ANATOMY OF THE DOG
An Illustrated Text

THIRD EDITION

Professor Dr Klaus–Dieter Budras

Medical Illustrator
Wolfgang Fricke
Institute of Veterinary Anatomy
Free University of Berlin

Dr Patrick H McCarthy
Department of Veterinary Anatomy, University of Sydney

Contributors
Professor Dr Ekkehard Henschel
Institute of Veterinary Anatomy
Free University of Berlin

Dr Cordula Poulsen Nautrup
Center of Anatomy
Medical School of Hanover

M Mosby-Wolfe

London Baltimore Bogotá Boston Buenos Aires Caracas Carlsbad, CA Chicago Madrid Mexico City Milan Naples, FL New York Philadelphia St. Louis Sydney Tokyo Toronto Wiesbaden

CONTRIBUTION TO THE GERMAN AND ENGLISH EDITION

Scientific Contribution: Frau Dr. Monika Höftmann, Frau Dr. Anita Wünsche
Design, Layout, Reprography: Herr Klaus Korpel
Editorial contribution: Frau Helga Schubert, Frau Dr. Beate Tetzlaff, Ms L. Hicks (Sydney)
Figures and Tables:
Prof. Dr. Dirk Berens v. Rautenfeld, Med. Hochschule Hannover: Fig. p.67 and to VIII.1. and to 14.1.
Herr Thomas Goller*: Fig. p. 50A, to 40.1., 41.12. and 43.2.
Herr Roland Hagen*: Textfig. p. III
Prof. Dr. Klaus Hartung, Klinik für Pferdekrankheiten und allgemeine Chirurgie, Freie Universität Berlin: Fig. to 44.3.
Prof. Dr. Aaron Horowitz, University of Zimbabwe
Prof. Dr. Ronald Hullinger, Purdue Univ., Lafayette (U.S.A.): Fig. p. I, fig. p. VIIA, p. 49A and to 28.8.
Herr Dr. Axel Kremer*: Tables to myology (p. 56 — 65)
Frau Dr. Elke Laue*: Lymphology
Herr Christoph Mülling*: Textfig. p. 16 and p. 33
Mrs. Jane Redmann, Barrington (U.S.A.): Textfig. p. I, fig. p. VII A, p. 49A and to 28.8.
Frau Sabine Röck, Berlin: Fig. p. 52 — 54 and to III.9.

*All staff members belong to the Institut für Veterinär-Anatomie, Fachbereich Veterinärmedizin,
Freie Universität Berlin, Koserstr. 20, 14195 Berlin.

1. German Edition 1983
2. German Edition 1987
3. German Edition 1991
1. Spanish Edition by BUDRAS/FRICKE/SALAZAR 1989

This is an international edition issued by Schlütersche, Hannover,
and Mosby-Wolfe, London, an imprint of Times Mirror International
Publishers Limited.

© 1994 Schlütersche Verlagsanstalt und Druckerei GmbH & Co.,
Hans-Böckler-Allee 7, 30173 Hannover.

Printed in Germany

ISBN 07234-19205

For full details of all Times Mirror International Publishers Limited
titles please write to: Times Mirror International Publishers Limited,
Lynton House, 7 — 12 Tavistock Square, London, WCIH 9LB, Eng-
land.

TABLE OF CONTENTS

PREFACE TO THE ENGLISH EDITION

This English edition of 'Atlas der Anatomie des Hundes' by Professor Dr. K.-D. Budras and Herr W. Fricke (and collaborators) will place their work before a wider scientific community than previously.

To reiterate what has been stated in the prefaces of previous German editions, this volume should rightfully be called a book-atlas. Its main thrust has been a closer integration of theoretical and practical knowledge of the anatomy of the dog and its further integration with clinical, surgical, radiographic and ultrasonographic correlative anatomy. Not only is such applied anatomy of intrinsic importance; it also demonstrates the importance of the basic subject and why one has to persevere with a discipline of great complexity. This is of psychological benefit to students. Indeed, this volume is a blue-print upon which integrated courses of veterinary anatomy can be fashioned according to the requirements of the veterinary course being served.

This book-atlas with its abundance of illustrations, is not only the result of patient, detailed research and design on the part of the authors. Constructive student criticism has been sought and tested over several years. Synoptic tables in the mid-section of the book-atlas provide summarized data on each of the anatomical systems dealt with in lectures and seen in detail in dissection laboratory classes. Such information has proved invaluable in achieving an overall view of the subject and also serves to encapsulate data prior to examinations.

For the practising veterinarian the book-atlas is a quick source of reference for anatomical information on the dog at the preclinical, diagnostic, clinical and surgical levels. It will also serve to deepen one's knowledge at a specialist level.

The authors wish to thank all collaborators featured on the title page of the present volume as well as those mentioned for their contributions in the prefaces to previous German editions. The present edition is a contribution of veterinary anatomists, illustrators and technicians from three continents.

Special thanks should be given to Frau H. Schubert and Frau Dr. B. Tetzlaff of the Institute of Veterinary Anatomy, Free University, Berlin and Ms L. Hicks of the Department of Veterinary Anatomy, University of Sydney, Australia for their secretarial help and skills in word processing and Professor Dr. A. Horowitz, University of Zimbabwe and Professor Dr. R. L. Hullinger, Purdue University, Indiana, U.S.A. for their translation of parts of the General Anatomy. Without such assistance this volume could not have reached fruition.

Berlin, summer 1994   The authors

INTRODUCTION TO ANATOMY

The concept of anatomy originates from the Greek word 'anatemnein', to dissect or cut apart. The distinguished Austro-Hungarian anatomist Hyrtl (1811-1894) has already spoken of the art of dissection in a logical, consistent manner. Nowadays the original significance of the word persists although it has incorporated other distinct aspects. Anatomy in the modern sense, is not only limited to simple description. It also serves a confirmed role in the relationship between form and function as well as in the application of purely morphological knowledge to clinical sciences. Now, as previously, the greater part of our knowledge and understanding of anatomy is acquired by dissecting the animal body in the dissection laboratory. There, veritably, one discovers the 'naked truth'. By this activity, one also acquires a fluency and a dexterity which will be useful in his or her future professional activity especially in the surgical area. It should be pointed out, furthermore, that there is no limit to the desire to investigate specimens one has dissected personally. On the contrary, neither can the best collection of prepared specimens replace individual practical work in the dissection laboratory. However, the thorough study of anatomical dissections in whatever form, is indispensable, and the use of texts and atlases complementary to this. Today, every avenue of assistance is more important than ever, since much less time is available for individual dissections than previously. The study of anatomy is also very important in the learning and comprehension of medical terminology; indeed, many terms designating diseases and methods of treatment are derived from the anatomical vocabulary.

Anatomy, as a whole, is divided into macroscopic and microscopic anatomy, and embryology. Notwithstanding, these subdivisions are interconnected in an intimate manner, and form a unity always emphasized and defended by the eminent anatomist and former director of the Institute of Veterinary Anatomy, Free University of Berlin, Professor Dr F. Preuss. The oldest and most complete part, macroscopic anatomy, is considered frequently to embrace the whole discipline. However, where observation is not served by conventional aids such as the naked eye and lens, one moves progressively to microscopic anatomy (histology and cytology) employing the microscope as the medium. The boundary between macroscopic and microscopic anatomy is known as mesoscopic, a term becoming more important day by day. Within these divisions the same material is studied and the same objectives are in mind, only the techniques vary. To the third division, embryology, is entrusted ontogenesis or individual development before and after birth. It utilizes techniques proper to the field as well as macroscopic, microscopic and mesoscopic methods.

Macroscopic anatomy, as with the other two divisions, is considered from different viewpoints depending on the emphasis given to its different aspects. Nevertheless, the classical study does not change.

Systematic descriptive anatomy deals with the animal body divided into well defined structural and organic systems. These include skeletal, muscular, circulatory, nervous, and visceral systems as well as sense organs and skin. Occasionally descriptions are complex and dwell too much on certain details while forgetting the more substantial. These descriptions, however, are a necessary prerequisite to comprehend other observations explained subsequently.

Systematic anatomy is also divided into general and specific anatomy. General anatomy (see pages I to IX) deals with given facts valid for the general structural or organic system. For example, the organization of the periosteum is the same for all bones.

Specific anatomy gives specific data applying to a specific structure, for example one bone only.

Comparative anatomy emphasizes anatomical correlations be they conformities or divergencies, among individual animal species including man. At times anatomical comparisons between or among different species are necessary and assist greatly in establishing homologies and determining the functions of certain structures. Goethe, for example, used the principles of comparative anatomy to discover the incisive bone of man. This bone is a constant feature in adult domestic animals and only at times is it present in man. In his study of the human skull, Goethe came upon one specific example with a well developed incisive bone and by comparing it with that of the animal skull, was able to homologize and identify it.

Topographic anatomy emphasizes the positional relationship of anatomical structures and underlines areas of clinical application. In different parts of the body these structures are analysed in layers, always having overall body structure in mind.

The anatomy of the live animal is, undoubtedly an important part of the whole subject. It shows the body in its natural state and certain difficulties in the study of its gross anatomy are obviated as compared with those of the cadaver. These include post mortem tissue variations in color, consistency and characteristics, as well as artificial changes with tissue fixation. One cannot consider live animal anatomy here for obvious reasons; notwithstanding, it is very informative when learning under the direction of an anatomist with clinical experience.

Radiographic anatomy (and also, more recently, ultrasonographic anatomy) is the direct link between anatomy and clinical sciences. During the interpretation of radiographic plates of the normal body in the dissection laboratory, one should gain the first perceived results of anatomical instruction. The illustration of abnormalities such as a persistent right-side aorta or even pathological changes, should excite the interest of veterinary students and 'add seasoning' to the anatomical instruction.

The present text-atlas of anatomy is considered to be very useful to combine and coordinate different methods of anatomical presentation and different ways of thinking. The textual portion is presented in compendium or compressed form since the majority of details can be seen in the adjacent coloured plates. These afford an ideal presentation of topographical dissections. With this atlas the demands of comparative veterinary anatomy are also realized since, in many aspects, the body of the dog is taken as a 'foundation stone'. Thus one can speak with a comparative perspective and in a more detailed manner about complicated anatomical concepts in other domestic mammals.

The intimate relationship between art and anatomy is apparent on any visit to a museum. The artist devotes himself or herself to inspire through the beauty of the animal body, and teachers and students of anatomy take advantage of meticulous art presentation. The dual relationship was demonstrated with genius by Leonardo da Vinci whose complex anatomical drawings were the result of basic detailed dissections of cadavers of several animal species. What fascination emanates from the portrayal of anatomy in art and literature as proved by the illustrations, descriptions, and contributions to research by the gifted greats of world history such as Aristotle, Leonardo da Vinci and Goethe. Aristotle, for example, published among many other things, anatomical descriptions of senile sex reversal in birds and a description of the horse's hoof with regard to thrush. The commitment to research was formulated appropriately by Goethe who felt 'inexpressible pleasure' at his discovery of the human incisive bone.

GENERALLY ANATOMY

OSTEOLOGY: MEMBRANOUS AND CARTILAGINOUS OSTEOGENESIS; BONE GROWTH IN LENGTH AND THICKNESS

a) **MEMBRANOUS (DESMAL) OSTEOGENESIS** refers chiefly to the formation of the flat bones of the neurocranium, that is the skull enveloping the brain. In this process the embryonic **connective tissue or mesenchymal cell (A)** is transformed into a **bone producing cell or osteoblast (B)** which gives rise to **primary trabecular bone (C)**. The osteoblast synthesizes the organic osseous matrix or osteoid, that is, the fundamental osseous substance and collagen fibers. The osteoblast becomes enveloped by the osteoid and, as it progressively mineralizes (ossifies), it is subsequently called an **osteocyte (D)**. This primary bone tissue is organized initially as trabeculae and the osteoid is mineralized by an orderly deposition of calcium phosphate crystals (hydroxyapatite) in the interstices of collagen. The transformation of primary trabecular bone into secondary compact bone is accomplished by a layer of osteoblasts apposed to the surface of the trabecular bone. The neurocranium increases in circumference by the deposition of bony tissue at its external surface with a simultaneous breakdown and resorption of bony tissue at its internal surface due to the activity of **osteoclasts (E)**. The apposed margins of the bones of the skull have not yet formed a bony union at birth and are joined by fibrous tissue. The largest of the spaces between these bones are termed fontanelles. This lack of bony fusion allows for a compression of the skull as it passes through the pelvic apertures at birth.

Membranous osteogenesis

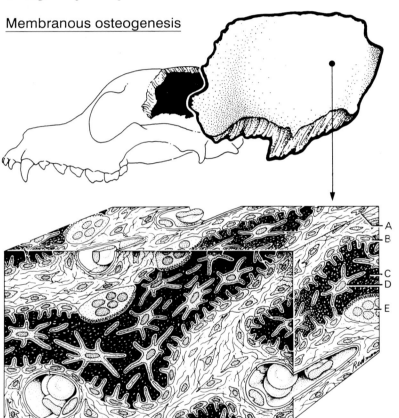

A Mesenchymal cells	C Primary trabecular bone	D Osteocyte
B Osteoblast		E Osteoclast

1 b) CARTILAGINOUS (ENDOCHONDRAL, ENCHONDRAL) OSTEOGENESIS refers to those bones which develop as replacement of a cartilaginous, primordial skeleton, for example vertebrae, long bones and those of the base of the skull. This formation of bones involves a small amount of membranous/perichondral ossification but is mainly enchondral. Perichondral ossification is present only at the periphery of the **diaphysis (4)** between the **epiphyses (1)** of the bone concerned. It first appears as a perichondral **osseous ring (3)** developing in the osteogenic layer of the **periosteum (6)** formerly perichondrium. This osseous ring provides stability. Endochondral ossification begins **3** in the diaphysis (see below) in the **primary ossification center** (see radiograph 1.3.), where the first structural evidence is the penetration of a perforating **blood vessel (21)** into the cartilage. The osseous ring is thus perforated, creating the anlage of major bone nutrition via the **nutrient canal (22)**. The penetrating vascular buds and the accompanying perivascular mesenchymal connective tissue enter and occupy the **primary (marrow) medullary cavity (7)**, subsequently forming the secondary bone marrow. This primary medullary cavity progressively enlarges as the cartilage is resorbed in the **zona resorbens cartilaginea (16)**. Resorption is a result of activity of hypertrophic chondrocytes, their degeneration, removal of their cell fragments and removal of chondroid, **2** the organic matrix of cartilage. Accompanying this there is activity of

chondroclasts (2). The osteoblast (5) differentiates from, among other cells, a perivascular mesenchymal cell and adheres to the surface of a trabecula of the remaining mineralized **cartilaginous matrix (19)**. It eventually becomes an **osteocyte (8)**. Thus osteocytes appear firstly embedded in the **primary (osseous) trabecula (18)** which, initially, has a core of effete chondroid. Mineralization of the chondroid and the osteoid makes possible the early detection of the ossification centers due **3** to the related density of the calcium salt in the matrices. The times of appearance of ossification centers varies with the bone. The pattern of appearance in anyone bone, however, is so typical and consistent that it may be used for age determination of the fetus and juvenile puppy. As presently used in textbooks of anatomy, 'primary' and 'secondary' with reference to centers of ossification have variously a 'temporal' (vis-a-vis birth or other centers) or a 'significance' (size – major/minor) usage.

With age and subsequent development, blood vessels also penetrate the epiphyses of long bones. As in the diaphysis, this signals the development of another ossification center, the **secondary center of ossification (12)**. Expanding from this center, the endochondral ossification progresses centrifugally, but most markedly at its diaphyseal interface. In the growing long bone, the epiphyses remain separated from the diaphysis by a plate of **epiphyseal cartilage (13)**. Until its mineralization, **4** this plate contrasts with the radiographic opacity of the diaphysis and epiphyses, making detection by radiography easy (see radiography I.2.). From the **zona reservata (14)** of the epiphyseal cartilage chondrocytes proliferate (proliferative zone) and are oriented in **columns (15)** parallel with the long axis of the bone. The most recently formed chondrocytes are located proximal to the plate, while the oldest are hypertrophic and are situated nearer the approaching zona resorbens. The **metaphysis (17)** is the diaphyseal region of transition between epiphyseal cartilage and diaphyseal bone and joining epiphysis to diaphysis. As the pup approaches physical maturity, the diaphyseal ossification advances in an epiphyseal direction and the lesser rate of epiphyseal ossification progresses in a diaphyseal direction. These advancing ossification 'fronts' are no longer compensated by proliferation of chondrocytes. Consequently, the epiphyseal plate becomes progressively thinner until it is finally obliterated by bone. Thereafter no growth in length of the bone is possible. Beam-like structures in the metaphysis which demonstrate an envelope of bone around a core of cartilaginous matrix or chondroid are called primary osseous trabeculae. They are soon destroyed by the activity of **osteoclasts (9)**. In turn, osteoblastic activity, depositing bone on remnants of these beams, forms a second generation of trabeculae, lacking a chondroid component and called secondary osseous trabeculae (see p. II). The osteocytes and collagen fibers of the ossified bone of the secondary trabeculae are more regularly arranged. It is in the metaphysis that the primary trabeculae appear and here their transformation to secondary trabeculae begins. Bone transformation is greatest in the metaphysis (see growth in circumference).

The replacement of cartilage and deposition of bone at the periphery of the cartilage model leads to compact bone formation. The compact bony tissue is organized as thin concentric lamellae around single blood vessels (see p. IIA).

Apophyses (10) are bony processes receiving many muscular insertions (other definitions are also used). The apophysis develops an **apophy- 4 seal cartilage (11)** and also a center of ossification.

c) **GROWTH IN LENGTH** of long bones is assured as long as the zone **5;6** of proliferating chondrocytes remains active. When the degeneration of hypertrophic chondrocytes and their replacement with trabecular and compact bony tissue is no longer compensated by chondrocyte proliferation, the zones of ossification approach each other, consuming the entire growth plate cartilage. The zones fuse and eliminate the possibility of further growth in length of the long bones. This osseous tissue replacing the radiolucent growth plate cartilage appears on a radiograph as a radiopaque epiphyseal line.

d) **GROWTH IN CIRCUMFERENCE AND THICKNESS** results primarily from the apposition of bony lamellae on the inner aspect of the thin osteogenic layer of the periosteum. This external apposition of new bony tissue coupled with an internal resorption of osseous tissue at the margin of the medullary (marrow) cavity, result in an enlargement of the cavity while increasing the circumference of the cortical bone. The relative thickness of the cortical compact bone is regulated by the degree of inner resorption. The growth in thickness in the metaphysis area follows expansion of the diameter of the epiphyseal plate. Herein the apposition of new bony tissue to the inner surface is accompanied by resorption at the external surface. (Healing of bone, see appendix p. **7,8** 69 1.7., 1.8.)

I

Cartilaginous osteogenesis

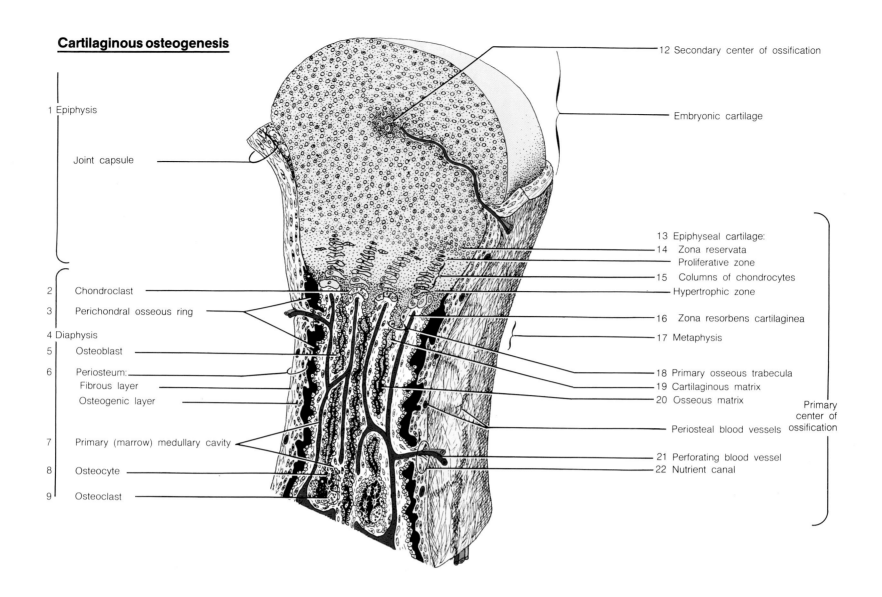

1 Epiphysis

Joint capsule

2 Chondroclast
3 Perichondral osseous ring
4 Diaphysis
5 Osteoblast
6 Periosteum:
  Fibrous layer
  Osteogenic layer

7 Primary (marrow) medullary cavity
8 Osteocyte
9 Osteoclast

12 Secondary center of ossification

Embryonic cartilage

13 Epiphyseal cartilage:
14 Zona reservata
  Proliferative zone
15 Columns of chondrocytes
  Hypertrophic zone
16 Zona resorbens cartilaginea
17 Metaphysis
18 Primary osseous trabecula
19 Cartilaginous matrix
20 Osseous matrix
  Periosteal blood vessels
21 Perforating blood vessel
22 Nutrient canal

Primary center of ossification

Femur (juvenile dog)

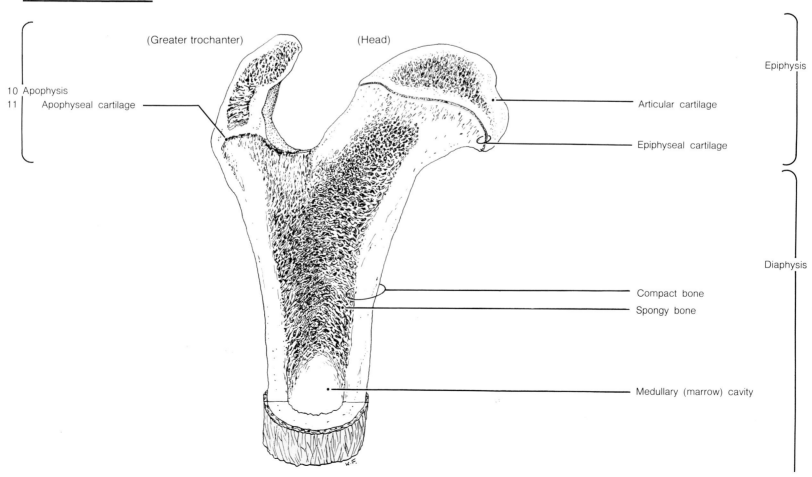

(Greater trochanter)   (Head)

10 Apophysis
11 Apophyseal cartilage

Epiphysis

Articular cartilage

Epiphyseal cartilage

Compact bone
Spongy bone

Diaphysis

Medullary (marrow) cavity

A

OSTEOLOGY: FORM AND STRUCTURE OF BONE

a) **BONY TISSUE** comprises bone cells (osteocytes, osteoblasts, and osteoclasts) and mineralized or ossified bone matrix, rich in collagenous fibers and hydroxyapatite crystals. The 'dry substance' of a mature bone is about one-third organic and two-thirds inorganic.

b) The **HARD COMPONENTS OF BONE** as an organ are the compact substance and the spongy substance.

I. The cortical (bark) part of a bone is composed of **compact substance (26)** whereas the internal part often is traversed by or contains spongy substance. Compact substance refers to a dense mesoscopic organization. Most bony tissue is deposited as lamellae, compact bone being deposited as a mantle of varying thickness to form the contour of the bone. In a region such as the diaphysis of a long bone, where the force vectors of weight-bearing act in parallel to the long axis of the bone, the dense, compact bone is organized on the basis of the **osteon (13)**, the microscopic structural unit of bone. In regions such as epiphyses, crests and carpal bones, where these forces cannot be represented by a single vector, the compact bony tissue is arranged irregularly. Osteons of the compact cortical substance may approach 2 cm in length and 0,5 mm in diameter. The central axis of an osteon is defined by a canal containing one or more capillaries and periangial reticular connective tissue. During osteogenesis, osteoblasts acting locally and in synchrony produce **lamelli osteoni (15)**. The **collagenous fibers (14)** polymerized in the osteoid are deposited in parallel and helically along the axis of the osteon, the inclination of which varies from one lamella to the next. These lamellae, when mineralized during ossification, are analogous to reinforced concrete: rigidity = hydroxyapatite salts (concrete) with limited flexibility + collagenous fibers (steel). By further analogy, if plywood could be wrapped into a very tight cylinder, making the lumen of it quite small, its organization would approximate that of an osteon. In the model the many laminae are each composes of wood, the grain of which alternates with the laminae.

Between osteons there remain portions of older osteons, the interstitial
1 **lamellae (7)** presenting morphological evidence of ongoing bone remodelling. Remodelling occurs throughout life. In the period of growth bone regeneration prevails; with maturity of the skeleton removal and replacement are balanced; and with decline due to age, destruction of osteoid may prevail, resulting in an increased fragility.

II. Spongy substance **(5)** forms the internal scaffolding beneath the cortical bone where the forces vary in their application, as with synovial articulations and tendon attachments. Its generally trabecular network is organized along **lines of tension (10)** or in some cases as lamellae or tubules along planes of tension. The interstices of the trabecular network, the cellulae medullares, are small compartments of medullary hemopoietic tissue. A small amount of spongy substance sandwiched between the cortical part of flat bones of the skull, is called diploë. Little or no spongy substance develops in some bones such as the petrous temporal bones of most species, the mandible of elephants, and pneumatic bones.

c) The **SOFT COMPONENTS OF BONE** include periosteum, endosteum, articular cartilage, blood vessels and bone marrow.

2 I. The **periosteum (17)** is composed of a superficial **fibrous layer (19)** from which **perforating fibers (18)** penetrate into the compact substance, and a deep cellular **osteogenic layer (20)** or cambium which is rich in periosteal nerves and vessels. As the periosteum passes around an articulation, its equivalent separates from its position next to the
3 bone to continue as the articular capsule. The **endosteum (9)** forms the inner lining of the medullary cavity, that is the surface of the trabeculae, and mesoscopic cellulae medullares of the spongy substance. It is composed of a single layer of cells, mostly osteoprogenitor cells, osteoclasts, and osteoblasts.

II. The **articular cartilage (1)** is composed of three zones, superficial, intermediate and deep. The **superficial zone (2)** sends collagen fibrils into the **intermediate zone (3)**. These collagen fibrils grow to fibers, change their orientation from tangential to the surface, to become parallel with the long axis of the bone, and then continue into the **deep zone (4)**. This deep zone manifests a calcified matrix on its inner aspect and forms an uneven border with the underlying bony tissue of the epiphysis.

III. **BLOOD SUPPLY OF BONES** is provided by **osseous vessels**. A **4** distinction can be made between small periosteal vessels and larger nutrient vessels. **Periosteal arteries (21)** and **veins (23)** form an extensive plexus and continue as **perforating rami (24)** into the compact bone (some authors deny the presence of arterial perforating rami). **Nutrient vessels** enter the bone at the **nutrient foramina** and continue in **nutrient canals** to supply the interior of the bone. Branches of these vessels supply the hemopoietic tissues of the bone marrow and enter the endosteal surface. The capillaries anastomose with branches of the periosteal vessels penetrating the compact substance from the periosteal surface. The plexus formed in the compact substance is oriented primarily as a parallel set of vessels each passing in the **central canal (6)** of an osteon. These central vessels are interconnected by perforating vessels, each passing in a **perforating canal (16)**. The **periosteal veins (23)** receive the blood from the capillaries via delicate branches. Long bones are supplied by diaphyseal, metaphyseal, epiphyseal and periosteal arterioles (see appendix II.4.). In young animals the supply of the epiphysis is relatively independent of the others due to a plate of epiphyseal cartilage. After this cartilaginous plate has been consumed, anastomoses occur.
Periosteal nerves (22), mainly sensory, accompany the larger blood vessels.

IV. **BONE MARROW** fills the **medullary cavity (12)** and the **cellulae medullares (8)** of the spongy bone. In the fetus and young animal, the marrow is red and supports an hemopoietic function. With aging, the red marrow is only present in short and flat bones, while in long bones, it is replaced by a fatty yellow marrow. This tissue may revert to hemopoietic tissue with stress occuring, for example, with blood loss and anemia. With old age or emaciation it is transformed irreversibly to a gelatinous fatty tissue.

d) **FORM AND SHAPE OF BONES** vary as with long bones, short **5** bones (carpus, tarsus) irregular bones (vertebrae), and flat bones (of the skull). The shape of the medullary cavity also varies with the shape and size of the bone and increases with age in long bones. Shape of bones also determines the mode of ossification. Development of bones involves both an endochondral (epi- and diaphyseal) and a perichondral (diaphyseal) ossification. Ossification of short bones is mainly endochondral, and that of flat bones is either a desmal or perichondral. Irregular bones ossify at epiphysis and diaphysis (located between two epiphyses) in a similar manner to long bones.

e) **SESAMOID BONES** develop in association with synovial joints. **6** They lack an osteogenic layer of periosteum since they develop within tendons, for example the patella, or as a support structure over which a tendon may glide, as, for example, the proximal sesamoid bones (see p. 7A). Some sesamoid bodies consist only of cartilage and are referred to as sesamoid cartilages.

f) **ORGANIC BONES** are present within an organ, for example the os penis of the dog and the os cordis of ruminants.

g) **CARTILAGE** is an avascular supporting tissue consisting of cartilage cells or chondrocytes and an intercellular matrix. This contains connective tissue fibers and a more or less compact ground substance containing, in particular, chondroitin-sulphate. Superficially, cartilage is covered by a vascular perichondrium which is lacking in articular cartilage. In accordance with sound building principles, cartilage is a vesiculated supporting or turgor tissue which provides turgidity. This is achieved by supporting the turgid properties of the cells, including the surrounding cartilage ground substances, with the functionally reciprocal tensile strength of the connective tissue fibers. In particular, collagen, but also elastic fibers, are present and these vary in morphology, quantity, course and orientation depending on body posture and the three different types of cartilage concerned.
Collagen fibers are arranged predominantly in a criss-cross fashion. In hyaline articular cartilage (illustrated sectional magnification upper left) of the knee, the collagen fiber bundles tend to run from the boundary between bone and cartilage in a curved manner to the surface. There they course parallel, returning again in an arched form into the bone. In the meshes between collagen fiber bundles, the cartilage cells surrounded by cartilage capsule and lacuna are situated in cell groups or territories. Collagen fibers occur in all three cartilage types namely hyaline, elastic and fibrous.

Femur (Homo sapiens)

(Head)

(Greater trochanter)

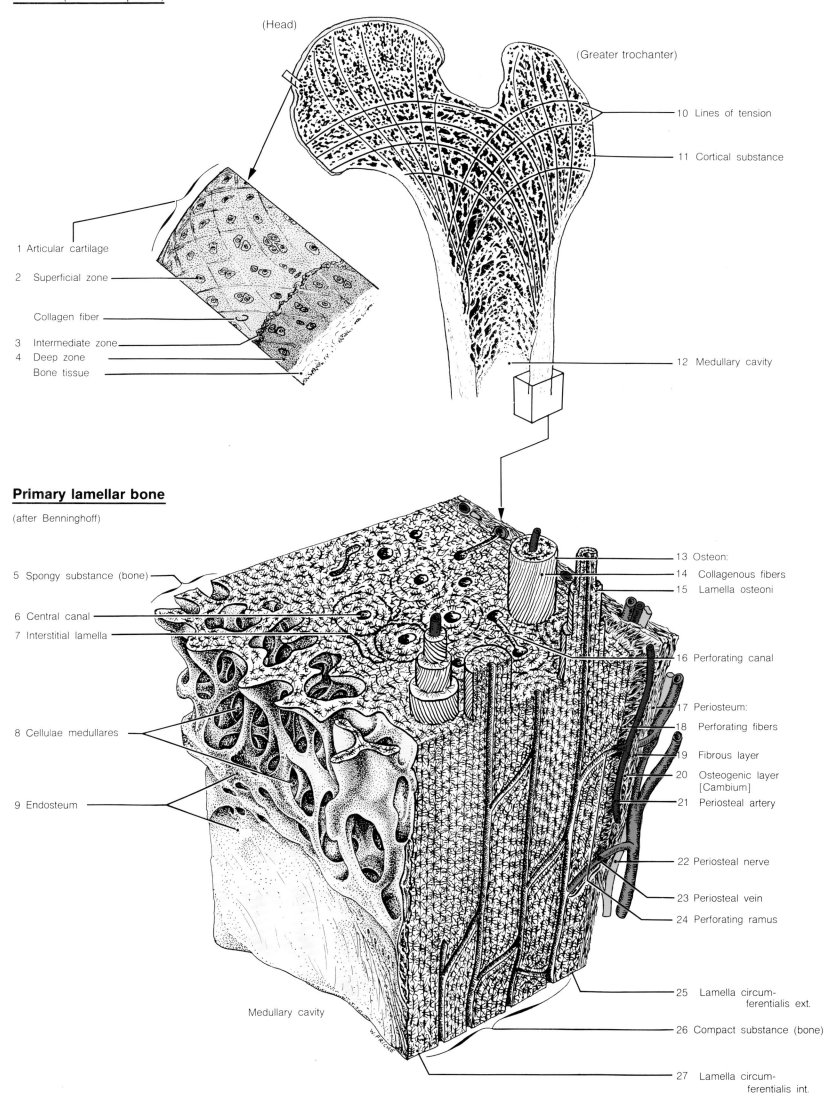

10 Lines of tension

11 Cortical substance

1 Articular cartilage

2 Superficial zone

Collagen fiber

3 Intermediate zone
4 Deep zone
Bone tissue

12 Medullary cavity

Primary lamellar bone

(after Benninghoff)

5 Spongy substance (bone)

6 Central canal
7 Interstitial lamella

8 Cellulae medullares

9 Endosteum

Medullary cavity

13 Osteon:
14 Collagenous fibers
15 Lamella osteoni

16 Perforating canal

17 Periosteum:
18 Perforating fibers
19 Fibrous layer
20 Osteogenic layer [Cambium]
21 Periosteal artery

22 Periosteal nerve

23 Periosteal vein
24 Perforating ramus

25 Lamella circum-ferentialis ext.

26 Compact substance (bone)

27 Lamella circum-ferentialis int.

A

1 a) **BONY ARTICULATIONS** include those without a joint cavity or synarthroses, and those with a joint cavity or diarthroses. Synarthroses include fibrous joints, cartilaginous joints and synostoses, while diarthroses are also termed synovial or true joints.

I. Fibrous joints include: 1. syndesmoses such as the interosseous membrane of the leg; 2. sutures such as those between bones of the neurocranium; and 3. gomphoses, the dentoalveolar articulations anchoring teeth in the bony alveoli.

II. Cartilaginous joints include: 1. synchondroses formed by the meeting of hyaline cartilages such as the slow-to-ossify portions of the sternum, for example the manubriosternal synchondrosis; 2. symphyses formed by the meeting of fibrocartilages such as the pelvic symphysis in young sexually mature females and the intervertebral symphysis formed by an intervertebral disc between two vertebral bodies.

III. Synostoses are found at the fusion of bony elements such as those forming the sacrum.

IV. True, freely moveable **synovial joints** are characterized by a joint space separating the ends of two bones covered with hyaline cartilage. Simple joints involve two bones such as the shoulder joint, and composite joints several bones such as the knee joint (see p. IIIA). The **joint**
2;3 **cavity (1)** with its synovial fluid is surrounded by a **joint capsule (2)** reinforced by **ligaments (4, 5)**. **Joint cartilage (3)** covers the ends of the
4;5 bones forming the articulation. The fibrous layer of the joint capsule is
6 a continuation of the outer layer of the periosteum. Deep to the capsule lies the synovial layer which is a modified continuation of the inner, cellular osteogenic layer of the periosteum. The synovial layer does not extend onto the articular cartilage, but projects as folds or plicae, and villi, into the joint cavity. It presents two types of epithelioid synovial cells: 1. secretory cells which add mucin to an ultrafiltrate of blood plasma forming a fluid which nourishes the non-vascularized joint cartilage and separates and lubricates the articular cartilages; and 2. phagocytic cells which phagocytose cells, for example erythrocytes after hemorrhage as in hemarthrosis, or after exfoliation of cartilage fragments as in degenerative arthritis. The synovial membrane presents more or less folding depending upon the location within a joint or the
7 range of movement of the joint. Outpouchings of the joint capsule and contained synovial fluid may surround tendons of origin, for example the sheath around the tendon of origin of m. extensor digitalis longus at the knee.

8 b) **SYNOVIAL BURSAE AND SYNOVIAL SHEATHS** comprise the same tissues as the fibrous joint capsule and synovial lining (see above). Synovial bursae lie between bones and the tendons which glide over them, such as the **distal infrapatellar (subtendineal) bursa (7)** or between bones and skin such as the subcutaneous calcaneal bursa (see p. VA-9 and p. 33). Synovial sheaths anchor tendons and allow them to glide as they follow a longer course over a bone (see p. V).

c) **FIBROCARTILAGINOUS MENISCI** or **DISCS** are suspended in many complex joints, functioning as shock absorbers, and presenting one of three variations in form:

I. A pair of **articular menisci (6)** extending from lateral and medial aspects of the incongruent knee joint, each with the shape of an orange section, and located within the joint cavity. They partition the synovial cavity incompletely, each is congruent with its respective bony articular surfaces, and is attached by meniscal ligaments;

II. A single **articular disc** subdividing the articular cavity into two compartments, as in the temporomandibular articulation;

9 **III. Intervertebral discs** which completely fill what would otherwise be the spaces between vertebral bodies. In this case each vertebral articulation is called a false joint. True joints do exist at the facets of the articular processes of the vertebrae (see p. 4A-16, 7) and between the first and second cervical vertebrae. In the latter case there is no intervertebral disc. The intervertebral disc comprises an outer capsule chiefly of fibrous cartilage, called the anulus fibrosus, and an inner core of gelatinous connective tissue, the nucleus pulposus. The latter is a remnant of the embryonic notochord.

d) Depending on the number of **COMPONENT BONES** forming the joint it is considered to be a simple articulation when comprising two bones and composite when more than two are involved.

e) **JOINT SHAPE** (see also text illustrations below)

I. Flat, sliding or **plane joints** occur between the relatively flat articular processes of the vertebrae. The sacroiliac articulation, termed amphiarthrosis, is a special case, in that it has very short ligaments, uneven articular surfaces, and a small range of gliding motion.

II. Ball-and-socket or **spheroidal joints** involve a sphere or convex surface articulating with a concave socket, for example the shoulder joint.

III. An **enarthrosis** is also a spherical form of the ball-and-socket joint, in which the socket extends beyond the equator of the ball, for example the hip joint of man. The canine hip joint is not an enarthrosis but a simple ball-and-socket.

IV. Pivot or **trochoid joints** involve a central cylindrical peg and a ring rotating relative to each other as, for example, the dens of the atlantoaxial articulation.

V. Ellipsoidal joints have an elliptical or oval head articulating with a reciprocal cavity such as the atlantooccipital joint.

VI. Saddle or **sellar joints** are biaxial and combine two surfaces, each of which are maximally convex in one direction, and maximally concave in a second at right angles to the first. An example of this is the distal interphalangeal articulation.

VII. Condylar joints occur where two condyles or knuckles or a single transversely-oriented 'roller' articulate with a congruent concave socket such as with the temporomandibular articulation. Special types of condylar joints are as follows (VIII — XI):

VIII. Hinge joints or **ginglymi** where ridges and grooves are reciprocal on the opposing members, restricting the pendulum-like movement to one plane as in the elbow joint;

IX. Cochlear joints where ridges and grooves are oriented obliquely, allowing movement in one plane, for example in the tarsocrural joint;

X. Spiral joints where the outer contour of the roller mechanism shows a partial spiral when viewed lateromedially (see p. IIIA above, left). The eccentricity is due to the shape of the articulating members and the insertion points of the collateral ligaments. The latter which are stretched more during extension and flexion of the joint than in the middle position, provide a 'braking' action (femorotibial articulation);

XI. Sled or **gliding joints**, for example the femoropatellar articulation where the patella glides in the groove between the cranial parts of the femoral condyles.

f) The **FUNCTION** of a joint depends upon the shape of the articulating members, and the properties of the capsule and associated structures, namely the extent of the capsule and the length and tension of the joint ligaments; **Amphiarthroses** have a small range of movement due to short ligaments and a tight capsule. Incongruent joints produce a deepening of the joint into which penetrates a joint eminence or protuberance.

According to the **degree of movement** joints may allow **one, two,** or **three axes of motion.** A given axis allows flexion and extension, the extension angle being less than 180°. Only in hyperextension does the angle become greater than 180°. During flexion the angle becomes smaller on its inner curvature or "hollow". With two axes of motion the joint enables flexion and extension as well as abduction and adduction as in saddle joints. With more than two axes of movement the articulation is termed a free joint, that is the movement is in multiple directions.

Classification of joints

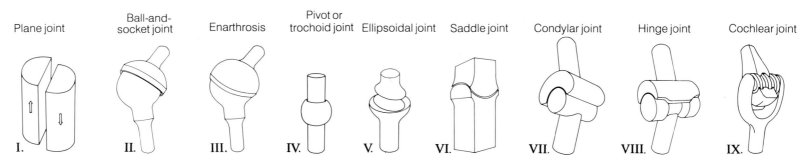

Plane joint	Ball-and-socket joint	Enarthrosis	Pivot or trochoid joint	Ellipsoidal joint	Saddle joint	Condylar joint	Hinge joint	Cochlear joint
I.	II.	III.	IV.	V.	VI.	VII.	VIII.	IX.

Synovial joints

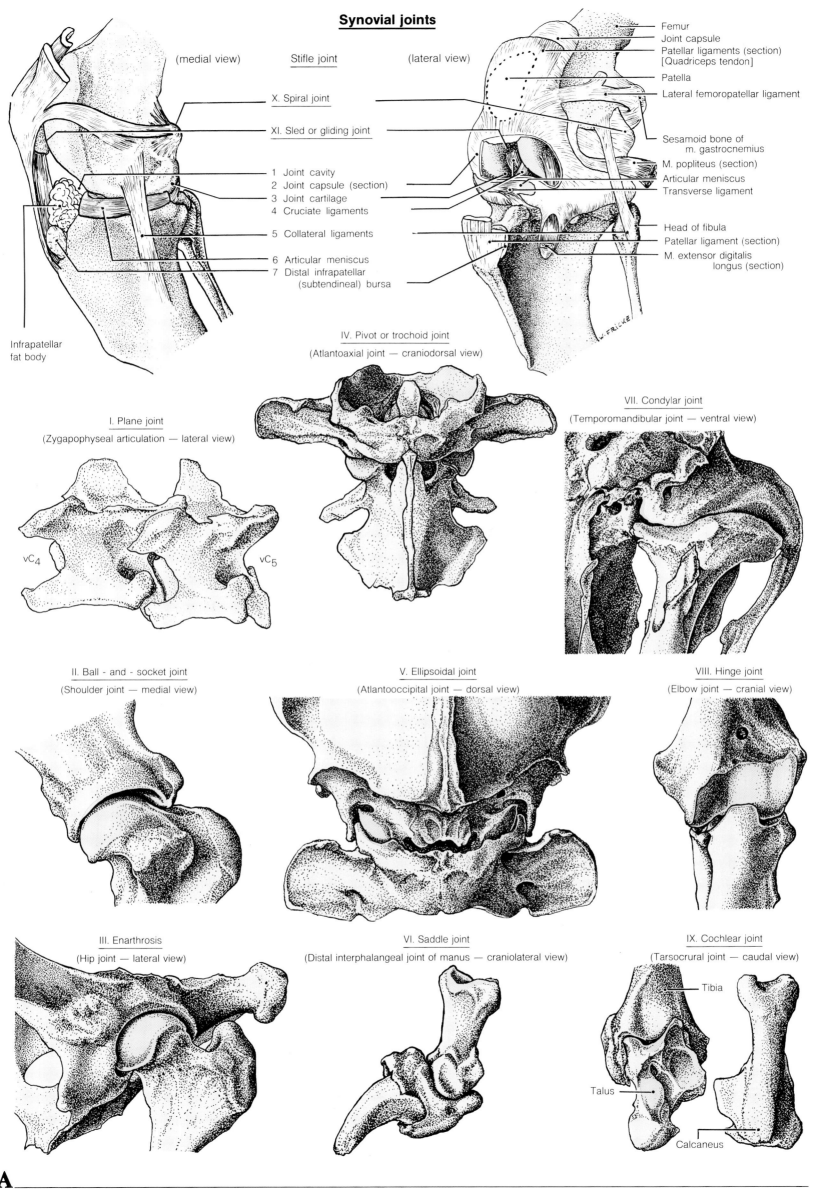

(medial view)   Stifle joint   (lateral view)

Femur
Joint capsule
Patellar ligaments (section)
[Quadriceps tendon]
Patella
Lateral femoropatellar ligament

X. Spiral joint
XI. Sled or gliding joint

Sesamoid bone of
m. gastrocnemius
M. popliteus (section)
Articular meniscus
Transverse ligament

1 Joint cavity
2 Joint capsule (section)
3 Joint cartilage
4 Cruciate ligaments
5 Collateral ligaments

6 Articular meniscus
7 Distal infrapatellar
(subtendineal) bursa

Head of fibula
Patellar ligament (section)
M. extensor digitalis
longus (section)

Infrapatellar
fat body

W. FRICKE

IV. Pivot or trochoid joint
(Atlantoaxial joint — craniodorsal view)

VII. Condylar joint
(Temporomandibular joint — ventral view)

I. Plane joint
(Zygapophyseal articulation — lateral view)

vC₄   vC₅

II. Ball - and - socket joint
(Shoulder joint — medial view)

V. Ellipsoidal joint
(Atlantooccipital joint — dorsal view)

VIII. Hinge joint
(Elbow joint — cranial view)

III. Enarthrosis
(Hip joint — lateral view)

VI. Saddle joint
(Distal interphalangeal joint of manus — craniolateral view)

IX. Cochlear joint
(Tarsocrural joint — caudal view)

Tibia

Talus

Calcaneus

I A

MYOLOGY I

The basic components of muscle tissue are the myocytes and the closely enveloping connective tissue stroma. Myocytes have as their principal characteristic, the ability to contract. Cytoplasmic myofilaments glide relative to one another in accordance with fine-structural and physiological properties of the muscle cell and tissue type. Three types of muscle are differentiated as follows:

I. Muscle Type	Location	Color	Cell Length	Cell Diameter	Innervation	Muscle Shape
A Striated	skeletal	red	up to 15 cm	approx 0.1 mm	somatic	variable/multiform
B Smooth	visceral	grayish	approx 0.1 mm	approx 0.01 mm	autonomic	layered
C Striated	cardiac	red-brown	approx 0.1 mm	approx 0.05 mm	intrinsic/autonomic	layered
						(approx = approximately)

Comments on the tables:

The **striations** of skeletal and **cardiac myocytes** are due to the regular architecture of the contractile filaments arranged as myofibrils, and their properties in polarized light. This does not occur in the **smooth myocytes** of visceral muscle.

The **distribution** of striated skeletal muscle (A) is not limited to the skeletal. Striated muscle of the viscera occurs also in the pharynx and esophagus. The skeletal muscle of the skin also has no connection with the skeleton, being anchored in the superficial fascia and skin giving it tone and movement. The **visceral smooth muscle (B)** in some cases is connected to the skeleton as, for example, the m. rectococcygeus extending from the wall of the rectum to vertebrae of the tail (see p. 29A). Smooth muscles also occur in the walls of blood vessels and the arrector muscles of hair follicles. **Striated cardiac muscle (C)** presents a further specialization as contractile and conductile myocytes (see p. 18), and cardiac myocytes extend beyond the heart into the initial portion of the pulmonary vein.

1 The **color** of skeletal muscle is more or less red, smooth muscle is grayish-yellow giving the typical color to the intestinal wall, and the continually active cardiac muscle is red-brown.

The **length** of the muscle cells is different between and within the three types. Skeletal myocytes can be extremely long, up to 15 cm with many hundreds of nuclei per cell. As the myocyte grows in size, nuclei are added by fusion with adjacent blastemic cells. The result is a multinu-

2 cleated syncytium often called a muscle fiber. Smooth myocytes vary in size depending upon physiological conditions, as in the gravid uterus where the length of the myocyte may increase ten-fold. The cardiac myocyte is defined at its junctions with other cells by intercalated discs. A few myocytes are binucleated. The cell-to-cell boundary is a more recent discovery. Formerly cells were thought to be longer and to form a structural syncytium. Cellular communication occurs at the discs through gap junctions and thus there exists a functional syncytium as well.

Innervation of skeletal muscle is by rapidly conducting myelinated nerves terminating on the myocytes at the motor endplates. This is voluntary control with the exception of esophageal skeletal muscle which

3 is innervated by the vagus nerve. Smooth muscle contracts slowly. The diameter of the lumen of each of the tubular organs is determined by the degree of contraction (tone) of the smooth muscle of its wall. Innervation is by unmyelinated neurons of the autonomic nervous system. An exception is smooth muscle of the uterus which contracts in response to oxytocin of the neurohypophysis. (Heart innervation and conductile system, see p. 18.) Each skeletal myocyte has a nerve contact via a motor endplate. In contrast, the smooth myocytes and cardiac myocytes receive only a few direct contacts with nerve fibers. The depolarization impulse spreads instead through the tissue by cell to cell 'touching', for example, at gap junctions.

The **shape of muscle** varies considerably and depends upon topography and function. There are about 250 skeletal muscles. Smooth visceral muscle and cardiac muscle are arranged in layers.

II. Skeletal muscle is described mesoscopically and macroscopically as follows: Each skeletal myocyte is limited by a **plasmalemma (12)** and a glycocalyx or cell coating in which are embedded small **collagen (13)** and **reticular fibers (14)**. The reticular fibers extend as a net around and between the myocytes as an **endomysium (11)**. Several myocytes together with this interstitial connecting tissue or endomyseal stroma comprise a **primary muscle bundle (10)**. A primary muscle bundle can be resolved with the unaided eye and is enveloped by **perimysium internum (9)**. Several primary bundles are enveloped by a fibrous sheath, the **perimysium externum (8)** forming an envelope around a **secondary muscle bundle (3)**. All myocyte bundles are enveloped by a thick layer of collagen fibers, the **epimysium (7)**. The muscle is bound to adjacent structures by tight **muscle fascia (6)**, a dense irregular collagenous and elastic tissue.

The **function** of this fibrous muscle stroma is multiple: 1. It supports and gives shape. 2. It enables a shear/gliding mechanism. The muscle is never fully active and following the principle of energy conservation, only some of the primary fiber bundles are in action, in proportion to loading or weight-bearing. The compartmentalization of the muscle allows this separation of function and optimizes energy expenditure. 3. These stromal structures also convey nerves, arteries, veins, and lymph vessels. 4. The muscle fascia is fixed locally to the underlying skeleton and therefore forms a tunnel serving as a directional determinant.

The **myotendinous junction (15)** is extensive and elaborate. The tendon collagen expands to envelop the digitiform ending of each myocyte. Collagen fibers of the tendon are also bound by reticular fibers. These extend along and around the digitiform processes, becoming embedded in the glycocalyx and continuous with the endomysium. 4

The **macrostructure** of muscle, especially the mode of cell and tissue insertion into the tendon, allows an evaluation of its muscle power. Where the muscle consists of simple origins and insertions, joined by a few, long myocytes extending the length of the muscle, it is classified as a **simple pennate** or **unipennate muscle (D)**. Each tendon and half the length of all myocytes correspond to the 'vane of a feather (penna)'. With a **bipennate (E)** or multipennate muscle, the tendons of origin and insertion are subdivided and extend within the muscle, varying their distances to connect with many but short myocytes. Unipennate muscles develop a larger lifting or up-and-down strike and multipennate muscles a lesser one, because the stroke grows in proportion to the length of the myocyte which can shorten 30-40%. Power of the muscle is the reverse, that is it is proportional to the number of myocytes and is greater in multipennate muscles. This becomes obvious when comparing the **anatomical diameter (4)** with the **physiological diameter (5)**. The former bisects the thickest part of the muscle belly perpendicular to the long axis of the muscle. The physiological diameter bisects myocytes perpendicular to the axes of their origin or insertion. In both muscle types of equal size the anatomical diameters are equal. In unipennate muscles both diameters are equal. In multipennate muscles the physiological diameter is greater than the anatomical.

The **insertion of the tendon (1)** is dependent upon the shape of the tendon. Where a great surface area of insertion occurs the collagen fibers radiate and are distributed into the fibrous layer of the periosteal surface. In turn, the fibrous layer is anchored into the cortical bone by perforating collagen fibers (of Sharpey). Very strong round **tendons (2)** appear to perforate the periosteum and insert directly in the osseous tissues. This takes place where periosteum is lacking as, for example, the calcaneal tendon inserting on the calcaneus. In such a case a great deal of power is concentrated on a small surface area of bone. As a result a bone fracture may result at the site of insertion. In sports medicine this is an important Achilles' tendon injury.

The discontinuity between the muscle cells and the collagen fibers of its tendon and the formation of collagen fibers in the shape of microwaves protect the muscle from laceration, allowing an elastic tensing during contraction and a gentle shock absorption.

Myology I

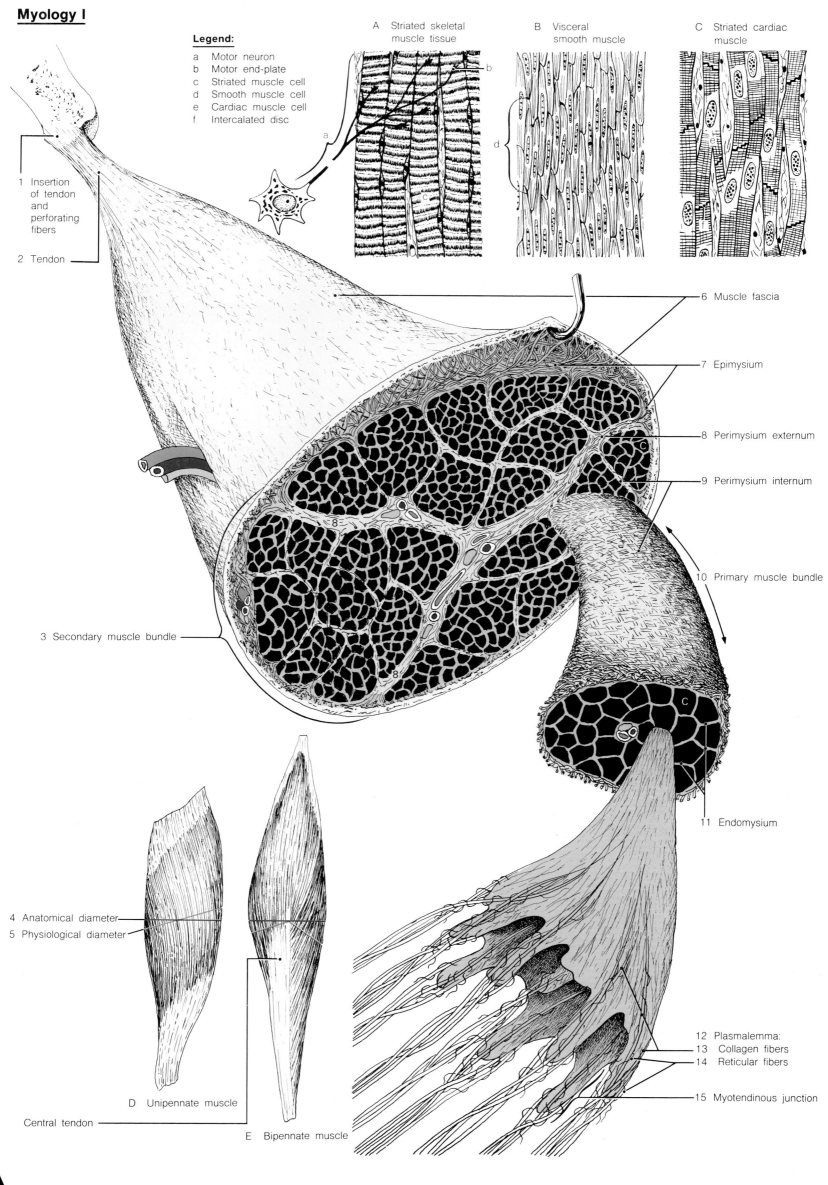

Legend:

a Motor neuron
b Motor end-plate
c Striated muscle cell
d Smooth muscle cell
e Cardiac muscle cell
f Intercalated disc

A Striated skeletal muscle tissue

B Visceral smooth muscle

C Striated cardiac muscle

1 Insertion of tendon and perforating fibers
2 Tendon

6 Muscle fascia
7 Epimysium
8 Perimysium externum
9 Perimysium internum
10 Primary muscle bundle

3 Secondary muscle bundle

11 Endomysium

4 Anatomical diameter
5 Physiological diameter

12 Plasmalemma:
13 Collagen fibers
14 Reticular fibers
15 Myotendinous junction

D Unipennate muscle

E Bipennate muscle

Central tendon

Muscle and tendon are treated together because of their close correlation. The typical spindle-shaped or **fusiform muscle,** such as **m. extensor carpi radialis (A)** can be compared to the shape of a small mouse; indeed the literal translation of the term 'musculus' means 'little mouse'. Similarly the muscle has a **head** or **origin (2),** a fleshy part, the **belly (3)** and a tail, the **attachment, termination** or **insertion (5).** With limb muscles, the tendon of origin is attached at a more fixed point on the skeleton, usually proximally on the limb or on the trunk. The tendon of insertion usually ends distally on a more moveable point of the skeleton of the limb. With some muscles, such as those of the vertebral column, the origin and insertion of the muscle are interchangeable. The origin and insertion of a muscle are readily determined from an anatomical preparation. Considering the two components, origin and insertion in a common sense manner, the function of a muscle is easily revealed. The functions of different muscles can complement one another as with synergistic muscles; for example the strong m. biceps brachii is synergistic to the more slight m. brachialis. Antagonistic muscles have opposing actions. In a harmonic movement synergists and antagonists act together. For example, ventral flexion of the trunk is initiated by the ventral abdominal muscles and is 'braked' or restricted by the action of the dorsal vertebral muscles.

The **tendon (4)** is usually adapted to the form of the muscle belly. Spindle-shaped muscles possess round tendons, whereas broad muscle bellies containing many parallel fibers likewise have wide, flat tendons with many parallel tendon fibers. Wide, sheet-like tendons are called **aponeuroses (1).** (This concept of an aponeurosis should not be confused with a fascia, the fibers of which cross one another in a scissors-like lattice-work.) The **central tendon** (see p. IVA), seen in multipennate muscles, should also be mentioned here. The central tendon is within the fleshy part of the muscle and lessens the quality of meat in food-producing animals. The superficial tendon is silvery and glistening and has a mirrorlike quality. Muscle bellies belonging to the one muscle may be in series one behind the other and are connected by **tendinous intersections (6).** Muscles with a large surface of origin or insertion are observed macroscopically as having a 'fleshy' origin or a 'fleshy' insertion. With microscopic examination, however, one can establish that there are many very short, fine tendons. The form of muscles and tendons as well as muscle function is, in many cases, embodied in the names of the muscles concerned (for example, **m. serratus — B — =** the serrated muscle; **m. digastricus — C — =** the two-bellied muscle; **m. bi-** or **quadriceps — D — =** two or four-headed muscle; **m. extensor — A — =** extensor muscle; **m. orbicularis — E — =** the ring-shaped muscle; its functional name: m. sphincter).

The **accessory structures of a muscle,** including its tendons, are the **synovial bursa** and the **tendon sheath.** The space within bursa and sheath is like the joint cavity, being lined internally with a **synovial layer (10)** and containing synovial fluid or synovia. The outer envelope of the bursa and tendon sheath is the **fibrous layer (11).** Bursa and tendon sheath, like the sesamoid bones, serve to protect a tendon which can be crushed or its tendon fibers frayed by pressure or rubbing especially when the underlying bone is hard and pointed. If the mechanical insult is only from one direction, it is on that side of the tendon that the synovial bursa is present. Tendon sheaths protect the entire surface of the tendon and are found especially at the carpus and tarsus.

Synovial bursae are found either in a subcutaneous position (**subcutaneous synovial bursa -9),** in which case they are usually acquired; or they are subtendinous (**subtendineal synovial bursa -15)** or submuscular in position and are then usually congenital or inherited.

The tendon sheath facilitates the movement of the tendon in relation to the underlying bone. The sheath consists of a **parietal part (12),** which is connected to the **tendinous part (13)** by means of a mesentery-like fold, the **mesotendon (14).** The tendinous part is directly applied to the tendon and receives its blood and nerve supply from vessels and nerves which pass in the mesotendon. The fibrous layer of the tendinous part is thin and identical to the connective tissue envelope of the tendon. Common tendon sheaths encompass several tendons and provide a smooth, slippery surface for their movement.

The term **fascia** is given to a sheet of stretched connective tissue, with specified fasciae enveloping muscles (see p. IVA) and larger body cavities. Fibers within the connective tissue cross one another, the spaces between the fibers having a somewhat rhombic configuration. With contraction of a muscle surrounded by its fascia, there is a consequent increase in muscle girth, a change in orientation of the connective tissue fibers of the fascia and of the rhombic spaces which the fibers define. Larger fasciae envelop muscle groups, are sometimes termed group fasciae, and separate them from neighbouring structures. In as much as these larger sheets of fasciae connect with periosteum of the underlying bone, osteofibrous chambers are formed which establish the position and direction of the muscles passing within the chambers. Muscles may also arise or insert at areas of reinforcement of the fasciae involved. The large fascial layers envelop the body and cavities of the trunk. They form the fascia trunci externa and the fascia trunci interna, being exter-

nal and internal, respectively, to the muscle layer which attaches to the skeleton of the trunk. The fascia trunci externa is subdivided into the fascia trunci superficialis and — profunda. The fascia trunci superficialis corresponds to the fibrous layer of subcutis and is connected with the cutaneous muscles where these are developed in trunk, neck, and head. The fascia trunci profunda lies in the body wall directly upon the muscle layer attached to the skeleton. This fascia separates the individual trunk muscles with fascial lamellae. For example, the thoracolumbar fascia gives off lamellae which pass between the muscles of the vertebral column. On the limbs, the antebrachial and crural fasciae and other of the large fascial sheets of fascia profunda binding the muscles, form reinforcing bands, the **retinacula (8),** for tendons passing beneath them and thus fixed in position.

The fasciae are relatively impermeable to fluids such as extravasates or pus, which then advance along fascial planes and may 'break through', far removed from the site of the disease.

The **cutaneous muscles** are usually poor in myoglobin and pale variants of skeletal muscles. Most have no direct connection to the skeleton but lie mainly between the lamellae of the fascia trunci superficialis and insert into the skin by short tendons. In this way, their contraction moves or stretches the skin.

The **blood vessels** and **nerve** to the muscle enter together at the **hilus of the muscle (7)** (the word 'hilum' meaning 'a small thing' is grammatically correct). The branches of these vessels run mainly in a longitudinal direction within the muscle, especially in the perimysium internum and — externum. Subsequent branches are encountered chiefly in the endomysium where 4 — 8 capillaries surround a muscle cell. The blood vascular supply is variable according to the type of muscle and, in association with the high energy requirement of skeletal muscle cells, is more intensive than the sparse vascularization of tendon. Besides the transport of metabolites the intense vascularization regulates temperature. The veins, according to the principle of a tubular cooling system similar to the watercooling system of an automobile, carry away the heat energy of muscular work. This heat is given off at the body surface by the cutaneous veins.

The **innervation** of most muscles is by a single nerve. Only the long muscles of the trunk, such as m. longissimus thoracis and m. longissimus lumborum, arising in the embryo from many myotomes, are multiply supplied. Also muscles, which, in the course of phylogenesis have proceeded from the fusion of two individual muscles show a dual innervation. For example in the dog the mm. pectineus et adductor longus are fused and in the cat completely separate. The nerve of a muscle ramifies just as the large blood vessels, in the peri- and endomysium. It is made up of medullated large (alpha-) and small (gamma-) motor neurons, non-medullated autonomic and medullated sensory fibers. A large motor neuron (comprising a large multipolar nerve cell with its dendrites from the ventral column of the spinal cord, and its process or axon) and the muscle cells innervated by it, form together a motor unit. In muscles which effect fine muscle movement such as the ocular muscles, only a few small muscle cells are incorporated in the motor unit. On the other hand, in the larger muscles of the extremities, several thousand muscle cells of considerable size belong to the motor unit. The small gamma motor neurons innervate the modified muscle cells of the neuromuscular spindle. The nonmedullated fibers of the nerve innervate the blood vessels within the muscle. The sensory fibers have their origin in pain receptors, in the neuromuscular spindles, and in the neurotendinous spindles (Golgi tendon organs), both spindle types mediating information concerning the length, tension, and rate of change in length of the muscle.

Muscle and nerve form a functional unit. The contraction of every muscle is brought about by a motor nerve acting via the motor endplates of the muscle cells of the muscle innervated. With regular, rigorous, muscle activity, the muscle fibers increase in diameter physiologically, resulting in muscle hypertrophy. The opposite would be the atrophy of inactivity. The functional and anatomical relationship of a muscle and its nerve is so close that it has scarcely been altered during the phylogenetic development of higher animals (conservative conduct). From a comparative anatomical viewpoint, with the different species, the muscle retains its specific nerve. For that reason the nerve is an important criterion in determining the homology of muscles since the connection between muscle and nerve, once entered into, is rarely lost.

In a phylogenetic regard, homologous muscles are quite variable in shape. From species to species the same muscle can appear very different and may no longer be recognized with certainty.

Functionally, muscles serve chiefly for movement of the body. Further, muscles have an auxiliary function of maintaining body processes such as respiration, defecation and micturition. In addition muscles enable the animal to stand by stabilizing and giving rigidity to the joints. With shivering due to cold, muscles produce heat for thermoregulation.

Myology II

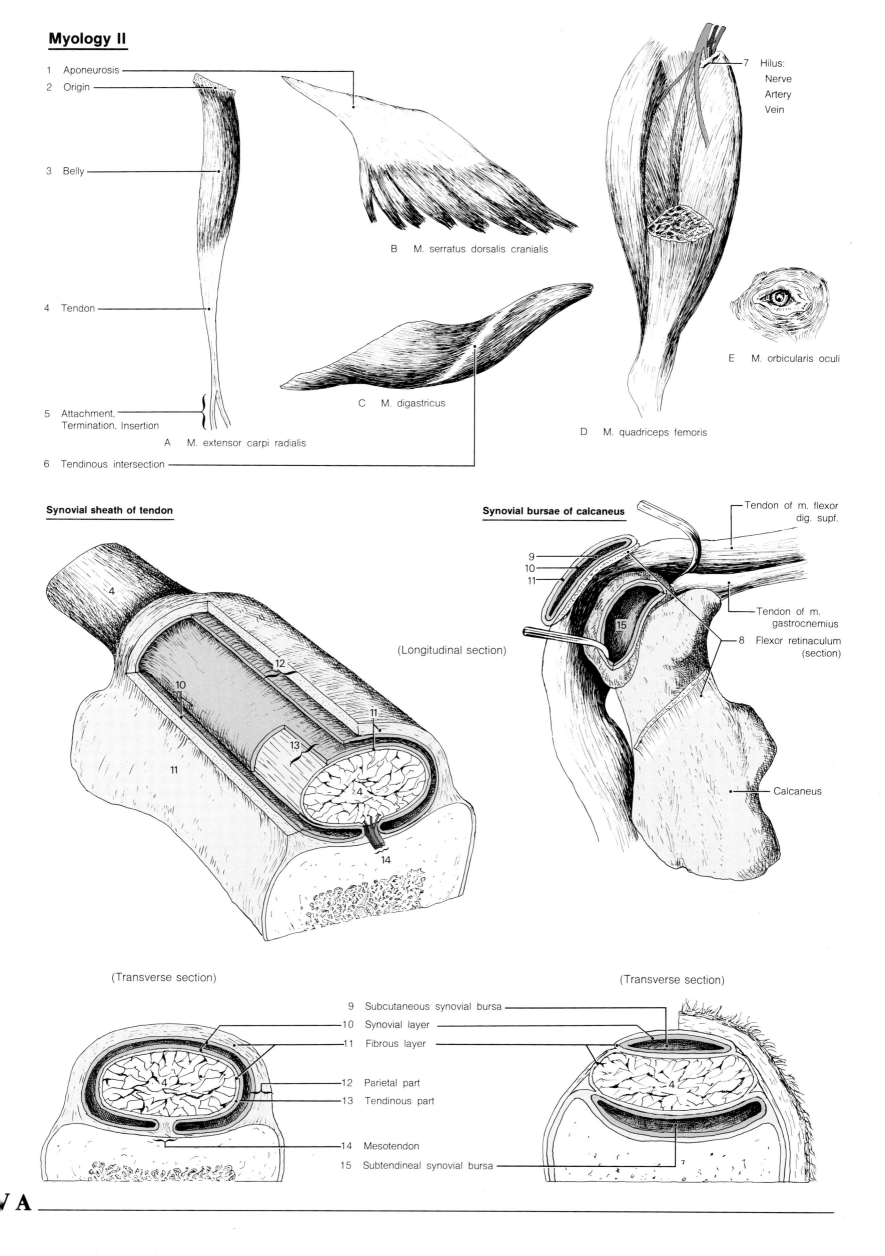

1 Aponeurosis
2 Origin
3 Belly
4 Tendon
5 Attachment, Termination, Insertion
6 Tendinous intersection

A   M. extensor carpi radialis

B   M. serratus dorsalis cranialis

C   M. digastricus

D   M. quadriceps femoris

7 Hilus:
Nerve
Artery
Vein

E   M. orbicularis oculi

Synovial sheath of tendon

(Longitudinal section)

Synovial bursae of calcaneus

Tendon of m. flexor dig. supf.

Tendon of m. gastrocnemius

8 Flexor retinaculum (section)

Calcaneus

(Transverse section)

(Transverse section)

9 Subcutaneous synovial bursa
10 Synovial layer
11 Fibrous layer
12 Parietal part
13 Tendinous part
14 Mesotendon
15 Subtendineal synovial bursa

1 The **STRUCTURAL UNIT** of nervous tissue is the nerve cell with its processes. This is also called a neuron and is supported by neuroglial cells, referred to as nerve cell 'glue', a supporting tissue. The neuron has the capacity to receive an impulse, to convey the impulse as an 'excitatory state', to modify and 'store' the impulse as well as to transmit it in an altered form. In this way the nervous system controls the coordination of internal body functions, especially those of the internal organs, and makes possible the animal's analysis of its environment. The latter is achieved by the reception of the stimulus and its further conduction from cutaneous receptors and other sense organs to the central nervous system. The conduction of impulses to the CNS is via the afferent limb of the reflex arc, whereas the response to the stimulus is via the efferent limb which proceeds from the central nervous system to the musculature.

The **DIVISION** of the nervous system is according to topographic and functional criteria. From a **topographical viewpoint**, it is divided into the central nervous system (CNS) and the peripheral nervous system (PNS). From a **functional viewpoint**, it is divided into the somatic and the autonomic (vegetative) nervous system, both of which belong partly to the CNS and partly to the PNS.

According to neurotransmitter substances, origin and topography, the autonomic nervous system is further divided into the sympathetic nervous system and the parasympathetic nervous system, which function predominantly in an opposing manner. Both contain chiefly efferent but, in addition, afferent portions (see p. 19).

The somatic nervous system makes possible an analysis of the environment. This analysis proceeds subconsciously but as it is partly conscious, it can be so influenced at any time.

The **CENTRAL NERVOUS SYSTEM** is composed of the brain (see pp. 45 − 47) and the spinal cord (see p. 44). In both brain and spinal cord **gray matter (15)** can be clearly distinguished from **white matter (16)**. The cell bodies of the nerve cells lie in the gray matter surrounded by neurophils. The white matter is made up chiefly of the conducting pathways formed by neuronal processes, the phospholipid medullary sheaths of which are markedly white. In the cerebrum and cerebellum the gray matter lies largely peripherally as the cerebral or cerebellar cortex, whereas the white matter is situated centrally as medullary substance. In the spinal cord the reverse is true. The gray matter forms the typical central 'butterfly' figure and is surrounded by the peripheral white matter, in which the conducting pathways, tracts or fasciculi, extend to and from the brain. In the white matter of the CNS, medullary 2 sheaths are formed by oligodendrocytes, that is glia cells of the CNS. As a rule one oligodendrocyte envelops several nerve cell processes (see VI.2.).

The **PERIPHERAL NERVOUS SYSTEM** embraces all neural portions which lie outside the membrana limitans gliae superficialis of the CNS (12 cranial-, 8 cervical-, 13 thoracic-, 7 lumbar-, 3 sacral-, and approx. 5 caudal nerves). The PNS chiefly integrates parts of the somatic nervous system. Most peripheral nerves, however, also convey autonomic modalities.

The **peripheral nerves** are formed from bundles of nerve fibers. The **nerve fiber (7)** consists of a cell process (axon, dendrite) and its sheaths. The nerve cell with its processes is called a **neuron (3)**. The **nerve cell body (4)** usually sends out several **dendrites (5)** and one neurite or **axon (6)**. The dendrite of the sensory neuron functions as a receptor or receives the excitatory state from receptors. The dendrite conveys the excitatory state only in one direction, toward the nerve cell body then via the axon to the spinal cord. The axon of the motor neuron arises at the **axon hillock (4)**, which is free of chromatophilic substance and conducts the excitatory state to the effector organ. Most nerve fibers, in some of which the associated nerve cell process may be up to one meter long, are each surrounded by a myelin sheath and are known as **medullated** or **myelinated fibers (25)**. The myelin sheath is formed by a peripheral **Schwann cell** or **neurolemmocyte (23)**. This is comparable to the opening of a sardine tin, the turning key of which represents the nerve cell process (see VI.2.). In the proximal and distal ends of the Schwann cell, the myelin sheath of the nerve cell process or neurofiber 3 is lacking in the region of the **nodes of Ranvier (8)** which are functional in saltatory conduction of the nerve impulse. The endoneurial sheath 4 and **endoneurium (20)** follow the nerve fiber as it extends peripherally. Formed by loose connective tissue with many fine reticular fibers, the sheath passes between individual nerve fibers and surrounds them.

4 The nerve fibers form bundles which are surrounded by the **perineurium**, a continuation of spinal arachnid (see p. 44), which exhibits an external **fibrous part (21)** and an internal neurothelium or **epithelioid part (22)**. Several bundles of nerve fibers are united to form a peripheral nerve which is clothed superficially by **epineurium (19)**, a continua-
4 tion of dura mater. From the epineurium, loose connective tissue with fat tissue radiates out as paraneurium, attaching the nerve to neighbouring structures. With a **nonmedullated nerve fiber (24)** several

axons invaginate into one peripheral glial cell (Schwann cell) but no medullary coat is 'rolled up'. A nerve can consist of medullated or nonmedullated fiber types, or both can be present.

Regarding the spinal nerves, the nerve cell bodies of the motor neurons of the **SOMATIC NERVOUS SYSTEM** are in the ventral horn of the 5 gray matter of the 'butterfly' configuration observed on cross-section of the spinal cord. With the cranial nerves, the cell bodies of the motor neurons are ventral in the brain stem. In the case of the spinal nerves, the conduction of the excitatory state to skeletal muscle is by a path passing from the spinal cord via the **ventral root (13)** to a motor endplate on individual skeleton muscle cells.

The **afferent neurons of the somatic nervous system** conduct nerve impulses from peripheral nerve endings in the skin and other organs or from receptors in those structures. Such neurons are found in sensory nerves. The path of the nerve impulses is by way of dendrites, usually quite long, to their nerve cell bodies lying in each of the **spinal ganglia (18)** near the spinal cord in the intervertebral foramina. From there, the impulse passes along a short axon into the **dorsal root (14)** of the spinal nerve to the dorsal horn of the gray matter of the spinal cord. The sensory portions of the cranial nerves have their nerve cell bodies in ganglia near the foramina through which sensory fibers pass into the skull.

Spinal nerves can be divided into segmental nerves and plexus nerves. **Segmental nerves (9)** of the somatic nervous system, for example the cranial iliohypogastric n., do not enter into a plexus formation with neighbouring nerves but pursue an isolated course and run approximately parallel to one another. The segmental nerves divide into a dorsal branch (d) and a ventral branch (v). Each of these branches divides further into a dorsomedial branch (dm) and a dorsolateral branch (dl) on the one hand and a ventromedial branch (vm) and a ventrolateral branch (vl) on the other.

Plexus nerves proceed from **nerve plexuses (2)**, such as the brachial plexus (p. 8A) and the lumbosacral plexus (p. 25A). The nerve plexuses arise by the intermixing of fibers of the **ventral branches** of several spinal nerves.

After exchange of fibers of the ventral nerve branches which contribute to the plexus, there arise from the plexus the individual plexus nerves such as the **genitofemoral n. (1)**. These plexus nerves are composed of nerve fibers from several spinal cord segments. For example, the genitofemoral n. proceeds from lumbar spinal nerves 3 and 4, and, correspondingly, the individual spinal cord segments give origin to nerve fibers of several plexus nerves. The third lumbar (L3) spinal nerve gives nerve fibers to the lateral cutaneous femoral and genitofemoral nn. (see p. 20). The dorsal branches take their course as typical segmental nerves.

Efferent neurons of the **AUTONOMIC NERVOUS SYSTEM** (here the sympathetic nervous system is used as an example) innervate, among other things, the smooth muscle of the organs involved and of the blood vessels leading to those organs. Such autonomic nerves regulate and coordinate the function of the internal organs and the process takes place subconsciously, that is the animal is generally unaware of it. A common feature of the autonomic nervous system is the formation of nerve plexuses. These autonomic plexuses are preferentially arranged around large arterial trunks and consist, predominantly, of medullated and nonmedullated sympathetic and parasympathetic nerve fibers. The plexuses contain multiple ganglia such as the solar plexus at the origin of the celiac and cranial mesenteric aa. containing ganglia of like name (see p. 44). The efferent limb of the autonomic nervous system consists of two neurons. The nerve cell bodies of the proximal neurons are found in the CNS; and the preganglionic medullated fibers extend in the **white ramus communicans (12)** to the autonomic ganglia such as the **ganglia** of the **sympathetic trunk (17)**. Here or further distally at the prevertebral ganglia such as the **cranial mesenteric ganglion (10)**, synaptic transmission of the nerve impulse to the postganglionic nonmedullated fiber takes place. The nerve cell body of the postganglionic fiber lies in the autonomic ganglion. The postganglionic nonmedullated or nonmyelinated fibers extend to the effector organs. If the effector organ, such as smooth muscle of the hair follicles, sweat and sebaceous gland cells, is in the skin, the postganglionic fiber will pass in the **gray ramus communicans (11)** to the peripheral spinal nerve to be distributed with its cutaneous nerves.

The **afferent (visceral sensory) neurons of the autonomic nervous system** lead from the periphery to the CNS. They mediate, for example, visceral pain. Their cell bodies lie near the CNS, for the most part in the spinal ganglion.

The **intramural nervous system** can also be accounted as part of the **autonomic nervous system**. The intramural system lies in the wall of hollow organs, for example, in the gut (see p. 44A). (More detailed information can be found in textbooks of histology.) Its nerve cell bodies belong predominantly to the parasympathetic nervous system. The intramural nervous system functions in the hollow organs as an independent regulatory system for the proper motility of the gut, the movements of which can continue even after it is isolated from the body.

V

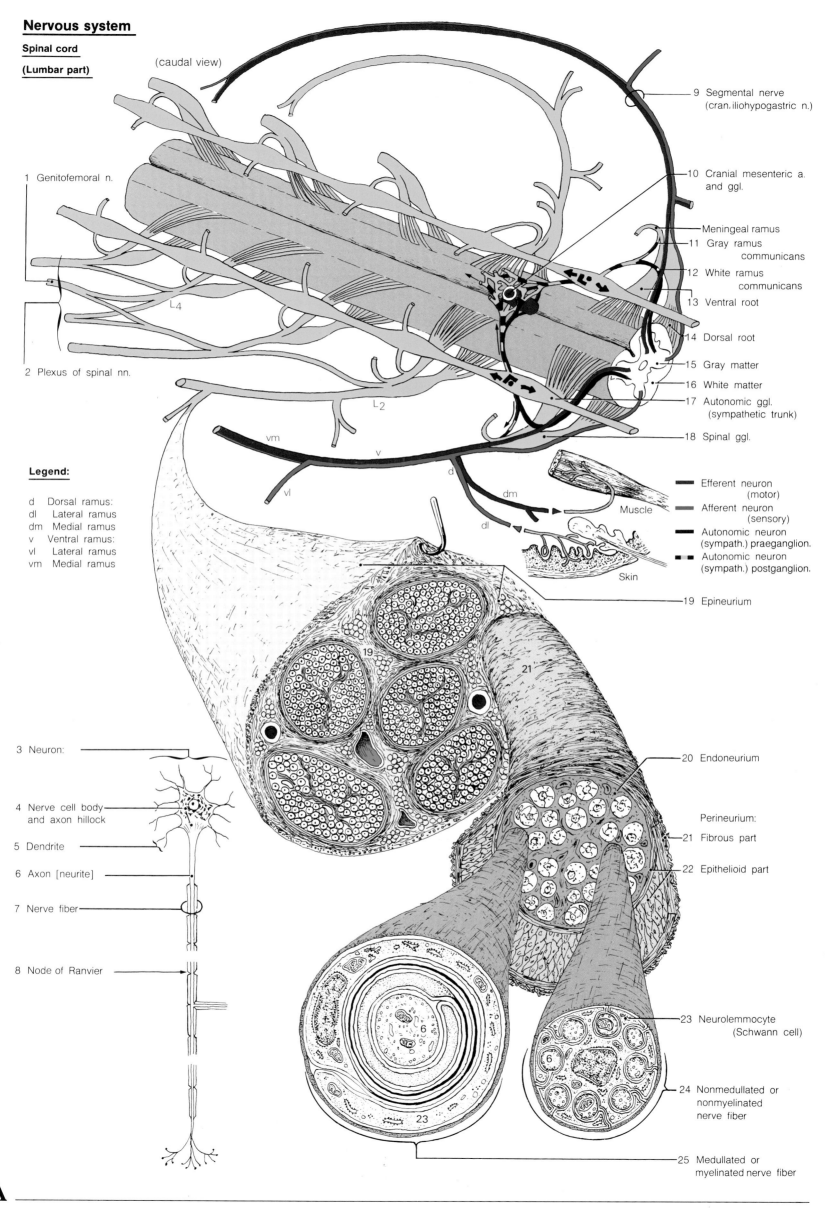

Nervous system

Spinal cord

(Lumbar part)

(caudal view)

1 Genitofemoral n.

L₄

2 Plexus of spinal nn.

Legend:

d   Dorsal ramus:
dl   Lateral ramus
dm Medial ramus
v   Ventral ramus:
vl   Lateral ramus
vm Medial ramus

9 Segmental nerve
(cran. iliohypogastric n.)

10 Cranial mesenteric a.
and ggl.

Meningeal ramus
11 Gray ramus
communicans
12 White ramus
communicans
13 Ventral root

14 Dorsal root

15 Gray matter
16 White matter
17 Autonomic ggl.
(sympathetic trunk)

18 Spinal ggl.

Efferent neuron
(motor)
Afferent neuron
(sensory)
Autonomic neuron
(sympath.) praeganglion.
Autonomic neuron
(sympath.) postganglion.

Muscle

Skin

19 Epineurium

20 Endoneurium

Perineurium:
21 Fibrous part

22 Epithelioid part

23 Neurolemmocyte
(Schwann cell)

24 Nonmedullated or
nonmyelinated
nerve fiber

25 Medullated or
myelinated nerve fiber

3 Neuron:

4 Nerve cell body
and axon hillock

5 Dendrite

6 Axon [neurite]

7 Nerve fiber

8 Node of Ranvier

vm

v

vl

d

dm

dl

L₂

I A

BLOODVASCULAR SYSTEM

The heart, the muscular pump of the **CARDIOVASCULAR SYSTEM,** pumps blood through all body regions. Thus it maintains the supply of nutrients, oxygen, water, hormones and antibodies as well as the transportation of waste products and the regulation of body heat.

The cardiovascular system is divided into: a **systemic circulation** conducting blood from the left ventricle, throughout the body, and via veins, back to the right atrium; a **pulmonary circulation** which pumps the deoxygenated blood to the lungs via pulmonary arteries, and the oxygenated blood via pulmonary veins to the left atrium; and a **portal system** which supplies the liver through two capillary systems connected by the portal vein (see p. 23).

Major organs such as heart, liver and lungs, have a **dual blood supply** by **vasa privata** and **vasa publica**. For example, in the lungs the pulmonary vessels pass through the lungs for overall oxygen gain (vasa publica) and the bronchial vessels supply the bronchial tree (vasa privata).

Blood vessels are classified according to the structure of their walls, which depends significantly upon blood pressure. Thin-walled veins, for example, veins of the leg, gradually develop a wall typical of arteries,for example coronary aa when grafted into the arterial system. The wall of large vessels has three tunics: a **tunica interna** or **intima (2)** consisting of an endothelium, a subendothelial layer of collagenous fibrils, and an internal elastic lamina or membrane; a **tunica media (3)** consisting of smooth muscle cells, an elastic lamina and fibers, and an external elastic lamina or membrane demarcating the tunica media; and a **tunica externa** or **adventitia (4)** comprised of collagenous and elastic fibers.

The **arteries** arise from the heart, and except for the pulmonary aa., carry oxygenated blood. Arteries are classified as **elastic (1)**, **muscular (11)**, and mixed types according to the structure of the tunica media. Elastic arteries are more proximal to the heart, for example, aorta and brachiocephalic trunk. Their tunica media contains smooth muscle cells but also very numerous and thick fenestrated elastic laminae contributing to the yellow appearance of these vessels. The coronary aa. and arteries more distant from the heart belong to the muscular type. Their media contains chiefly smooth muscle cells and fewer elastic laminae. **Terminal arteries**, for example, the cerebral and coronary aa., are of great clinical significance, since they have no **collateral arteries (7)** and usually no arterio-arterial anastomoses. Occlusion of any of these terminal arteries leads to ischemia and necrosis of the tissue supplied.

Convoluted arteries, for example helicine aa. of penis, are named according to their highly convoluted and sometimes spiral-like course. During penile erection they become engorged with blood and lose their convolutions while extending to their full length. Unlike the helicine aa., the morphology of the convoluted aa. of the mesovarium is independent of volume fluctuations. Convoluted arteries have portions which are occlusive in morphology, such as portions of the helicine aa. of the penis, in that their lumina become narrowed by **myoid cushions (6)** of the internal tunic or intima. Thus blood flow is regulated functionally, for example, diverting blood through the capillary system during erection (see appendix 28.6.). **Arteriovenous anastomoses (5)** also regulate blood flow. A few organs such as kidney and brain, require a large and consistent blood supply. However, for most organs, such as the gastrointestinal tract, blood volume depends upon functional demands. It is the arterial portion of anastomoses which regulates blood flow and when it narrows, blood passes on into the capillary system. If the arterial part is 'open' some blood may by-pass the capillary bed by diversion through the anastomoses. **Anastomoses arteriovenosae glomeriformes (8)** occur in the digital organ and in the skin. They have a tuft of capillaries midway along their course which is supplied extensively with autonomic nerves and contains nests of endocrine cells. The arterial portions of these anastomoses also function as regulators of blood flow.

Arterioles (9) are less than 10 mm in diameter and in transverse section their tunica media contains but a few spirally arranged smooth muscle cells. The precapillary arteriole has only one layer of smooth muscle cells.

The wall of the **capillaries (10)** comprises only a remnant of the tunica intima, namely an endothelium, a basal lamina, and a discontinous layer of pericytes. Capillaries measure up to 3 mm in length and their diameter varies between 3 and 10 μm. They may be classified into three types according to their endothelium and specific functions. **Non-fenestrated endothelial cells (13)** form a continuous lining of the capillary lumen, and may be found in musculature, skin, and connective tissue. **Fenestrated endothelial cells (12)** have intracellular pores which are numerous in organs such as gastrointestinal tract, kidney and endocrine glands. A discontinuous endothelium with **intercellular apertures (14)** and a discontinuous basal lamina occur in sinusoids of the liver, spleen, and bone marrow. The apertures facilitate the passage of macromolecular substances.

A **capillary rete** is an anastomosing capillary which occurs within the arterial portion. It is supplied and drained by arterioles as in the case of the coiled capillaries of the renal corpuscles (see 24.2.).

Venules (16) in the initial segment of the blood venous portion of the circulation measure up to 50 μm in diameter. Their tunica media lacks muscle cells, they have pericytes, and their tunica externa contains collagenous fibrils, leucocytes and macrophages. **Postcapillary venules (17)** of lymphatic organs have a typical cuboidal or columnar endothelium which facilitates the passage of lymphocytes and monocytes from the lumen into the perivascular, lymphoreticular tissue.

The wall of the **veins** is less distinct in organization than the arterial wall. In the tunica media of medium-large veins longitudinal bundles of smooth muscle cells lie chiefly in the collagenous connective tissue. In large veins, such as the caudal vena cava, the tunica externa contains the greater amount of smooth muscle cells, dominating the tunica media which contains only a few.

Venous valvules (18) are semilunar folds of tunica intima protruding into the lumen. A fibrous stroma of connecting tissue forms the core of the valves which regulate the blood flow. With back flow, the valves are forced to expand and contact each other centrally, thereby occluding the lumen and preventing further back flow.

Occlusive veins are a special type of vein occurring in endocrine glands, uterus, oviduct, cavernous bodies and liver. They are so-termed because their lumina can be narrowed by intimal cushions. This apparently retards the blood flow in the post-capillary system.

Venous plexuses occur in the form of networks such as the testicular vein forming the pampiniform plexus. This participates in thermoregulation (see p. 25A).

Venous sinuses are essentially expansions of veins, such as the sinus venarum cavarum, and the coronary sinus. The venous sinuses of the dura mater lack tunica media, tunica externa and valves (see p.45). Venous sinuses may have quite thick walls with smooth myocytes, as in the cavernae of the genital organs, or thin walls as in the tunica submucosa of the turbinates. Sinusoids are large capillaries which may have a discontinuous endothelium, attenuated endotheliocytes and an incomplete basal lamina. Their large lumen renders them 'sinus-like'.

Vasa vasorum (15) are nutritive blood vessels within the walls of large arteries, veins, and lymphatic trunks furnishing blood to the tunica externa and media.

Nerves of the blood vessels (15) are predominantly unmyelinated and belong to the autonomic system. Vasoconstriction is regulated by the sympathetic nervous system.

Bloodvascular system

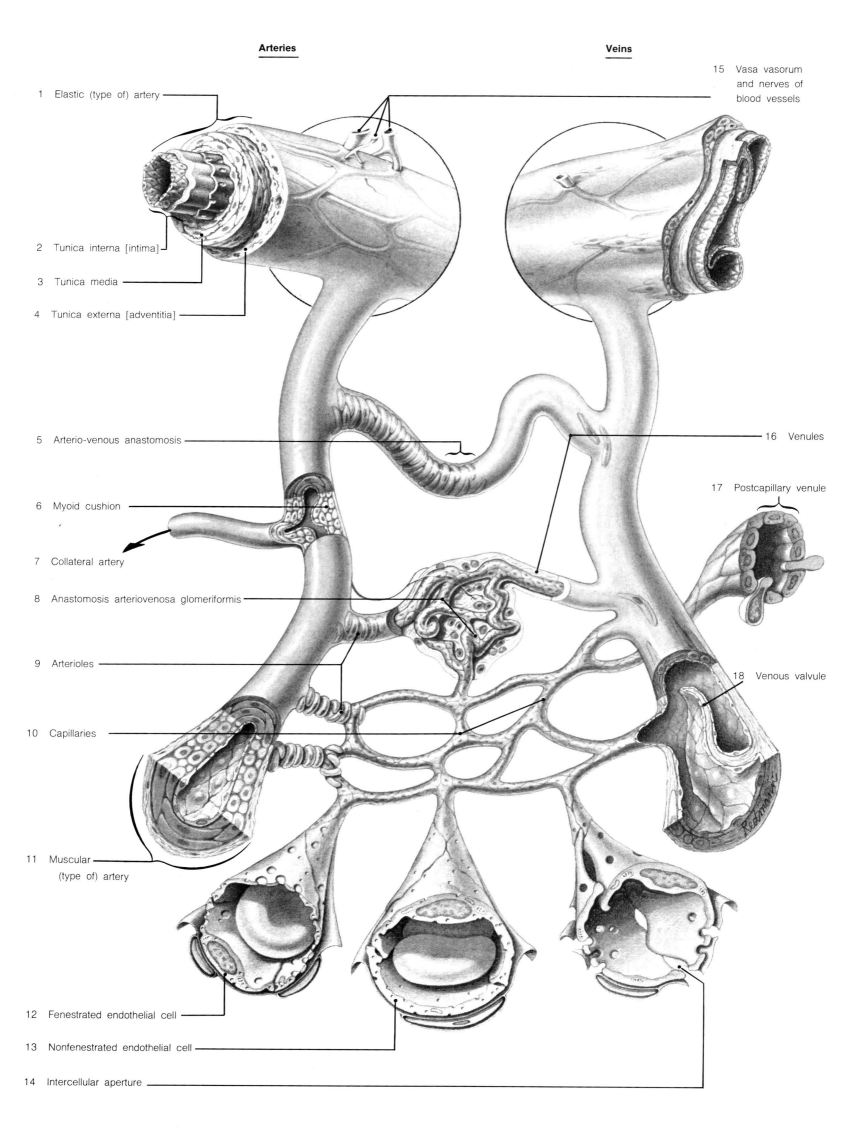

1 Elastic (type of) artery

2 Tunica interna [intima]

3 Tunica media

4 Tunica externa [adventitia]

5 Arterio-venous anastomosis

6 Myoid cushion

7 Collateral artery

8 Anastomosis arteriovenosa glomeriformis

9 Arterioles

10 Capillaries

11 Muscular
    (type of) artery

12 Fenestrated endothelial cell

13 Nonfenestrated endothelial cell

14 Intercellular aperture

15 Vasa vasorum
    and nerves of
    blood vessels

16 Venules

17 Postcapillary venule

18 Venous valvule

LYMPHATIC SYSTEM

The **LYMPHATIC SYSTEM** includes structures of the defense (immune) system and the lymph vascular system, both structural types being integrated in function. The defense system, that is, lymph nodules, lymph nodes and lymph organs recognizes and eliminates foreign bodies.

The **lymph vascular system** resorbs interstitial fluids, conducts lymph to lymph nodes and transports lymph to the blood vascular system from the lymph nodes. The lymph vascular system is not a complete circuit and possesses no lymph heart in mammals. It begins as blind **lymph capillaries** and **lymph capillary nets (1)** at the periphery and ends at the cervical-thoracic border at the 'venous angle' formed by the left internal and external jugular vv. Due to the absence of a lymph heart (pump) — found only in lower vertebrates and some birds — lymph volume transport per unit of time is only 0,03% of that for the blood vascular system.

Lymph is typically water-clear to amber in colour, while lymph passing from the gastrointestinal tract may be periodically milky due to suspended fat droplets. For the most part lymph forms from a protein-rich transudate arising from blood capillaries and accumulating in interstitial tissues. These tissue fluids also suspend or dissolve local metabolites and reenter the circulation via blood capillaries as blood plasma, or lymph capillaries as lymph plasma. Lymph capillaries are absent from the central part of the nervous system, non-glandular epithelia, cartilage, bone marrow, thymus and cornea. By routine light microscopy it is difficult to distinguish lymph and blood capillaries. Lymph capillaries normally lack erythrocytes. Lymph capillaries are distinguished using electron microscopy, the basis for much of the following structural/functional discussion.

The **endotheliocyte** of the **lymph capillary (22)** is characterized by overlapping cell boundaries and **anchoring filaments (24)**. The latter pass from the **collagen fibers (25)** to foci over the basal surface of the cell. Lymph capillaries often appear collapsed. When there is an abnormal accumulation of fluid (edema) in the interstitial tissue, a greater pressure results, the anchoring filaments tense, and the capillaries retain their lumina. At the cell-to-cell boundary the overlapping endothelio-
1 cytes form an **'open junction' (23)** that is valve-like, preventing the reflux of lymph into the interstitium while allowing the entrance of fluid to the lymph capillary. These 'open junctions' are quite large and extensive, allowing colloids, cellular debris, microorganism, tumor cells, as well as ameboid blood cells to enter. The colloids include hormones and long-chained fatty acids, which enter the lymphatics of the gut. Lymph capillaries are also more or less 'open' under normal physiologic conditions. This is especially true for areas such as the gut, and some endocrine organs. In the pleural and peritoneal serosal membranes of the diaphragm unique stomata occur in the lymph capillaries which open and close as a passive 'gulping' related to respiratory movements, assisting the uptake and thus transport of lymph.

The **postcapillary lymphatic vessels (26)** are also called conducting lymphatics. They have a larger lumen than the capillaries, possess many valves, and anastomose freely, forming a network.

The larger **muscular lymphatic vessels (27)** function as transporting lymphatics. These vessels possess a **tunica interna** or **intima (28)** of endothelium, a **tunica media (29)** of spirally-disposed myocytes, and a **tunica externa** or **adventitia (30)** of collagenous connecting tissue. The fibrous ring at the base of the valve leaflets or **segmental valvules (36)** is relatively non-expansive. The muscular lymphatic segment, the **afferent lymphatic vessels (2)** between the valves are distended and contract with the flow of lymph. When filled with lymph these vessels are beaded. Where tributary lymphatics flow into larger vessels, **entrance valves (35)** are present.

The major **lymphatic trunks (20)** include the thoracic duct, cysterna chyli, and the lumbar, intestinal and jugular trunks. They receive lymph from their tributary, muscular lymphatics. The jugular trunks empty their lymph into the blood of the external jugular veins at their respective 'venous angles', the single thoracic duct entering at the same point on the left side only. The wall of these vessels presents three layers. Smooth muscle fascicles are disposed longitudinally in the tunica externa as well as **vasa vasorum (21)** and **nerves of lymph vessels (21)**. Movement of lymph is due to several factors. A pressure is generated peripherally at the lymph capillary by the formation of lymph.

Together with the 'open junction' valve, this formation initiates the proximally directed lymph flow. This influence of valves is continued in the postcapillary lymphatics. Smooth musculature of the larger lymphatics, together with valves, are the bases of peristaltic lymph flow. Other influences support the movement of lymph. These are: the massage of lymphatics by the periodic contractions of neighbouring skeletal muscles; pulsation and fluid movements in an adjacent artery and vein bound to a lymphatic by connecting tissue (see text figure below); and respiratory and gastrointestinal visceral movements. Removal of a lymph node or other surgical intervention temporarily disrupts lymph drainage. Distal to this site the region becomes edematous due to the accumulation of lymph. The region feels cold and after digital palpation pits and folds of the skin disappear slowly. Lymph flow is reestablished by histogenesis of the endothelium and vascular wall. Lymph nodes do not regenerate. Postsurgically in man, lymph flow is supported by gentle massage in the direction of lymph flow. Together with dilation, massage stimulates contraction of the muscular lymphatics.

Effect of arterial pulse on lymphatic and venous transport

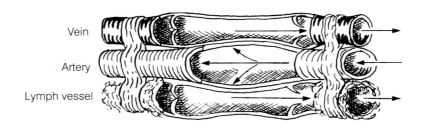

Lymph nodes (8) are named according to location, for example, the 2 mandibular lymph node. Body regions are drained by lymphatics and nodes. Several nodes which share an overlapping tributary region together comprise lymphocenters. The 'primary lymph' from a tributary region flows to the lymph node by **afferent lymphatic vessels (9)**. Several of these converge at the lymph node. From 2-4 **efferent lymphatic vessels (15)** exit the lymph node at its hilus. 'Secondary lymph' having about 10 times more lymphocytes than 'primary lymph' flows in the efferent lymphatics. Lymph nodes occur in series and secondary lymph is thus repeatedly filtered.

Lymph nodes are bean-shaped, have a firm, smooth surface, and are embedded in fat. They have a connective tissue **capsule (10)** perforated peripherally by afferent lymphatic vessels and at the **hilus (14)** by efferent lymphatics and blood vessels. From the capsule, **trabeculae (11)** project centrally. Lymph flow through the node is from afferent lymphatics to **subcapsular sinus (12)**, to **intermediate** or **cortical sinus (13)**, to **medullary sinus (16)**, and to the efferent lymphatics.

The **lymph sinus (3)** is bridged by many **reticular fibers (5)**. All sinus surfaces and reticular fibers are lined by littoral cells or **endotheliocytes (4)**. A reticular cell or **reticulocyte (32)** of the interstitial tissue may send phagocytic processes through the littoral cells. **Reticulocytes (6)** and **macrophagocytes (7)** reside in the sinus lumen and avidly phagocytose particles such as carbon, microorganisms, and cellular debris.

The lymph node parenchyma manifests a **cortex, paracortex (18)**, and **medulla**. The **lymphoreticular tissue (31)** is comprised of **reticulocytes (32)** and **reticular fibers (33)**. Some reticulocytes bind and 'present' antigen to the **lymphocyte (34)**. An accumulation and proliferation of 3 B-lymphocytes around such reticulocytes in the cortex produces the characteristic **lymphonodule (19)**. The paracortex is a site of accumulation and proliferation of T-lymphocytes, arriving from the thymus and exiting from the blood at post-capillary venules. The medullary parenchyma is an extension from the **paracortex (18)** as **(medullary) cords (17)** which contain B-lymphocytes, some T-lymphocytes and many transformed B-lymphocytes, that are plasma cells.

Lymphatic organs are discussed with the appropriate topographic anatomy and in the appendix: thymus p. 16 and appendix 16.6; spleen p. 22 and appendix 22.4; solitary and aggregate lymphonodules — appendix VIII. 1.-3. and 21.6.

Lymphatic System

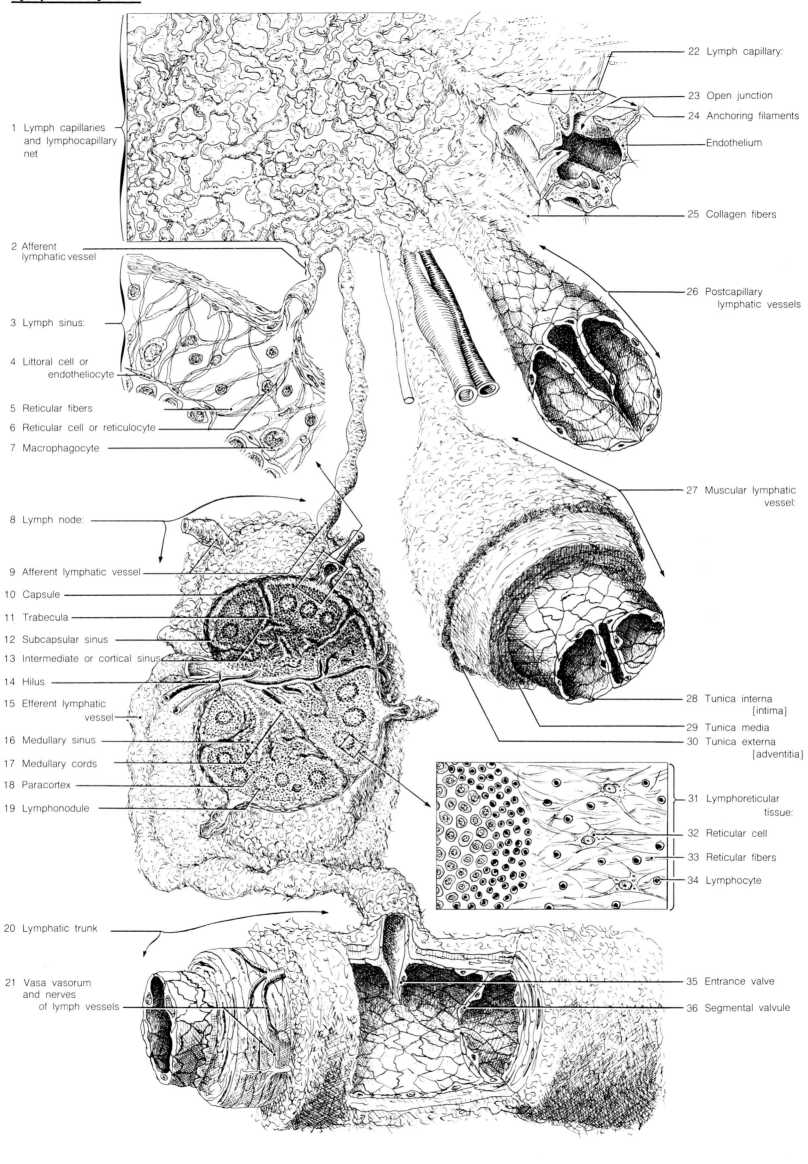

1 Lymph capillaries and lymphocapillary net

2 Afferent lymphatic vessel

3 Lymph sinus:

4 Littoral cell or endotheliocyte

5 Reticular fibers

6 Reticular cell or reticulocyte

7 Macrophagocyte

8 Lymph node:

9 Afferent lymphatic vessel

10 Capsule

11 Trabecula

12 Subcapsular sinus

13 Intermediate or cortical sinus

14 Hilus

15 Efferent lymphatic vessel

16 Medullary sinus

17 Medullary cords

18 Paracortex

19 Lymphonodule

20 Lymphatic trunk

21 Vasa vasorum and nerves of lymph vessels

22 Lymph capillary:

23 Open junction

24 Anchoring filaments

Endothelium

25 Collagen fibers

26 Postcapillary lymphatic vessels

27 Muscular lymphatic vessel:

28 Tunica interna [intima]

29 Tunica media

30 Tunica externa [adventitia]

31 Lymphoreticular tissue:

32 Reticular cell

33 Reticular fibers

34 Lymphocyte

35 Entrance valve

36 Segmental valvule

A

GLANDS, MUCOUS MEMBRANES AND SEROUS MEMBRANES

The **ACTIVITY OF GLANDS** is called secretion. The secretions, such as saliva and hormones have their effect outside of the cell. To prepare the secretion, the gland cell takes up the substances required, synthesizes the secretion within the cell organelles, and then releases it in a manner which is characteristic for that gland cell (see below). Glandular secretion is controlled chiefly by hormones and the autonomic nervous system. Different criteria are used to classify glands. Where **secretion** is discharged into **blood** or **lymph capillaries**, we are dealing with **endocrine glands**. The secretion, a hormone, is carried away by the vascular system to **target organs**, namely those receptive to that hormone. Where secretion is given off onto the **internal** or **external body surface** we are dealing with **exocrine glands**. The exocrine glands are treated below (see also pp. 2-3).

Exocrine glands may be classified according to the **shape of the secretory endpieces**. Tubular, acinar (shaped like a berry) and alveolar (in the form of a vesicle) are recognized. The endpieces of the tubular glands can be: 1. simple tubular, where the secretory endportion is an extended tube without branching; 2. simple coiled tubular, where the secretory endportion is a simple coiled tube; and 3. simple branched tubular where the secretory endpiece is branched. The compound tubular gland has branched efferent ducts.

Exocrine glands may be classified according to the **manner of the secretion**. Eccrine or merocrine gland cells discharge their secretion by exocytosis. In the case of apocrine glands, a part of the gland cell with its plasmalemma is cast off with the secretion. With holocrine gland cells, the gland cell is transformed entirely and forms the secretion.

Exocrine glands are classified according to the **character of the secretion** as: serous, a thin, watery, fluid secretion; mucous, a more viscous, slippery secretion; and mixed, having both serous and mucous portions.

Glands may be classified according to their **composition** as intraepithelial **unicellular**, for example, mucous producing goblet cells, and **multicellular**. The first named are without efferent ducts and open directly into the adjoining lumen. The multicellular glands lie either in the wall of the organ concerned and open into the lumen of the organ by means of a short efferent duct (wall glands); or they form larger aggregates which lie outside the wall of the organ and open into its lumen usually by a longer efferent duct (accessory glands).

The **MUCOUS MEMBRANE (TUNICA MUCOSA)** is the internal lining of those hollow organs which communicate with the exterior. This group of organs includes the alimentary tract, the respiratory system, the urinary system and those parts of the genital system which exhibit a lumen. With the exception of the urinary system, the lumen is covered by mucous which has its origin from the glands of the mucous membrane, namely goblet cells, wall glands, or accessory glands.

The **mucous membrane of the gut** consists of a **columnar epithelium** (-7, E, E') and a **lamina propria mucosae (6)**. The lamina propria mucosae is a connective tissue layer which attaches the epithelium to underlying tissue and contains a **lamina muscularis mucosae (5)**. In some parts of the gut a loose displaceable layer, the **tela submucosa (4)**, and a **tunica muscularis (2, 3)** are present. External to the submucosa is a **visceral part (13)** of the tunica serosa or a tunica adventitia which connects the gut to surrounding structures. In the individual organs the mucous membrane is adapted morphologically to its functional requirements.

I. In the **digestive apparatus** which begins at mouth and ends at anus, the functional differentiation of the mucous membrane is particularly distinct. At both ends of the digestive tube, the transition to the external skin is formed by a cutaneous mucous membrane of **stratified squamous epithelium (A — oral cavity)** (see histology). In those areas of the oral cavity subject to the highest mechanical stress such as the hard palate and those lingual papillae having a mechanical function, the stratified squamous epithelium is actually cornified (keratinized). The striated muscle of the tunica muscularis of the oral cavity and pharynx is continued well into the esophagus and assures the rapid transport of prehended food. Food transport is also enhanced by the lubricating secretions of the salivary glands of the oral cavity and pharynx.

The lamina muscularis mucosae is a delicate layer of smooth muscle forming part of the mucous membrane of the gut. It begins in the esophagus and increases slightly in strength caudally. As smooth muscle, it permits a certain autonomous mobility of the mucous membrane which is independent of the contraction of the tunica muscularis. Likewise, the tunica muscularis is formed predominantly of smooth muscle and is separated from the tunica muscularis mucosae only by the tela submucosa.

At and distal to the entrance of the stomach a **columnar epithelium (E — small intestine)** lines the lumen. This epithelium has less of a protective function and more of a secretory and absorptive one. These functional differences are expressed by the different types of epithelial cells. In the intestinal region the **secretion of mucus** is maintained chiefly by **goblet cells (9)**. The goblet cells are scattered in the luminal epithelium and increase in number from the duodenum through the large intestine

and rectum. The digestive juices are secreted by the wall glands of the intestine and by accessory glands such as the pancreas.

For **absorption** of digestive products, it is necessary to have a very large mucosal surface in contact with the intestinal content. This increase in surface area is obtained by **circular folds (1)** of the intestinal mucous membrane, by **intestinal villi (E', a)**, by **intestinal crypts (E', b)**, and by the **microvilli (8)** of the microvillous border on the luminal surface of the absorptive epithelial cells.

II. In the **respiratory tract** the **pseudostratified ciliated epithelium (B — trachea)** containing scattered goblet cells, produces and carries away mucus. In this way fine dust particles adherent to the mucus are carried to the exterior. The respiratory mucous membrane lines the air passages namely, nasal cavity, paranasal sinuses, nasal pharynx, and auditory tube including middle ear, larynx, trachea and bronchi. (The term 'pseudostratified' means falsely giving the appearance of stratification. In fact each cell has contact with the basal lamina but not all cells reach the luminal surface.)

The **mucous membrane of the genital tract**, for example that of uterine tube (see also 28.9.) and uterus, is clothed by a **simple columnar epithelium (C — uterine tube)** bearing cilia and microvilli. The epithelial cells lining parts of the male genital tract, such as the **pseudostratified columnar epithelium of the deferent duct (D)**, exhibit microvillous cell processes which are long, branched, and non-motile. These are the so-called stereocilia.

III. The **mucous membrane of the urinary tract** is free of glands in the dog. Its epithelium is a **transitional epithelium (F — ureter)** which lines the urinary tract, renal pelvis, ureter, urinary bladder and urethra. The transitional epithelium is a pseudostratified type. All of its cells have contact with the basal lamina but not all reach the luminal surface. According to another view transitional epithelium is a stratified epithelium that is, layers of cells where only the basal layer is in contact with the basal lamina. The transitional epithelial cells abutting the luminal surface possess special mechanisms for protection against the deleterious effects of urine.

The **SEROUS MEMBRANE (TUNICA SEROSA — G)** lines serous cavities and covers the external surface of many organs. It consists of the **mesothelium (14)**, a single layer of squamous cells of mesodermal origin, resting upon a layer of connective tissue, the **lamina propria serosae (15)**. A **tela subserosa (16)** functions as a displaceable interlayer between the lamina propria and the deeper tunica muscularis of the gut. It can serve as a fat storage depot which, especially in swine, is well developed as subserosal fat. A tela subserosa is present only on those organs which are subject to considerable variation in size, such as stomach and intestines.

The **function of the serosa** consists chiefly in the creation of a smooth, moist surface which reduces the friction between the serosal membranes. The serosa accomplishes this by producing serous fluid which is given off into the serous cavities. The fluid also provides the serosa with its glistening character. The amount of serous fluid within the cavities is normally constant due to an equilibrium between **transudation** and **resorption**. The **transudate** is formed by the passage of fluid from blood capillaries to lamina propria and then into the spaces between the serosal epithelial cells.

The serous cavities are spaces which, in life, contain only a little serous fluid. They are the pleural cavities, the pericardial cavity and the peritoneal cavity the serosal lining of which is known as pleura, serous pericardium and peritoneum respectively. Organs sink into the serous cavities drawing serosal folds with them. The serosal covering of the body wall, the **parietal part (11)** continues over the folds as the **intermediate part (12)** to cover the organs as the **visceral part (13)**.

The **relationship** of organs to peritoneum is of particular surgical significance. With surgical intervention this relationship determines whether the peritoneal cavity must be opened and a greater surgical risk taken. An **intraperitoneal organ*** is surrounded by peritoneum up to the mesenteric attachment of the organ, for example, the small intestine. A **retroperitoneal organ** is covered by peritoneum only on that surface facing the abdominal cavity, for example, the kidneys. Such an organ lies in the **retroperitoneal space (10)**. An **extraperitoneal organ** lies near the peritoneal cavity but is not covered by peritoneum, for example, the cisterna chyli, a cistern for the collection of lymph. Some organs which lie on the longitudinal axis of the body, such as the vagina, proceed from an intraperitoneal position, to a retroperitoneal position and to an extraperitoneal position.

The **sensory innervation** of the parietal layer of the peritoneum is by **segmental nerves** namely the lumbar and thoracic spinal nerves. The peritoneum on the abdominal side of the diaphragm receives its sensory innervation by way of the phrenic n. The sensory innervation of the visceral part is via nerves following the pathway of autonomic nerve fibers.

---

* The term 'intraperitoneal' in its restricted sense, refers to substances within the peritoneal cavity and would be limited to mean the small amount of serous fluid present there. In the liberal meaning given here, the term refers to all organs which receive a covering of peritoneum which is complete except for the narrow border at which peritoneum passes onto the organ.

Tunica mucosa and Tunica serosa

Small intestine (sectional view)

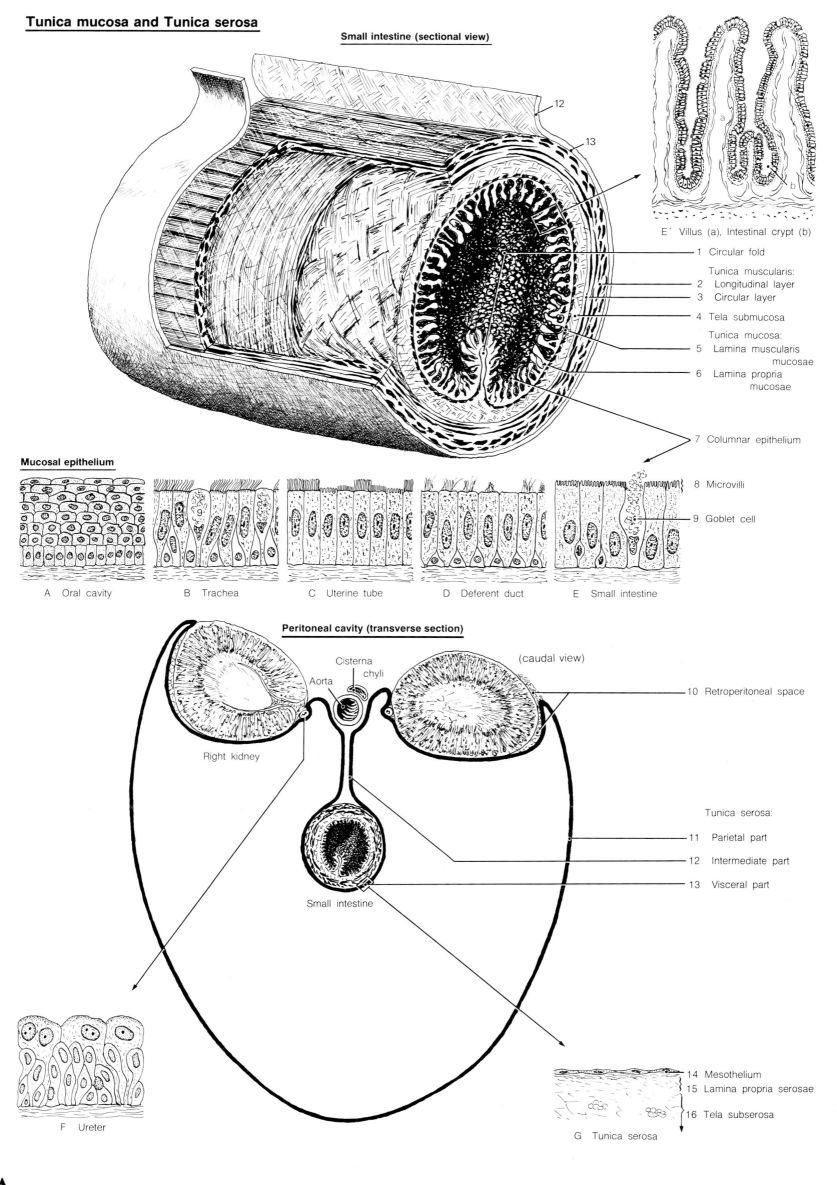

E´ Villus (a), Intestinal crypt (b)

1 Circular fold

Tunica muscularis:
2 Longitudinal layer
3 Circular layer
4 Tela submucosa

Tunica mucosa:
5 Lamina muscularis mucosae
6 Lamina propria mucosae

7 Columnar epithelium

8 Microvilli

9 Goblet cell

Mucosal epithelium

A Oral cavity   B Trachea   C Uterine tube   D Deferent duct   E Small intestine

Peritoneal cavity (transverse section)

Cisterna chyli

Aorta

(caudal view)

Right kidney

10 Retroperitoneal space

Tunica serosa:
11 Parietal part
12 Intermediate part
13 Visceral part

Small intestine

14 Mesothelium
15 Lamina propria serosae
16 Tela subserosa

F Ureter

G Tunica serosa

TOPOGRAPHIC ANATOMY

CHAPTER 1: BODY SURFACE AND AXIAL SKELETON

C1. S1: DIVISION OF THE ANIMAL BODY

a) DIVISION OF THE BODY
Longitudinal lines and body planes serve to orientate the body and its surface. The **dorsal median line (a)** and the **ventral median line (b)** are the median longitudinal lines of the dorsal and ventral parts of the body.

The **median plane (A)** is the plane between both longitudinal lines and divides the body in half in the midline. The **sagittal planes (B, also known as paramedian)** are planes parallel to the median. They also divide the body longitudinally but into unequal parts. **Transverse planes (C)** divide the body in planes perpendicular to those which are longitudinal. **Dorsal planes (D)** are parallel to the dorsal surface and cut the body vertical to the median and paramedian planes.

b) TERMS USED FOR POSITION AND RELATIONSHIP are
derived partly from portions of the body such as **caudal (c)** meaning towards the tail, and partly from directions on the animal surface, for example **sagittal (d)**. The terms internal and external are used when describing hollow or visceral organs. Furthermore, terms such as left,

right, short, long, deep, superficial, longitudinal and transverse, as well as collateral, lateral and medial, are also employed. The term towards the head **(cranial -e)** must not be used in relation to the head itself and the term **rostral (f)** is substituted. The term towards the dorsum **(dorsal- -g)** relates to the dorsal part of the trunk, but it is also used when considering the distal part of the limbs. There it refers to the dorsum of manus and pes. The word **ventral (h)** is not used with reference to the distal parts of the limbs, where one speaks of **palmar (i)** and **plantar (k)** for manus and pes respectively.

The terms **proximal (l)** and **distal (m)** are used relative to the spinal column, that is, the axis of the body. The term **abaxial (n)** relates to positions away from an axis, **axial (o)** referring to those towards an axis. Such usage can be applied for example to either manus or pes where the axis lies between digits 3 and 4. The general terms anterior, posterior, superior and inferior are used extensively in human anatomy. To avoid confusion in quadrupeds, however, these are not used and remain limited in veterinary anatomy to the head (superior and inferior eyelids, anterior and posterior planes of the ocular bulb).

c) PARTS AND REGIONS OF THE BODY divide the body and its
surface. The body parts are head, trunk, including neck, body and tail, and the limbs. The body regions divide the body surface and can be divided further into subregions as indicated in the adjoining table.

REGIONS OF THE BODY

Regions of Cranium
1 frontal
2 parietal
3 occipital
4 temporal
5 auricular

Regions of Face
6 nasal
6' dorsal nasal
6" lateral nasal
6‴ narial (of naris)
7 oral
7' of upper lip
7" of lower lip
8 mental (of chin)
9 orbital
9' upper palpebral
9" lower palpebral
10 zygomatic
11 infraorbital
12 temporomandibular articulation
13 masseteric
14 buccal
15 maxillary
16 mandibular
17 intermandibular

Regions of Neck
18 dorsal neck
19 lateral neck
20 parotid
21 pharyngeal

22 ventral neck
22' laryngeal
22" tracheal

Regions of Dorsum
23 thoracic vertebral
23' interscapular
24 lumbar

Pectoral Regions
25 presternal
26 sternal
27 scapular
28 costal
29 cardiac

Abdominal Regions
30 cranial abdominal
30' of hypochondrium
30" xiphoid
31 middle abdominal
31' lateral abdominal
31" paralumbar fossa
31‴ umbilical
32 caudal abdominal
32' inguinal
32" pubic and preputial

Pelvic Regions
33 sacral
34 gluteal
35 of tuber coxae

36 ischiorectal
37 of ischiatic tuberosity
38 caudal
38' of root of tail
39 perineal
39' anal
39" urogenital
40 scrotal

Regions of Thoracic Limb
41 shoulder
42 axillary
42' axillary fossa
43 brachial
44 tricipital
45 cubital
46 olecranon
47 antebrachial
48 carpal
49 metacarpal
50 phalangeal

Regions of Pelvic Limb
51 hip
52 femoral
53 knee
53' patellar
54 popliteal
55 crural
56 tarsal
57 calcaneal
58 metatarsal
59 phalangeal

Body regions and terms of site and direction in relation to parts of the body indicated

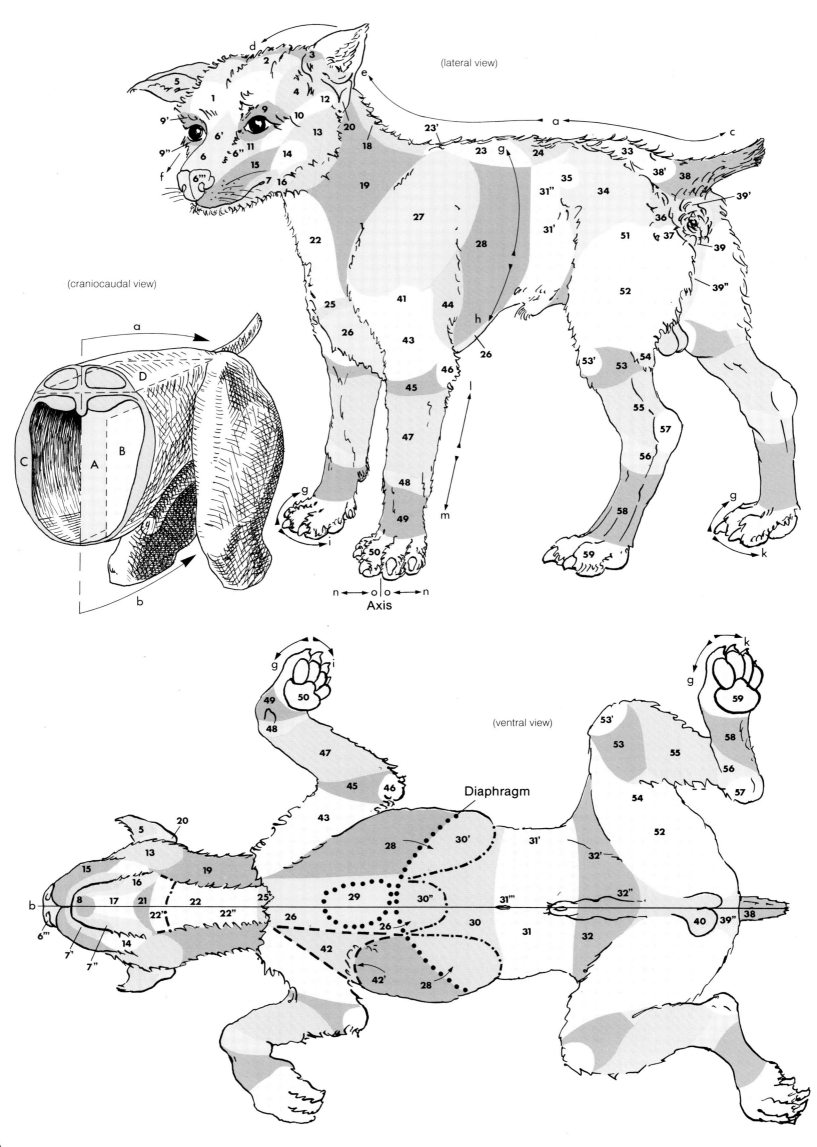

(lateral view)

(craniocaudal view)

Axis

Diaphragm

(ventral view)

1 A

C.1. S.2: SKIN

The **skin** forms the external body surface and consists of three layers, namely epidermis, dermis or corium, and subcutis. The latter is divided into an adipose layer and a fibrous one, the superficial fascia of the trunk.

1 The **epidermis (1)** consists of a squamous epithelium of several layers which is keratinized on its surface. Thickness and keratinization depend on utilization. The deep vital layers of epidermis guarantee the production of cells (stratum germinativum, **stratum basale — 27*** and **stratum spinosum — 26**), an intermediate layer **(stratum granulosum — 25)** and a keratinized layer **(stratum lucidum — 24 and stratum corneum — 23.)** (Besides the cells of the epidermis, melanocytes, Langerhans cells and sensory cells of Merkel are present particularly in the stratum germinativum.)

The degree of keratinization is different in the diverse parts of the body. On the digital cushions of manus and pes and also in other skin zones keratinization is termed soft or callous. Hard keratinization occurs on the claws. With soft keratinization, keratinized cells with an adhesion of 'membrane coating material' in the stratum disjunctium come away as cornified scales (see page 2A lower left). On the other hand, in keratinization of the claw and its distal growth, desquamation does not exist due to strong adhesion to the solid cornified mass (see page 2A lower right). The individual keratinized cell of the claw is much harder than that of the skin. Epidermis in areas of callus formation has a stratum granulosum (so-called because of its content of keratohyalin granules which fuse to form keratin), and a stratum lucidum. Microscopically this layer is lustrous and contains eleidin, a product of the transformation of the keratohyalin granules. In areas of formation of hard keratin these layers are lacking, the cells of the stratum spinosum keratinizing directly. The superficial configuration of the keratinization with its epidermal grooves and folds or papillae is determined by the form of the papillary layer of dermis lying deep to it.

The **function of the epidermis** is the continued production of keratinized cells for the protection of the body against radiations, excess hydration and dehydration, parasitic penetration and trauma.

The **corium** or **dermis (6)** consists of a thin **stratum papillare (2)** the papillae of which are dentiform and present a concave epidermal configuration, and an extensive **stratum reticulare (7)**. In particular, the stratum papillare contains thin collagen fibers. The stratum reticulare contains a mesh of specifically directed collagen fibers, hard, and resistant to traction. In both layers elastic fibers are also present and these act in the recovery of the typical configuration after change. (Regarding other cells present, particularly fibrocytes, fibroblasts, mastocytes, plasmacytes, macrophages and pigment cells, see an histology text.) The dermis is an effective protection for the body and is also used in the production of leather.

2 The **subcutis** or **tela subcutanea (10)** comprises loose connective tissue and adipose tissue. This is traversed by thick retinacula which fasten the skin to the deep fascia and to the periosteum. Functionally the subcutis and the panniculus adiposus store energy and water, and are used for heat insulation and shock-absorption. The connective tissue also permits slippage. Where the subcutis is not well developed (for example lips, cheeks and eyelids), no slippage occurs and musculature ends directly in the dermis.

Innervation is due to the presence of sensory and sympathetic fibers (sympathetic plexuses invest the blood vessels and act in regulating temperature and blood pressure). Skin can be considered as an expansive sensory organ. **Free nerve endings (16)** and terminal nerve corpuscles, for example the **tactile corpuscles (17)** of Meissner and **lamellated corpuscles (22)** of Vater-Pacini act as sensory receptors.

Granted that free nerve endings lose the myelin sheaths at definite points they also penetrate the epidermis and act as receptors of pain and temperature.

Blood circulation of the skin is due to the presence of large arteries and veins of the subcutis which traverse it in a sinuous pattern to allow for slippage of the skin. These vessels send branches to the dermis where they form two vascular nets. The **dermal arterial net (9)** is adjacent to the subcutis and the **subpapillary arterial net (3)** lies between the stratum papillare and stratum reticulare and sends subepidermal capillaries to the papillary bodies. The corresponding venous plexuses have more or less the same disposition. Other vascular plexuses deep to the fascia irrigate the subcutis. Blood supply can be regulated by **arteriovenous anastomoses (4)** to avoid capillaries and thus regulate cutaneous vascularization. The stratum papillare is the best irrigated. These vessels dilate to provide heat, and contract to maintain it, and in this way also act in thermoregulation in conjunction with sweat glands. The venous plexus is equally a storage place for blood.

Lymph circulation is by means of lymph capillary nets which begin in the subepidermis and are disposed around hair follicles and cutaneous glands.

The **external panniculus adiposus** is a layer of fat in the subcutis. Adipose tissue is divided into that containing white and that containing brown fat. Both present morphological and marked functional variations. White adipose tissue is characterized by large 'signet-ring' cells, each with a diameter up to 100 μm, having the nucleus displaced peripherally and containing only one fat droplet (except in formative and reformative stages). Subserous fat tissue is encountered, to greater or lesser extent in mesentery as well as in the retroperitoneal space where it acts as an energy reservoir, a reservoir of liquid, and a heat insulator. Subcutaneous fat tissue shapes body proportions typical of sex and age (such as 'baby fat'). Structural fat, including that associated with nutritional status, is localized in the digital cushion, coronary arteries, renal periphery, orbital fossa where it is a shock absorber, and in the heart which it moulds. In muscles it acts as a layer allowing slippage, as in muscles of the ocular bulb. White fat acts in a subordinate role as in the juvenile mamma or as replacement fat, for example, in the involution of thymus and the medulla of bone.

Hair covers almost the whole body surface with the exception of the rostral portion of nose, anus, lips of vulva, and digital cushions. Hairs are thread-like keratinized structures produced by cornification of epidermal cells. In each, one can differentiate the **scapus of the hair (15)** which extends beyond the body surface, and the **root of the hair (21)** penetrating obliquely into the dermis and widening at its termination in the **bulb of the hair (8)**. Root and bulb are contained in an epithelial root sheath. This has an external part in continuous union with the most superficial epidermis and an internal part which is lost by keratinization at the mouth of the sebaceous gland. The connective tissue and epithelium of the root sheath form the hair follicle. The structural parts of the hair are the **medulla (12)**, the **cortex (13)** and, more superficially, **the cuticle (14)** which consists of very fine keratinized cells. The **m. arrector pili (5)** begins deep to the mouth of the sebaceous gland and oblique to the hair follicle. During muscle contraction, as in 'goose flesh', hairs are erected, sebaceous glands emptied (see page 3) and airspaces between the hairs increased. Two hair types occur, namely **outer cover hairs (11)** and deeper **woolly hairs (pili lanei)**. Dog's hair is disposed in groups or clusters. Each group contains a large principal hair and up to ten complementary hairs. The central principal hair is very strong and erect and the complementaries fine and undulated. The finest of these are the wool hairs which are very wavy and in a majority of breeds constitute the undercoat. In some breeds such as the Puli, complementary hairs are superficial to the erstwhile cover hair and provide a superficial woolly coat.

Tactile hairs (19) are very special hair forms disposed in particular around the oral cleft. For the perception of tactile stimuli each tactile hair is surrounded by a **blood sinus (20)** within the hair follicle. This sinus contains many terminal sensory nerves, and can perceive the finest stimuli.

Hair direction. The flow of the hair characterizes the coat. Hairs converge towards or diverge from a center in a vortex. Hairs also converge towards or diverge from a convergent or divergent hair line. Converging hair lines can also form a cross.

* See Illustration p. 2A for Anglicized terms.

Common integument

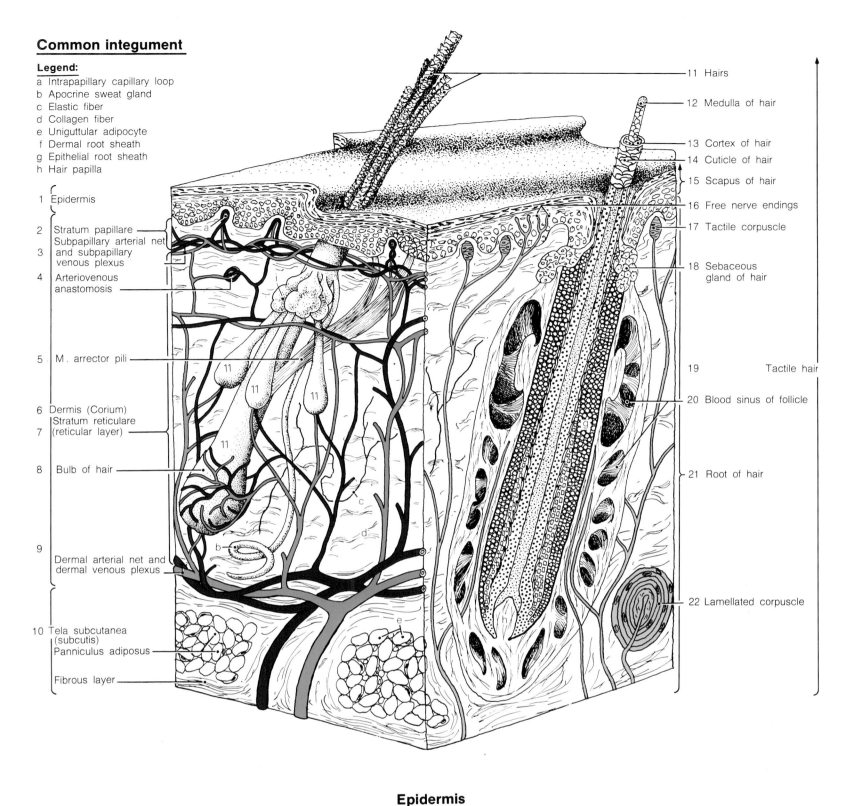

Legend:
a Intrapapillary capillary loop
b Apocrine sweat gland
c Elastic fiber
d Collagen fiber
e Uniguttular adipocyte
f Dermal root sheath
g Epithelial root sheath
h Hair papilla

1 Epidermis
2 Stratum papillare
3 Subpapillary arterial net and subpapillary venous plexus
4 Arteriovenous anastomosis
5 M. arrector pili
6 Dermis (Corium)
7 Stratum reticulare (reticular layer)
8 Bulb of hair
9 Dermal arterial net and dermal venous plexus
10 Tela subcutanea (subcutis)
Panniculus adiposus
Fibrous layer

11 Hairs
12 Medulla of hair
13 Cortex of hair
14 Cuticle of hair
15 Scapus of hair
16 Free nerve endings
17 Tactile corpuscle
18 Sebaceous gland of hair
19 Tactile hair
20 Blood sinus of follicle
21 Root of hair
22 Lamellated corpuscle

Epidermis

Epidermis of digital pad

Epidermis of wall of claw

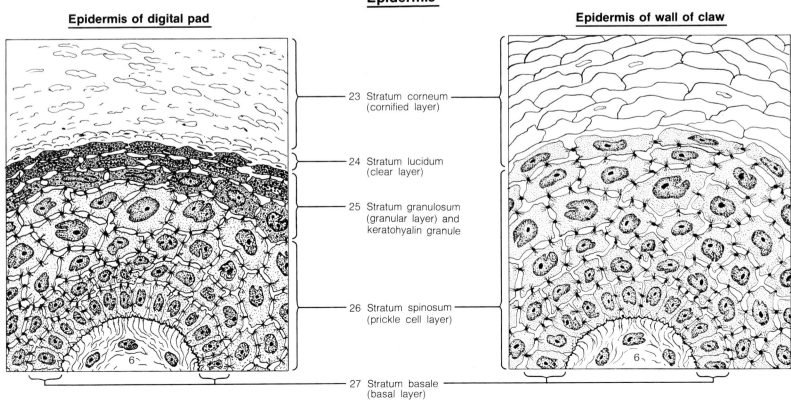

23 Stratum corneum (cornified layer)
24 Stratum lucidum (clear layer)
25 Stratum granulosum (granular layer) and keratohyalin granule
26 Stratum spinosum (prickle cell layer)
27 Stratum basale (basal layer)

A

The **SKIN GLANDS** include the sebaceous glands and sweat or sudoriferous glands as well as the mammary gland which is a modified sweat gland.

The lobulated **sebaceous glands** (see p 2) open into the hair follicle or they can be unassociated, as occurs in several sites of the body at the transition of skin and cutaneous mucosa (for example at lips and anus). The peripheral cells have a high rate of mitosis and the daughter cells are pushed centrally towards the lumen. There, the largest and oldest cells are destroyed (holocrine secretion) so that the sebum thus liberated reaches the gland lumen, its short secretory duct, and then the infundibulum of the hair follicle. The sebum keeps the skin soft and pliable and gives the hair a natural shine.

The **sweat or sudoriferous glands** are subdivided into merocrine (eccrine) and apocrine or scent glands. The classification is based on a presumed apocrine function of the scent glands which, however, has been refuted. (Both types of sweat glands have a merocrine or eccrine mode of secretion — see Histology).

The **merocrine sweat glands** are mostly coiled, unbranched, tubular glands. In dogs they exist solely on the digital pads (see below). Some authors however, consider these to be apocrine. In man, purely merocrine (eccrine) sweat glands are developed over large areas of the skin surface.

The **apocrine sweat glands or scent glands** (see p 2A) are present over large areas of the skin surface but they are relatively poorly developed. These tubular glands generally open by their secretory ducts into hair follicles, their secretion being alkaline and responsible for body odour which is specific to the individual. (In man they are well developed but present only on a few parts of the body namely the anal, urogenital, and axillary regions.)

Specifically modified skin glands include the ceruminous or wax secreting glands of the external auditory meatus, the circumanal glands, the glands of the paranal sinuses and the caudal or tail glands (see p 26 and 21.11 - 21.13), glands of the eyelid and the mammary gland.

The **ceruminous glands** consist principally of sebaceous glands and fewer apocrine sweat glands. The brown, greasy secretion is known as cerumen.

The **circumanal glands** surround the anus in a zone where hair is either sparse or absent. In the dog they are modified sebaceous glands whereas in other domestic mammals they are apocrine sweat glands. Individual glands of this nature are situated superficially and open into the mouths of the hair follicles. Because of their similarity histologically to liver tissue the more deeply situated glands are also known as hepatoid glands. They possess no secretory ducts and their function remains in doubt.

The **glands of the paranal sinuses** (see 21.12) are apocrine sweat, and sebaceous in type.

The **glands of the dorsal organ of the tail** contain sebaceous and apocrine sweat glands, and are mentioned in 21.13.

Glands of the eyelid (see 40.1).

The **mammary gland** (see p 14 and 14.1) is regarded as a modified sweat gland because it originates as such during development.

Prenatally, denser epidermal buds sprout first of all from the tips of each teat primordium into the deep skin layers, a process which continues also in the male without any recognizable difference. A specific sex difference becomes evident only with puberty and the first estrus. Particularly under the influence of estrogen, partial duct organization and ramification occur in the epidermal buds. In addition, estrogen causes adipose tissue to be deposited. This depot fat is displaced further with the development of the epidermal buds during pregnancy.

During pregnancy the epidermal buds, duct formation and ramification within the gland increase greatly, induced chiefly by a raised estrogen blood level. Vesicular alveoli develop in the second half of pregnancy particularly through the influence of progesterone on the ends of the mammary canaliculi.

During the approximately thirty-day-long lactation period, the alveoli guarantee milk secretion. From an individual embryonic epidermal bud, as a consequence, a large gland develops with alveoli, alveolar lactiferous ducts, lactiferous ducts, a lactiferous sinus and a papillary duct (see illustration p 14A). Approximately 8 - 20 individual glands open on the apex of a teat or papilla, by means of a similar number of ducts. The mammary papilla and its associated glands constitute a mammary complex.

The **hormonal-nervous regulation** of postnatal development of the mammary gland is complex and its details are not yet completely elucidated. Besides estrogen and progesterone, glucocorticoids from the adrenal cortex, and prolactin and growth hormone from the hypophysis (pituitary gland) also participate.

After the lactation period the alveoli in particular and a large part of the duct system regress.

SKIN MODIFICATIONS include the nasal plane or planum nasale and the digital pads.

The **nasal plane** or **planum nasale** (see p 40) is either unpigmented or deeply pigmented, depending on breed. The corium or dermis forms distinct papillae while the overlying epidermis is remarkably thin. Its superficial stratum corneum consists of hard keratinized tissue on which it is possible to recognize many small polygonal areas or plaques. Glands are absent on the planar surface and it is kept moist by lacrimal flow (see p 40) and secretion of the lateral nasal glands. These lie in the maxillary recesses deep within the nasal cavity.

The **pads** present in the dog are those of the **digit** (c) at the level of the distal interphalangeal articulation, those of **metacarpus** (b) and **metatarsus** at the level of metacarpo- and metatarsophalangeal articulations respectively, and the **carpal pad** (a) on the mediopalmar aspect of the carpus. The thick subcutis of the pad has an abundant adipose tissue (panniculus adiposus largely removed in illustration 3A) and contains sweat glands. The subcutis is subdivided into minute chambers by radially arranged net-like retinacula of collagen and elastic fibers; these are very sensitive to pressure increases caused by inflammation. The retinacula radiate from the dermis into the subcutis and fix the pads to the underlying fascia and bone. The metatarsal and metacarpal pads also possess well developed **tension bands** or **tractus tori** (12) that fix the pads to the metatarsal and metacarpal bones respectively (see p 33A). The dermis produces very tense or rigid connective tissue bands and forms very high papillary bodies associated with conical tubercles. The epidermis is up to 2 mm thick and forms corresponding configurations from the soft keratinized tissue. The pads have an abundant supply of blood vessels, lymphatics and nerves.

The **digital end organ** is the osseous distal phalanx surrounded by modified external skin (see p.7). With the exception of the digital pad it lacks a subcutis. The dermis is developed in the form of papillae, villi or laminae or at times results in a flat surface. The inner surface of the epidermis also produces a corresponding configuration (epidermal pegs, crypts, laminae or a smooth surface). Both, dermis and epidermis, fit tightly into each other like a die or patrix (dermis) into its impression or matrix (epidermis).

The **claw or unguicula** developes in the form of a cone and surrounds the **unguicular process** (11). The differentiation of dermis and epidermis in the claw is similar to that in the human fingernail.

The (osseous) unguicular crest is covered by a cutaneous **vallum** or **limbus** (7). The outer lamella of the vallum is populated with hair while the inner lamella is free of hair, its epidermis forming a thin soft **horn of vallum** or **eponychium** (1) over the surface of the claw wall. In the horse this glossy eponychial layer is known as 'hoof varnish'; a similar layer present on the human fingernail is removed with manicuring.

In the deeper part of the (osseous) unguicular groove that is a specific characteristic of the distal phalanx, a folded tissue segment is present. Its dermis bears **dermal papillae** (10). The overlying epidermis forms a tubular horn, the **horn of the unguicular groove** or **mesonychium** (2), that constitutes an important part of the unguicular horn. This corresponds to the mesonychium of the human fingernail.

In the lateral field of the unguicular process, dermis and epidermis bear **dermal lamellae** (9) and the epidermis forms corresponding **epidermal (nonkeratinized) laminae**. The latter is known also as 'binding' horn (with the dermis); the **lateral parietal horn** or **lateral hyponychium** (4) forms the inner lining of the conical claw. Proximolateral and dorsal to the (osseous) unguicular process, a smooth **dorsal dermal torus** (8) is in contact with a corresponding smooth concave area of the inner surface of the epidermis. In this way a special characteristic of the claw is formed, namely the **dorsal parietal horn** or **dorsal hyponychium** (3). On the palmar (or plantar) surface of the unguicular process lies an area of sole, the dermis of which also possesses papillae. There, an epidermal tubular **horn of the sole** (6) is formed. Around the tip of the unguicular process, a soft **terminal horn** or **terminal hyponychium** (5) completes the terminal part of the claw cone.

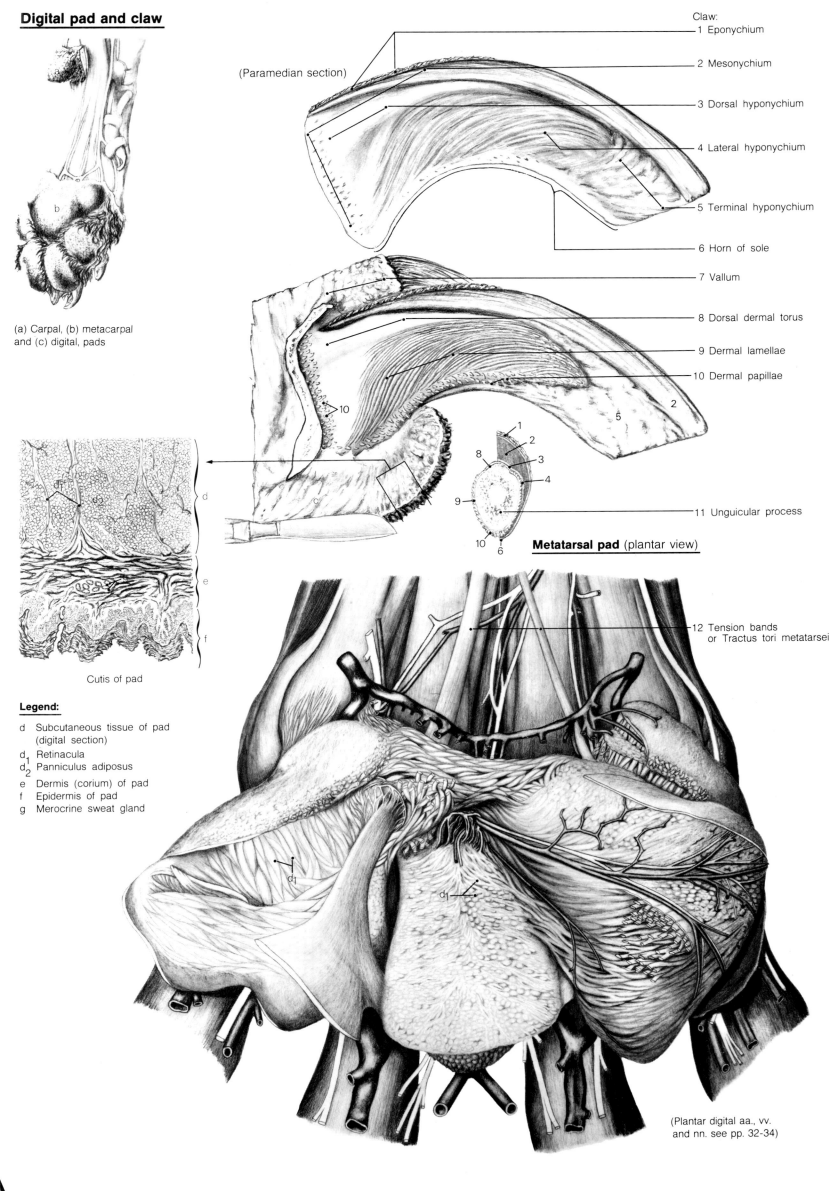

Digital pad and claw

(a) Carpal, (b) metacarpal
and (c) digital, pads

Cutis of pad

Legend:

d Subcutaneous tissue of pad
    (digital section)
d_1 Retinacula
d_2 Panniculus adiposus
e Dermis (corium) of pad
f Epidermis of pad
g Merocrine sweat gland

Claw:
1 Eponychium
2 Mesonychium
3 Dorsal hyponychium
4 Lateral hyponychium
5 Terminal hyponychium
6 Horn of sole
7 Vallum
8 Dorsal dermal torus
9 Dermal lamellae
10 Dermal papillae
11 Unguicular process

(Paramedian section)

Metatarsal pad (plantar view)

12 Tension bands
   or Tractus tori metatarsei

(Plantar digital aa., vv.
and nn. see pp. 32–34)

1 *One uses both an articulated and a disarticulated skeleton to make a general survey of the normal S-shaped vertebral column, its ventral convexity being referred to as lordosis and its ventral concavity, kyphosis. From the viewpoint of forensic anatomy a special aim is the identification of individual vertebrae so that comparisions may be made between the different segments of the vertebral column.*

2 a) The **VERTEBRAL COLUMN** consists of seven cervical vertebrae (vC 1-7), thirteen thoracic vertebrae (vT 1-13), seven lumbar vertebrae (vL 1-7), three sacral (vS 1-3) fused to form the sacrum, and about twenty coccygeal or caudal vertebrae (vCy 1-20).

3 I. **Vertebrae** (see also text illustration) have three basic elements, the body and its constituent parts the arch and processes, which, according to their functional requirements, vary from region to region.

4;5 The body of the vertebra (1) possesses a ventral crest (2) (distinct in the cervical region) and a cranial (3) and a caudal extremity (4). On the thoracic vertebrae, caudal (5) and cranial (6) costal foveae together form articular facets for the heads of the ribs (see below). In vivo, the vertebral canal (7) contains the spinal cord and cauda equina and lies between the body and arch of each vertebra.

6 At its base, the vertebral arch (8) has a pedicle, and dorsally a flattened lamina. The intervertebral foramina (9), bounded by cranial (10) and caudal (11) vertebral notches respectively of contiguous vertebrae, afford passage to the spinal nerves. An exception to this is the first spinal nerve (see below).

Of the vertebral processes, the spinous (12) is the most prominent except on the first cervical and all caudal vertebrae. The transverse processes (13) are well developed on the cervical and lumbar vertebrae and on the thoracic vertebrae each bears a costal fovea (14), an articular facet for the tubercle of the rib (see below). Transverse foramina (15) are present at the bases of the transverse processes of the first six cervical vertebrae. Aligned in series, these form the transverse canal containing the vertebral a. v. and n.

7 The cranial (16) and caudal (17) articular processes form true articulations between the vertebrae. A costal process (18) appears at the ventrocranial end of the bipartite transverse process between the third and sixth cervical vertebrae inclusive. On the lumbar part of the vertebral column the ends of the transverse processes exhibit costal processes which can grow to fully developed lumbar ribs since they include costal rudiments. An accessory process (19), absent on each of the vertebrae of the caudal part of the lumbar vertebral column, is fully developed as an independent vertebral process on the cranial part of the series. The process loses its independent character at the transition of the thoracic part of the vertebral column, by merging with the caudal contour of the transverse process. On each of the lumbar vertebrae, the mamillary process (20) is mounted on the cranial articular process and could therefore be termed mamilloarticular. Likewise at the transition of the thoracic part of the column it is displaced to the cranial contour of the transverse process. Hemal processes (21) are fully developed from the fourth caudal vertebra but gradually become less distinct more caudally. Those of the fourth and fifth caudal vertebrae can close to form hemal arches (22). The interarcuate spaces open dorsally, the lumbosacral (23) and sacrococcygeal (24) spaces being particularly large. This is significant for epidural anesthesia. Likewise the atlantooccipital space is suitable for puncture of the subarachnoid space, which is filled with cerebrospinal fluid (not illustrated).

Special features are present in the following cervical vertebrae: The first cervical vertebra, namely the atlas (25), bears a broad flat, lateral process (26) which is known as the wing of the atlas. The alar notch (27), which in other domestic species is an alar foramen, is situated cranially at the attachment of the wing of the atlas and is occupied by the ventral branch of the first cervical nerve. In contradistinction to the other spinal nerves, the first cervical nerve does not pass through an intervertebral foramen but through the lateral vertebral foramen (28). Besides having a dorsal or vertebral arch (29) the atlas, as a unique vertebra, employs its body as a ventral arch (30). This term is justified due to the displacement of the major part of the embryological primordium of its vertebral body to the dens of the axis. As stated, the second cervical vertebra, the axis (31), **8** contains the displaced body of the atlas in its dens (32). The last cervical vertebra differs from the remaining six due to its long spinous process,

its caudal costal fovea for the articulation of the head of the first rib, and the absence of a transverse foramen.

II. The **sacrum** results from the fusion of the three sacral vertebrae. Laterally, it bears the wing of the sacrum (33) of which the auricular surface (34) forms the strong sacroiliac articulation with a congruent ear-shaped surface of the ilium. The median sacral crest (35) originates as a result of the incomplete fusion of the spinal processes. The lateral extremities of the transverse processes form the lateral sacral crest (36) while the intermediate sacral crest (37) originates from the craniocaudal series of mamilloarticular processes. The promontory (38) is the cranioventral contour of the sacrum and participates in delimiting the cranial pelvic aperture. From the vertebral canal, the sacral nerves pass through the intervertebral foramina and after their division into dorsal and ventral nerve branches, they leave the vertebral canal by way of the dorsal (39) and ventral (40) sacral foramina respectively.

Lumbar vertebra

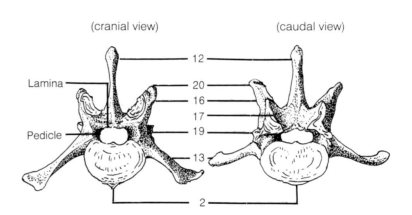

(cranial view)   (caudal view)

Lamina — 12
20
16
17
Pedicle — 19
13

2

b) Of the 13 **RIBS** of each side, the first nine are termed sternal ribs (41, see p. 4A upper left) because they articulate with the sternum. Ribs 10 to 12 inclusive are termed asternal ribs (42), do not articulate with the sternum, and because of their movement, play a part in respiration. Due to the serial placement of costal cartilages, a costal arch occurs on each side of the body. As a rule the last rib, as a floating rib (43), ends freely, does not participate in the costal arch, and terminates in the musculature of the body wall. Ribs, sternum and thoracic part of vertebral column together form the thorax containing the thoracic cavity. Its cranial thoracic aperture is bounded by the first pair of ribs, its caudal thoracic aperture by the costal arch. The dorsal part of the rib is osseous (44), its head (45) bearing the articular surface of the head of the rib (46). This articulates between two vertebral bodies, not only with the cranial costal fovea of the vertebral body of like numeration, but also with the caudal costal fovea of the preceding body. An indistinct neck of the rib (47) unites the head to the body of the rib (48). The tubercle of the rib (49), situated proximodorsally, bears an articular surface (50) for articulation with the costal fovea of the transverse process. The angle of the rib (51) is indistinct. The costal cartilage (52) begins at the costochondral articulation and slightly distal to this it inclines cranially at the genu costae (53). In the other domestic mammals this feature lies at the costochondral articulation.

c) The **STERNUM** consists of a manubrium (54) cranially, a body (55) with its six sternebrae (56), and a xiphoid process (57) terminating caudally at the expanded xiphoid cartilage. The first pair of ribs articulates with manubrium, the second pair with the border between the first intersternebral cartilage and the body, and the third to the seventh pair inclusive with successive intersternebral cartilages. The eighth and ninth pairs together, articulate with the border between the last intersternebral cartilage and the xiphosternum.

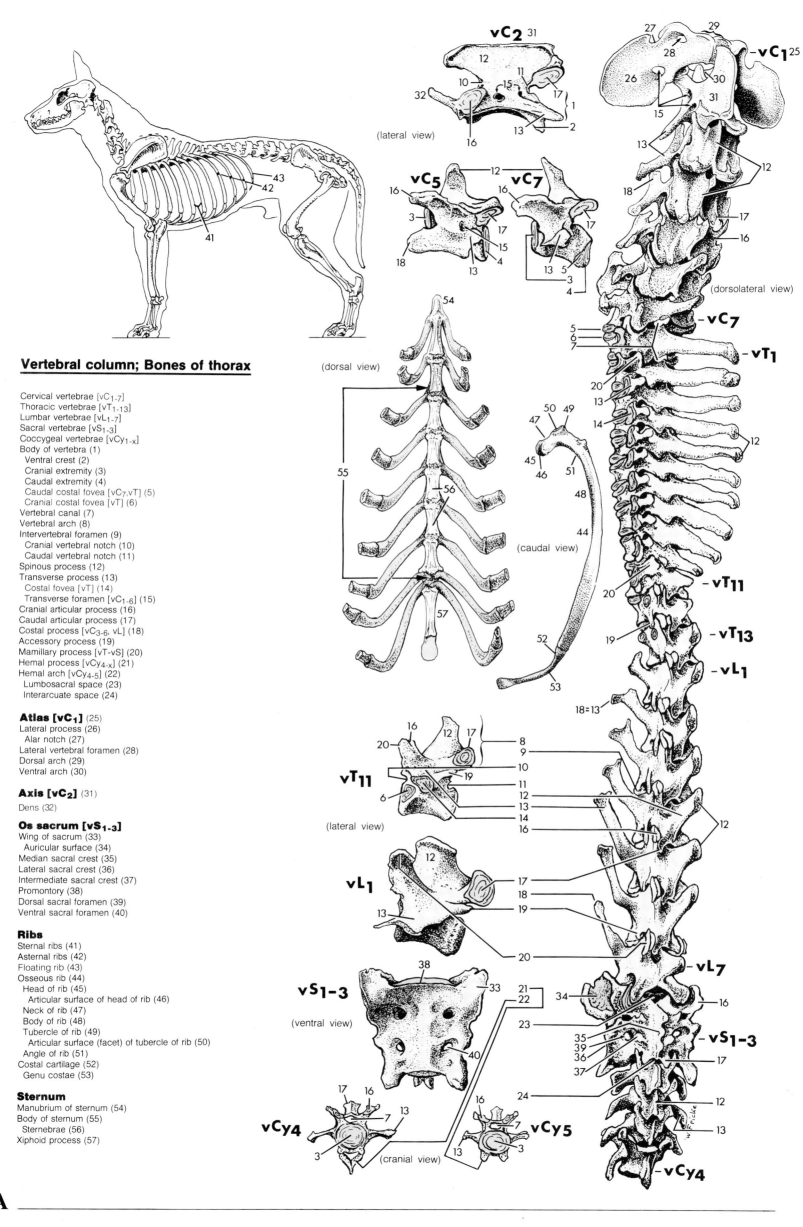

Vertebral column; Bones of thorax

Cervical vertebrae [vC$_{1-7}$]
Thoracic vertebrae [vT$_{1-13}$]
Lumbar vertebrae [vL$_{1-7}$]
Sacral vertebrae [vS$_{1-3}$]
Coccygeal vertebrae [vCy$_{1-x}$]
Body of vertebra (1)
  Ventral crest (2)
  Cranial extremity (3)
  Caudal extremity (4)
  Caudal costal fovea [vC$_7$,vT] (5)
  Cranial costal fovea [vT] (6)
Vertebral canal (7)
Vertebral arch (8)
Intervertebral foramen (9)
  Cranial vertebral notch (10)
  Caudal vertebral notch (11)
Spinous process (12)
Transverse process (13)
  Costal fovea [vT] (14)
  Transverse foramen [vC$_{1-6}$] (15)
Cranial articular process (16)
Caudal articular process (17)
Costal process [vC$_{3-6}$, vL] (18)
Accessory process (19)
Mamillary process [vT-vS] (20)
Hemal process [vCy$_{4-x}$] (21)
Hemal arch [vCy$_{4-5}$] (22)
  Lumbosacral space (23)
  Interarcuate space (24)

Atlas [vC$_1$] (25)
Lateral process (26)
  Alar notch (27)
Lateral vertebral foramen (28)
Dorsal arch (29)
Ventral arch (30)

Axis [vC$_2$] (31)
Dens (32)

Os sacrum [vS$_{1-3}$]
Wing of sacrum (33)
  Auricular surface (34)
Median sacral crest (35)
Lateral sacral crest (36)
Intermediate sacral crest (37)
Promontory (38)
Dorsal sacral foramen (39)
Ventral sacral foramen (40)

Ribs
Sternal ribs (41)
Asternal ribs (42)
Floating rib (43)
Osseous rib (44)
  Head of rib (45)
    Articular surface of head of rib (46)
  Neck of rib (47)
  Body of rib (48)
  Tubercle of rib (49)
    Articular surface (facet) of tubercle of rib (50)
  Angle of rib (51)
Costal cartilage (52)
  Genu costae (53)

Sternum
Manubrium of sternum (54)
Body of sternum (55)
  Sternebrae (56)
Xiphoid process (57)

4A

> *To demonstrate the cutaneous muscles (see p. 5A, upper part) a longitudinal skin incision is made on the left side of the cadaver from the base of the ear to the ventral end of the last rib, passing over the midlength of the scapula. The underlying cutaneous musculature must be preserved in its entirety. At each end of the parent incision, a superficial transverse incision is made and the resultant skin flaps reflected to the dorsal and ventral midlines, respectively. To avoid unintentional injury and contamination of the field, the external jugular and omobrachial veins are then dissected carefully away from the skin.*

a) The CUTANEOUS MUSCLES terminate by very fine fibers in the skin. They produce movement possibly as a defence against insects. The fibers of the m. cutaneus trunci (11) converge towards the axillary region and the linea alba ventromedially and are perforated by very fine cutaneous nerve branches. The motor innervation to the muscle is derived from the lateral thoracic n. (12) the branches of which are visible on the ventral half of the cutaneous muscle of the trunk. They are directed parallel to the muscle fibers.

1 The platysma colli (1) is displayed from its linear origin on the dorsal midline of the neck to the boundary between head and neck where it is continuous with the m. cutaneous faciei, the platysma of the head. The nerve ramus to the platysma colli (2) is derived from the caudal auricular n., a branch of the facial or seventh cranial nerve. The ramus crosses under the fiber bundles of platysma in a dorsoparamedian course at approximately the level of the ends of the transverse processes of the cervical vertebrae. By separating the large muscle fiber bundles one is able to determine the nerve branch.

The m. sphincter colli superficialis (3) is situated on the ventral part of the neck, its fibers oriented transversely. Only with great effort can its muscle fiber bundles be detached from the skin.

> *To demonstrate the cutaneous nerves of the neck the dorsal linear origin of the platysma is severed and the muscle reflected cranially as far as the cranial skin incision. To demonstrate the cutaneous nerves of the thorax the m. cutaneous trunci is severed along the caudal transverse skin incision at the level of the last rib and at the torus anconeus; it is reflected ventrally as far as the linea alba. During further dissection in the ventral abdominal and thoracic areas, it is essential to preserve the aponeurosis of the m. obliquus externus abdominis (see p. 5A-35).*

b) The CUTANEOUS NERVES supply the outer skin chiefly with sensory innervation and are therefore visible as individual portions of subcutaneous nerves. Each of the spinal nerves (for example nC4, see p. 19A) divides as it leaves the intervertebral foramen into a dorsal branch (d) and a ventral branch (v), each of which divides further into a medial (dm, vm respectively) and a lateral branch (dl, vl respectively). With the exception of the dorsal neck region the deeply situated medial branches convey motor innervation either completely or almost so. The lateral branches supplying the skin are mainly sensory. Of the eight cervical nerves, nC1 passes through the lateral vertebral foramen, nC2 to 7 appear cranial to their respective segmental cervical vertebrae, and nC8 caudal to the last cervical vertebra. The dorsomedial branch (nC1 dm) of the first cervical nerve does not reach the skin of the neck. The greater occipital n. (nC2 dm) runs from under the m. cervicoauricularis superficialis to the occipital region. The subsequent four dorsomedial branches (nC3 dm to nC6 dm inclusive) are often fully duplicated and must not be identified individually. The dorsomedial branches of the seventh and eighth cervical nerves (nC7 dm and nC8 dm respectively) are so delicate that they do not, as a rule, reach the skin, but remain in the thick musculature. The innervation of the dorsal cervical skin areas through the dm-branches demonstrates a divergence from the rule for the other body regions. There, the skin is supplied by lateral branches and the musculature by medial branches. The divergence becomes distinct due to the opposite arrangement of the passage of nerve branches in the dorsal cervical and thoracic regions.

I. The series of dorsal cutaneous branches of the cervical nerves courses to the skin surface with concomitant blood vessels in the dorsal midline, and is formed from dorsomedial nerve branches.

II. The series of dorsal cutaneous thoracic nerves appears with accompanying blood vessels several centimeters dorsosagittally, that is, almost laterally; in normal circumstances the series is formed from dl-branches. Each of the thirteen thoracic nerves (see p. 19A) leaves the vertebral canal through the intervertebral foramen caudal to the vertebra of like numeration, and divides into a dorsal and a ventral branch. The ventral branch extends distally between the ribs as the intercostal n. At the approximate midlength of the intercostal space this nerve gives off a proximal ventrolateral branch, vl (prox.) and at the ventral end of the space, a distal ventrolateral branch, vl (dist.).

III. The series of lateral cutaneous branches of the cervical nerves are formed by consecutive branches from nC2 vl to nC5 vl. The nC2 v to nC5 v communicate with one another and deep within the musculature form the cervical plexus. The nC2 vl provides the great auricular n. (5) to the ear region and the transverse cervical n. (7) to the ventral neck region. The major portions of the nerve roots from nC6 v to nT2 v contribute to the brachial plexus (see p. 8A) from which the nerve supply to the thoracic limb arises. (The nerves of the plexus are not divided into vm- and vl-branches!)

IV. The series of lateral cutaneous branches of the thoracic nerves are formed from the previously mentioned vl (prox.) branches, and are known as the lateral cutaneous branches of the intercostal nn.

V. A series of ventral cutaneous branches of the thoracic nerves are formed from very slender vl (dist.) branches and are known as the ventral cutaneous branches of the intercostal nn.

C. 2. S. 2.: DORSAL EXTRINSIC MUSCLES OF THORACIC LIMB OR 'DORSAL TRUNK-LIMB MUSCLES'

> *A knowledge of the osseous pectoral girdle is necessary before undertaking the dissection (see p.7). The muscle origins/insertions on skull, cervical and thoracic vertebrae, ribs and sternum associated with the trunk, and those on the bones of the thoracic limb, determine the second name assigned to these muscles. When an insertion occurs on a part of the pectoral girdle, the term muscle of the pectoral girdle is also justified. Of these muscles the m. serratus ventralis produces the fleshy connection between trunk and thoracic limb, the rotational filed of which lies on the facies serrata of the scapula. To continue the dissection, the m. cleidocervicalis and m. trapezius are severed along the course of the dorsal ramus of the accessory n. (XI) which innervates them, and reflected.*

Both parts of the m. trapezius (13) (three parts according to Donat et al. 1967) arise at the dorsal midline from the spinous processes of cervical and thoracic vertebrae. The thoracic part (15) is directed cranioventrally to insert along the dorsal third of the spine of the scapula. The cervical part (14) courses caudoventrally to its insertion on the dorsal two-thirds of the spine. In spite of the difference in direction of the muscle fibers, both parts function as protractors of the limb since the thoracic part inserts dorsal to the field of rotation of the fleshy connection (m. serratus ventralis) between trunk and thoracic limb.

According to Donat et al. (1967) the m. cleidocervicalis (9) is classified as the third clavicular part of the m. trapezius. It is directed between the clavicular intersection (10) and the dorsal midline of the neck.

The dorsal ramus of the accessory n. (6) innervating the m. cleidocervicalis and m. trapezius, appears at the apex of a triangle bounded by the former muscle and the cervical part of the latter. One pursues it further at the beginning of the incision through the m. trapezius (see lower p. 5A).

The m. omotransversarius (8), as its name suggests, extends between the acromion of the scapula and the transverse process of the atlas. Innervation: nC4 vn. The superficial cervical lymph node lies under the dorsomedial surface of the muscle and should be preserved.

The m. latissimus dorsi (17) arises from the broad thoracolumbar fascia (16) and ends on the brachial fascia and the crests of the greater and lesser tubercles of the humerus. In this way a wider axillary arc results.

The thoracodorsal n. (see p. 8A-2) and accompanying blood vessels course along the medial surface of the muscle and are seen by lifting its ventral border.

The m. rhomboideus (4) is covered by the m. trapezius and consists of the mm rhomboideus capitis (nC vm), cervicis (nC vm) and thoracis (nT vm). These take origin from the nuchal crest and the dorsal mid-line and insert on the scapular cartilage.

Function: Fixes the scapula, raises and retracts the thoracic limb, and raises the neck.

Cervical and pectoral regions (lateral view)

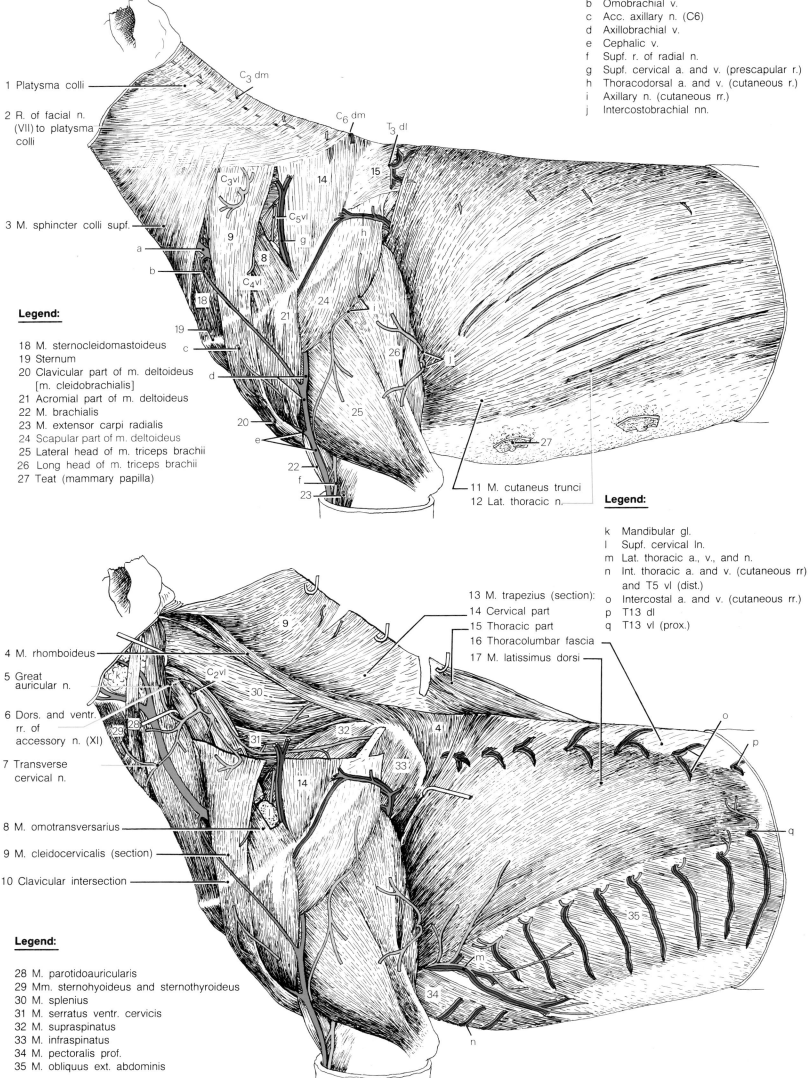

1 Platysma colli

2 R. of facial n. (VII) to platysma colli

3 M. sphincter colli supf.

C₃ dm
C₆ dm
T₃ dl
C₃ vl
C₅ vl
C₄ vl

Legend:

a Ext. jugular v.
b Omobrachial v.
c Acc. axillary n. (C6)
d Axillobrachial v.
e Cephalic v.
f Supf. r. of radial n.
g Supf. cervical a. and v. (prescapular r.)
h Thoracodorsal a. and v. (cutaneous r.)
i Axillary n. (cutaneous rr.)
j Intercostobrachial nn.

Legend:

18 M. sternocleidomastoideus
19 Sternum
20 Clavicular part of m. deltoideus [m. cleidobrachialis]
21 Acromial part of m. deltoideus
22 M. brachialis
23 M. extensor carpi radialis
24 Scapular part of m. deltoideus
25 Lateral head of m. triceps brachii
26 Long head of m. triceps brachii
27 Teat (mammary papilla)

11 M. cutaneus trunci
12 Lat. thoracic n.

Legend:

k Mandibular gl.
l Supf. cervical ln.
m Lat. thoracic a., v., and n.
n Int. thoracic a. and v. (cutaneous rr) and T5 vl (dist.)
o Intercostal a. and v. (cutaneous rr.)
p T13 dl
q T13 vl (prox.)

13 M. trapezius (section):
14 Cervical part
15 Thoracic part
16 Thoracolumbar fascia
17 M. latissimus dorsi

4 M. rhomboideus

5 Great auricular n.

6 Dors. and ventr. rr. of accessory n. (XI)

7 Transverse cervical n.

8 M. omotransversarius

9 M. cleidocervicalis (section)

10 Clavicular intersection

C₂ vl

Legend:

28 M. parotidoauricularis
29 Mm. sternohyoideus and sternothyroideus
30 M. splenius
31 M. serratus ventr. cervicis
32 M. supraspinatus
33 M. infraspinatus
34 M. pectoralis prof.
35 M. obliquus ext. abdominis

C. 2. S. 3: VENTRAL EXTRINSIC MUSCLES OF THORACIC LIMB OR 'VENTRAL TRUNK-LIMB MUSCLES'

> *The extrinsic muscles of the thoracic limb function not only in the movement of head, spinal column and limb but also in the suspension of the two limbs from the trunk. The wide muscles situated dorsally are predominant in locomotion and take over the suspension of the limb. In particular the broad ventral extrinsic muscles act in the suspension of the trunk and are therefore infiltrated extensively by tendinous tissue. As the dissection progresses the superficial and deep pectoral muscles are transected 2 cm lateral to the thoracic linea alba which is situated midsagittally.*

The **m. pectoralis superficialis** (16, innervated by cranial and caudal pectoral nn) forms a lateral pectoral groove in conjunction with the clavicular part of the m. deltoideus. This contains the cephalic v. The wide **m. pectoralis transversus** (17) has a linear origin extending from the cranial end of the body of the sternum to the manubrium of the sternum. The narrow **m. pectoralis descendens** (18) lying on the surface of the latter muscle, also arises from the manubrium. The superficial pectoral muscles insert onto the crest of the greater tubercle of the humerus.

The **m. pectoralis profundus** (19, cranial and caudal pectoral nn) consists of two portions, a **principal part** (20) which supports a narrower, **accessory part** (21) situated laterally. The m. pectoralis profundus takes origin from the manubrium and body of the sternum and inserts onto the greater and lesser tubercles of the humerus. The accessory part inserts onto the brachial fascia. Nerves supplying the muscle are visible on its cut surface.

The **m. serratus ventralis** (34) is subdivided into the m. serratus ventralis cervicis (nC vm) and the m. serratus ventralis thoracis (**long thoracic n. —35**). In the region of the cranial thoracic aperture where their fibers are directed transversely, the subdivisions fuse completely. They arise from the transverse processes of the cervical vertebrae and the ribs respectively, and both parts insert on the facies serrata of the scapula.

The **m. sternocleidomastoideus**[1] (3, ventral ramus of accessory n.) consists of three individual muscles. The **mm cleidomastoideus** (6) and sternomastoideus (4) fuse cranially, while the m. sternomastoideus and the **m. sternooccipitalis** (5) do so caudally. The lateral surface of the m. sternocleidomastoideus forms the jugular groove containing the external jugular v. The innervating ventral ramus of the accessory n. becomes evident at the caudal border of the mandibular gland. There the **accessory n.** (see p. 5A —6) lies between the mm sternomastoideus and sternooccipitalis and is seen when the muscles are artificially separated from each other. The nerve is divided into the elongated dorsal ramus which communicates with nC2 and the short venral ramus.

The **m. deltoideus** (13, axillary n.; accessory axillary n. for the clavicular part), because of its **clavicular part** (14, m. cleidobrachialis of texts) as opposed to its scapular and acromial parts, belongs to the musculature of the pectoral girdle. After raising the lateral or medial border of the clavicular part, it becomes evident that the **accessory axillary n.** (nC6 —15) enters the medial surface of the clavicular part about 4 cm distal to the clavicular intersection and supplies it with motor innervation. Together with its sensory skin branches, the accessory axillary n. then pierces the clavicular part (m. cleidobrachialis).

The **m. sternohyoideus** (8, nC lvm) and the **m. sternothyroideus** (2, nC lvm), do not belong to the ventral extrinsic muscles of the thoracic limb but rather to the muscles of the hyoid apparatus (long hyoid muscles). The right and left mm. sternohyoidei are in contact along their medial borders in the ventral midline of the neck. The adjacent m. sternothyroideus lies laterally.

C. 2. S. 4: ARTERIES, VEINS, NERVES AND LYMPHATIC VESSELS OF NECK; AND VISCERA OF NECK

> *The jugular groove and its contained external jugular vein were exposed previously. To demonstrate the structures of the neck, both mm. sternohyoidei are separated from each other midventrally or are transected in common with the m. sternothyroideus.*

a) The **EXTERNAL JUGULAR V.** (9) arises from the brachiocephalic v. at the level of the cranial thoracic aperture as well as the subclavian and internal jugular vv. In caudocranial sequence the external jugular v. discharges the cephalic, superficial cervical and omobrachial vv. On the caudal aspect of the mandibular gland the external jugular v. divides into a dorsal **maxillary v.** (23) and a **linguofacial v.** (22) situated ventrally. Initially the **cephalic v.** (12) runs along the lateral pectoral groove to the thoracic limb. The **superficial cervical v.** (11) ramifies in the extrinsic limb muscles in company with the artery of like name from the thoracic cavity. The **omobrachial v.** (10) is directed laterally over the m. deltoideus to the arm.

b) Of the **ARTERIES, VEINS AND NERVES OF THE NECK** the **internal jugular v.** (25) is directed along the lateral aspect of the trachea, sending branches to brain, thyroid gland, larynx and pharynx. The left and right common carotid arteries arise from the brachiocephalic trunk at the level of the cranial aperture of the thorax (see p. 19A —11). The **common carotid a.** (27) runs cranially along the dorsolateral surface of the trachea and also sends branches to thyroid gland, larynx and pharynx. The large **vagosympathetic trunk** (26), a part of the autonomic nervous system, is situated dorsal to the common carotid a. It conveys a sympathetic constituent from the thoracolumbar part of the sympathetic trunk to the head (see p. 19A). Parasympathetic portions of the vagus n. (cranial nerve X) also extend from the head by means of the vagosympathetic trunk, chiefly to the body cavities. After its separation from the sympathetic n. within the thoracic cavity, the vagus n. gives off the recurrent laryngeal n. (see p. 19A —4) and subsequently carries parasympathetic fibers exclusively. The **recurrent laryngeal n.** (30) containing predominantly motor and sensory fibers, travels back up the neck. It lies lateral to the trachea within surrounding connective tissue, the esophagus and trachea both receiving nerve branches from it. The nerve itself is easily detected on the medial surface of the thyroid gland.

c) Concerning the **LYMPHATIC SYSTEM**, lymphatic trunks and nodes relating to the neck are discussed at this stage. The jugular trunk of its respective side is the largest in the neck. It begins at the medial retropharyngeal lymph node, receives lymph from superficial and deep cervical nodes, and opens at the venous angle resulting from the confluence of external and internal jugular vv. Shortly before its opening, the **terminal part of the left jugular trunk** (32) unites with the **end of the thoracic duct** (33) which conveys lymph from the body cavities. The **medial retropharyngeal lymh node** (1) lies on the m. sternothyroideus at its insertion. It receives lymph from the head region. The **superficial cervical lymph node** (31) lies deep to the dorsal border of the m. omotransversarius on the m. serratus ventralis. Its drainage areas include the superficial neck regions and also parts of head, trunk and thoracic limb. The deep cervical lymph nodes lie close to the trachea and consist of an inconstant cranial, and constant middle and caudal groups. The drainage area is the immediate vicinity in the neck.

d) The **VISCERA OF THE NECK** include esophagus, trachea, thyroid gland and parathyroid glands. At the midlength of the neck, the cervical part of the **esophagus** (29) lies dorsal to the trachea and at the cranial aperture of the thorax it is positioned sinistrodorsally. Its reddish colour is produced by infiltration of striated muscle. This visceral type of striated muscle is innervated by the vagus n., nonmyelinated fibers (!) of which reach the muscle cells. The **trachea** (7) consists of c-shaped cartilaginous rings. These are closed dorsally by membranous walls which contain the m. trachealis (transversus) and they are connected one to the other by annular ligaments. The patency of the tracheal lumen is maintained by the cartilaginous rings which are slightly distorted by fibroelastic tissue. The 'tension effect' brings about alterations in length during breathing and swallowing and is responsible for the typically round cross-section of the tracheal lumen. This can be compressed somewhat by the tension of the m. trachealis.

The **thyroid gland** (28) lies at the cranial end of the trachea; occasionally its right and left lobes are united ventrally by a fragile isthmus.

The two round **parathyroid glands** (24) of each side have diameters of about 3 mm. They lie on the medial and lateral surfaces of the associated thyroid lobe or within its parenchyma.

---

[1] There is no m. sternocleidomastoideus in the official nomenclature; the m. cleidomastoideus is the mastoid part of m. cleidocephalicus and mm. sternomastoideus and sternooccipitalis are mastoid and occipital parts of m. sternocephalicus.

6

Ventral region of neck and ventral pectoral region

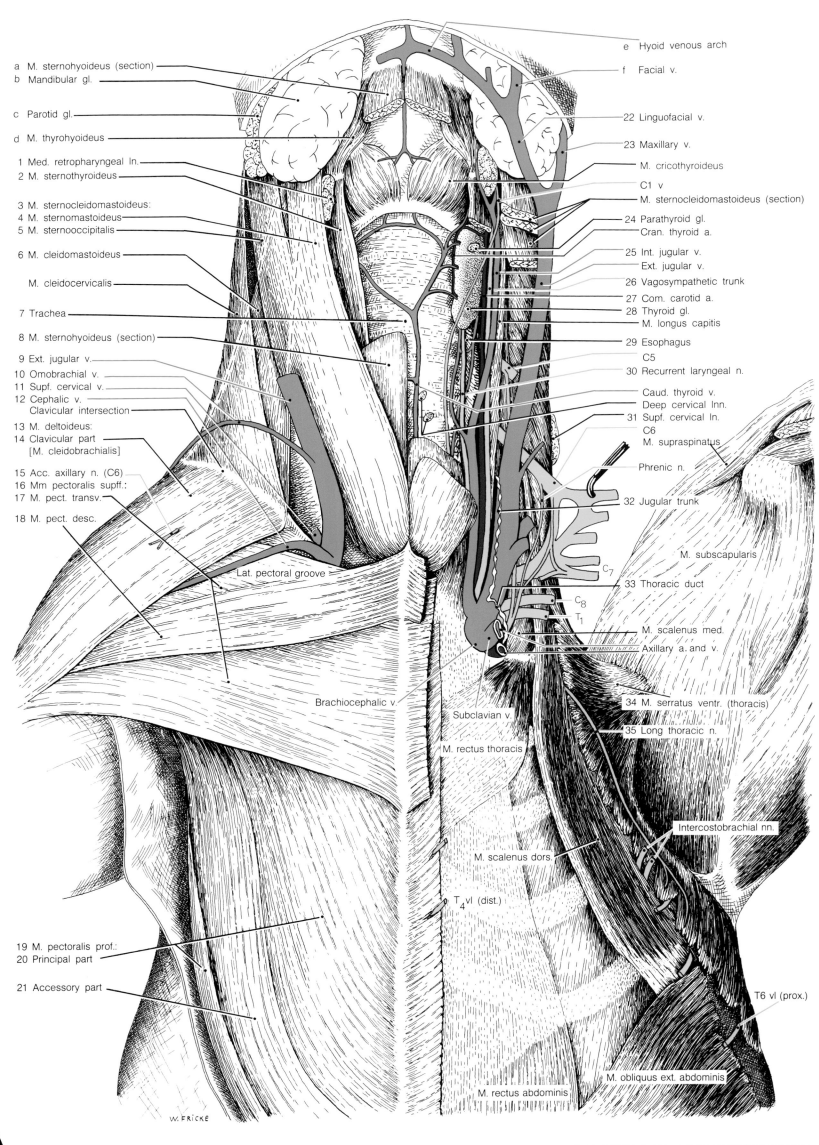

a M. sternohyoideus (section)
b Mandibular gl.

c Parotid gl.

d M. thyrohyoideus

1 Med. retropharyngeal ln.
2 M. sternothyroideus

3 M. sternocleidomastoideus:
4 M. sternomastoideus
5 M. sternooccipitalis

6 M. cleidomastoideus

M. cleidocervicalis

7 Trachea

8 M. sternohyoideus (section)

9 Ext. jugular v.
10 Omobrachial v.
11 Supf. cervical v.
12 Cephalic v.
Clavicular intersection

13 M. deltoideus:
14 Clavicular part
[M. cleidobrachialis]

15 Acc. axillary n. (C6)
16 Mm pectoralis supff.:
17 M. pect. transv.

18 M. pect. desc.

Lat. pectoral groove

Brachiocephalic v.

Subclavian v.

M. rectus thoracis

M. scalenus dors.

T₄ vl (dist.)

19 M. pectoralis prof.:
20 Principal part

21 Accessory part

M. rectus abdominis

e Hyoid venous arch
f Facial v.

22 Linguofacial v.
23 Maxillary v.
M. cricothyroideus
C1 v
M. sternocleidomastoideus (section)
24 Parathyroid gl.
Cran. thyroid a.
25 Int. jugular v.
Ext. jugular v.
26 Vagosympathetic trunk
27 Com. carotid a.
28 Thyroid gl.
M. longus capitis
29 Esophagus
C5
30 Recurrent laryngeal n.
Caud. thyroid v.
Deep cervical lnn.
31 Supf. cervical ln.
C6
M. supraspinatus
Phrenic n.
32 Jugular trunk
M. subscapularis
C₇
33 Thoracic duct
C₈
T₁
M. scalenus med.
Axillary a. and v.
34 M. serratus ventr. (thoracis)
35 Long thoracic n.
Intercostobrachial nn.
T6 vl (prox.)
M. obliquus ext. abdominis

W. FRICKE

6A

C. 3. S. 1: SKELETON OF THORACIC LIMB

The pectoral girdle consists of scapula, coracoid and clavicle which are fully developed as individual bones in animals other than mammals, for example birds. In domestic mammals a far reaching reversal in development results in a coracoid process of the scapula and an extensive clavicular intersection formed by connective tissue. Occasionally very small osseous remnants of the clavicle persist and are visible on radiographs at the medial end of the intersection.

a) The **PECTORAL GIRDLE** has the scapula as its principal element. The costal surface (1) of the scapula is subdivided into a facies serrata (2), the surface of insertion of the m. serratus ventralis, situated dorsally, and a ventral subscapular fossa (3), conforming to the insertion of the m. subscapularis. The lateral surface (4) is divided by the spine of the scapula (5) into a supraspinous (6) and an infraspinous fossa (7) affording origin to the mm supraspinatus and infraspinatus respectively. The acromion (8) and hamate process (9) are at the ventral end of the spine. The caudal border (10) of the scapula is almost a straight line. The curved cranial border (11) has a scapular notch (12) distally, while the dorsal border (13) bears a narrow scapular cartilage (14). Of the three angles, caudal (15), cranial (16) and ventral (17), the latter possesses a shallow, oval, articular surface, the glenoid cavity (18). Caudodistal to the neck of the scapula (19) is the infraglenoid tubercle (20) and craniodistally, the supraglenoid tubercle (21). Craniomedially, the coracoid process (22) is associated with the supraglenoid tubercle.

b) The **HUMERUS** bears a head (23) which is the convex articular surface of the shoulder joint or scapulohumeral articulation. The head of the humerus is well defined only in its caudal part due to the presence of the neck of the humerus (24). From the cranial aspect of the greater tubercle (25) the crest of the greater tubercle (26) extends distally, and from its caudal aspect the tricipital line (27) does likewise. The intertubercular groove (28) houses the tendon of origin of the m. biceps brachii and distinctly limits the lesser tubercle (29) situated medially. The crest of the lesser tubercle (30) extends distally from the latter feature and is continuous with the lateral supracondylar crest. At the junction of its proximal and middle thirds the body of the humerus (31) possesses the deltoid tuberosity (32) on its lateral aspect. From there the crest of the humerus (33) continues distally as far as the medial epicondyle. Cranially the crest limits the spiralling brachial groove (34) which houses the m. brachialis. The humeral condyle (35) consists of a large trochlea of the humerus (36) medially, for articulation with the ulna, and a small capitulum of the humerus (37) laterally, articulating with the radius. The condyle of the humerus possesses epicondyles. From the lateral epicondyle (38) (containing tuberculi for the origins of extensor muscles of carpus and digits), the prominent lateral supracondylar crest (38') proceeds proximally. The medial epicondyle (39) is the prominence for the origins of flexor muscles of carpus and digits. The deep olecranon fossa (40) caudally, and the shallow radial fossa (41) cranially, communicate through the supratrochlear foramen (42).

c) The **BONES OF THE FOREARM OR ANTEBRACHIUM** are the radius and ulna.

I. **Radius.** Slightly distal to the surface articulating with the humerus, the head of the radius (43) has a medial cylindrical articular elevation, the articular circumference (44). This articulates with the radial notch of the ulna proximally. On its medial aspect the indistinct neck of the radius (45) features an insignificant radial tuberosity (46) for attachment of the tendon of the m. biceps brachii. The body of the radius (47) merges with the trochlea of the radius (48) articulating with the carpal bones distally. The trochlea participates in the distal articulation with the articular circumference of the ulna by means of the ulnar notch (49) laterally. Medially it terminates in the medial styloid process (50).

II. **Ulna.** Proximally the ulna towers above the head of the radius by means of the olecranon (51) which is thickened at the tuber of the olecranon (52). The trochlear (semilunar) notch (54) begins both medially and laterally at the hooklike anconeal process (53) and extends distally to the medial (55) and lateral (56) coronoid processes. The radial notch (57) lies at the transition between olecranon and body of ulna (58). The distal (!) head of the ulna (59) has an articular circumference (60) situated medially and terminates distally at the lateral styloid process (61). The antebrachial interosseous space (62) is particularly wide in the distal third of the antebrachium.

d) Embryologically the **BONES OF THE CARPUS** are disposed in three rows and postnatally they are reduced to two. The radial carpal bone (63) medially, contains the intermediate carpal bone of the proximal row and the central carpal bone of the middle row. The ulnar carpal bone (64) distal to the ulna and the accessory carpal bone (65) projecting laterally, complete the proximal row. Carpal bones I to IV (66) form the distal row and articulate with the metacarpal bones.

e) Each of the **METACARPAL BONES I TO IV** possesses a base (67) with an articular surface proximally, a long body (68) and a head (69) situated distally (!). Metacarpal bone I may be absent or even divided into two parts, the proximal of the two fusing with carpal bone I.

f) The **BONES OF EACH FULLY DEVELOPED DIGIT OF THE MANUS** include a proximal, a middle and a distal phalanx. The first digit usually lacks a middle phalanx. The proximal phalanx (70) and the middle phalanx (71) each contain a base (72) proximally, a body (73), and a head (74) distally. The indistinct flexor tuberosity (75) for the attachment of the superficial flexor tendon, is situated on the proximopalmar surface of the middle phalanx. The distal phalanx (76) has an articular surface (77) proximally, and an ill-defined extensor process (78) dorsally for the insertion of the digital extensor tendon. A distinct flexor tuberosity (79) is present on the proximopalmar surface for insertion of the deep digital flexor tendon. The sharply angled unguicular crest (80) overlies the unguicular groove (81) and the base of the process (82) of like name which bears the claw.

The sesamoid bones of the manus are as follows: the sesamoid bone of the m. abductor digiti I (83) situated medially on carpal bone I; the paired proximal sesamoid bones (84) on the palmar surface of each metacarpophalangeal articulation; the unpaired distal sesamoid bone (85) on the palmar aspect of each distal interphalangeal articulation; and the unpaired dorsal sesamoid bones (86) on the dorsal aspect of each metacarpophalangeal and each proximal interphalangeal articulation respectively. Some sesamoid bodies are either always (85) or occasionally cartilaginous (86).

Thoracic limb (lateral view)

(Artésien Normand Basset)

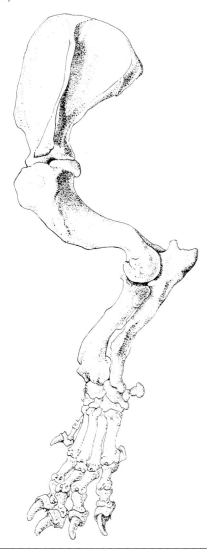

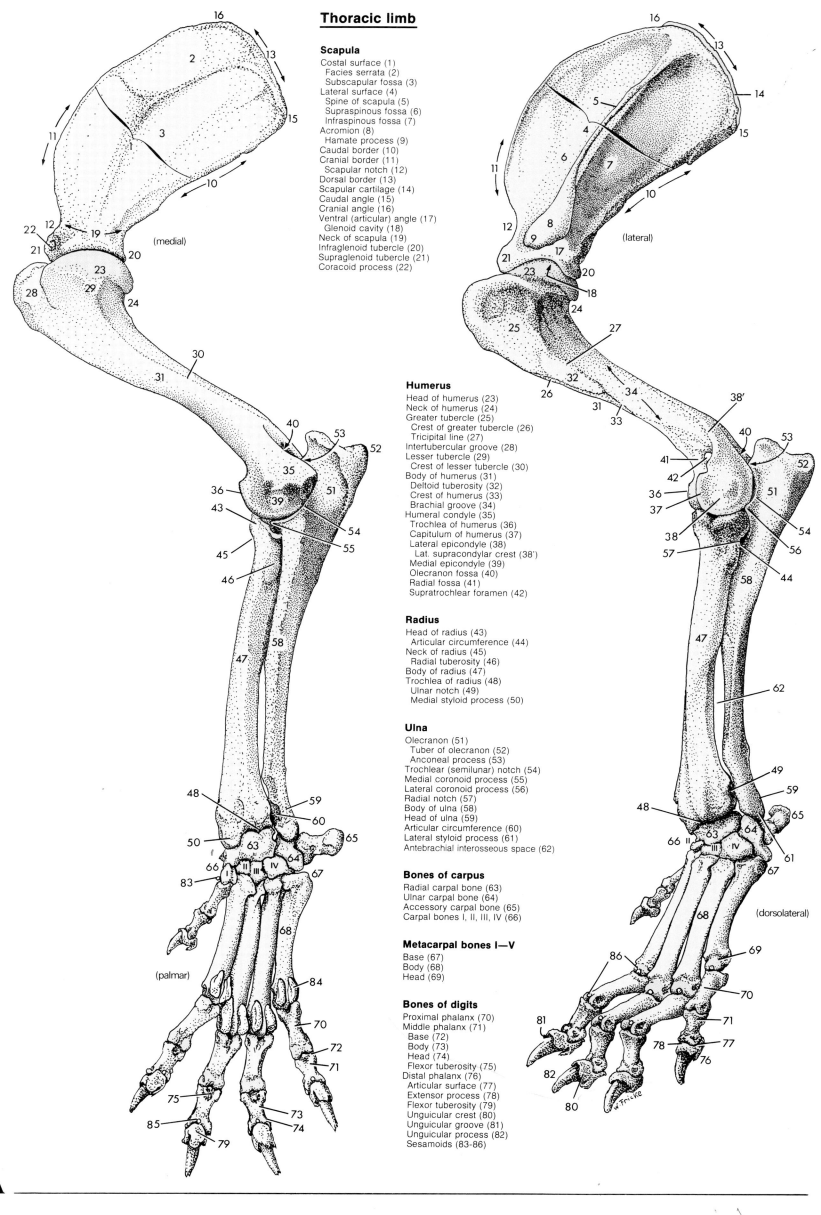

Thoracic limb

Scapula
Costal surface (1)
  Facies serrata (2)
  Subscapular fossa (3)
Lateral surface (4)
  Spine of scapula (5)
  Supraspinous fossa (6)
  Infraspinous fossa (7)
Acromion (8)
  Hamate process (9)
Caudal border (10)
Cranial border (11)
  Scapular notch (12)
Dorsal border (13)
Scapular cartilage (14)
Caudal angle (15)
Cranial angle (16)
Ventral (articular) angle (17)
  Glenoid cavity (18)
Neck of scapula (19)
Infraglenoid tubercle (20)
Supraglenoid tubercle (21)
Coracoid process (22)

Humerus
Head of humerus (23)
Neck of humerus (24)
Greater tubercle (25)
  Crest of greater tubercle (26)
  Tricipital line (27)
  Intertubercular groove (28)
Lesser tubercle (29)
  Crest of lesser tubercle (30)
Body of humerus (31)
  Deltoid tuberosity (32)
  Crest of humerus (33)
  Brachial groove (34)
Humeral condyle (35)
  Trochlea of humerus (36)
  Capitulum of humerus (37)
  Lateral epicondyle (38)
    Lat. supracondylar crest (38')
  Medial epicondyle (39)
  Olecranon fossa (40)
  Radial fossa (41)
  Supratrochlear foramen (42)

Radius
Head of radius (43)
  Articular circumference (44)
Neck of radius (45)
  Radial tuberosity (46)
Body of radius (47)
Trochlea of radius (48)
  Ulnar notch (49)
  Medial styloid process (50)

Ulna
Olecranon (51)
  Tuber of olecranon (52)
  Anconeal process (53)
Trochlear (semilunar) notch (54)
Medial coronoid process (55)
Lateral coronoid process (56)
Radial notch (57)
Body of ulna (58)
Head of ulna (59)
  Articular circumference (60)
  Lateral styloid process (61)
Antebrachial interosseous space (62)

Bones of carpus
Radial carpal bone (63)
Ulnar carpal bone (64)
Accessory carpal bone (65)
Carpal bones I, II, III, IV (66)

Metacarpal bones I—V
Base (67)
Body (68)
Head (69)

Bones of digits
Proximal phalanx (70)
Middle phalanx (71)
  Base (72)
  Body (73)
  Head (74)
  Flexor tuberosity (75)
Distal phalanx (76)
  Articular surface (77)
  Extensor process (78)
  Flexor tuberosity (79)
  Unguicular crest (80)
  Unguicular groove (81)
  Unguicular process (82)
Sesamoids (83-86)

(medial)

(lateral)

(palmar)

(dorsolateral)

W. Fricke

7A

Further dissection necessitates the separation of thoracic limb from trunk. Transect the extrinsic muscles of the thoracic limb a few centimetres proximal to the clavicular intersection and proximal to the other muscle insertions on the thoracic limb. The external jugular vein is severed cranial to the branching of the omobrachial v., and the brachiocephalic v. is transected slightly caudal to its division into subclavian and external jugular vv. Segmental nerves (nCv6 to nTv2) are severed a short distance dorsal to their contribution to the brachial plexus. In so doing one preserves the three branches of the phrenic n. (nCv5 to 7) running to the cranial thoracic aperture. The skin is then removed from the isolated limb as far distally as the metacarpophalangeal articulations, leaving the carpal, metacarpal and digital pads intact for further inspection. Care should be taken to leave superficial veins and accompanying subcutaneous nerve branches intact, particularly those relating to the flexor muscles of elbow and dorsal surface of antebrachium. The m. pronator teres (46) is transected medial to the elbow to display adjacent veins.

a) VEINS are defined on the basis of the areas they supply. The sequence in which veins ramify serves only as a guide since they vary markedly from one individual to the other. There is, however, little variance with arteries and nerves. At the level of the first rib, the very short subclavian v. continues as the **axillary v. (21)**. As its first branch this gives off the **external thoracic v. (22)** which is often duplicated. The next vein, the **lateral thoracic v. (4)**, together with the concomitant artery and nerve, courses along the lateral border of the m. pectoralis profundus. The vein is in contact with the **axillary (21)** and the **accessory axillary lymph node (3)**, at its origin and at the level of the second intercostal space respectively. Then from the axillary v. the **subscapular v. (5)** proceeds to the appropriate muscle and the **thoracodorsal v. (2)** goes to the medial surface of the m. latissimus dorsi. The subscapular v. gives off the **caudal circumflex humeral v. (18)**, which immediately goes deeper, is directed in an arch around the humerus, and anastomoses with the **cranial circumflex humeral v. (23)** (see p. 11A). The latter is a very weak vessel arising from the axillary v. and proceeding to the hilar region of the m. biceps brachii. After giving off the axillobrachial v. (see p. 11A —33) which may also arise from the caudal circumflex humeral v., the axillary v. continues as the **brachial v. (6)**. The **superficial brachial v. (10)** arises on the surface of the elbow. Its continuation, the **median cubital v. (27)** arises from the cephalic v. The brachial v. gives off the **common interosseous v. (12)** deep to the transected m. pronator teres and then continues as the **median v. (13)**.

1 **b) SPINAL NERVES nCv6 to nTv2** form the roots of the brachial plexus where the fibers of several spinal nerve segments meet. From the plexus, nerves proceed to the thoracic limb. In the case of the sequential identification of muscles and nerves, nerves serve as guiding structures for formulating muscle homologues and reciprocally, nerves are identified on the basis of their fields of innervation.

The **axillary n.** (nC7 and 8 —17) sends a branch to the **m. teres major (1)** which arises proximally on the caudal border of the scapula and inserts on the teres tuberosity. This slight prominence lies on the medial aspect of the humerus at the junction between its proximal and middle thirds. In addition, the axillary n. innervates the caudal border of the m. subscapularis and the shoulder joint. At the caudal border of the m. subscapularis, the nerve courses deeply and eventually appears on the lateral surface of the shoulder (see p. 9). The motor component of the **accessory axillary n.** (nC6 —15) (which is not derived from the plexus), ramifies in the clavicular part of the m. deltoideus, while its cutaneous sensory component pierces the muscle. Two branches of the **subscapular n.** (nC6 and 7 —16) enter the **m. subscapularis (16)** which arises on the subscapular fossa and inserts on the lesser tubercle of the humerus. The **suprascapular n.** (nc6 and 7 —14) crosses the junction between middle and distal thirds of the cranial border of the scapula and goes between the mm subscapularis and supraspinatus to the lateral surface of the shoulder (see p. 9). The **musculocutaneous n.** (nC6 to nT1 —25) lies cranial to the brachial a. Proximal muscle branches of the musculocutaneous nerve supply the mm coracobrachialis and biceps brachii and it then communicates with the median n. about 2 cm proximal to the elbow. Deep to the m. biceps brachii the musculocutaneous n. divides into a distal muscle branch supplying the m. brachialis and (yet again) the m. biceps brachii, and the **medial cutaneous antebrachial n. (11)**. This runs subcutaneously after passing between the brachial v. and the end of the m. biceps brachii. The short spindle-shaped **m. coracobrachialis (20, B)** arises at the coracoid process and inserts onto the crest of the lesser tubercle of the humerus between its proximal and middle thirds. The dense tendon of origin of the elongated **m. biceps brachii (26, A)** arises in the vicinity of the supraglenoid tubercle of the scapula and lies initially within the capsule of the shoulder joint. This forms a synovial sheath around the tendon which is held in place in the intertubercular groove by a transverse ligament. The tendon of insertion of the m. biceps brachii (see text illustration) forms a Y-shaped division distal to the flexor surface of the elbow, the tendon parts so formed inserting onto the proximomedial aspects of radius and ulna. The tendon of insertion of the **m. brachialis (C)** passes through the Y-shaped division described above, to insert on the proximal part of the ulna. The muscle itself arises caudal to the head of the humerus and spirals in the groove of

the m. brachialis first caudally, then laterocranially and medially, to its insertion. Distal to the thick tendon of insertion of the m. teres major, the **radial n.** (nC7 to nT2 —19) is directed deeply to innervate the m. 2 triceps (see p. 9). Proximal to this, on the proximomedial aspect of the arm, the radial n. gives off a slender branch to the thin **m. tensor fasciae antebrachii (7)**. This muscle takes origin at the insertion of the m. latissimus dorsi and radiates into the antebrachial fascia at the level of the olecranon. The **ulnar** (nC8 to nT2 —8) and **median** (nC8 to nT2 —24) nn arise in common from the brachial plexus. They separate from each other at the level of the shoulder joint and accompany the brachial v. before reaching the muscles of the antebrachium. Here the median n. lies medially and the ulnar n. caudally. At the junction of the middle and distal thirds of the brachium the ulnar n. gives off the **caudal cutaneous antebrachial n. (9)** which proceeds towards the olecranon. (According to current nomenclature, the cranial and caudal pectoral nn, the long thoracic n., the thoracodorsal n. and the lateral thoracic n. are also considered to be nerves of the brachial plexus.)

Mm biceps brachii, coracobrachialis, brachialis

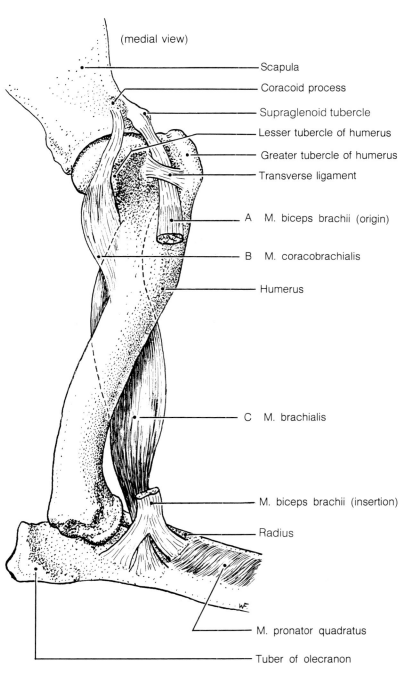

(medial view)

- Scapula
- Coracoid process
- Supraglenoid tubercle
- Lesser tubercle of humerus
- Greater tubercle of humerus
- Transverse ligament
- A M. biceps brachii (origin)
- B M. coracobrachialis
- Humerus
- C M. brachialis
- M. biceps brachii (insertion)
- Radius
- M. pronator quadratus
- Tuber of olecranon

8

Thoracic limb (caudomedial view)

Legend:

a Phrenic n.
b Deep brachial a. and v.
c Collateral ulnar a. and v.
d Transverse cubital a. and v.
e Recurrent ulnar a. and v.
f Ulnar a. and v.
g Dors. rr. of ulnar a. and n.
h Caud. interosseous a. and v.
i Bicipital a. and v.
j Deep brachial a. and v.
k Cephalic v.
l Medial r. of supf. cran. antebrachial a.
   and supf. r. of radial n. (medial r.)
m Acc. cephalic v.
n Radial a. and v.
o Dors. carpal r.
p Abaxial palmar digital n. I
q Com. palmar digital aa. and nn.
r Supf. palmar arch
s Com. palmar digital vv.

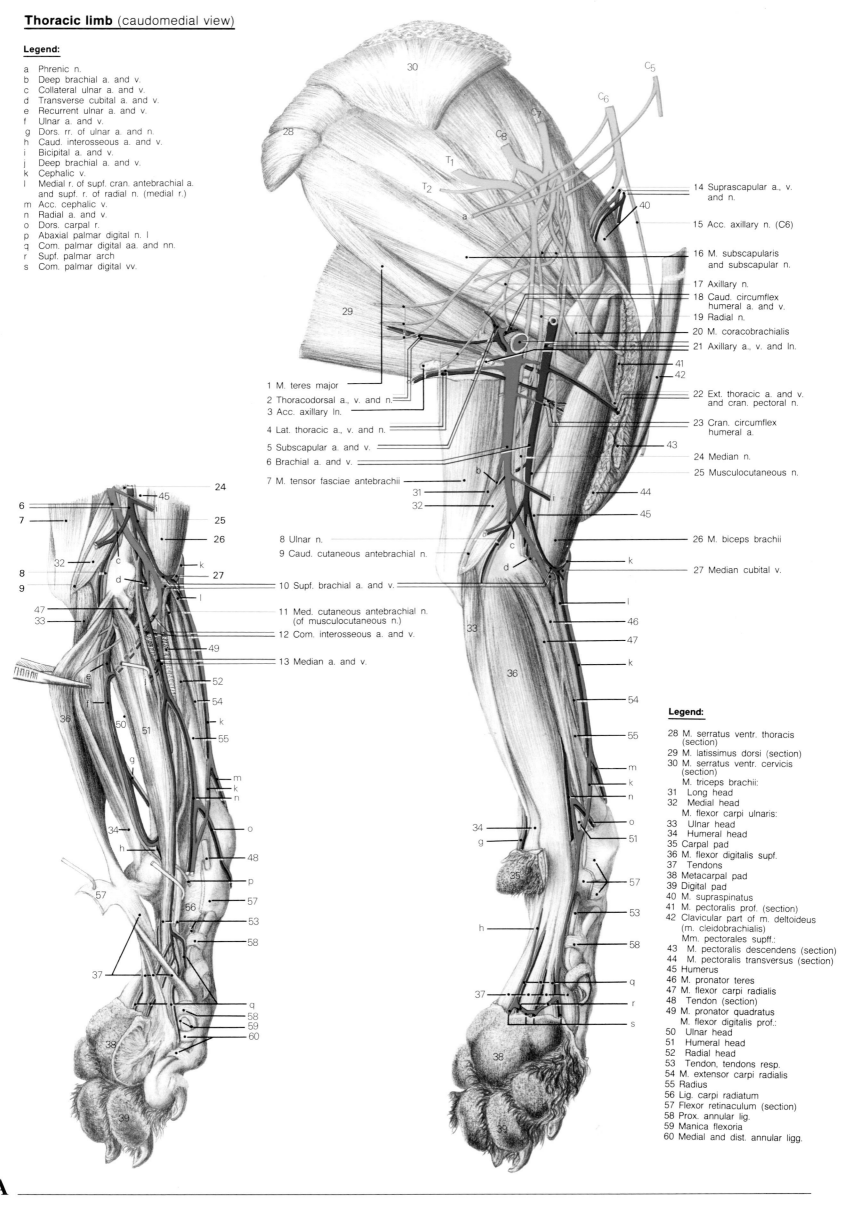

1 M. teres major
2 Thoracodorsal a., v. and n.
3 Acc. axillary ln.
4 Lat. thoracic a., v. and n.
5 Subscapular a. and v.
6 Brachial a. and v.
7 M. tensor fasciae antebrachii

8 Ulnar n.
9 Caud. cutaneous antebrachial n.
10 Supf. brachial a. and v.
11 Med. cutaneous antebrachial n.
   (of musculocutaneous n.)
12 Com. interosseous a. and v.
13 Median a. and v.

14 Suprascapular a., v.
   and n.
15 Acc. axillary n. (C6)
16 M. subscapularis
   and subscapular n.
17 Axillary n.
18 Caud. circumflex
   humeral a. and v.
19 Radial n.
20 M. coracobrachialis
21 Axillary a., v. and ln.
22 Ext. thoracic a. and v.
   and cran. pectoral n.
23 Cran. circumflex
   humeral a.
24 Median n.
25 Musculocutaneous n.
26 M. biceps brachii
27 Median cubital v.

Legend:

28 M. serratus ventr. thoracis
   (section)
29 M. latissimus dorsi (section)
30 M. serratus ventr. cervicis
   (section)
   M. triceps brachii:
31 Long head
32 Medial head
   M. flexor carpi ulnaris:
33 Ulnar head
34 Humeral head
35 Carpal pad
36 M. flexor digitalis supf.
37 Tendons
38 Metacarpal pad
39 Digital pad
40 M. supraspinatus
41 M. pectoralis prof. (section)
42 Clavicular part of m. deltoideus
   (m. cleidobrachialis)
   Mm. pectorales supff.:
43 M. pectoralis descendens (section)
44 M. pectoralis transversus (section)
45 Humerus
46 M. pronator teres
47 M. flexor carpi radialis
48 Tendon (section)
49 M. pronator quadratus
   M. flexor digitalis prof.:
50 Ulnar head
51 Humeral head
52 Radial head
53 Tendon, tendons resp.
54 M. extensor carpi radialis
55 Radius
56 Lig. carpi radiatum
57 Flexor retinaculum (section)
58 Prox. annular lig.
59 Manica flexoria
60 Medial and dist. annular ligg.

To demonstrate the proximal anastomoses between the medial deep and the lateral superficial venous system, and the muscle branches of the axillary n., the scapular and acromial parts of the m. deltoideus are separated at their origins on the spine of the scapula (see text illustration). To display the muscle branches of the radial n., the lateral head of the m. triceps is transected at its midlength (see p. 9A —14).

a) The CUTANEOUS VEINS SITUATED LATERALLY run in the shoulder and brachial regions unaccompanied by arteries of like name.
1 The **cephalic v. (20)** courses in the lateral pectoral groove to the craniolateral aspect of the brachium. There it receives the large **axillobrachial v. (7)**. Before this vein passes deep to the scapular part of the m. deltoideus it receives the **omobrachial v. (6)** described above. On the flexor aspect of the elbow the cephalic v. gives rise to the **median cubital v. (21)** which is directed transversely. The continuation of the latter, known as the superficial brachial v., reaches the brachial vein situated medially. In its course along the cranial aspect of the antebrachium the cephalic v. is accompanied by medial and lateral rami of the superficial ramus of the radial n. and the cranial superficial antebrachial a., the distal continuation of the slender superficial brachial a. About 6 cm proximal to the carpus, the cephalic v. curves medially and reaches the palmar aspect of the manus. The straight course adopted by the cephalic v. on the cranial aspect of the thoracic limb is continued by the **accessory cephalic v. (22)**. On the dorsal aspect of the manus this gives rise to dorsal common digital veins.

b) The NERVES supplying the LATERAL MUSCLES of shoulder and brachium arise from the brachial plexus medial to the shoulder, where the initial parts of the brachial plexus have been identified (see p. 8A).
The **axillary n. (see p. 8A —17)** gives muscle branches to the medial surface of the scapular and acromial parts of the m. deltoideus and to the spindle-shaped m. teres minor. Subsequently, the axillary n. terminates in the **craniolateral cutaneous brachial n. (10)** and the **cranial cutaneous antebrachial n. (11)**. These proceed subcutaneously on the ventral aspect of the m. deltoideus between its scapular and acromial parts in company with the axillobrachial v. The **m. deltoideus (3, C)** arises along the spine of the scapula by means of its **scapular part (4)** and at the acromion by its **acromial part (5)**. Both parts insert onto the deltoid tuberosity of the humerus. The clavicular part (m. cleidobrachia-lis of texts) runs between the clavicular intersection and the distal end of the crest oft the humerus and is innervated by the accessory axillary n. (nC6). The **m. teres mibor (D)** arises on the infraglenoid tubercle and the distal part of the caudal border of the scapula and inserts at the deltoid tuberosity. The **suprascapular n. (2)** identified previously, innervates the m. supraspinatus and is directed lateral to the scapula and ventral to the base of the acromion to enter the m. infraspinatus. The **m. supraspinatus (1, A)** takes origin from the supraspinous fossa of the scapula and inserts cranially onto the greater tubercle of the humerus. The **m. infraspinatus (9, B)** lies deep to the scapular part of the m. deltoideus, its origin being in the infraspinous fossa. It inserts caudally onto the greater tubercle of the humerus by passing over the subtendinous bursa of the m. infraspinatus. The tendons of insertion of the m. infraspinatus and the m. subscapularis assume the function of medial and lateral contractile 'tension bands' of the shoulders joint, since proper collateral ligaments are absent. Deep to the transected lateral head of the m. triceps brachii, proximal muscle branches of the **radial n. (16)** enter the muscle and also the m. anconeus. The continuing radial n. then passes over the lateral supracondylar crest and divides into a **deep ramus (18)** innervating extensors of carpus and digits, **and a superficial ramus (17)**. The medial and lateral branches of this ramus accompany the cephalic and accessory cephalic vv. disally on their respective sides. Subsequently they provide dorsal common digital nn to the dorsal aspect of the manus. The **m. triceps brachii (12**, see also small illustration) has the following heads and sites of origin: the **long head (13)** from the caudal border of scapula, the **accessory head (15)** caudally from the neck of humerus, the **lateral head (14)** caudally from the tricipital line, and the **medial head (8)** proximomedially from the humerus. A subtendinous bursa lies under the common tendon of insertion of the m. triceps brachii and slightly laterally. The inconstant subcutaneous bursa of the olecranon is situated on the olecranon tuber. The **m. anconeus (19)** arises around the border of the olecranon fossa and terminates laterally on the olecranon by a fleshy attachment.

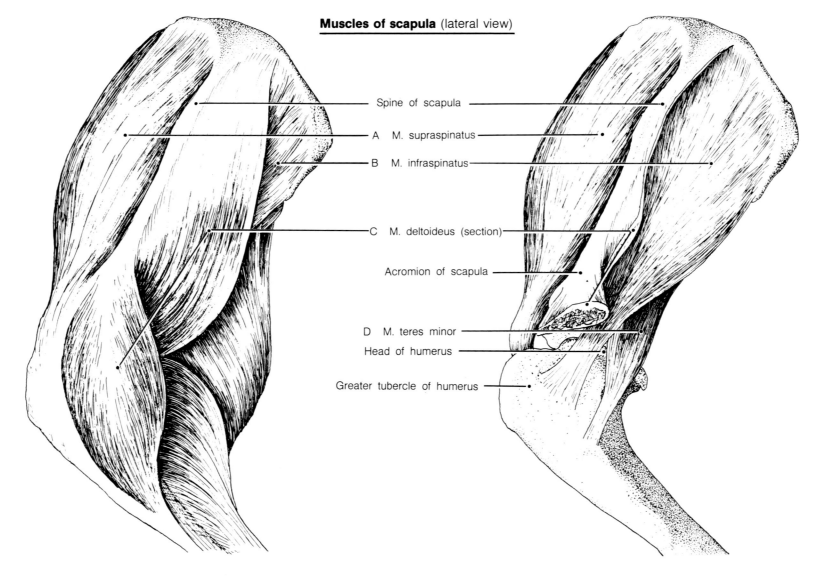

Muscles of scapula (lateral view)

Spine of scapula

A   M. supraspinatus

B   M. infraspinatus

C   M. deltoideus (section)

Acromion of scapula

D   M. teres minor

Head of humerus

Greater tubercle of humerus

Thoracic limb (craniolateral view)

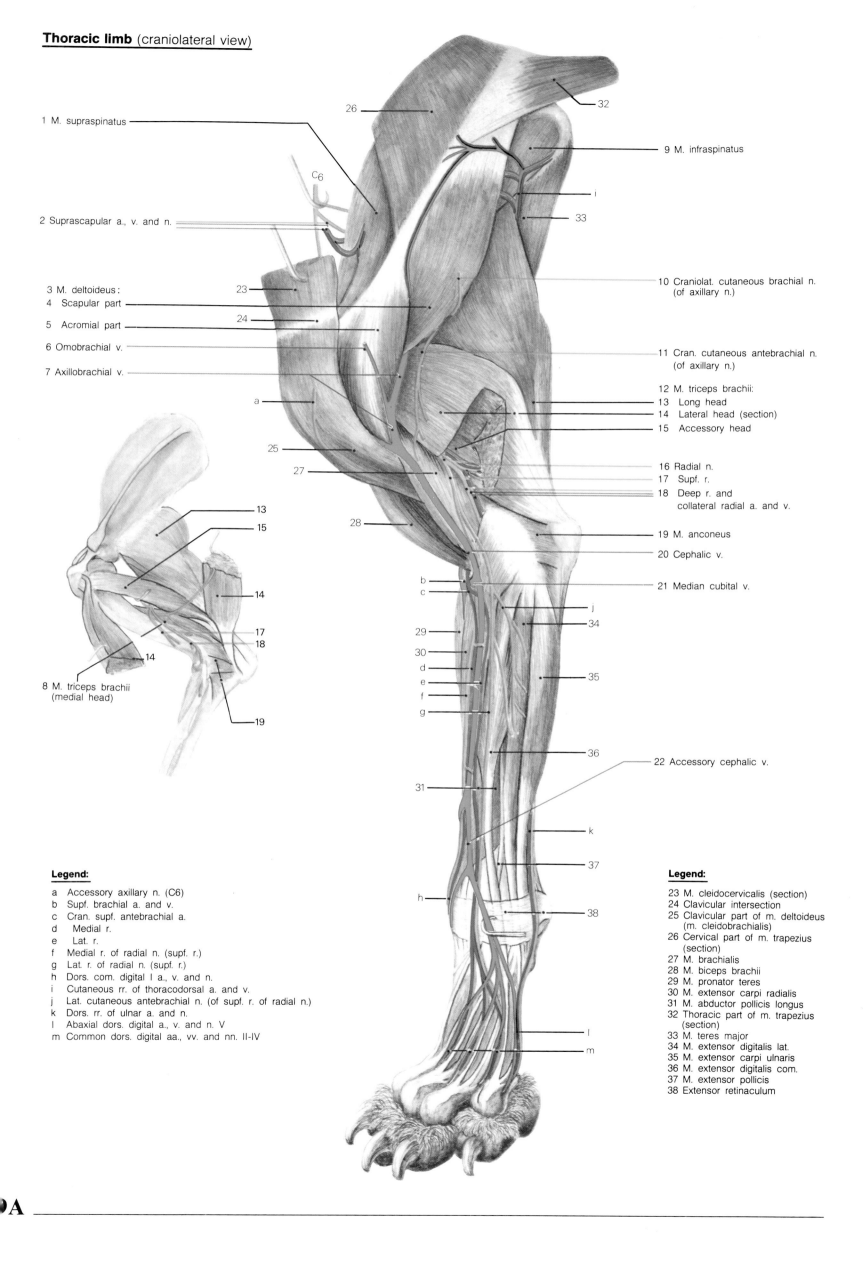

1 M. supraspinatus

2 Suprascapular a., v. and n.

3 M. deltoideus:
4 Scapular part
5 Acromial part
6 Omobrachial v.
7 Axillobrachial v.

8 M. triceps brachii
   (medial head)

26

C6

23

24

25

27

28

29

30

31

a

b
c

d
e
f
g

h

13
15
14
17
18
14
19

32

9 M. infraspinatus

i

33

10 Craniolat. cutaneous brachial n.
   (of axillary n.)

11 Cran. cutaneous antebrachial n.
   (of axillary n.)

12 M. triceps brachii:
13   Long head
14   Lateral head (section)
15   Accessory head

16 Radial n.
17   Supf. r.
18   Deep r. and
     collateral radial a. and v.

19 M. anconeus
20 Cephalic v.

21 Median cubital v.

j
34

35

36

22 Accessory cephalic v.

k

37

38

l

m

Legend:

a Accessory axillary n. (C6)
b Supf. brachial a. and v.
c Cran. supf. antebrachial a.
d Medial r.
e Lat. r.
f Medial r. of radial n. (supf. r.)
g Lat. r. of radial n. (supf. r.)
h Dors. com. digital I a., v. and n.
i Cutaneous rr. of thoracodorsal a. and v.
j Lat. cutaneous antebrachial n. (of supf. r. of radial n.)
k Dors. rr. of ulnar a. and n.
l Abaxial dors. digital a., v. and n. V
m Common dors. digital aa., vv. and nn. II-IV

Legend:

23 M. cleidocervicalis (section)
24 Clavicular intersection
25 Clavicular part of m. deltoideus
   (m. cleidobrachialis)
26 Cervical part of m. trapezius
   (section)
27 M. brachialis
28 M. biceps brachii
29 M. pronator teres
30 M. extensor carpi radialis
31 M. abductor pollicis longus
32 Thoracic part of m. trapezius
   (section)
33 M. teres major
34 M. extensor digitalis lat.
35 M. extensor carpi ulnaris
36 M. extensor digitalis com.
37 M. extensor pollicis
38 Extensor retinaculum

> *To distinguish unequivocally between the individual flexor muscles of carpus and digits their tendons of insertion must be identified. To this end, one opens the carpal canal which acts as a tunnel conveying both digital flexor tendons, and arteries, veins and nerves on the palmar aspect of the carpus. Firstly, one transects the superficial lamina of the flexor retinaculum between accessory carpal bone and medial styloid process. Subsequently the superficial digital flexor tendon is lifted out of the carpal canal and then one severs the deep lamina of the flexor retinaculum between superficial and deep digital tendons. After elevating the deep flexor tendon, the deep boundary of the carpal canal is visible through the capsule of the carpal articulation. Then the skin of the second digit is completely removed to demonstrate the insertions of digital extensor and flexor tendons. The manica flexoria of the superficial digital flexor tendon is incised laterally so that the passage of the deep flexor tendon can be followed to better advantage.* **Hint on use of illustration on p. 10A:** *Structures not. designated are illustrated on pp. 8A and 9A.*

a) The CAUDOMEDIAL ANTEBRACHIAL MUSCLES include two digital flexors, two carpal flexors and two pronators of the radioulnar articulations.

Nerve supply is derived from the **median (5)** and **ulnar (10) nn**, the muscles situated more medially (the mm pronator teres, pronator quadratus, flexor carpi radialis, and radial head of m. flexor profundus) are innervated solely by the median n., while the more caudal muscles (the m. flexor carpi ulnaris and ulnar head of the m. flexor digitalis profundus) are supplied solely by the ulnar n. An exception is the m. flexor digitalis superficialis which, as a muscle situated further caudally, is supplied by the median n. In the centre, lying between both fields of innervation, the three bellies of the humeral head of the m. flexor digitalis profundus receive a double innervation from both ulnar and median nn. In the proximal third of the forearm the ulnar n. divides into a palmar ramus and a **dorsal ramus (15 — dorsal n. of manus)**. The latter proceeds to the caudolateral aspect of the m. extensor carpi ulnaris distally, and terminates on the dorsal aspect of the manus as abaxial dorsal digital n. V. The **palmar ramus of the ulnar n. (18)** and the median n. traverse the carpal canal on the deep and superficial surfaces of the deep digital flexor tendon respectively. Both nerve branches continue distally on the palmar aspect of the manus (Details: see p. 11).

The **muscle origins** are situated predominantly on the medial epicondyle of the humerus. A sole exception is the m. pronator quadratus. Its fibers, directed horizontally, occupy the antebrachial interosseous space. Besides an origin from the medial epicondyle of the humerus, the two 'multiheaded' muscles, namely the mm flexor carpi ulnaris and flexor digitalis profundus have additional heads from the bones of the antebrachium. The former muscle has an ulnar head in addition; the latter, a muscle with three heads, has additionally an ulnar and a radial head. Corresponding to their names, these heads originate proximally on the respective bones of the antebrachium.

The **muscle insertions** of both digital flexors are situated on the bones of the digits, those of the flexors of carpus on carpus or on the proximal metacarpus and those of the pronators on the bones of the antebrachium (considered when identifying the muscles).

I. Of the two **digital flexor muscles**, the **m. flexor digitalis superficialis (11)** continues as a tendon on the distal third of the forearm. On the proximal part of the metacarpus, this subdivides into four tendons of insertion, one for each digit. A short distance proximal to the insertion on the flexor tuberosity of the middle phalanx, the attachment of each tendon is transformed into a tunnel or sleevelike **manica flexoria (21)** for the passage of the tendon of insertion of the m. flexor digitalis profundus. In the distal third of the forearm, the tendons of the weak lateral **ulnar head (8)**, the strong middle **humeral head (9)** and the weaker medial **radial head (7)** of the **m. flexor digitalis profundus (6)** form the deep flexor tendon. Each of its five tendons of insertion ends on the flexor tuberosity of the respective distal phalanx after passage through the tunnel-like manica flexoria.

II. Of the two **flexor muscles of the carpus**, the **m. flexor carpi ulnaris (16)** is divided into two heads, the predominantly tendinous ulnar head and mainly fleshy humeral head, both of which insert onto the accessory carpal bone. The two tendons of insertion of the **m. flexor carpi radialis (2)** are attached to the proximopalmar aspect of metacarpal bones II and III.

III. Of the two **pronators** of the radioulnar articulations, **m. pronator teres (A)** inserts craniomedially onto the radius. The **m. pronator quadratus (4)** is situated in the interosseous space of the antebrachium between radius and ulna (see also text illustration, p. 8).

b) The CRANIOLATERAL ANTEBRACHIAL MUSCLES include two digital extensors, two muscles of digit I, two extensors of the carpus and two supinators of the radioulnar articulations.

The **innervation** is from the radial n., the **deep ramus (1)** of which sends branches to all craniolateral antebrachial muscles. The lateral and medial rami of the superficial ramus accompany the cephalic vein distally, one to either side, the superficial ramus itself giving rise to the lateral cutaneous brachial n. Both medial and lateral rami accompany the accessory cephalic v. onto the dorsal aspect of the manus where they divide into the dorsal **common digital nn (20)**, providing in turn the dorsal proper digital nn.

The **origins** of the craniolateral antebrachial muscles are situated mainly on the lateral epicondyle of the humerus. Exceptions are, the mm. extensor carpi radialis and brachioradialis arising from the lateral supracondylar crest further proximally, and both muscles of digit I (the pollex), coming from radius and ulna somewhat more distally.

The muscle insertions are considered for the identification of the craniolateral antebrachial muscles. Of the four muscle groups, in turn two individual muscles, the digital extensors, insert distally onto the extensor processes of the distal phalanges, the muscles of digit I between metacarpus and distal phalanx, the carpal extensors onto the metacarpus directly distal to the carpus, and the supinators onto the radius further proximally.

I. Of the two **digital extensor muscles**, the **m. extensor digitalis communis (13, H)** has tendinous insertions from the second to the fifth digit and similarly the **m. extensor digitalis lateralis (17, I)** inserts by tendons onto digits III to V inclusive.

II. Of the two **muscles of the first digit (pollex)**, **m. abductor pollicis longus (12, D)** terminates proximally on metacarpal bone I, and the **m. extensor pollicis (14, E)** goes by two weak tendons to the first and second digits.

III. Of the two **extensors of the carpus**, the two tendons of the **m. extensor carpi radialis (3, C)** insert proximally onto metacarpal bones II and III. The **m. extensor carpi ulnaris (19, G)** inserts onto the end of metacarpal bone V, and terminates on the accessory carpal bone by a transverse attachment. Therefore it is said to function as an abductor in appropriate positions of the articulation.

IV. Of the two **supinators** (see text illustration), the **m. supinator (F)** lies deep to the muscle origins of the digital extensors and ends on the proximocranial part of the radius. The **m. brachioradialis (B)** courses superficially over the flexor aspect of the elbow to terminate in the antebrachial fascia at the level of the midlength of radius.

Antebrachial muscles (craniolateral view)

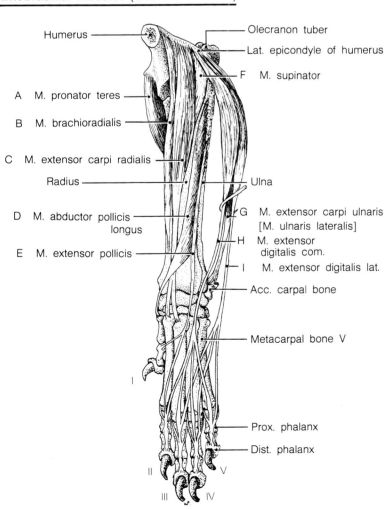

Humerus — Olecranon tuber — Lat. epicondyle of humerus — F M. supinator

A M. pronator teres — B M. brachioradialis — C M. extensor carpi radialis — Radius — Ulna — G M. extensor carpi ulnaris [M. ulnaris lateralis] — H M. extensor digitalis com. — I M. extensor digitalis lat. — D M. abductor pollicis longus — E M. extensor pollicis — Acc. carpal bone — Metacarpal bone V — Prox. phalanx — Dist. phalanx

Antebrachium or Forearm

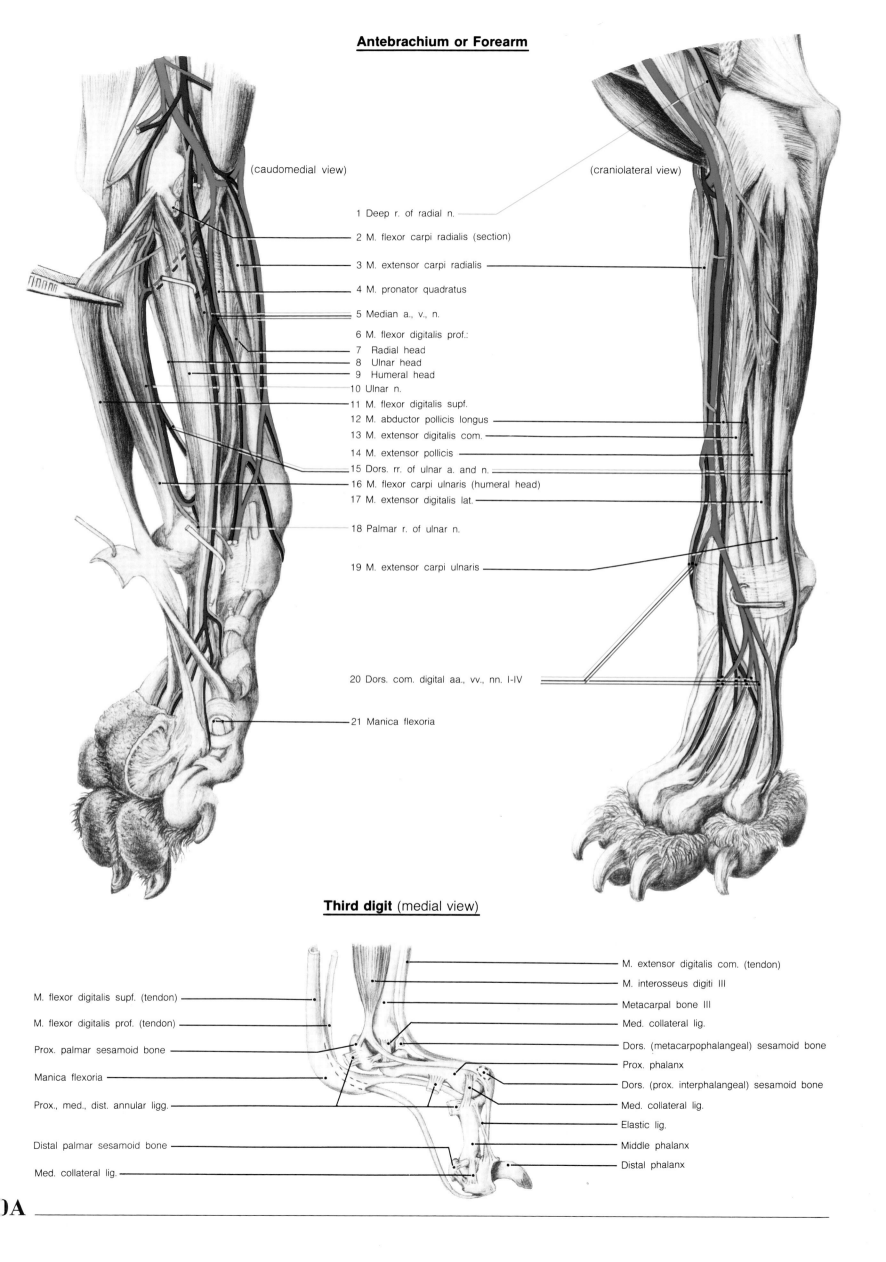

(caudomedial view) (craniolateral view)

1 Deep r. of radial n.

2 M. flexor carpi radialis (section)

3 M. extensor carpi radialis

4 M. pronator quadratus

5 Median a., v., n.

6 M. flexor digitalis prof.:
7 Radial head
8 Ulnar head
9 Humeral head
10 Ulnar n.
11 M. flexor digitalis supf.
12 M. abductor pollicis longus
13 M. extensor digitalis com.
14 M. extensor pollicis
15 Dors. rr. of ulnar a. and n.
16 M. flexor carpi ulnaris (humeral head)
17 M. extensor digitalis lat.

18 Palmar r. of ulnar n.

19 M. extensor carpi ulnaris

20 Dors. com. digital aa., vv., nn. I-IV

21 Manica flexoria

Third digit (medial view)

M. flexor digitalis supf. (tendon)

M. flexor digitalis prof. (tendon)

Prox. palmar sesamoid bone

Manica flexoria

Prox., med., dist. annular ligg.

Distal palmar sesamoid bone

Med. collateral lig.

M. extensor digitalis com. (tendon)

M. interosseus digiti III

Metacarpal bone III

Med. collateral lig.

Dors. (metacarpophalangeal) sesamoid bone

Prox. phalanx

Dors. (prox. interphalangeal) sesamoid bone

Med. collateral lig.

Elastic lig.

Middle phalanx

Distal phalanx

DA

> *During the dissection of the arteries of the thoracic limb, accompanying veins and nerves of like name are to be taken into account as well. The blood supply of the thoracic limb comes from only one artery, the axillary. On the other hand three veins are concerned with venous return from the limb namely the axillary v. situated medially, and the cephalic and omobrachial vv laterally.*

a) The **SHOULDER, ARM, AND FOREARM** are supplied by blood vessels and nerves which run chiefly on the medial and lateral aspects of the limb.

The **axillary a. and v. (15)** lie on the flexor aspect of the shoulder joint superficial to the axillary n. At the level of the first rib they are continuous with the long subclavian a. and the very short subclavian v. respectively. The axillary a. and v. give rise to the external thoracic, the lateral thoracic and the subscapular aa and vv; proximal to their continuation as the brachial a. and v., the axillary vessels discharge the cranial circumflex humeral a. and v. The **external thoracic a. and v. (16)** and accompanying cranial pectoral nn penetrate the mm. pectorales. The **lateral thoracic a., v. and n. (3)** run along the lateral border of the m. pectoralis profundus, the blood vessels supplying the axillary and the inconstant accessory axillary lymph node and the thoracic mammae. The **subscapular a. and v. (1)** travel along the caudal border of the m. subscapularis, the artery providing the **thoracodorsal a. (2)** caudally. The thoracodorsal v. (2) arises from the axillary v. (see p. 8). Thoracodorsal a., v. and n. all enter the medial surface of the m. latissimus dorsi. As further branches of the scapular vessels, the **caudal circumflex humeral a. and v. (4)** take a deep arcuate course around the humerus laterally, to anastomose with the smaller cranial circumflex a. and v. respectively. The caudal circumflex humeral a. and v. provide the collateral radial a. and v. as vessels accompanying the radial n. The **cranial circumflex humeral a. and v. (17)** branch from the distal end of the axillary a. and v. cranially, proceeding deep to the m. biceps brachii at the level of its hilus. As a continuation of the axillary vessels, the **brachial a. and v. (5)** discharge the deep brachial, bicipital, collateral ulnar, superficial brachial and transverse cubital aa and vv. Distal to the elbow, the brachial a. and v. continue as the median a. and v. after giving off the common interosseous a. and v. In the distal third of the brachium, the **deep brachial a. and v. (6)** run caudally to the m. triceps brachii and the **bicipital a. and v. (18)** proceed cranially to the m. biceps brachii. The **collateral ulnar a. and v. (7)** accompany the ulnar n. to the tuber of the olecranon and anastomose distally with the ulnar a. and v. by way of the recurrent ulnar vessels. The **superficial brachial a. (8)** arises in the distal third of the brachium and the vein at the elbow. Then artery and vein proceed superficially and are directed transversely at the flexor aspect of the elbow. Following this, both vessels take a different course. The superficial brachial a. continues distally as the cranial superficial antebrachial a., the branches of which accompany the cephalic at first and then more distally the accessory cephalic v., to reach the dorsal aspect of the manus. On the flexor aspect of the elbow, the very short **superficial brachial v. (8)** is continuous with the median cubital v., a branch of the cephalic v. Both the superficial brachial and median cubital vv, directed transversely and merging one into the other, form the horizontal part of a venous configuration resembling the letter H. Its medial vertical limb is represented by the brachial v., the lateral by the cephalic v. On the flexor aspect of the elbow, the **transverse cubital a. and v. (9)** are directed transversely and deep to the terminal part of the m. biceps brachii. Near their origin, the **common interosseous a. and v. (10)** send off the **ulnar a. and v. (11)** which accompany the ulnar n. distally. The common interosseous a. and v. then divide into cranial and caudal interosseous aa and vv respectively. Both sets of vessels run distally in the antebrachial interosseous space. In the proximal third of the antebrachium, the **median a. and v. (12)**, the continuations distally of the brachial a. and v., give rise to the **deep antebrachial a. and v. (13)** caudally supplying antebrachial muscles. About two centimeters distally, the **radial a. and v. (14)** arise cranially and course along the radius. Distally and after the termination of its accompanying vein, the median a. passes through the carpal canal to reach the surface of the deep flexor tendons on the palmar surface of the manus.

b) **BLOOD SUPPLY AND INNERVATION OF MANUS** are derived from deep and superficial aa, vv and nn. On the dorsal and palmar aspects of the manus, deeply situated aa, vv and nn are referred to by the terms dorsal and palmar metacarpal respectively. Similar superficial structures are known as dorsal or palmar common digital aa, vv and nn. In turn, these divide into either dorsal proper digital or palmar proper digital aa, vv and nn respectively.

I. On the **dorsal aspect of the manus** dorsal common digital aa I to IV arise from both branches of the cranial superficial antebrachial a. which is a continuation distally of the superficial brachial a. The common dorsal digital vv I to IV arise from the accessory cephalic v. The common dorsal digital nn I to IV are a continuation of both medial and lateral rami of the superficial ramus of the radial nn. The dorsal ramus of the ulnar n. terminates as the abaxial dorsal n. of digit V on the dorsal aspect of the manus. The deeply situated dorsal metacarpal aa and vv I to IV arise from the dorsal arterial and venous carpal retia respectively. The venous rete is formed from dorsal carpal rami, originating in turn from accessory cephalic and radial vv. The arterial rete is also supplied by dorsal carpal rami derived from caudal interosseous, ulnar and radial aa.

II. On the **palmar aspect of the manus** common palmar digital aa and vv I to IV take origin from their superficial palmar arterial and venous arches. The superficial palmar arterial arch arises distal to the carpus, its medial limb from the median and radial aa and its lateral limb from the confluence of the ulnar and caudal interosseous aa. The superficial palmar venous arch lying more distally at the digital pads, is the result of a medial confluence of cephalic and radial vv as well as a lateral fusion of ulnar and caudal interosseous vv. The palmar metacarpal aa and vv I to IV arise slightly distal to the flexor aspect of the carpus from deep palmar arterial and venous arches. The arterial arch is formed from the radial a. medially and a mergence of ulnar and caudal interosseous aa laterally. The venous arch arises from respective veins of like name and the cephalic v. as an addition, medially. The palmar common digital nn I to III originate from the median n. while the palmar common digital n. IV arises from the superficial ramus of the palmar ramus of the ulnar n. The palmar metacarpal nn I to IV stem from the deep ramus of the palmar ramus of the ulnar n.

c) The **LYMPHATIC DRAINAGE** from the thoracic limb arises 1 from superficial and deep lymphatic vessels. In the main, the superficial lymphatic vessels accompany the superficial lateral cutaneous veins. They convey the lymph to the superficial cervical lymph node (see p. 6A —31), and subsequent to nodal transit, to the venous angle between internal and external jugular vv. The deep lymphatic vessels accompany the deeper blood vessels (situated medially) as far proximally as the axillary and the accessory axillary lymph node (see p. 8A). These nodes also receive drainage from the thoracic wall and the three cranial mammae. The disc-shaped axillary lymph node with a diameter of approximately 2 cm is accessible to palpation in the angle of divergence between axillary a. and v. on the one hand and lateral thoracic a. and v. on the other. The accessory axillary lymph node (see p. 8A —3) lies one intercostal space caudally on the course of the lateral thoracic a. and v. Lymphatic vessels from the axillary and accessory axillary lymph node arrive, likewise, at the venous angle.

Legend p. 11A

19 Circumflex a. and v. of scapula	35 Recurrent ulnar a. and v.
20 Brachiocephalic v.	36 Cran. interosseous a. and v.
21 Subclavian v.	37 Caud. interosseous a. and v.
22 Collateral radial a. and v.	38 Interosseous r.
23 Ulnar n.	39 Dorsal r. of ulnar a. and n.
24 Caud. cutaneous antebrachial n.	40 Deep palmar arch
25 Palmar r.	41 Musculocutaneous n.
26 Deep r.	42 Med. cutaneous antebrachial n.
27 Supf. r.	43 Radial n.
28 Abaxial palmar digital a. and v.	44 Supf. r.
29 Supf. palmar arch	45 Lat. r.
30 Palm. and dors. com. digital aa., vv., nn. I-IV	46 Med. r.
31 Palm. and dors. prop. digital aa., vv., nn. resp.	47 Deep r.
32 Phrenic n.	48 Supf. cran. antebrachial a.
33 Axillobrachial v.	49 Lat. r.
34 Median n.	50 Med. r.
	51 Axillary n.

52 Cran. cutaneous antebrachial n.	
53 Suprascapular n.	
54 Acc. axillary n. (C6)	
55 Ext. jugular v.	
56 Cephalic v.	
57 Median cubital v.	
58 Acc. cephalic v.	
59 Omobrachial v.	
60 Palmar metacarpal aa, vv and nn I-IV	
61 Abaxial palm. and dors. digital nn I resp.	
62 Dors. carpal r. of radial a. and v.	
63 Dors. carpal r. of caud. interosseous a. (interosseous r.)	
64 Dors. carpal r. of acc. cephalic v.	
65 Dors. carpal r. of ulnar v.	
66 Dorsal rete of carpus	
67 Dors. metacarpal aa. and vv. I-IV	
68 Abaxial dors. digital a., v. and n. V	

Arteries, Veins, Nerves of thoracic limb

(Artésien - Normand Basset)

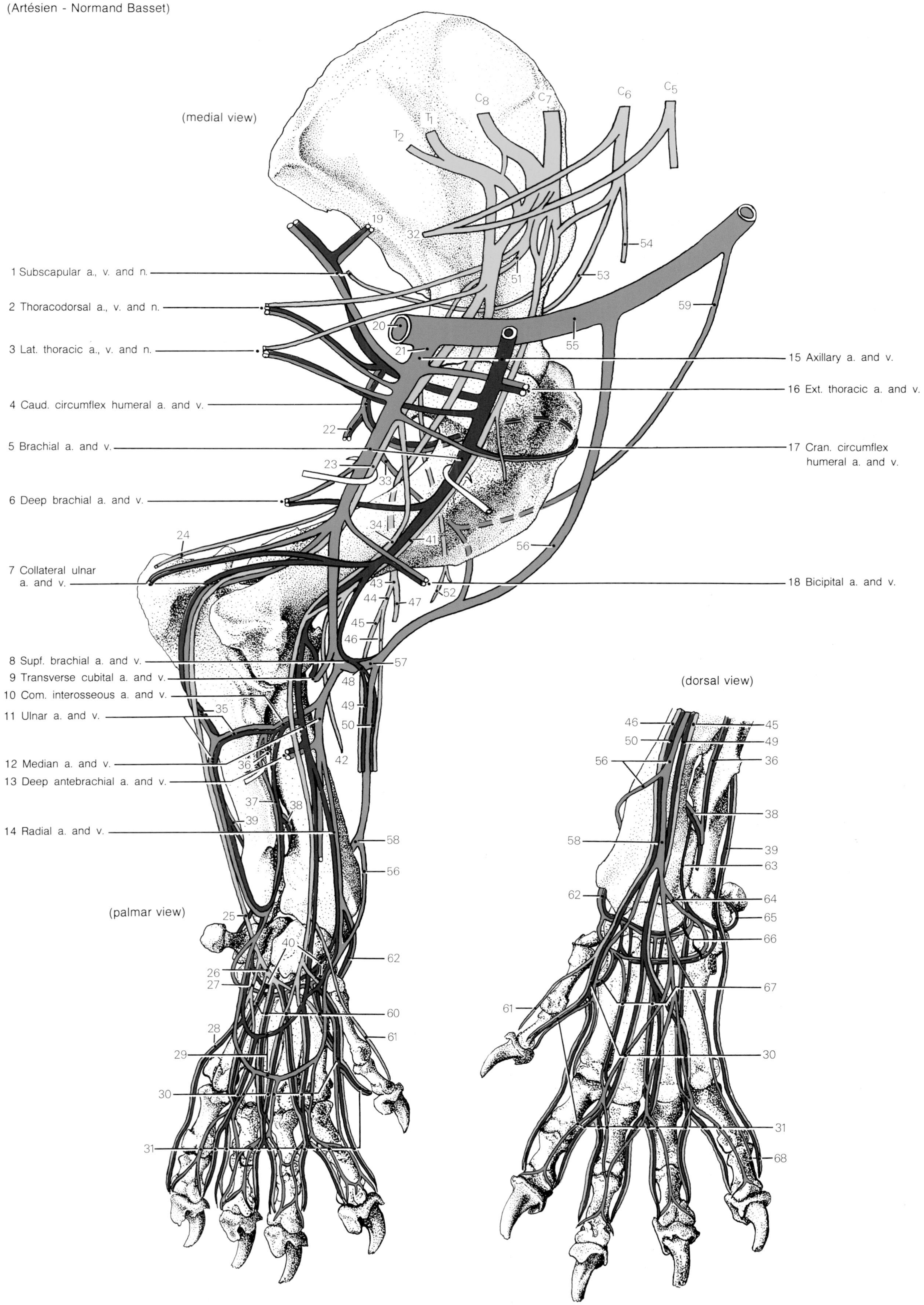

(medial view)

1 Subscapular a., v. and n.

2 Thoracodorsal a., v. and n.

3 Lat. thoracic a., v. and n.

4 Caud. circumflex humeral a. and v.

5 Brachial a. and v.

6 Deep brachial a. and v.

7 Collateral ulnar a. and v.

8 Supf. brachial a. and v.
9 Transverse cubital a. and v.
10 Com. interosseous a. and v.
11 Ulnar a. and v.

12 Median a. and v.
13 Deep antebrachial a. and v.

14 Radial a. and v.

15 Axillary a. and v.

16 Ext. thoracic a. and v.

17 Cran. circumflex humeral a. and v.

18 Bicipital a. and v.

(palmar view)

(dorsal view)

4A

> *To demonstrate the muscles of the vertebral column, the skin is removed from the lateral body wall, the dorsum, and the sacral region as far as the caudal end of the sacrum. Then one removes the remnants of the extrinsic muscles of the thoracic limb. The m. serratus dorsalis cranialis is detached from its rib insertions and reflected dorsally. The* **thoracolumbar fascia (22**, *see also text illustration p. 13) is incised longitudinally along a line parallel to and 2 cm from the dorsal midline. In the lumbosacral area, the underlying* **lumbodorsal tendon (39)** *is incised longitudinally at the same level as the fascia, and a transverse incision is made at the caudal end of the parent cut. Then the tendon is detached from the underlying musculature. The lumbodorsal tendon divides at the lateral border of the m. longissimus lumborum. The deep lamina is directed between the mm. iliocostalis and longissimus lumborum as an intermuscular septum (see text illustration p. 13) and the superficial lamina runs over the m. iliocostalis. In the neck, the* **m. splenius (1)** *and m. semispinalis capitis are transected after being exposed. The sequence to be followed in the dissection of the muscles corresponds to the numeration in the table below.*

1 a) The **MUSCLES OF THE VERTEBRAL COLUMN** are subdivided into a dorsal and a ventral group, and a specific epaxial group which moves the head. All the dorsal vertebral muscles (Nos. I to X) function in the extension and lateral movement or inclination of the vertebral column. The **innervation** of all dorsal vertebral muscles and the epaxial group of 'head movers' (Nos. XIV to XVII) results from the dorsal rami of the spinal nerves. The ventral vertebral muscles (Nos. XI to XIII) flex the vertebral column and incline it laterally. Their **innervation** is derived from ventral rami of segmental spinal nerves. The muscles situated ventral to the lumbar part of the vertebral column, namely the mm. quadratus lumborum, psoas major and psoas minor, belong to the sublumbar or inner loin muscles and are dealt with on p. 20.

MUSCLES OF THE VERTEBRAL COLUMN

Dorsal (autochthonous) muscles of vertebral column		Ventral muscles of vertebral column	Epaxial muscle group moving head
I. M. splenius	VI. **Mm. multifidi**	XI. Mm. scaleni	XIV. M. rectus capitis dors. major
II. M. iliocostalis	VII. M. sacrococcygeus dors. med.	XII. M. longus capitis	XV. M. rectus capitis dors. minor
III. M. longissimus	VIII. M. sacrococcygeus dors. lat.	XIII. M. longus colli	XVI. M. obliquus capitis caud.
IV. M. semispinalis capitis	IX. Mm. interspinales		XVII. M. obliquus capitis cran.
V. M. spinalis et semispinalis thoracis et cervicis	X. Mm. intertransversarii		

I. The **m. splenius (1)**, previously transected at its mid-length, extends from the spinous processes of the first three thoracic vertebrae to the nuchal crest on the skull. II. The **m. iliocostalis** arises from the wing of the ilium and inserts onto the lumbar transverse processes (**m. iliocostalis lumborum —24**), the angles of the ribs, and the transverse processes of the last two cervical vertebrae (**m. iliocostalis thoracis —16**). III. Extending caudally as far as the same level, the m. longissimus is divided along its length into the **mm. longissimus lumborum (23)**, **thoracis (17)**, **cervicis (12)** and **capitis (2)**. Corresponding to their range, these muscles insert onto lumbar transverse processes, tubercles of ribs, cervical transverse processes and the mastoid process of cranium respectively. IV. The **m. semispinalis capitis** lies in the neck region dorsal to the m. longissimus and consists of a dorsal **m. biventer cervicis (4)**, characterized by tendinous intersections directed transversely, and a ventral **m. complexus (3)**. Both muscles extend from the cervicothoracic boundary to the cranium. V. The **m. spinalis et semispinalis thoracis et cervicis (15)** lies medial and adjacent to the m. longissimus and courses between the second cervical and eleventh thoracic vertebrae. VI. The multipennate **mm. multifidi** extend from axis to sacrum. In the caudal half of the neck, the **m. multifidus cervicis (10)** lies deep to the m. complexus and is traversed by dorsal branches of cervical nerves on its ventrolateral aspect. The m. multifidus thoracis is situated deeply, while the **m. multifidus lumborum (27)** lies superficial and adjacent to lumbar vertebrae and their spinous processes. VII. At the level of the seventh lumbar vertebra the **m. sacrococcygeus dorsalis medialis (26)** continues the oblique pennate m. multifidus by means of an approximating fiber-flow directed caudally. VIII. The **m. sacrococcygeus dorsalis lateralis (25)** begins acutely at the fourth lumbar vertebra and, as the caudomedial continuation of the m. longissimus, proceeds to the tail by a strong terminal tendon. IX. The **mm. interspinales** lie deeply between the spinous processes. X. The **mm. intertransversarii (8)** are superficial in the neck while in the thoracolumbar region they are situated deeply along the vertebral column. The mm. intertransversarii cervicis are located ventral to the line of insertion of the m. longissimus cervicis. XI. The **mm. scaleni** extend from fourth or fifth cervical vertebra to the eighth rib (**m. scalenus dorsalis — 14**) and the first rib (**m. scalenus medius —13,C**). The m. scalenus ventralis is absent in the dog. XII. The **m. longus capitis (9, A)** lying adjacent and ventromedial to the mm. scaleni, is situated ventrolateral to the cervical vertebrae. Arising from the sixth cervical vertebra it inserts onto the muscle process of the occipital bone. XIII. The **m. longus colli (B)**, which appears plaited, is situated ventromedially on the cervical and thoracic parts of the vertebral column. Arising on the first cervical vertebra it extends to the sixth thoracic. XIV. The **m. rectus capitis dorsalis major (6)** is the cranial continuation of the nuchal ligament. Between the spinous process of axis and the occipital bone it overlies the deeper **m. rectus capitis dorsalis minor (XV.)**. XVI. The **m. obliquus capitis caudalis (7)**, the oblique cranial continuation of the nuchal ligament, terminates on the wing of the atlas. XVII. The **m. obliquus capitis cranialis (5)** runs from the wing of the atlas to the occipital bone.

b) Midsagittally, the **NUCHAL LIGAMENT (11)** overlies the spinous processes of the cervical vertebrae. It connects the spinous processes of the second cervical and the first thoracic vertebra, caudal to which it is transformed into the supraspinous ligament. The yellow colour of the nuchal ligament, which is duplicated along its length, indicates a predominance of elastic fibers.

c) The **LUMBAR CUTANEOUS NERVES** form a dorsal, a lateral and a ventral cutaneous series of nerves, by means of their serial passage through the strata of skeletal muscles.
I. The **dorsal lumbar cutaneous nerve series** is formed from nL1 to 4 dl (cranial clunial nn) and becomes subcutaneous approximately 8 cm from the dorsal midline. As a rule the nL5 to 7 dl do not reach the cutaneous field of innervation.
II. The **lateral lumbar cutaneous nerve series** arises from branches of the cranial iliohypogastric (nL1 vl), caudal iliohypogastric (nL2 vl), ilioinguinal (nL2 and 3 vl) and lateral cutaneous femoral (nL3 and 4 vl) nn. The nerve transits through the abdominal muscles lie on a line directed caudodorsally from the ventral end of the last rib to the tuber coxae. The **cranial iliohypogastric n. (18**, with accompanying blood vessels), and the **caudal iliohypogastric n. (19)** pierce the m. obliquus externus abdominis. The **ilioinguinal n. (20**, sometimes absent) and the **lateral cutaneous femoral n. (21**, with accompanying blood vessels) become subcutaneous over the dorsal border of the muscle.
III. The **ventral lumbar cutaneous nerve series** reaches the skin in the vicinity of the linea alba, by means of minute nerve twigs (see p. 14A).

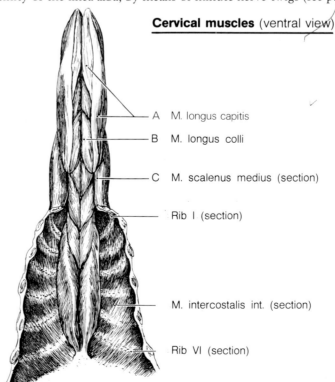

Cervical muscles (ventral view)

A M. longus capitis

B M. longus colli

C M. scalenus medius (section)

Rib I (section)

M. intercostalis int. (section)

Rib VI (section)

Cervical and thoracic muscles (lateral view)

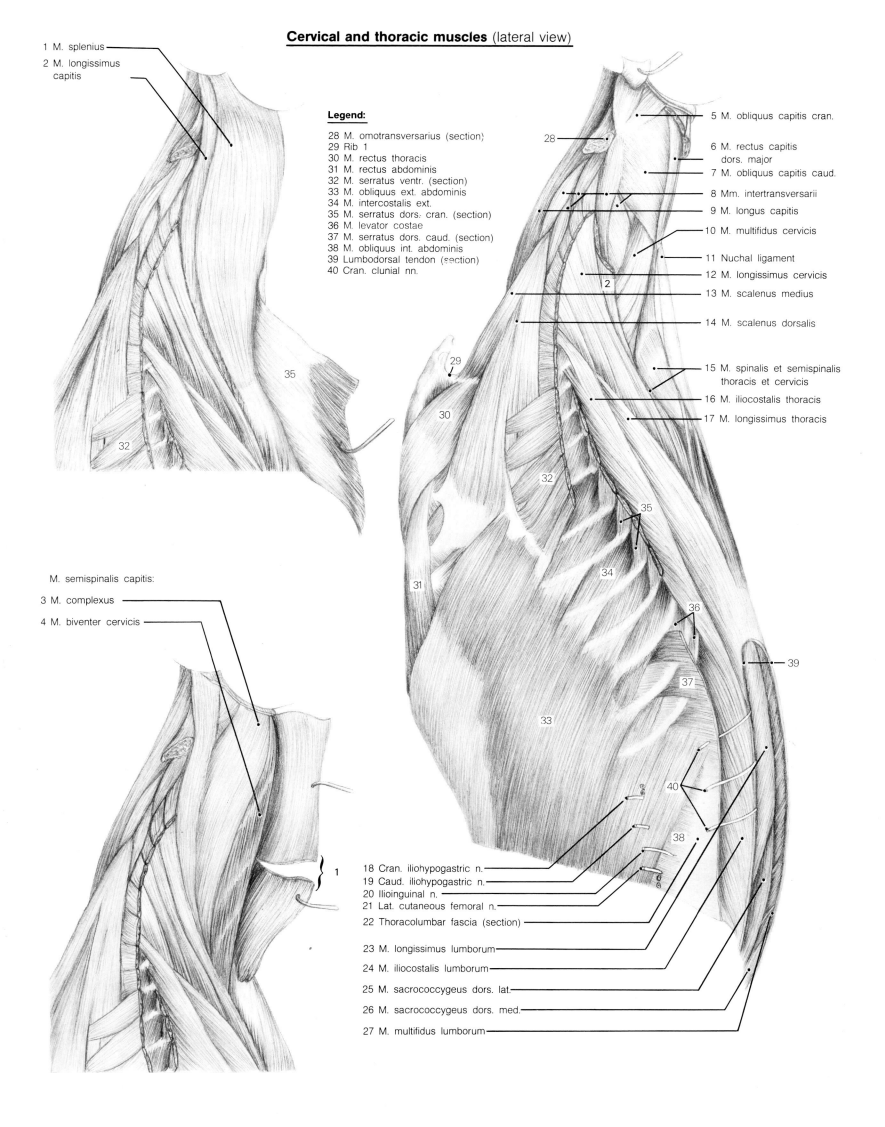

1 M. splenius

2 M. longissimus capitis

Legend:

28 M. omotransversarius (section)
29 Rib 1
30 M. rectus thoracis
31 M. rectus abdominis
32 M. serratus ventr. (section)
33 M. obliquus ext. abdominis
34 M. intercostalis ext.
35 M. serratus dors. cran. (section)
36 M. levator costae
37 M. serratus dors. caud. (section)
38 M. obliquus int. abdominis
39 Lumbodorsal tendon (section)
40 Cran. clunial nn.

5 M. obliquus capitis cran.
6 M. rectus capitis dors. major
7 M. obliquus capitis caud.
8 Mm. intertransversarii
9 M. longus capitis
10 M. multifidus cervicis
11 Nuchal ligament
12 M. longissimus cervicis
13 M. scalenus medius
14 M. scalenus dorsalis
15 M. spinalis et semispinalis thoracis et cervicis
16 M. iliocostalis thoracis
17 M. longissimus thoracis

M. semispinalis capitis:

3 M. complexus
4 M. biventer cervicis

18 Cran. iliohypogastric n.
19 Caud. iliohypogastric n.
20 Ilioinguinal n.
21 Lat. cutaneous femoral n.
22 Thoracolumbar fascia (section)
23 M. longissimus lumborum
24 M. iliocostalis lumborum
25 M. sacrococcygeus dors. lat.
26 M. sacrococcygeus dors. med.
27 M. multifidus lumborum

2A

The thoracic cavity is opened to display all the respiratory muscles (see pp. 16A and 17A). To facilitate this, the right thoracic limb and its extrinsic muscles are removed and the following structures are severed or transected on both sides of the body: The mm. obliquus abdominis externus and scalenus dorsalis at their tendinous margins, and the first to ninth ribs of the right side and the second to ninth of the left inclusive. Dorsally the ribs in question are transected along the lateral border of the m. iliocostalis and ventrally along that of the m. rectus abdominis. In this way the line of insertion of the diaphragm remains intact. This runs from the ninth costal cartilage, across the eleventh costochondral articulation and caudodorsally to the midlength of the thirteenth rib, and represents the boundary between the thoracic and pleural cavities on the one side and the abdominal and peritoneal cavities on the other. With the thoracic cavity opened, the five layers of the thoracic wall can be studied, namely skin, external fascia of the trunk, a musculoskeletal layer, internal fascia of the trunk, and pleura. To expose the m. intercostalis internus the m. intercostalis externus is removed from the tenth intercostal space.

1 The **RESPIRATORY MUSCLES** lie on the thorax and are known therefore as muscles of the thorax. From the functional viewpoint they are divided into an expiratory and an inspiratory group.

The **innervation** of the diaphragm is provided by the phrenic n. of its respective side and that of the remaining respiratory muscles by intercostal nn. The main respiratory muscle, the diaphragm, and the other obligatory respiratory muscles, are supported functionally by auxiliary respiratory muscles, which have been discussed in connection with other muscle groups, for example the muscles of the vertebral column. The mm. scalenus and serratus ventralis are auxiliary inspiratory muscles and the m. iliocostalis and abdominal muscles are auxiliary expiratory muscles, although some authors consider the m. scalenus to be an obligatory muscle. Keeping in mind their positional relationship to the thorax, the obligatory respiratory muscles are also divided into external, middle and internal respiratory muscles.

RESPIRATORY MUSCLES

Exspiratory Muscles		Inspiratory Muscles
I. M. serratus dorsalis caudalis	External respiratory muscles	IV. M. serratus dorsalis cranialis V. M. rectus thoracis
II. Mm. intercostales interni incl. mm. subcostales and retractor costae	Middle respiratory muscles	VI. Mm. intercostales externi incl. mm. levatores costarum
III. M. transversus thoracis	Internal respiratory muscles	VII. Diaphragm

a) The **EXSPIRATORY MUSCLES** run to the caudal borders of the ribs, their fibers being directed cranioventrally. They draw the ribs caudomedially and in so doing narrow the thorax.
I. The **m. serratus dorsalis caudalis (2)** takes origin from the thoracolumbar fascia (see text illustration) and its fibers course cranioventrally to insert on the caudal borders of the last three ribs.
II. The **mm. intercostales interni (5)** lie dorsal to the costochondral articulations and deep to the mm. intercostales externi. They also appear deep to the m. rectus abdominis in the crevices between the costal cartilages. The mm. subcostales and the **m. retractor costae (37)** also belong to the system of internal intercostal muscles. As a longer muscle portion, each of the mm. subcostales passes over the medial surface of a rib to insert onto the next or the next but one. The m. retractor costae extends from the transverse processes of the first three lumbar vertebrae to the caudal border of the last rib.
III. The **m. transversus thoracis (16)** situated cranial to the diaphragm, is the cranial continuation of the m. transversus abdominis. It arises on the deep surface of the sternum and inserts in a crenate manner onto the medial surface of each genu costae.

b) The **INSPIRATORY MUSCLES** run to the cranial borders of the ribs, the fiber bundles being directed caudoventrally. They draw the ribs craniolaterally and widen the thorax.
IV. The **m. serratus dorsalis cranialis (3)** takes origin from the supraspinous ligament dorsal to the first eight thoracic spinous processes and terminates by seven to nine fleshy insertions onto the cranial borders of ribs three to ten.
V. At the level of the fourth costal cartilage, the **m. rectus thoracis (6)** continues the m. rectus abdominis cranially. It runs obliquely over the aponeurotic origin of the latter as far as the first rib.
VI. The **mm. intercostales externi (4)** are situated mainly between the osseous ribs, and extend ventrally as far as the costochondral articulations at the approximate level of the lateral border of the m. rectus abdominis. Only sparse muscle bundles lie further ventrally. As the vertebral portions of the mm. intercostales externi, the **mm. levatores costarum (1)** can also be classified with the system of external intercostal muscles. Each of the mm. levatores costarum runs in approximately the same direction from the transverse process of one thoracic vertebra over the angle of the rib to the cranial border of the subsequent caudal rib. The mm. levatores are largely covered by the mm. iliocostalis and serratus dorsalis caudalis.
2 VII. The **diaphragm** is a tendo-muscular septum between the thoracic and abdominal cavities. Its tendinous cupola situated ventrally projects a considerable way into the thoracic cavity. The diaphragm functions as the main respiratory muscle, the contraction of which flattens the cupola laterally. The crown of the cupola is fixed at the caval foramen and during respiration, its position remains largely constant.
Innervation is provided by the **phrenic n. (12)** which arises by three roots from the fifth to seventh cervical n. At the level of the coronary groove the nerve passes over the heart. The right phrenic n. reaches the diaphragm by attachment to the plica venae cavae, the left phrenic n. in a short fold of the mediastinum. The diaphragm is divided into a peripheral muscular portion and a centrally placed tendinous portion, the central tendon. The muscular portion consists of a **sternal part (15)** inserting onto the sternum, a **costal part (13)** inserting onto ribs nine to thirteen inclusive, and a **lumbar part (10)**, the crura of which insert onto the third and fourth lumbar vertebrae. The free medial border of the weak **left crus (8)** and that of the stronger **right crus (7)** bound the **aortic hiatus (9)**. This affords passage to aorta, thoracic duct and right azygos v. The free dorsolateral borders of the crura form the lumbocostal arches over which the sympathetic trunk and the ramifying greater splanchnic n. are crossing. The slit like **esophageal hiatus (11)** provides a transit for esophagus and accompanying dorsal and ventral vagal trunks. It lies in the muscular part of the diaphragm bordering the central tendon.
The central tendon (14) is V-shaped and with the **caval foramen (12)** is present in the cupola region. This gives passage to the caudal vena cava and the accompanying right phrenic n.

Thoracolumbar fascia (transverse section)

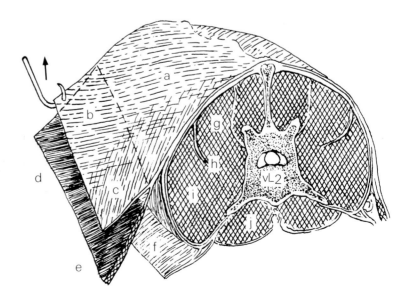

a	Thoracolumbar fascia	f	M. transversus abdominis
b	M. latissimus dorsi	g	M. longissimus lumborum
c	M. obliquus ext. abdominis	h	Intermuscular septum
d	M. serratus dors. caud.	i	M. iliocostalis lumborum
e	M. obliquus int. abdominis	j	M. quadratus lumborum and m. psoas minor

Thoracic muscles (left lateral view)

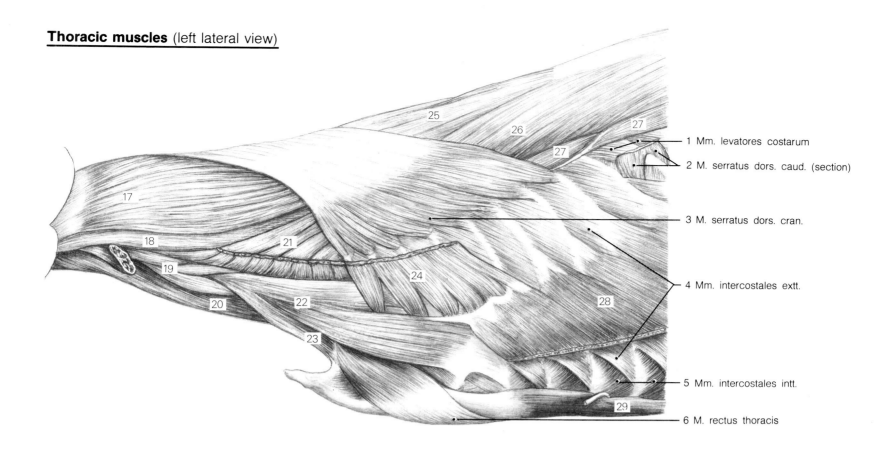

1 Mm. levatores costarum
2 M. serratus dors. caud. (section)
3 M. serratus dors. cran.
4 Mm. intercostales extt.
5 Mm. intercostales intt.
6 M. rectus thoracis

Diaphragm

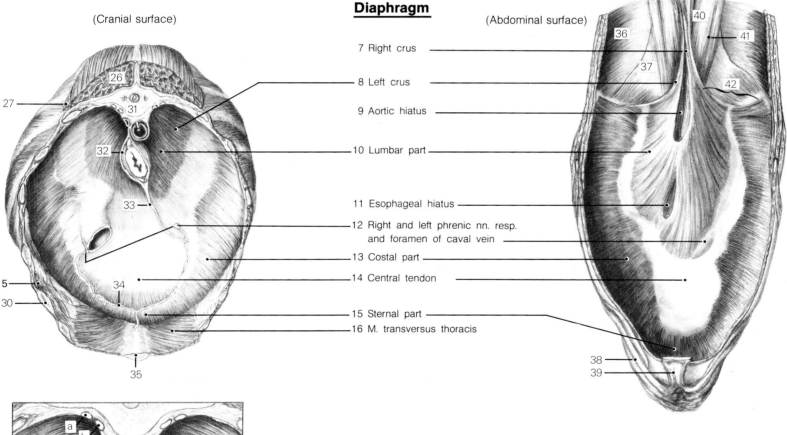

(Cranial surface)

(Abdominal surface)

7 Right crus
8 Left crus
9 Aortic hiatus
10 Lumbar part
11 Esophageal hiatus
12 Right and left phrenic nn. resp. and foramen of caval vein
13 Costal part
14 Central tendon
15 Sternal part
16 M. transversus thoracis

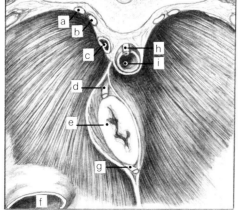

> *The remaining skin of the abdomen is removed from both sides of the body, keeping intact the mammae or the penis as the case may be. In the bitch, the mammae of the right side are preserved by incising the skin around the base of each teat. Beginning laterally, the mammae of the left side are removed after cutting through the left m. supramammarius and the suspensory ligament of the mamma. In the male one removes the hairy outer skin adjacent to the penis and its cranial continuation, the external lamina of prepuce, while keeping intact the m. preputialis cranialis and the suspensory ligament of penis.*

a) The **PREPUCE** (see p. 14A upper right) consists of an external lamina covered with hair, which is continous at the preputial ostium with the hairless cutaneous mucous membrane of the internal lamina. At the fundus of the prepuce, this lamina continues on the surface of the glans penis as the lamina of the penis.

b) The **MAMMAE** lie on both sides of the median intermammary groove and as a rule consist of five mammary complexes per side, namely a cranial and a caudal thoracic mammary complex, a cranial and a caudal abdominal mammary complex, and an inguinal mammary complex. The mammae of the male are characterized by insignificant mammary papillae or teats. A mammary complex (see p. 14A upper left) consists of a body of the mamma containing between sixteen and twenty mammary glands, and corresponding numbers of lactiferous sinuses, papillary ducts and ostia opening on the papilla. Each lactiferous sinus is divided into a glandular part situated in the body of the mamma and a papillary part in the teat. The fascial and muscular suspensory apparatus of, and the vascular, lymphatic and nerve supply to the mammae and prepuce, agree in many respects and are studied together.

As the continuation of external trunk fascia, the **suspensory ligament of penis or mamma (8)** separates from it at the linea alba and runs in the prepuce and around the penis (fascia of penis) or mammary complex (fascial relationship - see p. 15). The **m. preputialis cranialis** or the **m. supramammarius cranialis (6)** arises from the linea alba at the level of the xiphoid cartilage and runs to the bases of the abdominal mammae or the prepuce. The caudal muscles of like name are insignificant.

Regarding **vascular, lymphatic and nerve supply**, a cranial field of supply is present for both thoracic and the cranial abdominal mammary complexes. A caudal field supplies the caudal abdominal and inguinal mammary complexes, and the prepuce, external skin adjacent to penis, and scrotum of the male. Between the cranial and caudal fields anastomoses are present at the umbilicus.

I. The **blood vessels** of the cranial field originate from the **lateral thoracic a. and v. (3)** and the **internal thoracic a. and v. (4)**, perforating rami of the latter vessels appearing near the linea alba to supply both thoracic mammae. The internal thoracic a. and v. end as the **cranial superficial epigastric a. and v. (5)**. At the level of the costal arch these pierce the thoracic wall and after supplying the cranial abdominal mamma at the level of umbilicus, anastomose with caudal superficial epigastric a. and v. The blood vessels of the caudal field of supply originate from the **external pudendal a. and v. (12)**. After their passage through the inguinal space, these divide at the level of the inguinal teat into **ventral labial or scrotal rr (13)** and the **caudal superficial epigastric a. and v. (9)**, which, in turn give mammary or preputial rami.

II. The **lymphatic vessels** of the cranial field run to the **axillary (1)** and the **accessory axillary lymph node (2)**. The lymphatic vessels of the caudal field drain both caudal mammae and the prepuce, external skin adjacent to penis, and the scrotum of the male. They run to the **superficial inguinal lymph node (11)** situated at the base of the inguinal mamma where the external pudendal a. and v. divide into ventral labial or scrotal rami and the caudal superficial epigastric a. and v. On the other hand, deep lymphatic vessels run from the testis to the medial iliac lymph node (see below) or by circumventing it, to the lumbar aortic lymph nodes and the lumbar trunk.

III. The **sensory nerve supply** is derived from the intercostal nn cranially and the cranial and caudal iliohypogastric nn caudally. These approach the mammae by means of lateral cutaneous rami from the lateral series of thoracic and lumbar cutaneous nn and ventral cutaneous rami near the linea alba. The inguinal mamma is innervated by the **genitofemoral n. (10)** which passes through the inguinal space.

> *To expose the abdominal muscles and the sheath of the m. rectus abdominis, the m. obliquus externus abdominis and the overlying **deep fascia of trunk (7)** on the left side are transected 2 cm ventral and parallel to the origin of the oblique muscle from ribs and thoracolumbar fascia. The ventral muscle remnant is reflected ventrally as far as the lateral border of the m. rectus abdominis. Then the m. obliquus internus abdominis is transected 2 cm dorsal and parallel to the border between the muscle and its aponeurosis, and also reflected.*

Taking into consideration their sites of origin, the **ABDOMINAL MUSCLES** are subdivided into parts, namely the costal part taking origin from the ribs, the sternal part from sternum, the lumbar part from **thoracolumbar fascia** (see text illustration p. 73), and the inguinal part from the **inguinal ligament (20)**. With the exception of the m. rectus abdominis, they each insert onto the linea alba in the ventral midline by means of an aponeurosis or abdominal tendon. The m. obliquus internus abdominis also inserts along the costal arch by means of a costal tendon and the m. obliquus externus abdominis terminates by a pelvic tendon or lateral crus at the pecten ossis pubis. Lateral to their midventral insertions the aponeuroses of the abdominal muscles form the sheath of m. rectus abdominis. Ventrosagitally, the aponeuroses of the external lamina of the sheath form a meshlike zone which is anchored to the **tendinous intersections (15)** of the m. rectus abdominis. Midventrally, the **linea alba (21)** arises as a consecutive series of crisscrossings of tendon fibers. It begins at the mesosternum as a ventromedian 'anchoring raphe' for the aponeuroses of the abdominal muscles. At its widest, it encircles the **umbilical ring (22)** by means of two umbilical crura and terminates by tapering abruptly at the cranial end of the pelvic symphysis.

Innervation of the abdominal muscles comes from intercostal nn, **cranial (16)** and **caudal (17) iliohypogastric nn** and the **ilioinguinal n. (19)**. By means of their vm branches, all such nerves course across the lateral surface of the m. transversus abdominis, appearing as skin or mammary branches in the vicinity of the linea alba. Vl branches of the nerves run on the m. obliquus externus abdominis to the mammae.

The **m. obliquus externus abdominis (24)** arises as costal and lumbar parts which transform into an abdominal and a pelvic tendon. Known respectively as the **medial crus (26)** and the **lateral crus (25)** they bound the **external inguinal ring (27)**. The aponeurosis of the muscle participates throughout in forming the external lamina of the sheath of the m. rectus abdominis.

The **prepubic tendon** is represented by a stout tendinous mass extending ventrally from the brim of the pelvis. It consists mainly of the tendon of origin of the m. pectineus (et adductor longus) of its respective side. The pelvic tendons of the external and internal oblique abdominal muscles radiate into the prepubic tendon. The canine prepubic tendon differs from that of ungulates in one respect. It has two separate halves because the right and left pectineal tendons have their origins restricted to the ipsilateral pubis and body of ilium. In contrast, the origins of the ungulate pectineal tendons are from contralateral pubic bones. In the dog both right and left prepubic tendon complexes must be considered together as the homologue of the ungulate prepubic tendon.

The **m. obliquus internus abdominis (23)** has lumbar and inguinal parts. Its aponeurosis participates in the sheath of the m. rectus abdominis in three ways: 1) A cranial 2 cm-wide section of aponeurosis subscribes to the internal lamina only. 2) A subsequent 2-4 cm-wide section participates in forming both external and internal laminae in the umbilical region. 3) Caudally, the aponeurosis only passes to the external surface of the m. rectus abdominis.

The **m. transversus abdominis (18)** subscribes to the sheath in a similar manner to the m. obliquus internus abdominis. However, its contribution to both laminae occurs approximately one to two vertebral lengths more caudally. On the lateral surface of the muscle, ventromedial branches of thoracic and lumbar nn (cranial and caudal iliohypogastric and ilioinguinal nn) run ventrally and parallel to the muscle fibers.

The **m. rectus abdominis (14)** arises from the first rib and the first four sternebrae and could therefore be said to have a costal and a sternal part. It terminates on the pecten ossis pubis. Caudal to the costal arch it is enclosed in its sheath formed from the aponeuroses of the remaining **abdominal muscles and the internal and external fasciae of the trunk**. The external and internal laminae of trunk fascia contribute throughout to the respective external and internal laminae of the sheath.

Mammary region, Preputial region

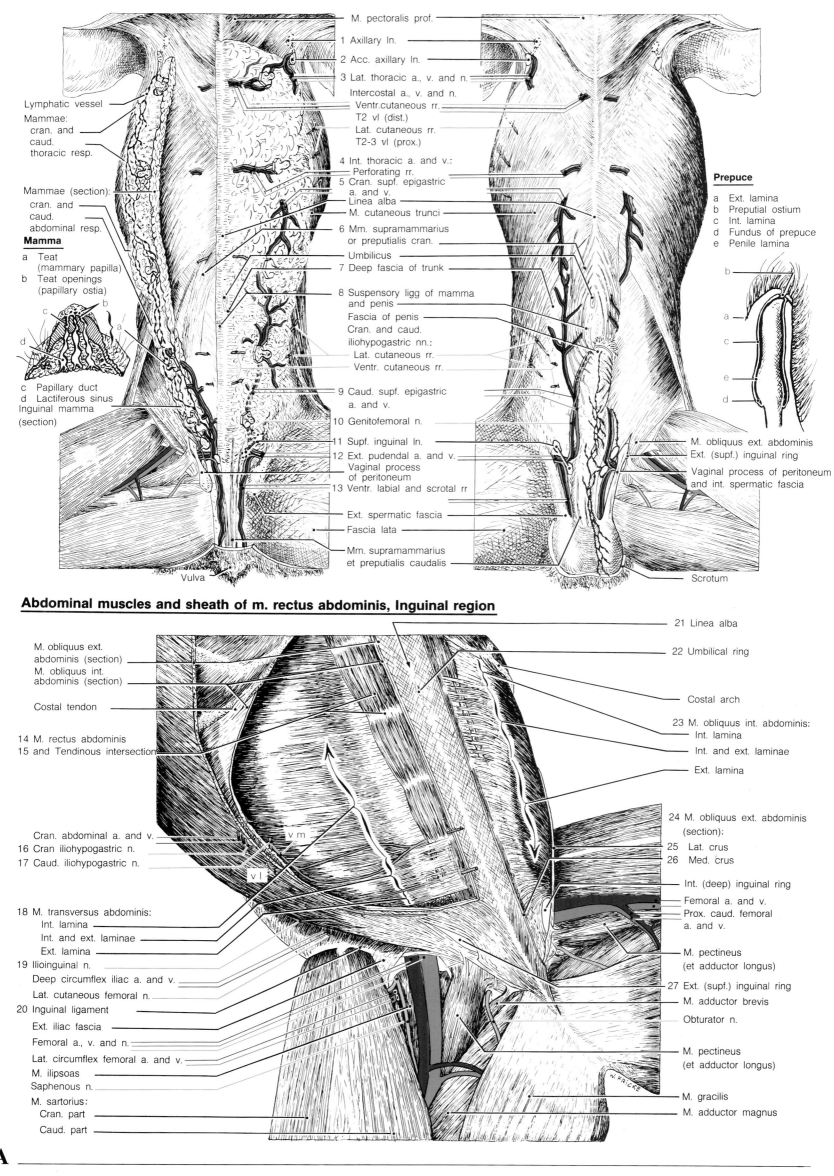

Lymphatic vessel
Mammae:
cran. and
caud.
thoracic resp.

Mammae (section):
cran. and
caud.
abdominal resp.

Mamma

a Teat
(mammary papilla)
b Teat openings
(papillary ostia)

c Papillary duct
d Lactiferous sinus
Inguinal mamma
(section)

M. pectoralis prof.
1 Axillary ln.
2 Acc. axillary ln.
3 Lat. thoracic a., v. and n.
Intercostal a., v. and n.
Ventr. cutaneous rr.
T2 vl (dist.)
Lat. cutaneous rr.
T2-3 vl (prox.)
4 Int. thoracic a. and v.:
Perforating rr.
5 Cran. supf. epigastric
a. and v.
Linea alba
M. cutaneous trunci
6 Mm. supramammarius
or preputialis cran.
Umbilicus
7 Deep fascia of trunk
8 Suspensory ligg of mamma
and penis
Fascia of penis
Cran. and caud.
iliohypogastric nn.:
Lat. cutaneous rr.
Ventr. cutaneous rr.
9 Caud. supf. epigastric
a. and v.
10 Genitofemoral n.
11 Supf. inguinal ln.
12 Ext. pudendal a. and v.
Vaginal process
of peritoneum
13 Ventr. labial and scrotal rr
Ext. spermatic fascia
Fascia lata
Mm. supramammarius
et preputialis caudalis
Vulva

Prepuce

a Ext. lamina
b Preputial ostium
c Int. lamina
d Fundus of prepuce
e Penile lamina

M. obliquus ext. abdominis
Ext. (supf.) inguinal ring
Vaginal process of peritoneum
and int. spermatic fascia

Scrotum

Abdominal muscles and sheath of m. rectus abdominis, Inguinal region

M. obliquus ext.
abdominis (section)
M. obliquus int.
abdominis (section)

Costal tendon

14 M. rectus abdominis
15 and Tendinous intersection

Cran. abdominal a. and v.
16 Cran iliohypogastric n.
17 Caud. iliohypogastric n.

18 M. transversus abdominis:
Int. lamina
Int. and ext. laminae
Ext. lamina
19 Ilioinguinal n.
Deep circumflex iliac a. and v.
Lat. cutaneous femoral n.
20 Inguinal ligament
Ext. iliac fascia
Femoral a., v. and n.
Lat. circumflex femoral a. and v.
M. ilipsoas
Saphenous n.
M. sartorius:
Cran. part
Caud. part

21 Linea alba
22 Umbilical ring
Costal arch
23 M. obliquus int. abdominis:
Int. lamina
Int. and ext. laminae
Ext. lamina
24 M. obliquus ext. abdominis
(section):
25 Lat. crus
26 Med. crus
Int. (deep) inguinal ring
Femoral a. and v.
Prox. caud. femoral
a. and v.
M. pectineus
(et adductor longus)
27 Ext. (supf.) inguinal ring
M. adductor brevis
Obturator n.
M. pectineus
(et adductor longus)
M. gracilis
M. adductor magnus

W. FRICKE

Before commencing the dissection, details should be provided on the **relationship of the fasciae.** *The superficial and deep fasciae of the trunk are classed as the external fascia of the trunk. The superficial fascia is closely united to the skin and ensheaths the cutaneous muscles of the abdomen by means of two laminae. The deep fascia is intimately united to the surface of the m. obliquus externus abdominis. Dorsally, in the lumbar region, it is known as the thoracolumbar fascia which courses over the muscles of the vertebral column between spinous and transverse processes. In the midline ventrally, the deep fascia fuses with the linea alba and there the suspensory ligament of penis or mamma, as the case may be, separates off from it. In the inguinal region the external fascia of the trunk (comprising the superficial and deep fasciae of trunk) continues as the external spermatic fascia. This envelopes the vaginal process of peritoneum externally and at the inguinal groove passes over onto the thigh as the fascia lata.*

By and large, the internal fascia of the trunk is adherent to the serosa and is known by different terms, depending on its location. In the thoracic cavity it is known as the endothoracic fascia, in the abdominal cavity as transverse fascia, on the ventral surface of the deep lumbar muscles as iliac fascia, and within the pelvic cavity as fascia of the pelvis. The internal fascia of the trunk, or more particularly the transverse fascia, continues as the internal spermatic fascia and invests the vaginal process of the peritoneum.

In the inguinal region three conduits are exposed, namely the inguinal space, the lacuna musculorum and the lacuna vasorum. Initially, the right inguinal mamma is reflected medially, and following this, one is able to study the participation of the five layers of the abdominal wall in the formation of the inguinal space. This begins at the internal or deep inguinal ring and ends at the external or superficial inguinal ring. Observe the continuation of the external fascia of the trunk as the tubelike external spermatic fascia. Externally, this envelopes the vaginal process and its covering of internal spermatic fascia. Then on the right side, external trunk fascia, the three expansive abdominal muscles, transverse fascia and peritoneum are transected and reflected.

This incision is made along the long axis of the external inguinal ring as far as the costal arch. In so doing, each individual abdominal layer is studied (see upper p. 15A).

1 a) The **INGUINAL SPACE** extends from the internal to the external inguinal ring. The caudal angles of both rings lie one above the other, whereas the cranial angle of the internal ring lies approximately 2 cm craniolateral to that of the external ring. This results in a corresponding lengthening of the inguinal space.

I. The **skin** does not participate in the formation of the inguinal space and merges with the integument of scrotum or lips of vulva.

II. The **external fascia of the trunk (1)** turns at the external inguinal ring to envelope the **vaginal process of peritoneum (23)** and its tubular content and is then known as the **external spermatic fascia (25).** In the male the process contains the spermatic cord, while in the bitch it houses the round ligament of the uterus and its enveloping body of fat. In contrast to most other female mammals, the bitch possesses a vaginal process of peritoneum and an accompanying external spermatic fascia. The **external pudendal a. and v. (21)**, the **genitofemoral n. (20)** and the **m. cremaster (externus)** pass through the space on the outside of the vaginal process and then enter the tubular external spermatic fascia. Within a few millimeters the two blood vessels and nerve pierce the fascia and the vessels branch into the **caudal superficial epigastric a. and v.** (see p. 14A —9) and the **ventral scrotal or ventral labial rami** (see p. 14A —13) in the vicinity of the superficial inguinal lymph node.

III. By a cleavage in its aponeurosis, the **m. obliquus externus abdominis (15)** forms the **lateral (17)** and **medial (16) crura** of the **external (superficial) inguinal ring (18).**
The free caudal border of the **m. obliquus internus abdominis (14)** contributes to the formation of the **internal (deep) inguinal ring (19)** together with the lateral border of the m. rectus abdominis and the deep surface of the lateral crus of the m. obliquus externus abdominis.
In rodents the **m. cremaster (externus —22)** is divided into a primary part derived from the m. transversus abdominis and a secondary part from the m. obliquus internus abdominis. To a large degree in domestic mammals, the united m. cremaster (externus) has lost its direct connection with the two abdominal muscles. The resultant independent muscle takes origin from the inguinal ligament and passes through the inguinal space, outside the vaginal process. It is distinct in the male whereas it seems weaker in the bitch.
The **m. transversus abdominis (2)** does not contribute to the formation of the inguinal space since its free caudal border is already on the level of the tuber coxae.

IV. The **internal fascia of the trunk (3,** transverse fascia) is adherent to the peritoneum. It protrudes through the inguinal space as the tubular **internal spermatic fascia (24)** and envelopes the vaginal process.
V. As stated above, the **peritoneum (4)** bulges out as a tubular-shaped vaginal process of peritoneum, penetrates the inguinal space and adheres to the enveloping internal spermatic fascia. The **vaginal ring (5)**, on the lateral part of the abdominal cavitiy, is the entrance into the vaginal process and does not belong to the inguinal space.

b) The **LACUNA MUSCULORUM** is the entrance to a passage for the m. iliopsoas and the femoral n. contained in it. At the level of the lacuna and the transition from lateral to dorsal body wall, the **transverse fascia (3)** extends over the **m. iliopsoas (11)** and is then known as the iliac fascia. The inguinal ligament is woven into this fascial covering and subdivides the iliac fascia into an **internal iliac fascia (13)** on the side of the abdominal cavity and an **external iliac fascia (7 -** iliac lamina) adjacent to the thigh. This fascia ends on the ilium. Occasionally absent, the very weak **inguinal ligament (6)** arises at the tuber coxae. In the region of the lacuna musculorum it is interwoven with the iliac fascia to provide a connective tissue reinforcement for the origin of the m. obliquus internus abdominis. At the level of the inguinal space more ventrally, the ligament unites with the lateral crus of the m. obliquus externus abdominis; both radiate into the prepubic tendon and end on the pecten ossis pubis. In common with the m. iliopsoas, the **femoral n. (8)** and its branch, the **saphenous n. (9)**, passes through the lacuna musculorum. This is bounded by the inguinal ligament ventrally and the ilium dorsally. Subsequently, the saphenous n. attains the **femoral triangle (10)** the three limiting sides of which are formed by the inguinal ligament, the m. pectineus (et adductor longus), and the caudal part of the m. sartorius. The femoral triangle limits the **femoral space (12,** femoral canal) which is bounded deeply by the m. iliopsoas and covered superficially by the fascia lata. Besides the saphenous n. the space houses the femoral a. and v.

c) The **LACUNA VASORUM** is the entrance to the passage for the blood vessels of the leg. It is bounded by the lacuna musculorum dorsolaterally, the body of ilium dorsomedially, and the inguinal ligament ventrally. The medial section of the lacuna vasorum not occupied by blood vessels is known as the femoral ring. Due to its covering of peritoneum and transverse fascia, the femoral ring is a self-contained **2** access to the femoral space or canal and is the site of femoral hernia.

PARTICIPATION OF LAYERS OF ABDOMINAL WALL IN THE INGUINAL SPACE

Abdomen	Rings		Process
I. Skin	—		Skin of scrotum
II. Ext. fascia of trunk	—		Ext. spermatic fascia
III. M. obl. ext. abdom.	Ext. inguinal ring	Inguinal	
III. M. obl. int. abdom.	Int. inguinal ring	space	
			M. cremaster (ext.)
III. M. transv. abdom.	—		
IV. Int. fascia of trunk (transverse fascia)	—		Int. spermatic fascia
V. Peritoneum	Vaginal ring		Vaginal process

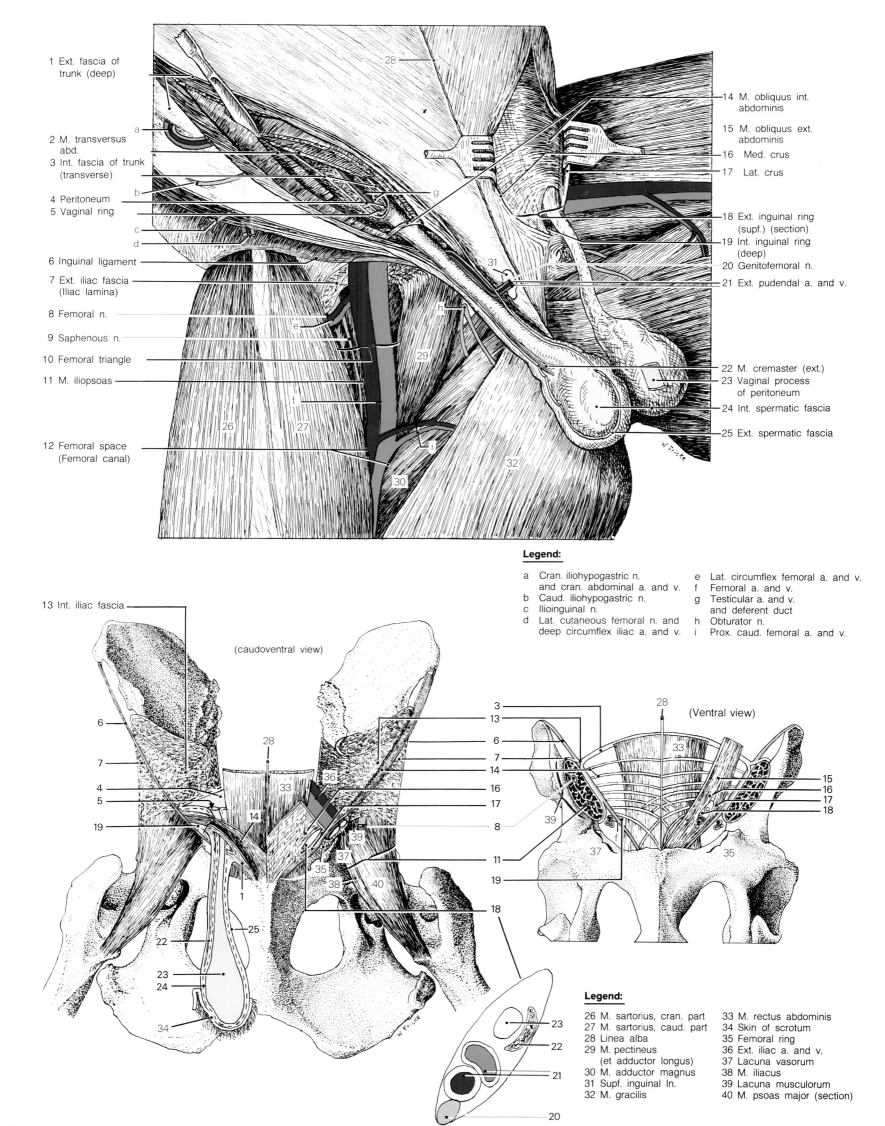

1 Ext. fascia of trunk (deep)

2 M. transversus abd.

3 Int. fascia of trunk (transverse)

4 Peritoneum

5 Vaginal ring

6 Inguinal ligament

7 Ext. iliac fascia (Iliac lamina)

8 Femoral n.

9 Saphenous n.

10 Femoral triangle

11 M. iliopsoas

12 Femoral space (Femoral canal)

14 M. obliquus int. abdominis

15 M. obliquus ext. abdominis

16 Med. crus

17 Lat. crus

18 Ext. inguinal ring (supf.) (section)

19 Int. inguinal ring (deep)

20 Genitofemoral n.

21 Ext. pudendal a. and v.

22 M. cremaster (ext.)

23 Vaginal process of peritoneum

24 Int. spermatic fascia

25 Ext. spermatic fascia

13 Int. iliac fascia

(caudoventral view)

Legend:

a Cran. iliohypogastric n. and cran. abdominal a. and v.
b Caud. iliohypogastric n.
c Ilioinguinal n.
d Lat. cutaneous femoral n. and deep circumflex iliac a. and v.

e Lat. circumflex femoral a. and v.
f Femoral a. and v.
g Testicular a. and v. and deferent duct
h Obturator n.
i Prox. caud. femoral a. and v.

(Ventral view)

Legend:

26 M. sartorius, cran. part
27 M. sartorius, caud. part
28 Linea alba
29 M. pectineus (et adductor longus)
30 M. adductor magnus
31 Supf. inguinal ln.
32 M. gracilis

33 M. rectus abdominis
34 Skin of scrotum
35 Femoral ring
36 Ext. iliac a. and v.
37 Lacuna vasorum
38 M. iliacus
39 Lacuna musculorum
40 M. psoas major (section)

a) The **THORACIC CAVITY** (see also text illustration), cranial to the diaphragm, is enclosed by the thorax comprising thoracic vertebrae ribs and sternum. Caudal to the diaphragm the thorax encloses the intrathoracic part of the abdominal cavity. The thoracic cavity is lined

1 with a serosal membrane, the pleura, which also covers the organs within
2 the cavity. The pleural lining of the thoracic wall is referred to as parietal, the parts of which, depending on their location, are termed **costal pleura (B)** and **diaphragmatic pleura (T)**. The **visceral pleura (C)** covers the lung and is therefore named pulmonary pleura. Portions of the parietal and the plumonary pleura come together in the midline as a 'middle portion' called the **mediastinum (5, D)**, corresponding to the mesentery of the abdominal cavity (see ahead). The mediastinum may be regarded as the mesentery of the thoracic part of the esophagus and inserts onto the vertebral bodies dorsally, the sternum ventrally, and the diaphragm caudally. There the line of insertion is pushed further to the left of the midline (see text illustration). At these positions the parietal pleura changes to **mediastinal pleura (E)**. The following structures pass through the mediastinum in a craniocaudal direction: **right azygos v. (F)**, **thoracic duct (3, G)**, **descending aorta (9, H)**, **dorsal vagal trunk (8, I)**, **esophagus (2, J)**, **ventral vagal trunk (7, K)** and **trachea (1)** and its bifurcation. The pulmonary ligament (see p. 17A) extends from the caudal lobe of the lung to the mediastinum, and thus connects the pulmonary pleura with the mediastinal pleura.

Within the thoracic cavity the pleura forms two pleural cavities which pass through the cranial thoracic aperture into the neck region as **pleu-**
3 **ral cupolae (A)**. That of the left is about two ribs widths in extent, the right about one. The most caudal part of each pleural cavity, the
4 **costodiaphragmatic recess (10, X)** lies immediately cranial to the line of diaphragmatic insertion. Functionally, the pleural cavities are fluid filled crevices of capillary dimension. Pulmonary and parietal pleurae are adjacent to each other and closely bound together by cohesive forces, so that the lung must follow the movement of the thoracic wall. As a result,

inspiration develops by a widening of the thorax. In addition, due to the asymmetrical caudal insertion of the mediastinum, the left pleural cavity has a **costomediastinal recess (6, V)**. In contrast, the right pleural cavity has a left **mediastinodiaphragmatic recess (W)**. The **caudal vena cava (L)** traverses the right pleural cavity in a proper **plica venae cavae (M)**, a folded part of the mediastinum. The latter produces a further niche, the **recess of the mediastinum (N)**, for the accessory lobe of the right lung. An additional cavity, the **serous cavity of the mediastinum (O)**, is situated within the mediastinum to the right of the esophagus and caudal to the bifurcation of the trachea. This is the result of a constriction of part of the peritoneal cavity during ontogenesis. The serous cavity of the mediastinum and accompanying tension in the pulmonary ligament are displayed by raising the caudal lobe of the right lung. The serous cavity of the mediastinum is opened by an incision directed ventroparallel to the line of insertion of the pulmonary ligament (see p. 17A), and one is able to probe its extent from the diaphragm to the root of lung with the index finger. The **cavity of the pericardium (P)**, 5 containing the heart, is situated in the mediastinum ventral to the esophagus and the bifurcation of the trachea. It is the fourth of the intrathoracic serous cavities and lies within the mediastinum. With the pericardium opened, one recognizes the **serous pericardium (Q)**, and its **parietal lamina (R)**. At the base of the heart this becomes the **visceral lamina (S)** or epicardium which invests the heart surface. The **fibrous pericardium (U)** is the outer enveloping connective tissue foundation of the pericardium. It is continuous with the **phrenicopericardiac ligament (Y)** and the **endothoracic fascia (Z)** which is the connective tissue base for the parietal pleura. The portion of mediastinum covering the pericardium is termed the pericardiac pleura.

b) The **THYMUS (4)** lies in the mediastinum cranial to the pericardium 6 and is the site of formation of the T-lymphocytes. After birth, its lymphoepithelial parenchyma is replaced gradually and almost completely by adipose connective tissue.

Thoracic cavity

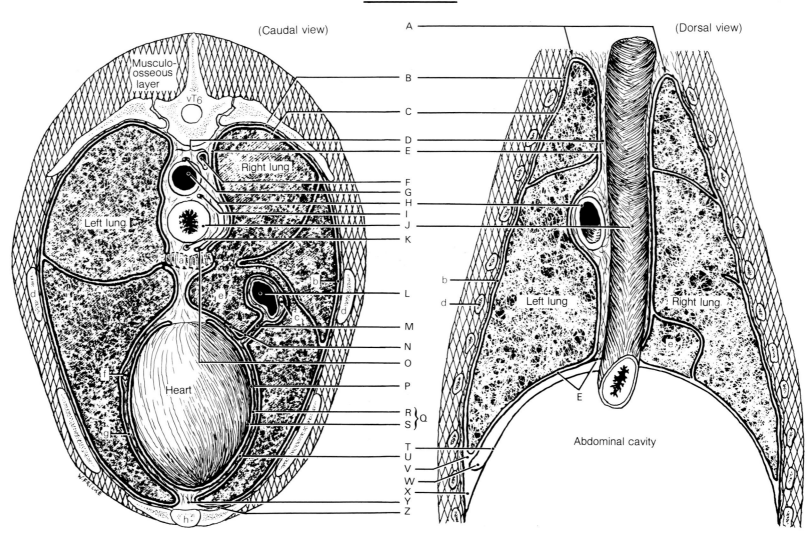

(Caudal view)

Musculo-osseous layer

vT6

Right lung

Left lung

Heart

Transverse section

(Dorsal view)

Left lung

Right lung

Abdominal cavity

Longitudinal section

Thoracic cavity (left side)

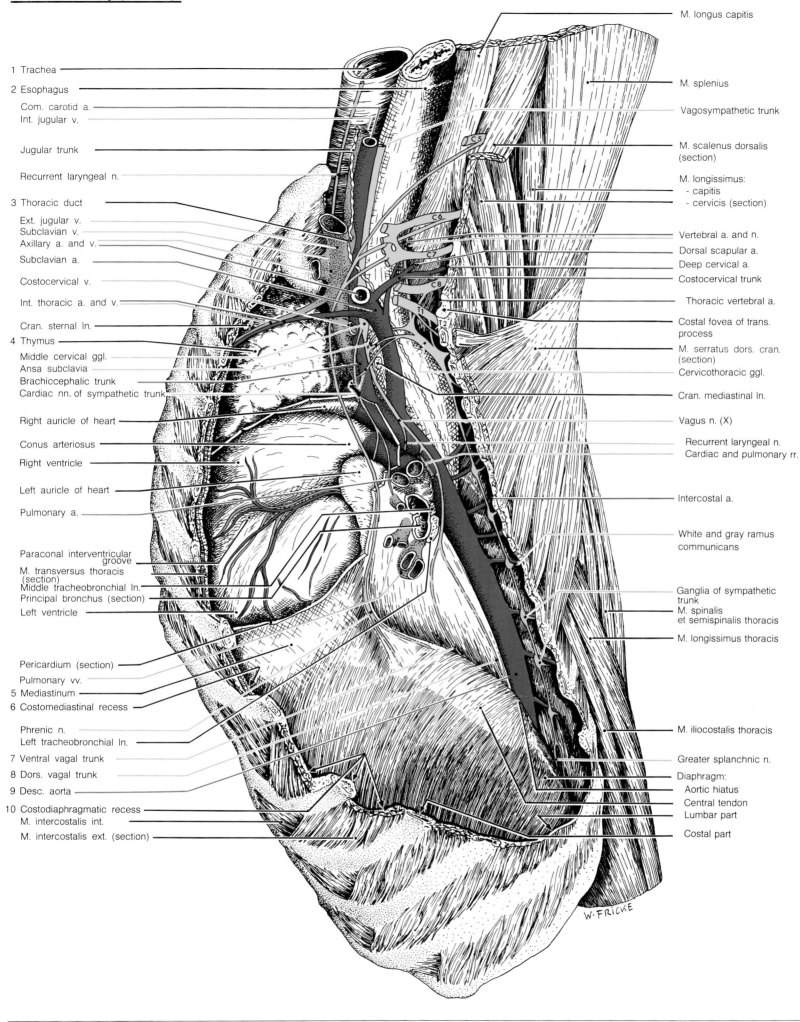

1 Trachea
2 Esophagus
Com. carotid a.
Int. jugular v.
Jugular trunk
Recurrent laryngeal n.
3 Thoracic duct
Ext. jugular v.
Subclavian v.
Axillary a. and v.
Subclavian a.
Costocervical v.
Int. thoracic a. and v.
Cran. sternal ln.
4 Thymus
Middle cervical ggl.
Ansa subclavia
Brachiocephalic trunk
Cardiac nn. of sympathetic trunk
Right auricle of heart
Conus arteriosus
Right ventricle
Left auricle of heart
Pulmonary a.
Paraconal interventricular groove
M. transversus thoracis (section)
Middle tracheobronchial ln.
Principal bronchus (section)
Left ventricle
Pericardium (section)
Pulmonary vv.
5 Mediastinum
6 Costomediastinal recess
Phrenic n.
Left tracheobronchial ln.
7 Ventral vagal trunk
8 Dors. vagal trunk
9 Desc. aorta
10 Costodiaphragmatic recess
M. intercostalis int.
M. intercostalis ext. (section)

M. longus capitis
M. splenius
Vagosympathetic trunk
M. scalenus dorsalis (section)
M. longissimus:
 - capitis
 - cervicis (section)
Vertebral a. and n.
Dorsal scapular a.
Deep cervical a.
Costocervical trunk
Thoracic vertebral a.
Costal fovea of trans. process
M. serratus dors. cran. (section)
Cervicothoracic ggl.
Cran. mediastinal ln.
Vagus n. (X)
Recurrent laryngeal n.
Cardiac and pulmonary rr.
Intercostal a.
White and gray ramus communicans
Ganglia of sympathetic trunk
M. spinalis et semispinalis thoracis
M. longissimus thoracis
M. iliocostalis thoracis
Greater splanchnic n.
Diaphragm:
Aortic hiatus
Central tendon
Lumbar part
Costal part

C 5
C 6
C 7
C 8
T1
T 2

W. FRICKE

Legend p. 16:

A Pleural cupolae
B Costal pleura
C Visceral pleura (pulmonary)
D Mediastinum
E Mediastinal pleura
F Right azygos v.
G Thoracic duct
H Desc. aorta
I Dors. vagal trunk
J Esophagus
K Ventr. vagal trunk

L Caudal vena cava
M Plica venae cavae
N Recess of mediastinum
O Serosal cavity of mediastinum
P Cavity of pericardium
Q Serous pericardium:
R Parietal lamina
S Visceral lamina (epicardium)
T Diaphragmatic pleura
U Fibrous pericardium

V Costomediastinal recess
W Left mediastinodiaphragmatic recess
X Costodiaphragmatic recess
Y Sternopericardiac lig.
Z Endothoracic fascia

a Bifurcation of trachea
b Pleural cavity
c Right phrenic n.
d Ribs
e Accessory lobe of right lung
f Left phrenic n.
g Pericardiac pleura
h Sternum

5A

a) The **BIFURCATION OF THE TRACHEA** (see text illustration —**B**) lies ventral to the esophagus approximately halfway between the cranial thoracic aperture and diaphragm. At the bifurcation, the trachea divides into the left and right principal bronchi each of which enters the hilus of the appropriate lung. There the left principal bronchus divides into two and the right principal into four lobar bronchi. (According to another viewpoint, the right lung has only three lobar bronchi since the bronchus of the accessory lobe is not considered to be equivalent to a lobar bronchus.)

Bifurcation of trachea (ventral view)

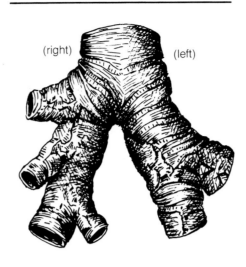

(right)　　　　　(left)

1;2　b) The **LUNG** (see text illustration) is auscultated within the 'field of the lung'. This is bounded by the ventral border of the m. iliocostalis, the tuber of the olecranon and a caudo-convex arch from the costochondral articulation of the sixth rib to the midlength of the eighth rib
3;4　and then to the vertebral end of the eleventh rib. The left lung consists

of a cranial and a caudal lobe corresponding to the number of left lobar bronchi. The right lung consists of cranial, middle, caudal and accessory lobes. The latter lobe lies in the mediastinal recess.

The **blood supply** of the lung comes from the **pulmonary aa and vv** 5 **(C)**, the vasa publica, and the **bronchoesophageal a. and v. (D)**, the vasa privata. The latter are derived from the aorta and right azygos v. respectively or from the fifth to the seventh intercostal vessels.

The **nerve supply** is from sympathetic and parasympathetic fibres 6 within the autonomic pulmonary plexus. Viscerosensory nerves, present at the root of the lung, are carried via autonomic pathways. The lung parenchyma, however, is not supplied with pain conducting fibers.

c) The **LYMPHATIC SYSTEM OF THE LUNG** includes the pulmonary lymph nodes within the lung parenchyma and the bronchial lymphocentre containing the **right (3, E)**, **middle (4, F)** and **left (G)** 7 **tracheobronchial lymph nodes.** These nodes are situated on the cranial aspect of the right principal bronchus, dorsal to the bifurcation of trachea, and caudodorsal to the left principal bronchus respectively. Their drainage areas are from lung and heart. After flowing through the cranial mediastinal lymph nodes, the efferent lymphatic vessels discharge directly or indirectly into the lymphatic trunk openings at the venous angle between internal and external jugular vv.

d) The **LYMPH NODES OF THE THORACIC CAVITY** include the **cranial mediastinal lymph nodes (1, A)** and the **cranial sternal lymph nodes (2).** The former lie in the precardial mediastinum, particularly adjacent to the walls of the large blood vessels or thoracic viscera, the latter on the cranial border of the m. transversus thoracis. Middle and caudal mediastinal and caudal sternal lymph nodes are absent in the dog. Inconstant intercostal lymph nodes are seen proximally in the fifth or sixth intercostal space. The drainage areas lie in the immediate vicinity of the lymph nodes in question. Their efferent lymphatic vessels either go directly to the venous angle from the cranial mediastinal lymph nodes or indirectly from cranial sternal and intercostal lymph nodes, after flowing through the cranial mediastinal lymph nodes.

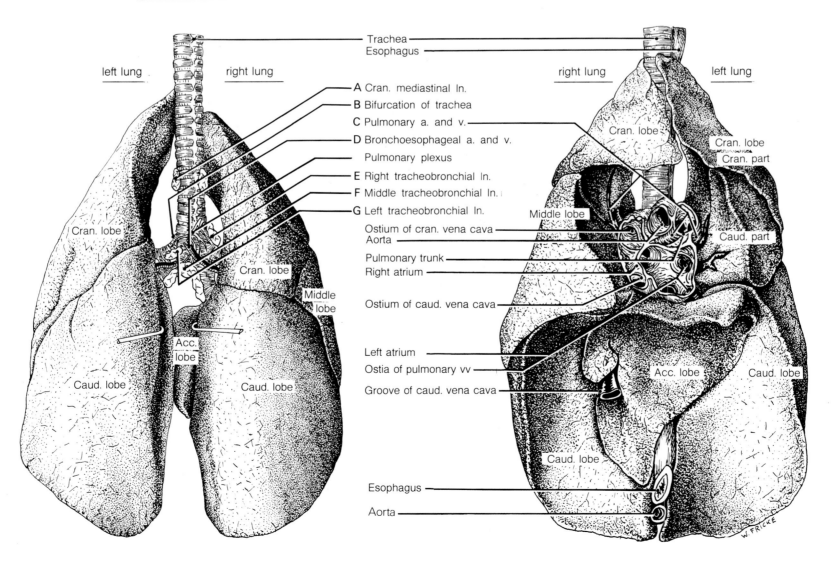

Lung (dorsal view)　　　　　　　　**Lung** (ventral view)

left lung　　right lung　　　　　right lung　　left lung

Trachea
Esophagus
A Cran. mediastinal ln.
B Bifurcation of trachea
C Pulmonary a. and v.
D Bronchoesophageal a. and v.
Pulmonary plexus
E Right tracheobronchial ln.
F Middle tracheobronchial ln.
G Left tracheobronchial ln.
Ostium of cran. vena cava
Aorta
Pulmonary trunk
Right atrium
Ostium of caud. vena cava
Left atrium
Ostia of pulmonary vv
Groove of caud. vena cava
Esophagus
Aorta

Cran. lobe
Cran. lobe
Middle lobe
Acc. lobe
Caud. lobe
Caud. lobe

Cran. lobe
Cran. lobe
Cran. part
Middle lobe
Caud. part
Acc. lobe
Caud. lobe
Caud. lobe

W. FRICKE

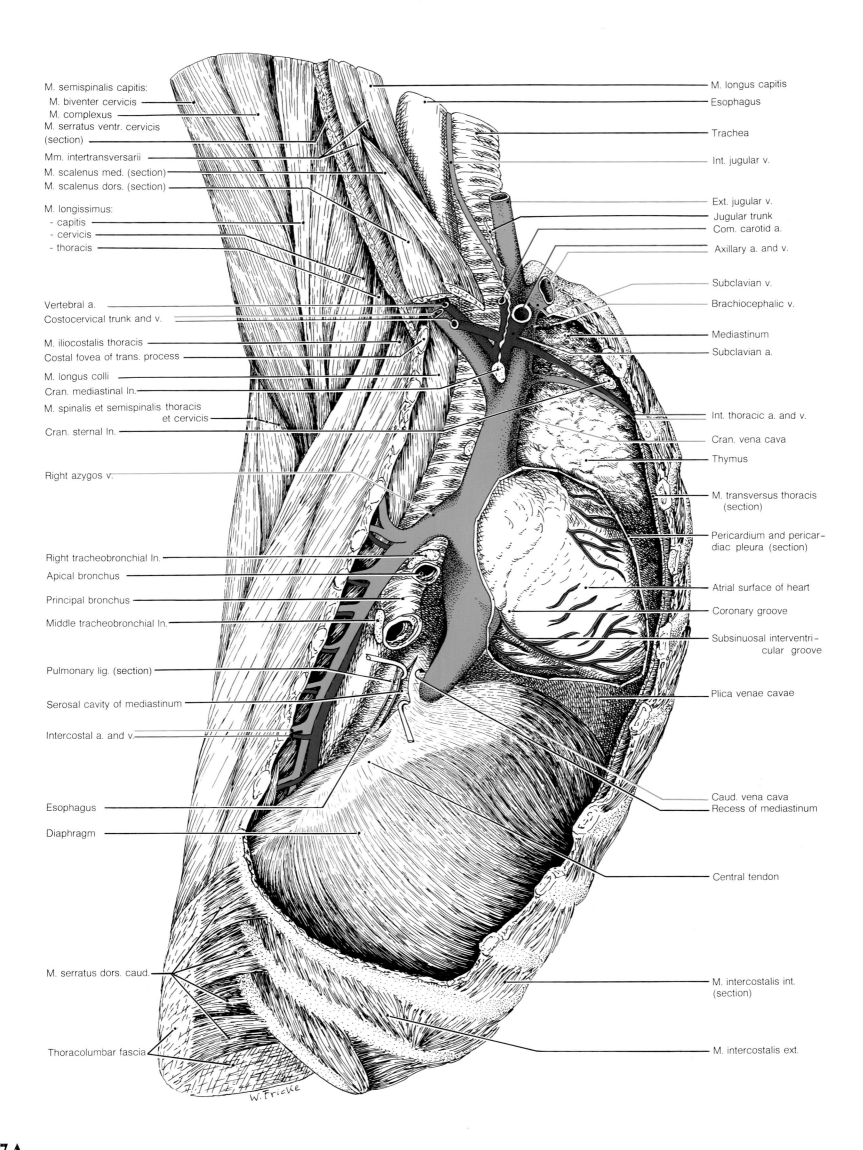

M. semispinalis capitis:
M. biventer cervicis
M. complexus
M. serratus ventr. cervicis (section)
Mm. intertransversarii
M. scalenus med. (section)
M. scalenus dors. (section)
M. longissimus:
- capitis
- cervicis
- thoracis

Vertebral a.
Costocervical trunk and v.

M. iliocostalis thoracis
Costal fovea of trans. process
M. longus colli
Cran. mediastinal ln.
M. spinalis et semispinalis thoracis et cervicis
Cran. sternal ln.

Right azygos v.

Right tracheobronchial ln.
Apical bronchus
Principal bronchus
Middle tracheobronchial ln.

Pulmonary lig. (section)
Serosal cavity of mediastinum
Intercostal a. and v.

Esophagus
Diaphragm

M. serratus dors. caud.

Thoracolumbar fascia

M. longus capitis
Esophagus
Trachea
Int. jugular v.

Ext. jugular v.
Jugular trunk
Com. carotid a.
Axillary a. and v.

Subclavian v.
Brachiocephalic v.

Mediastinum
Subclavian a.

Int. thoracic a. and v.

Cran. vena cava
Thymus

M. transversus thoracis (section)

Pericardium and pericardiac pleura (section)

Atrial surface of heart
Coronary groove

Subsinuosal interventricular groove

Plica venae cavae

Caud. vena cava
Recess of mediastinum

Central tendon

M. intercostalis int. (section)

M. intercostalis ext.

W. Fricke

The pericardium of the isolated heart is opened by a circular incision at the level of the coronary groove. The aorta is shortened until the aortic valve becomes visible. Likewise, the pulmonary trunk is sectioned piece by piece until the valve of the pulmonary trunk is also visible.

1;2 a) The **HEART SURFACE** is characterized by a **base (1)**, with the great vessels leaving and arriving at the organ, and an **apex (2)** with its muscle arranged like a vortex. The extremities of the **auricles (3, 4)** of the **right** and **left atria** respectively are used to denote the **auricular surface (5)** of the heart. The surface opposite this, where auricles are absent, is designated the **atrial surface (6)**. Externally, the left or arterial side of the heart is differentiated from the right or venous side by its great extension, it alone forming the apex. Apart from this it has a thicker, more compact ventricular wall. The **right (7)** and **left (8) ventricular borders** lie between the two heart surfaces in the region of the respective ventricles of the heart. Between both auricular extremities, the passage for the expulsion of blood from the right ventricle to the pulmonary trunk is curved to form the **conus arteriosus (9)**. At the level of the line of reflection of the pericardium the **ligamentum arteriosum (10)** travels between aorta and pulmonary trunk and is the **3** remnant of the embryonic ductus arteriosus. The **paraconal interventricular groove (11)** lies beside the conus arteriosus. The groove, which represents the boundary between the ventricles externally, contains interventricular rami or 'longitudinal limbs' of coronary blood vessels. The indistinct **subsinuosal interventricular groove (12)** runs from the coronary sinus, the site of opening of great and middle coronary vv., towards the heart apex ventrally. The **coronary groove (13)** containing the 'circular limbs' of the coronary blood vessels, is the detectable external boundary between atria and ventricles.

Using scissors, the right atrium is incised to study the internal relationships of the heart. The incision begins at the extremity of the right auricle, follows its ventral border and at the coronary groove is extended through the atrial wall to the atrial septum. The left atrium is opened in a similar manner. The ventricles are opened using different guidelines for the incisions. The right ventricle is displayed by a longitudinal incision beginning at the valve of the pulmonary trunk, proceeding parallel to the ventricular septum and terminating about halfway to the apex. From there, a transverse incision is made through the wall of the entire right ventricle, taking care to preserve the papillary muscles. To display the left ventricle, a longitudinal incision is made along the left border of the ventricle, thus bisecting the parietal cusps of the bicuspid valve.

b) The **INTERNAL RELATIONSHIPS OF THE HEART** are described according to the direction of blood flow. Within the **right atrium (A)** is a smooth walled region associated with the opening of both venae cavae, the **sinus venarum cavarum (sinus venosus) (a)**. This is separated indistinctly from the atrial region proper with its grooved, netlike internal relief. Of the coronary vessels, the **great coronary v. (15)** and the **middle cardiac v. (17)** open into the **coronary sinus (b)** ventral to the entrance of the caudal vena cava. The sinus produces a tubular protrusion of the right atrium. The **right cardiac vv (d)** and the **vv. cordis minimae** are located in the right atrium, between the **mm. pectinati (c)** which project into the atrial lumen in a comb-like manner. A fewer number of vv. cordis minimae appear in the left atrium and the left and right ventricles. Considering the **right ventricle (B)**, the **4** three cusps of the **tricuspid valve (e)** are anchored to the anulus fibrosus of the cardiac skeleton. Chordae tendineae go from the **parietal cusp (e' 1)** chiefly to the **m. papillaris magnus (e"1)**, from the **septal cusp (e'2)** mainly to the inconspicuous **mm. papillares parvi (e"2)**, and from the indistinct **angular cusp (e'3)** situated in the angle between the other two cusps, mainly to the **m. papillaris subarteriosus (e"3)** lying ventral to **5** the arterial pulmonary trunk. The **valve of the pulmonary trunk (f)** is characterized by having three cusps known as **right (f1)**, **left (f2)** and **intermediate (f3) semilunar valvules** depending on their positions relative to right and left sides of heart.

In the atrial region between both halves of the heart, is the **interatrial 6 septum (g)**. If this is held against a light source, the translucent **fossa 7;8 ovalis (h)** is discernible. Since the **epicardium (i)** and **myocardium (j) 9** are absent, the fossa ovalis consists of a duplication of **endocardium (k)** only. Blood flows into both ventricles through a right and a left **atrioventricular ostium (l)** respectively, each ventricle being separated completely from the other by a thick **interventricular septum (m)**. In **10** both ventricles, tendon-like **septomarginal trabeculae (n)** extend from septum to marginal region, particularly to the papillary muscles. They contain muscle tissue adapted to the conduction of electrical stimuli. **Trabeculae carneae (o)** particularly distinct in the region of the heart apex of both ventricles, project into the ventricular lumen as small fleshy struts or columns. The **chordae tendineae (p)** are tendinous threads which connect each cusp of both atrioventricular valves with two corresponding papillary muscles.

Five to eight **ostia of the pulmonary vv (q)** are present in the **left 11 atrium (C)**, while the **left ventricle (D)** has the **bicuspid valve (r)** at its entrance. Chordae tendineae connect its **parietal cusp (r' 1)** mainly with the **m. papillaris subauricularis (r" 1)**, and the **septal cusp (r' 2)** mainly with the **m. papillaris subatrialis (r" 2)**. The route of expulsion **12** of the blood to the **aortic valve (s)** lies between the septal cusp and the ventricular septum. The crescent-shaped valvules of the aortic valve are denoted according to their positions with reference to right and left sides of the heart and the ventricular septum. They are the **right (s 1)**, the **left (s 2)** and the **septal (s 3) semilunar valvule**.

13 The coronary arteries branch from the aorta immediately distal to the aortic valve. The **left coronary a. (14)** gives off the paraconal interventricular ramus, runs between the left ventricle and atrium as the circumflex ramus, and terminates in the subsinuosal interventricular ramus. The **right coronary a. (16)** runs in the coronary groove between the right ventricle and atrium.

c) The **CONDUCTION SYSTEM OF THE HEART** (see text **14** illustration) includes the sinoatrial node which is situated at the entrance of the cranial vena cava into the sinus venosus, and maintains the pacemaker function. From the node stimuli are conducted further over the 'working' myocardium to the atrial region and the atrioventricular node. Commencing there, the 'stimulus conducting' myocardium, namely the atrioventricular fasciculus or bundle of His, goes through the cardiac skeleton to the atrioventricular boundary. The skeleton acts as an **insulator between atrium and ventricle** and the 'working' atrial myocardium does not go beyond this boundary. Here the atrioventricular fasciculus divides into a right and a left crus each going to its appropriate ventricle. These crura run in the interventricular septum, over the heart apex to the marginal wall, and also send branches through the septomarginal trabeculae to the papillary muscles. Terminations of the 'stimuli conducting' myocardium come into contact with the 'working' ventricular myocardium.

d) The **AUTONOMIC NERVE SUPPLY** to the heart comes from **15** the cardiac plexus which is formed from parasympathetic and sympathetic nerve fibers and ganglion cells. The afferent fibers running in the autonomic nerves serve as pain mediators from the region of the heart.

Conduction system of heart

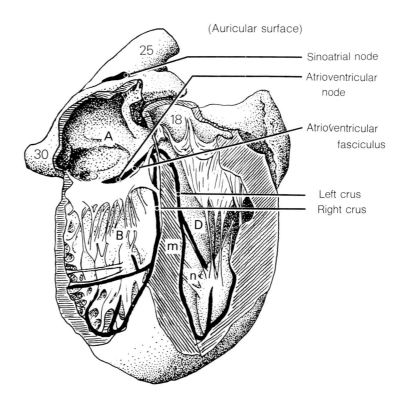

(Auricular surface)

Sinoatrial node
Atrioventricular node
Atrioventricular fasciculus
Left crus
Right crus

Base of heart

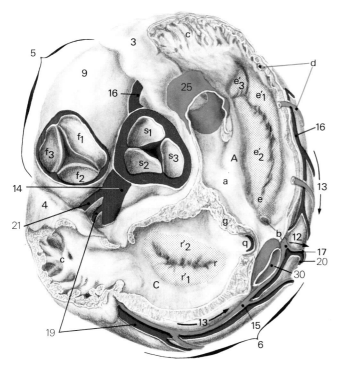

1 Base of heart
2 Apex of heart
3 Auricle of right atrium
4 Auricle of left atrium
5 Auricular surface
6 Atrial surface
7 Right ventricular border
8 Left ventricular border
9 Conus arteriosus
10 Lig.arteriosum
11 Paraconal interventricular groove
12 Subsinuosal interventricular groove
13 Coronary groove
14 Left coronary a.
15 Great coronary v.
16 Right coronary a.
17 Middle cardiac v.

Legend:

18 Aorta
    Left coronary a:
19   Circumflex r.
20   Subsinuosal interventricular r.
21   Paraconal interventricular r.
22 Brachiocephalic trunk
23 Left subclavian a.
24 Intercostal aa.
25 Cran. vena cava
26 Pulmonary trunk
27   Right pulmonary a.
28   Left pulmonary a.
29 Pulmonary vv.
30 Caud. vena cava

Left atrium and left ventricle (section)

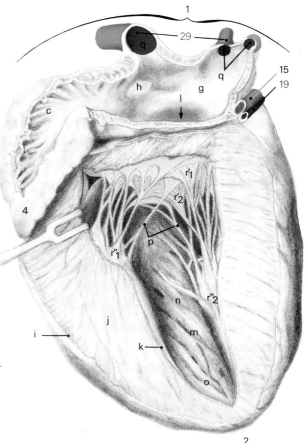

Coronary arteries and veins of heart

(Auricular surface)

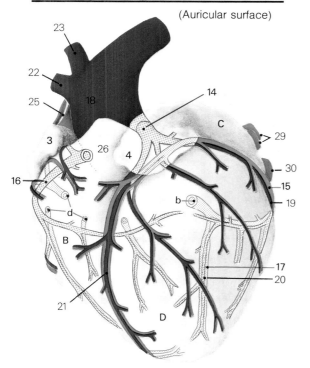

Auricular surface and right ventricle (section)

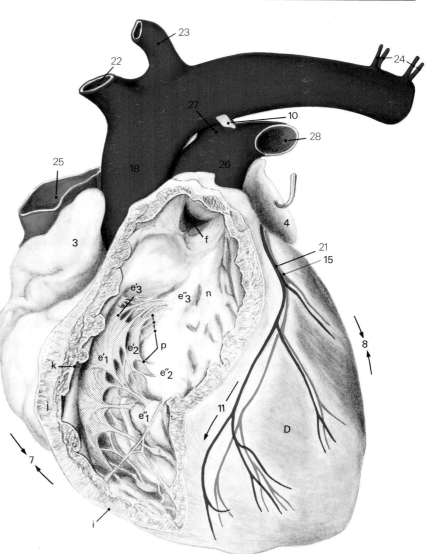

A	Right atrium	g	Interatrial septum
a	Sinus venarum cavarum (Sinus venosum)	h	Fossa ovalis
		i	Epicardium
b	Coronary sinus	j	Myocardium
c	Mm. pectinati	k	Endocardium
d	Right cardiac vv		

C	Left atrium
q	Ostia of pulmonary vv.

B	Right ventricle	D	Left ventricle
e	Tricuspid valve (Right atrioventricular valve)	r	Bicuspid (mitral) valve (Left atrioventricular valve)
e'1	Parietal cusp	r'1	Parietal cusp
e'2	Septal cusp	r'2	Septal cusp
e'3	Angular cusp	r''1	M. papillaris subauricularis
e''1	M. papillaris magnus	r''2	M. papillaris subatrialis
e''2	Mm. papillares parvi		
e''3	M. papillaris subarteriosus	s	Aortic valve
f	Valve of pulmonary trunk	s1	Right semilunar valvule
f1	Right semilunar valvule	s2	Left semilunar valvule
f2	Left semilunar valvule	s3	Septal semilunar valvule
f3	Intermediate semilunar valvule		

l	Atrioventricular ostium
m	Interventricular septum
n	Septomarginal trabeculae
o	Trabeculae carneae
p	Chordae tendineae

> *To expose the blood vessels, the dog cadaver is placed in dorsal recumbency and the manubrium of sternum and the next two sternebrae are removed. The thin, fragile autonomic nerves are preserved, particularly the vertebral n. and the ansa subclavia.*

1 a) Initially the **AORTIC ARCH** gives rise to the **brachiocephalic trunk (11)** which gives off the **right (3)** and the **left common carotid a. (3)** as well as the **right subclavian a. (7)**. The **left subclavian a. (7)** branches from the convexity of the aortic arch 2 cm distally. Right and left subclavian aa each discharge four arteries. The **vertebral a. (1)** crosses under the **costocervical trunk (9)** and passes along the transverse canal of the cervical vertebrae in company with the vertebral v. and n. The **internal thoracic a. (25)** passes under the m. transversus thoracis to proceed in a caudoventral direction. Its branches pass through the thoracic wall and send **mammary rami (26)** to the thoracic mammary complexes. At the level of the xiphoid process an artery is given off to the diaphragm before the parent artery subdivides into **cranial epigastric (27)** and **superficial cranial epigastric (28) aa.** These run along the deep and superficial surfaces respectively of the m. rectus abdominis and anastomose with the caudal vessels of like name. The **superficial cervical a. (5)** discharges a deltoid ramus to the lateral pectoral groove and ramifies on the side of the neck. The **bronchoesophageal a. (19)** is given off either directly from the descending aorta or from intercostal aa. V. to VII; its bronchial ramus supplies nutriment to the pulmonary tissue of the bronchial tree. The **intercostal aa.** (and accompanying vv. and nn.) run directly caudal to the ribs and join the aorta to the internal thoracic a.

b) The **CRANIAL VENA CAVA (8)** gives off the **right azygos v. (18)** near the heart, and consecutively in the direction of the cranial aperture of the thorax, the **costocervical (9)** and **internal thoracic (25)** vv. It then bifurcates into right and left **brachiocephalic vv. (6)**. Each brachiocephalic v. discharges a subclavian v. and an internal and an external jugular v. The **caudal vena cava (24)** possesses a distinct 'mesentery', the plica venae cavae, which is a branch of the mediastinum.

2

> *The autonomic (vegetative) nervous system regulates internal body functions. Being autonomic, such tissues as the smooth muscle of glands, blood vessels and organ systems, among others, are affected unconsciously. In this way, the internal milieu of the body is regulated and coordinated with regard to metabolic processes such as water and temperature conservation and different body functions such as circulation and respiration. Like the somatic nervous system, the autonomic system also has afferent sensory and efferent motor pathways, with the latter predominating. The afferent viscerosensory fibers mediate visceral pain and regulate the gradation and modification of peripheral stimuli. Their nerve cells are situated in spinal ganglia close to the spinal cord and their fibers run to it in the dorsal roots of the spinal nerves. Only at this location do they course separately from the efferent fibers. These leave the cord in the ventral roots of the spinal nerves and consist of two fibres in series. The two divisions of the autonomic nervous system are the sympathetic and parasympathetic. They differ histochemically regarding their transmitter substances, adrenalin or noradrenalin predominating in the sympathetic division, and acetylcholine in the parasympathetic. They also differ topographically.*
>
> *Functionally, both divisions of the system have a basic tendency to be antagonistic. Among other functions the sympathetic division has a capacity to activate the body, through an increase in energy turnover, blood pressure and heart frequency. On the other hand the parasympathetic division is concerned with the relaxation of the body, during which energy reserves are built up, for example by activating digestion. The autonomic nervous system tends to be formed from nerve plexuses. These contain ganglia of like name and in this regard differ unequivocally from plexuses of the somatic nervous system such as the brachial and lumbosacral plexuses. The ganglia contain the perikarya (nerve cell bodies), and synapsings occur there between myelinated preganglionic and non-myelinated postganglionic fibers.*

c) The **SYMPATHETIC DIVISION** commences, by means of preganglionic cell bodies, in the lateral horn of the thoracolumbar segment of the spinal cord (see p. 44A). The myelinated fibers leave the ventral root of the spinal cord and arrive at the sympathetic ganglia via white rami communicantes.
The sympathetic division includes the thoracic part of the sympathetic trunk which is formed by a subpleural, paravertebral, chain of **ganglia of the sympathetic trunk (20)**, joined longitudinally by **interganglionic rami (21)**. By way of gray and white rami communicantes the paravertebral chain has afferent and efferent pathways (details see p. 20). In common with the caudal cervical ganglion, the first three paravertebral ganglia of the thoracic part of the trunk form the **cervicothoracic ganglion (15)**, also known as the stellate. Two limbs of the **ansa subclavia (16)** begin from this ganglion and after surrounding the subclavian a., both join again at the indistinct **middle cervical ganglion (10)**. From there, sympathetic fibers are conducted cranially in the vagosympathetic trunk and through the cranial cervical ganglion (k) to the head. From the cervicothoracic ganglion the following three sets of branches arise: **cardiac nn (17)** which also receive roots from the ansa subclavia and the middle cervical ganglion and proceed to the cardiac plexus; the **vertebral n. (1)**, which, with its accompanying vessels of like name, runs in the transverse canal of the cervical vertebrae and gives off sympathetic branches to the cervical nerves; and weak **gray rami communicantes (14)** which join the brachial plexus to provide it with autonomic fibers.
From the thoracic part of the sympathetic trunk (before its transition to the lumbar part), the thick **greater splanchnic n. (23)** arises at the level of the twelfth or thirteenth thoracic vertebra. It runs over the lumbocostal arch of the diaphragm within the abdominal cavity to the solar plexus (see p. 20). Caudally, the thoracic part of the sympathetic trunk becomes very thin and is continuous with the lumbar part of the trunk which becomes thicker again (see p. 20A).

d) The **PARASYMPATHETIC DIVISION** includes a rostral portion derived from the brain, and a caudal sacral portion. Of the cranial nerves (nos. III, VII, IX and X) containing parasympathetic fibers, the **vagus n. (X —12)** runs into the body cavities. In the neck region, the vagus n. is contained in the vagosympathetic trunk, separating from the sympathetic division at the middle cervical ganglion (a sympathetic ganglion), and accompanying the trachea to the root of the lung. There it provides **cardiac** and **pulmonary rami (13)** to the cardiac and pulmonary plexuses. Caudal to the root of the lung the right and left vagi each bifurcate to give a long dorsal and a short ventral branch. Both dorsal and both ventral branches unite to form a **dorsal** and a **ventral vagal trunk (22)**. These accompany the esophagus on its dorsal and ventral aspects respectively, and pass through the esophageal hiatus of the diaphragm to supply the organs of the abdominal cavity as far as and including the transverse colon. In the thoracic cavity the vagus discharges the recurrent laryngeal n. After it has branched, the right recurrent laryngeal n. loops around the right subclavian a.; at the base of the heart the **left recurrent laryngeal n. (4)** winds around the aortic arch. Then each recurrent nerve courses up the neck, ventral to its appropriate common carotid a., as far as the larynx. There, each terminates as a caudal laryngeal n., tracheal and oesophageal branches having been given off along its course. After the origin of the recurrent laryngeal n. which carries both motor and sensory fibers, the vagus n. contains an exclusively parasympathetic motor portion besides a viscerosensory portion. The synapsings between myelinated preganglionic fibers and non-myelinated postganglionic fibers occur chiefly in intramural ganglia deep within the effector organs. (For the parasympathetic sacral subdivision see pp. 20 and 25.)

Legend p. 19A (lower illustration)

a	Ciliary ganglion	t	Caud. mesenteric ganglion
b	Pterygopalatine ganglion	u	Hypogastric n.
c	Lacrimal gland	v	Sacral splanchnic nn.
d	Mandibular and sublingual ganglia	w	Pelvic nn.
e	Sublingual gland	x	Pelvis plexus
f	Mandibular gland	y	Esophageal hiatus
g	Otic ganglion		
h	Zygomatic gland		
i	Parotid gland		
j	Distal (nodose) ganglion		
k	Cran. cervical ganglion		
l	Vagosympathetic trunk		
m	White ramus communicans		
n	Celiac ganglion		
o	Lesser splanchnic n.		
p	Adrenal r.		
q	Cran. mesenteric ganglion		
r	Abdominal aortic plexus		
s	Lumbar splanchnic nn.		

Cervical and thoracic arteries, veins and nerves (left view)

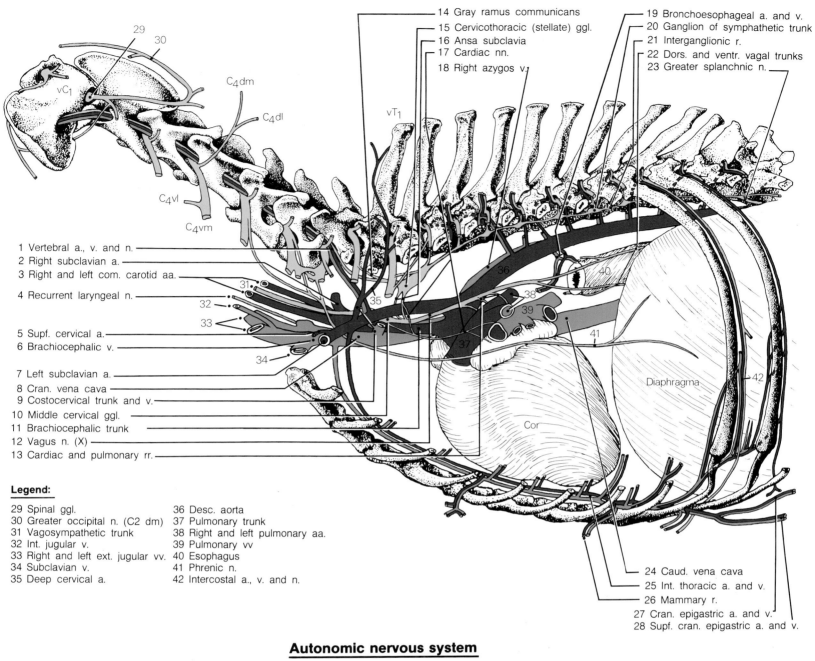

14 Gray ramus communicans
15 Cervicothoracic (stellate) ggl.
16 Ansa subclavia
17 Cardiac nn.
18 Right azygos v.

19 Bronchoesophageal a. and v.
20 Ganglion of sympathetic trunk
21 Interganglionic r.
22 Dors. and ventr. vagal trunks
23 Greater splanchnic n.

1 Vertebral a., v. and n.
2 Right subclavian a.
3 Right and left com. carotid aa.
4 Recurrent laryngeal n.
5 Supf. cervical a.
6 Brachiocephalic v.
7 Left subclavian a.
8 Cran. vena cava
9 Costocervical trunk and v.
10 Middle cervical ggl.
11 Brachiocephalic trunk
12 Vagus n. (X)
13 Cardiac and pulmonary rr.

24 Caud. vena cava
25 Int. thoracic a. and v.
26 Mammary r.
27 Cran. epigastric a. and v.
28 Supf. cran. epigastric a. and v.

Legend:

29 Spinal ggl.
30 Greater occipital n. (C2 dm)
31 Vagosympathetic trunk
32 Int. jugular v.
33 Right and left ext. jugular vv.
34 Subclavian v.
35 Deep cervical a.

36 Desc. aorta
37 Pulmonary trunk
38 Right and left pulmonary aa.
39 Pulmonary vv
40 Esophagus
41 Phrenic n.
42 Intercostal a., v. and n.

Autonomic nervous system

Parasympathetic division: ——— preganglionic fibers
                           - - - postganglionic fibers

Sympathetic division: ——— preganglionic fibers
                           - - - postganglionic fibers

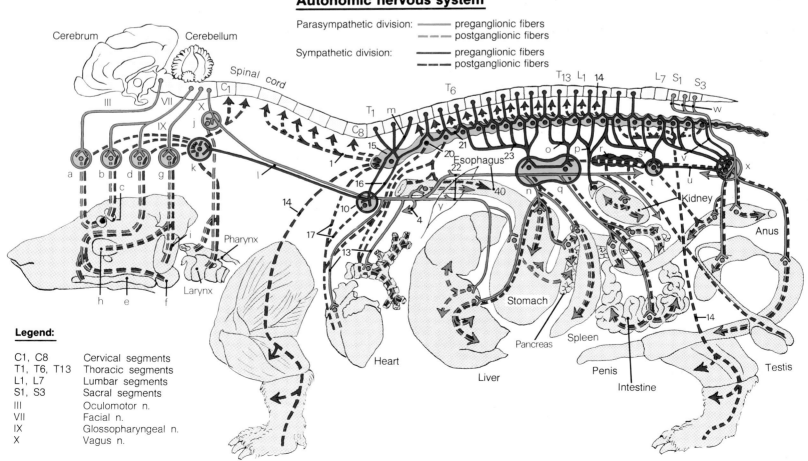

Legend:

C1, C8   Cervical segments
T1, T6, T13   Thoracic segments
L1, L7   Lumbar segments
S1, S3   Sacral segments
III   Oculomotor n.
VII   Facial n.
IX   Glossopharyngeal n.
X   Vagus n.

Without damaging the abdominal viscera, the abdominal cavity is opened by a paramedian incision as far as the pubis, and a transverse incision along the costal arch. The greater omentum is reflected cranially and the abdominal viscera is displaced to the right to display the lumbar part of the left sympathetic trunk and the abdominal aorta with its large unpaired intestinal arteries. In addition, the left kidney and its peritoneal covering are detached from underlying structures to reach the retroperitoneal space, the aorta and the lumbar part of the sympathetic trunk for dissection.

a) The **LUMBAR PART OF THE SYMPATHETIC TRUNK** lies in the lumbar region medial and adjacent to the m. psoas minor. Two or three thin nerves take origin from the first three lumbar ganglia and these form the lesser splanchnic n. In the sacral region, the lumbar parts of both sides, including sacral ganglia, fuse to give an unpaired part of the sympathetic trunk with its contained ganglion impar. The caudal continuation is the unpaired coccygeal part and its ganglia. In contrast to the initial section of the lumbar part of the sympathetic trunk, the last lumbar ganglion, and the sacral and coccygeal ganglia contain no segmental white rami communicantes. Sympathetic fibers run within the lumbar part of the sympathetic trunk, through the sacral to the coccygeal sympathetic trunk. In the ganglia of the lumbar part in particular or in the subsequent prevertebral ganglia (see below), synapsing takes place between myelinated preganglionic and unmyelinated postganglionic fibers. In contrast to the parasympathetic division, synapsing of postganglionic sympathetic fibers occurs preferably in the vicinity of the central nervous system. After fibers have synapsed in the ganglia of the sympathetic trunk, unmyelinated postganglionic fibers are present as gray rami communicantes, providing a second connection and sympathetic fibers to the segmental (somatic) spinal nerves. The smaller portion of myelinated preganglionic fibers which course through the ganglia of the sympathetic trunk without synapsing, runs in **lumbar splanchnic nn. (13)**, sacral splanchnic nn., **aortic (9)** and **adrenal (4) plexuses** to prevertebral plexuses containing ganglia of like name. The prevertebral ganglia are the **celiac ganglion (1)** and **cranial mesenteric ganglion (3)**, which, in common with their plexuses form the solar plexus, and the **caudal mesenteric ganglion (16)**. In these ganglia, situated at the origins of unpaired arteries of like name, synapsing results in unmyelinated postganglionic fibers. These form periarterial plexuses in the adventitia surrounding the arteries and reach the viscera in company with them. Lumbar splanchnic nn. from the lumbar sympathetic trunk join the caudal mesenteric ganglion caudoventrally. From there, sympathetic fibers reach the pelvic plexus within the pelvic cavity in the **hypogastric n. (25)**.

1 b) The **SACRAL PORTION OF THE PARASYMPATHETIC DIVISION** sends its fibers into the abdominal cavity in a reverse direction as far cranially as the transverse colon. The vagus n. takes over the parasympathetic innervation of abdominal organs cranial to the transverse colon. It enters the abdominal cavity by means of the dorsal and ventral vagal trunks and also supplies a branch to the solar plexus.

c) From its dorsal wall the **ABDOMINAL AORTA (10)** discharges segmentally paired lumbar aa. With their accompanying veins they proceed to the lumbar vertebral column and its adjacent areas. The common trunk of the caudal phrenic and cranial abdominal aa. arises from the lateral wall of the aorta as well as the renal a., the ovarian or testicular a., and deep circumflex iliac a., all accompanied by like-named veins. The unpaired celiac, cranial and caudal mesenteric aa. take origin from the ventral wall of the aorta. In the initial part of their course to the viscera these arteries have no accompanying veins. The **celiac a. (1)** arises directly caudal to the aortic hiatus at the level of the thirteenth thoracic vertebra. The **cranial mesenteric a. (3)** is situated slightly more caudal at the level of the first lumbar vertebra. The **cranial abdominal a. (5)** arises from the aorta in common with the caudal phrenic a. (not shown) at the level of the second lumbar vertebra. The **renal a. (6)** lies immediately caudal at the same level. The **ovarian a. (11)** or the testicular a., as the case may be, arises ventral to the third lumbar vertebra.

The **caudal mesenteric a. (16)** branches ventral to the fourth lumbar vertebra and is visible at the commencement of the rectum, when tension is applied to the colonic mesentery. The **deep circumflex iliac a. (18)** branches at right angles slightly caudal to the caudal mesenteric a. and ventral to the fourth lumbar vertebra. At the level of the fifth lumbar vertebra the abdominal aorta subdivides into the **external (19)** and the **internal iliac a. (21)**.

d) The **CAUDAL VENA CAVA (15)** lies to the right of the aorta and provides concomitant veins to the above-named arteries with the exception of the celiac, cranial and caudal mesenteric aa.

e) The **SUBLUMBAR MUSCLES** lie ventrolateral to the lumbar part of the vertebral column and are innervated by ventral branches of lumbar nn. The **m. quadratus lumborum (12)** takes origin from the bodies of the last three thoracic and all the lumbar vertebrae including lumbar transverse processes. Its insertion extends from the alar spine to the auricular surface of the ilium. The **m. psoas major (22)** arises from the vertebral ends of the last two ribs and the lumbar vertebrae and unites at the level of pelvis with the m. iliacus to give the m. iliopsoas (see lower p. 15A). The m. iliopsoas passes through the lacuna musculorum to insert on the lesser trochanter of the femur. The **m. iliacus (24)** takes origin from the sacropelvic surface of the ilium and the lateral surface of the tendon of insertion of the m. psoas minor. The **m. psoas minor (14)** begins at the last three thoracic vertebrae and the first four lumbar vertebrae where it overlies the m. psoas major on its ventral side. It ends by a flat tendon on the tubercle of m. psoas minor on the ilium.

f) The **LUMBAR PLEXUS** of the somatic nervous system (see also text illustration) is formed from the ventral branches of the lumbar nn from nLv 3 caudally. It connects with the sacral plexus to give the lumbosacral plexus. The nerve plexus lies within the sublumbar muscles and cannot be surveyed without their removal (see p. 25A). As segmental nerves, the **cranial (nLv 1 —2)** and **caudal iliohypogastric (nLv 2 —7)nn.** do not come from the lumbar plexus. The first plexus nerve is the **ilioinguinal (nLv 2 and 3 —8)**. All three nerves appear between the mm. psoas minor and quadratus lumborum. Running subperitoneally, the nerves pierce the m. transversus abdominis 3 cm lateral to m. quadratus lumborum, each dividing into a vl- and a vm-branch. The vl-branches pierce the abdominal muscles obliquely and supply the ventrolateral skin of the abdomen (see p. 14A). The vm-branches supply abdominal muscles and peritoneum. They run over the lateral surface of the m. transversus abdominis to the m. rectus abdominis, reaching the skin of abdomen and mammae in the vicinity of the linea alba. The **lateral cutaneous femoral n. (nLv 3 and 4 —17)** appears between the psoas muscles and runs transversely through the mm. obliquus abdominis and transversus abdominis in company with the deep circumflex iliac a. and v. The nerve supplies the skin of the kneefold. The **genitofemoral n. (nLv 3 and 4 —20)** appears medial to the tendon of the m. psoas minor, and accompanies the external iliac a. laterally, dividing ventral to it into a very weak femoral ramus and a strong genital ramus. The femoral ramus goes through the lacuna vasorum to the femoral space. After passing through the inguinal space, the genital ramus supplies scrotal and preputial skin or the inguinal mamma. The **femoral n. (nLv 4 to 5 —23)** courses in the m. iliopsoas through the lacuna musculorum, giving off an elongated branch, the saphenous n., before entering the m. quadriceps femoris. The **obturator n. (nLv 4 to 6 —26)** appears medial 2 to the m. iliopsoas, crosses the ilium medially, and after piercing the m. levator ani passes through the obturator foramen to the adductor muscles.

Lumbar plexus (lateral view)

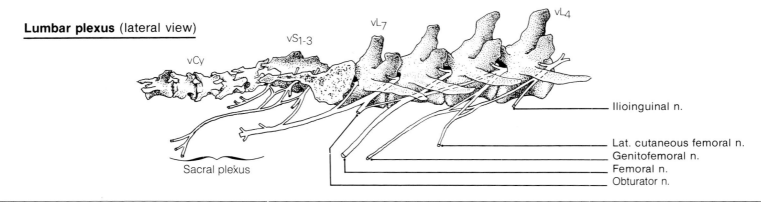

vCy   vS1-3   vL7   vL4

Ilioinguinal n.

Sacral plexus

Lat. cutaneous femoral n.
Genitofemoral n.
Femoral n.
Obturator n.

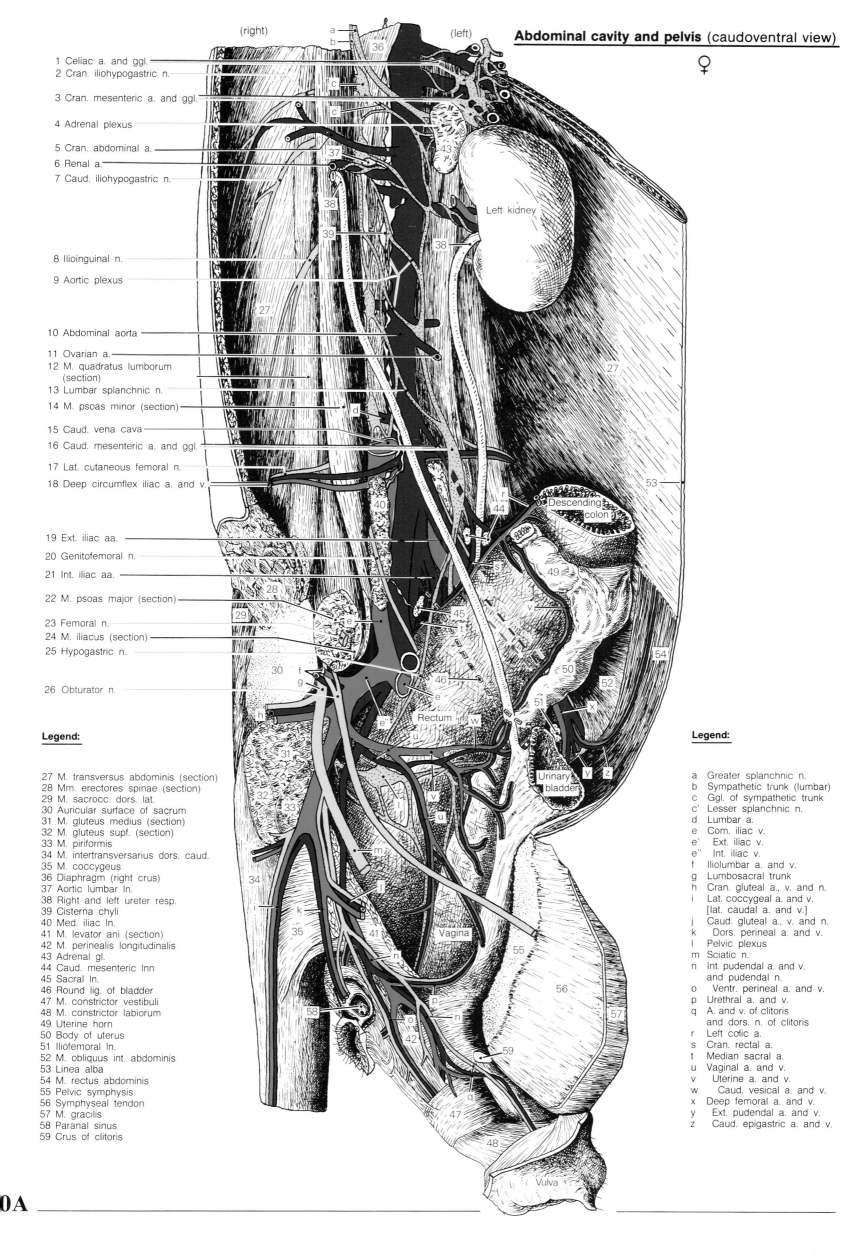

(right)   (left) ♀

1 Celiac a. and ggl.
2 Cran. iliohypogastric n.
3 Cran. mesenteric a. and ggl.
4 Adrenal plexus
5 Cran. abdominal a.
6 Renal a.
7 Caud. iliohypogastric n.
8 Ilioinguinal n.
9 Aortic plexus
10 Abdominal aorta
11 Ovarian a.
12 M. quadratus lumborum (section)
13 Lumbar splanchnic n.
14 M. psoas minor (section)
15 Caud. vena cava
16 Caud. mesenteric a. and ggl.
17 Lat. cutaneous femoral n.
18 Deep circumflex iliac a. and v.
19 Ext. iliac aa.
20 Genitofemoral n.
21 Int. iliac aa.
22 M. psoas major (section)
23 Femoral n.
24 M. iliacus (section)
25 Hypogastric n.
26 Obturator n.

Left kidney

Descending colon

Rectum

Urinary bladder

Vagina

Vulva

Legend:

27 M. transversus abdominis (section)
28 Mm. erectores spinae (section)
29 M. sacrocc. dors. lat.
30 Auricular surface of sacrum
31 M. gluteus medius (section)
32 M. gluteus supf. (section)
33 M. piriformis
34 M. intertransversarius dors. caud.
35 M. coccygeus
36 Diaphragm (right crus)
37 Aortic lumbar ln.
38 Right and left ureter resp.
39 Cisterna chyli
40 Med. iliac ln.
41 M. levator ani (section)
42 M. perinealis longitudinalis
43 Adrenal gl.
44 Caud. mesenteric lnn
45 Sacral ln.
46 Round lig. of bladder
47 M. constrictor vestibuli
48 M. constrictor labiorum
49 Uterine horn
50 Body of uterus
51 Iliofemoral ln.
52 M. obliquus int. abdominis
53 Linea alba
54 M. rectus abdominis
55 Pelvic symphysis
56 Symphyseal tendon
57 M. gracilis
58 Paranal sinus
59 Crus of clitoris

Legend:

a Greater splanchnic n.
b Sympathetic trunk (lumbar)
c Ggl. of sympathetic trunk
c' Lesser splanchnic n.
d Lumbar a.
e Com. iliac v.
e' Ext. iliac v.
e'' Int. iliac v.
f Iliolumbar a. and v.
g Lumbosacral trunk
h Cran. gluteal a., v. and n.
i Lat. coccygeal a. and v. [lat. caudal a. and v.]
j Caud. gluteal a., v. and n.
k Dors. perineal a. and v.
l Pelvic plexus
m Sciatic n.
n Int. pudendal a. and v. and pudendal n.
o Ventr. perineal a. and v.
p Urethral a. and v.
q A. and v. of clitoris and dors. n. of clitoris
r Left colic a.
s Cran. rectal a.
t Median sacral a.
u Vaginal a. and v.
v Uterine a. and v.
w Caud. vesical a. and v.
x Deep femoral a. and v.
y Ext. pudendal a. and v.
z Caud. epigastric a. and v.

a) The thoracic part of the **ESOPHAGUS** is situated in the mediastinum and passes through the esophageal hiatus of the diaphragm into the abdominal cavity. The very short **abdominal part (13)** goes immediately to the stomach. Both these parts have an outer serosal membrane and differ from the cervical part with its surrounding adventitia.

1;2 **b)** The simple **STOMACH** has one compartment possessing glands over the whole of its mucous membrane. The stomach wall, as well as that of the intestine, consists of a mucous membrane (tunica mucosa with its laminae epithelialis, propria and muscularis mucosae), a tela submucosa acting as a sliding layer, a tunica muscularis and an outer covering, the tunica serosa. The position and shape of the stomach vary considerably depending on how full it is. Almost empty, the stomach lies with its long axis practically transverse, and with its fundus pushing as far as the diaphragm. It is separated from the ventral body wall by 3 overlying liver. With marked filling, the stomach projects into the middle of the abdominal cavity and even as far caudally as the level of the third lumbar vertebra. At that stage the surface of the distended stomach lies against the lateral and ventral abdominal wall.

The **cardiac part (15)** with its contained **cardiac ostium (14)** continues 4 at the **greater curvature (18)** into the **fundus ventriculi (16)** and at the **lesser curvature (3)** without affording a recognizable boundary into the **body of the stomach (8)**. The **ventricular or gastric groove (17)** lies on the internal surface of the body of the stomach in the region of the lesser curvature where the **angular notch (4)** marks the boundary of the pyloric part. This consists of a thin-walled **pyloric antrum (5)** with a wide lumen, merging internally at a transverse fold, with the thick-walled, cone-shaped **pyloric canal (2)**. The **pylorus (1)** has multiple sphincter-like loops of muscle associated with it, and its ostium connects with the duodenum. The mucous membrane of the stomach is thrown into **gastric folds (6)**. In its cardiac part the membrane bears the narrow ring-shaped zone of cardiac glands, in the fundus and body it possesses zones of proper gastric glands while a zone of pyloric glands is situated in the pyloric part. Along the lesser curvature, the pyloric glands extend as far as the body of the stomach. The stomach contains an external longitudinal muscle layer merging with that of esophagus and duodenum, and a deeper circular muscle layer thickened at the cardiac and pyloric sphincters. Besides these two layers, oblique fibers are prominent in the gastric fundus and body. At the greater and lesser curvatures the **serous membrane (7)** is in continuity with greater and lesser omenta respectively.

5 **c)** The **SMALL INTESTINE** digests and resorbs nutrients, the area of its internal surface being increased considerably by circular folds, intestinal villi and crypts. The small intestine consists of duodenum, jejunum and ileum and extends from pylorus as far as the opening of the ileum into the large intestine. The small intestine is approximately three and a half times as long as the canine body.

I. The **duodenum** (see text illustration, left) is C-shaped and surrounds the pancreas. It begins with a **cranial part** which ascends to the right and dorsally as far as the porta of the liver. There, at the cranial flexure, it continues as the descending part. Internally at its commencement, the **descending part** bears the major duodenal papilla for the opening of the pancreatic and the bile duct. The minor duodenal papilla on which the accessory pancreatic duct opens, is situated approximately 6 cm caudally. At the caudal duodenal flexure, the descending part continues as the

transverse part. This is directed transversely and to the left of the midline, caudal to the cranial mesenteric a. There it changes direction to become the **ascending part.** At the duodenojejunal flexure the ascending part of the duodenum is continuous with the jejunum. The boundary between the two lies at the indistinct cranial margin of the **duodenocolic fold** where a sudden displacement of mesentery occurs.

II. The **jejunum**, by far the longest part of the small intestine, is suspended by a long mesentery, making possible a greater extension of the jejunal loops spatially between stomach and cranial aperture of pelvis.

III. The short **ileum (10)** begins at the indistinct free end of the 6 **ileocecal fold (11)**. The position at which the arterial and venous **antimesenteric ileal rami (12)** terminate, indicates the boundary between jejunum and ileum more clearly.

d) The **LARGE INTESTINE** resorbs water and dissolved salts arriving 7 in the small intestine with the digestive juices. Villi are absent on the inner surface of the large intestine and the crypts are particularly deep. Relative to the other domestic mammals the dog's large intestine is short and simple in form. It consists of cecum, colon, rectum and anal canal.

I. The **cecum (9)** is shaped like a cork screw and consists of a body and base limited by the ascending colon. It lies on the right side in the concavity of the C-shaped duodenum and can be regarded as a diverticulum of the colon. Cecum and ascending colon communicate through the cecocolic ostium adjacent to the ileal ostium. This also opens into the ascending colon.

II. The **colon** (see text illustration, right) is hook-shaped. It commences 8 at the cecocolic ostium on the right side as the short **ascending colon** and ascends to the level of the cranial mesenteric a. There at the right colic flexure it continues as the **transverse colon** to the left side of the body and cranial to the artery. At the left colic flexure this continues caudally as the **descending colon.**

III. The **rectum (19)** commences at the T formed by the terminal 9 division of the caudal mesenteric a. into the left colic and the cranial rectal a. It becomes wider at the ampulla of rectum, cranial to the anal canal.

IV. The **anal canal (21)** consists of three zones lying caudal to one another. The **columnar zone (22)** begins at the anorectal line where the 10 single layered, high prismatic glandular mucosa containing intestinal glands ends. Up to one centimeter high columnar-like arches of cutaneous mucose membrane are characteristic of the columnar zone. Solitary lymph nodes and anal glands opening between the mucosal arches, are situated in this zone. The columns cover blood vessels which occur lengthwise and thus form erectile bodies assisting in anal closure. The **intermediate zone (23)**, also called anocutaneous line is detected as an indistinct circular fold about 1 mm wide which is a transition between mucosa and skin. The **cutaneous zone (24)** is about 4 cm long, possesses 11 a few hairs increasing in number caudally, and contains circumanal glands. Each of the two paranal sinuses lies to the side of the anal canal 12 between it and the m. sphincter ani externus. Each sinus opens laterally between the intermediate and cutaneous zones.

Closure of the anus is produced in particular by the action of the mm. 13 sphincter ani internus and externus. The m. sphincter ani internus is a reinforced continuation of smooth circular muscle of the rectum. The striated **m. sphincter ani externus (20)** closes the anus and compresses the paranal sinuses, thus emptying them.

Intestine (ventral view)

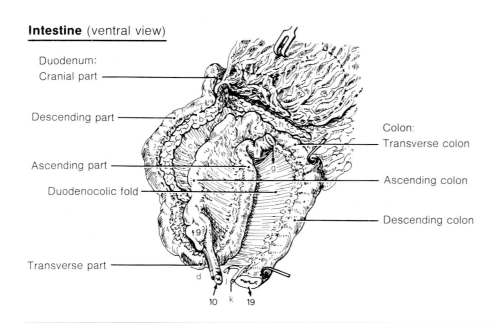

Duodenum:
Cranial part
Descending part
Ascending part
Duodenocolic fold
Transverse part

Colon:
Transverse colon
Ascending colon
Descending colon

Legend:
a   Cran. duodenal flexure
b   Right lobe of pancreas
c   Mesoduodenum
d   Caud. duodenal flexure
e   Mesocolon
f   Duodenojejunal flexure
g   Jejunum (section)
h   Left lobe of pancreas
i   Greater omentum
j   Caud. mesenteric a.
k   Cran. rectal a.

21

Stomach (sectioned visceral surface)

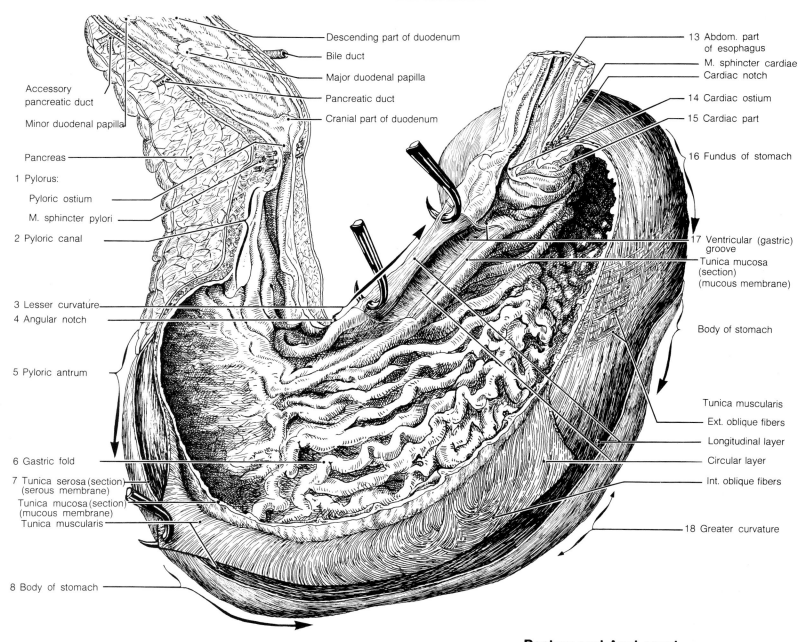

Descending part of duodenum

Bile duct

Major duodenal papilla

Pancreatic duct

Cranial part of duodenum

Accessory pancreatic duct

Minor duodenal papilla

Pancreas

1 Pylorus:

Pyloric ostium

M. sphincter pylori

2 Pyloric canal

3 Lesser curvature

4 Angular notch

5 Pyloric antrum

6 Gastric fold

7 Tunica serosa (section) (serous membrane)

Tunica mucosa (section) (mucous membrane)

Tunica muscularis

8 Body of stomach

13 Abdom. part of esophagus

M. sphincter cardiae

Cardiac notch

14 Cardiac ostium

15 Cardiac part

16 Fundus of stomach

17 Ventricular (gastric) groove

Tunica mucosa (section) (mucous membrane)

Body of stomach

Tunica muscularis

Ext. oblique fibers

Longitudinal layer

Circular layer

Int. oblique fibers

18 Greater curvature

Cecum and Ileum

Rectum and Anal canal

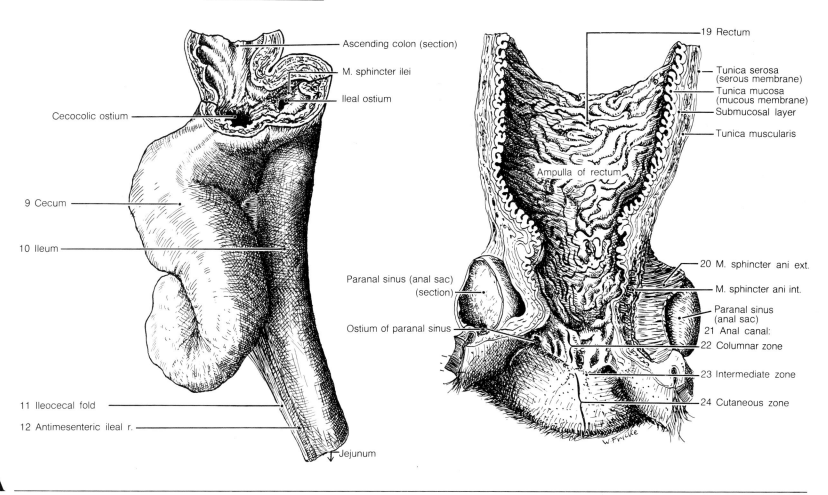

Ascending colon (section)

M. sphincter ilei

Ileal ostium

Cecocolic ostium

9 Cecum

10 Ileum

11 Ileocecal fold

12 Antimesenteric ileal r.

Jejunum

19 Rectum

Tunica serosa (serous membrane)

Tunica mucosa (mucous membrane)

Submucosal layer

Tunica muscularis

Ampulla of rectum

Paranal sinus (anal sac) (section)

Ostium of paranal sinus

20 M. sphincter ani ext.

M. sphincter ani int.

Paranal sinus (anal sac)

21 Anal canal:

22 Columnar zone

23 Intermediate zone

24 Cutaneous zone

W. Fricke

1A

1 a) The parietal **PERITONEUM** covers the peritoneal cavity, and the visceral peritoneum, the intraperitoneal organs. The peritoneal cavity commences at the diaphragm cranially and terminates in the pelvic cavity caudally where it forms three excavations or pouches lying one above the other. The dorsal **rectogenital pouch (6)** extends furthest caudally, being bounded by the rectum dorsally and the uterus or deferent duct as the case may be, ventrally. The **vesicogenital pouch (3)** lies between the named genital organs and the urinary bladder. The **pubovesical pouch (5)**, situated ventrally, lies between the urinary bladder and the pubis and is divided longitudinally by the **median ligament of the bladder (1)**. The vaginal process of peritoneum is an evagination of the peritoneal cavity, shaped like the finger of a glove, beginning at the **vaginal ring (7)** and passing through the inguinal space. During ontogenesis a part of the peritoneal cavity is constricted off cranially by the developing diaphragm to give the serous cavity of the mediastinum. Lying within the mediastinum, this cavity extends as far as the base of the heart. The **parietal peritoneum (20)** continues ventrally from the dorsal abdominal wall as a 'double lamina' of mesentery and as **visceral peritoneum (4)** covers the intraperitoneal organs such as the urinary bladder. On the other hand, organs such as the kidney which lie retroperitoneally, have no mesentery and are covered by peritoneum only on that side facing the peritoneal cavity.

b) The **MESENTERIES** include the dorsal and ventral mesogastrium (see also text illustration) and the common dorsal mesentery.

2 I. The **dorsal mesogastrium (19)**, also known as the greater omentum or epiploon, covers an extensive part of the intestinal tract ventrolaterally. Only the stomach, descending part of duodenum, descending colon and rectum remain uncovered. The **greater omentum (11)** is arranged into a parietal and a visceral layer. Each layer consists of two interfacing lamellae between which course blood and lymphatic vessels covered with large strands of adipose tissue. This results in each layer having a net-like appearance. The **visceral layer (12)** commences at the dorsal abdominal wall in common with the transverse mesocolon. It covers the intestinal convolutions ventrolaterally and continues as the **parietal layer (13)** at the cranial aperture of the pelvis. This layer lies on the ventral abdominal wall and inserts onto the greater curvature of the stomach. There the two lamellae separate, cover the stomach surface and come together again at the lesser curvature of the stomach to form the lesser omentum. The parietal and visceral layers form the partitions of the omental bursa surrounding the omental cavity. The left lobe of the pancreas develops between the two lamellae of the visceral layer and similarly the spleen develops within the lamellae of the parietal layer. The **epiploic foramen (10)**, which opens into the **vestibule of the omental bursa (22)** is situated caudal to the liver, ventral to the caudal vena cava, and dorsal to the portal vein. The vestibule lies between the abdominal wall and the lesser omentum and has access to the cavity of the omental bursa at the lesser curvature of the stomach. A specific section of the greater omentum is named the **gastrosplenic ligament (27)**. The **veil portion** (not illustrated) lies on the left side passing between spleen and descending mesocolon. It is the only portion of the greater omentum that does not participate in the formation of the omental bursa and the limitation of its cavity.

II. The liver develops in the **ventral mesogastrium** which is thus divided into a distal or parietal portion and a proximal or visceral

triangular and coronary ligaments (see p. 23) are secondary or accessory mesenteries, which, in contrast to those named above, do not contain blood vessels.

III. The **common dorsal mesentery (14)** extends from the beginning of the duodenum to the rectum, and is known by the names of the individual intestinal sections it suspends, namely mesoduodenum, — jejunum, — ileum, — cecum, — colon and — **rectum (17)**. Due to intestinal displacement and accompanying rotation during ontogenesis, the mesentery situated cranially at the origin of the cranial mesenteric a. also undergoes rotation and is known as the **root of the mesentery (18)**. The **descending mesocolon (15)** situated caudally, is attached to the dorsal abdominal wall in a linear manner. The ventral mesogastrium is not formed caudal to the cranial part of the duodenum.

c) The **LYMPH NODES** of stomach and intestines lie preferably in the mesenteries of those sections of viscera drained by the nodes in question. The **hepatic lymph nodes (21)**, also known as portal lymph nodes, are situated cranial to the pancreas on both sides of the porta of the liver. The **jejunal lymph nodes (8)** lie in series in the proximal third of the mesojejunum. Individual lymph nodes attain a length of up to 20 cm. The **splenic lymph nodes** (not illustrated) are grouped around the origin of splenic a. and v. The inconstant gastric lymph node is situated in the vicinity of pylorus on the lesser curvature of the stomach. The **colic lymph nodes (9)** lie in the region of the intestinal attachment of the ascending and transverse mesocolon while the **caudal mesenteric lymph nodes (16)** are found at the terminal division of the caudal mesenteric a. In the main, the efferent vessels from those lymph nodes mentioned, are united to the visceral trunk which opens into the cisterna chyli at the level of the kidneys.

d) The **CISTERNA CHYLI** (see p. 20A —39) is located dorsal to the 3 abdominal aorta, between the crura of the diaphragm at the level of the kidneys. As its chief afferent lymphatic vessels, the lumbar trunks conduct lymph from the pelvis and pelvic limb, and the visceral trunk from the viscera, to the cisterna. Cranially at the aortic hiatus of the diaphragm the cisterna chyli is continuous with the thoracic duct which proceeds to the venous angle. This has been mentioned previously (see p. 6A).

e) The position of the **SPLEEN** (2 - see also radiographs and text 4 illustration) is dependent on the fullness of the stomach with which it has a loose attachment due to the gastrosplenic ligament. In an almost empty stomach the **dorsal extremity** lies deep to the left costal arch. The **ventral extremity** is situated caudal to the left costal arch and can extend across the midline to the right side. With a full stomach the spleen is displaced a considerable way caudally, as far as the level of the third lumbar vertebra. Both lamellae of the parietal layer of the greater omentun are attached to the **hilus of the spleen (28)** and as a peritoneal covering envelop the organ superficially. Several branches of the splenic a. and v. (see p. 23A), autonomic nerves and lymphatic vessels, are directed to the hilus which runs lengthwise along the visceral surface of the spleen. On the cut surface of a fresh spleen one may study the red and white splenic pulp, splenic capsule and the trabeculae containing smooth muscle and blood vessels. Occasionally, accessory spleens result either from dispersed splenic primordia or from torsion of the spleen itself. Constricted splenic portions are supplied by their own blood vessel branches.

Mesogastrium (lateral view)

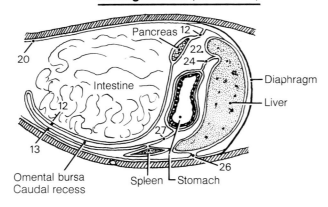

Pancreas 12
22
24
20
Intestine
12
Diaphragm
Liver
27
13
26
Omental bursa
Caudal recess
Spleen
Stomach

Spleen

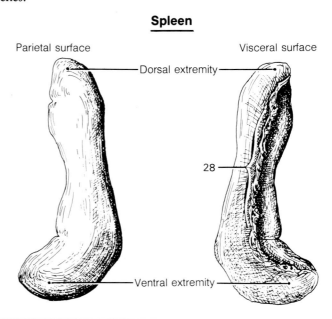

Parietal surface
Visceral surface
Dorsal extremity
28
Ventral extremity

portion. The proximal portion is the **lesser omentum (23)** which consists of **hepatogastric (24)** and **hepatoduodenal (25)** ligaments. The distal portion of the ventral mesogastrium is the **falciform ligament (26)** extending between liver and ventral body wall. In its free border it contains the round ligament of the liver, the obliterated umbilical vein of the fetus. The remaining hepatic ligaments, namely the

Mesentery and peritoneal cavity (right surface)

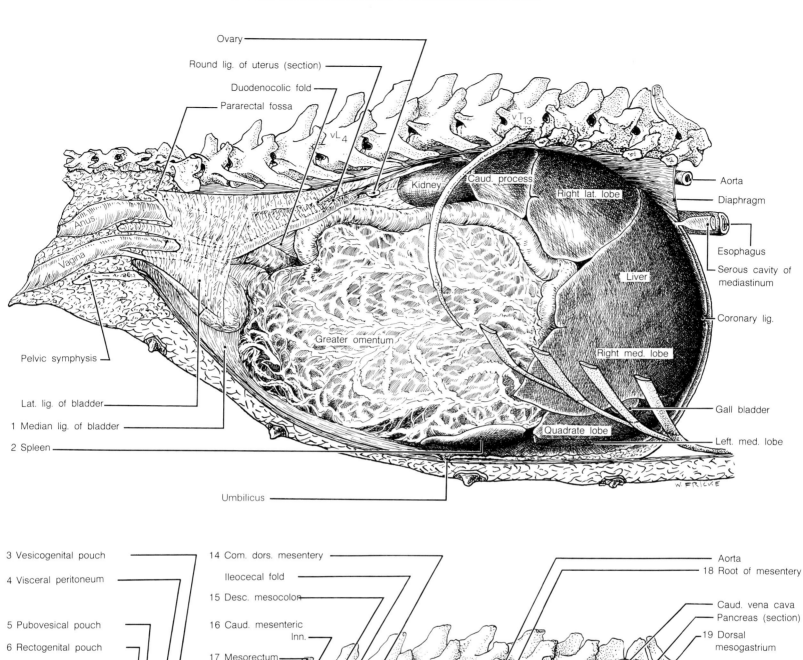

Ovary
Round lig. of uterus (section)
Duodenocolic fold
Pararectal fossa
vL₄
vT₁₃
Kidney
Caud. process
Right lat. lobe
Aorta
Diaphragm
Esophagus
Serous cavity of mediastinum
Liver
Anus
Vagina
Greater omentum
Coronary lig.
Pelvic symphysis
Right med. lobe
Lat. lig. of bladder
1 Median lig. of bladder
2 Spleen
Quadrate lobe
Gall bladder
Left. med. lobe
Umbilicus

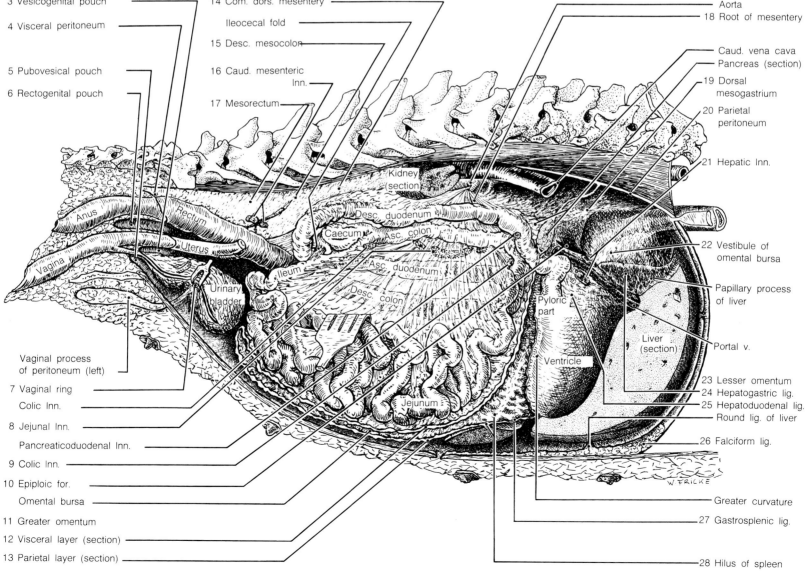

3 Vesicogenital pouch
4 Visceral peritoneum
14 Com. dors. mesentery
Ileocecal fold
15 Desc. mesocolon
16 Caud. mesenteric Inn.
17 Mesorectum
Aorta
18 Root of mesentery
Caud. vena cava
Pancreas (section)
19 Dorsal mesogastrium
20 Parietal peritoneum
5 Pubovesical pouch
6 Rectogenital pouch
Kidney (section)
21 Hepatic Inn.
Anus
Rectum
Desc. duodenum
Caecum
Asc. colon
Vagina
Uterus
Ileum
Asc. duodenum
Urinary bladder
Desc. colon
22 Vestibule of omental bursa
Papillary process of liver
Pyloric part
Liver (section)
Portal v.
Vaginal process of peritoneum (left)
Ventricle
7 Vaginal ring
Colic Inn.
8 Jejunal Inn.
Pancreaticoduodenal Inn.
9 Colic Inn.
10 Epiploic for.
Omental bursa
11 Greater omentum
12 Visceral layer (section)
13 Parietal layer (section)
Jejunum
23 Lesser omentum
24 Hepatogastric lig.
25 Hepatoduodenal lig.
Round lig. of liver
26 Falciform lig.
Greater curvature
27 Gastrosplenic lig.
28 Hilus of spleen

2A

1 a) The **LIVER** (see also text illustrations) possesses a smooth parietal or diaphragmatic surface, and a visceral surface with distinct impressions of the different adjacent abdominal organs. Dorsally the liver has an obtuse or a blunt shape and is limited by sharp borders laterally and ventrally. From the viewpoint of comparative anatomy, the liver is subsected by two imaginary 'lines of relationship' extending from its dorsal to its ventral border. The left line runs between the **esophagus (23)** dorsally and the **round ligament of the liver (N)** which is, however, generally absent in the adult dog; the right line runs between the **caudal vena cava (6, F)** dorsal, and the gall bladder ventrally. Between the two lines, the **quadrate lobe (M)** is situated ventrally while the **caudate lobe (A)** is positioned dorsal to the porta of liver. Its large **caudate process (B)** is attached to the cranial extremity of the right kidney by the hepatorenal ligament, while its small **papillary process (C)** projects to the left and is covered by lesser omentum. Lateral to the 'lines of relationship' and separated by **interlobar notches (L)**, the 'dual' hepatic lobes of left and right side are subdivided into **right** and **left medial** and **lateral lobes (E, G, I, K)** respectively. Accessory mesenteries resembling ligaments attach the liver to the diaphragm. The right and left **triangular ligaments (D)** anchor the lateral lobes of the respective sides, while medially the left, right, and intermediate **coronary ligaments (J)** fix the left and right medial lobes. Likewise the **hepatorenal ligament (O)** belongs to the accessory ligaments. The **falciform ligament (P)**, the distal part of the ventral mesogastrium, carries the round ligament of the liver in its free border from liver to umbilicus. The area nuda of the liver is a site of fusion with the diaphragm which is free of peritoneum. The following structures pass through the porta of the liver: the portal v., the hepatic a. (supplying hepatic tissue with nutriment), biliferous ducts, vagal and sympathetic rami, and lymphatic vessels to the hepatic or portal lymph nodes.

2 b) The **GALL BLADDER (1, H)** lies ventral to the porta of the liver in the fossa of the gall bladder. Bile flows through the **hepatocystic ducts (22)**, then either through the **cystic duct (2)** into the gall bladder or directly through the **bile duct (3)** to the duodenum. The bile duct opens on the major duodenal papilla located in the region of transition between the cranial and descending parts of the duodenum.

3 c) The **PANCREAS** is hook-shaped and consists of a **body of the pancreas (12)** adjacent to the cranial part of the duodenum, a prominent **left lobe (31)** running transversely in the visceral layer of the greater omentum, and the **right lobe (14)** in the descending mesoduodenum. The organ consists of an exocrine and an endocrine portion. The excretory ducts of the predominantly exocrine portion lie within the gland parenchyma. They form two ducts, the pancreatic duct opening on the major duodenal papilla and the accessory pancreatic duct opening a few centimeters caudally on the minor duodenal papilla. One excretory duct may be absent. The minute islets of the pancreas are only pin-point-sized and form the endocrine portion of the pancreas, the hormones being carried by the vascular system.

d) The **ARTERIES OF STOMACH AND INTESTINES** originate from the celiac, cranial and caudal mesenteric aa.
The **celiac a. (30)** has three main branches. After giving off an esophageal ramus, the **left gastric a. (25)** supplies the left section of the lesser curvature of the stomach to the side of the cardia. The **splenic a. (29)** provides pancreatic rami, **splenic rami (26)**, **short gastric aa. (24)** and the **left gastroepiploic a. (28)** for the greater curvature of the stomach. The **hepatic a. (13)** runs in the direction of the porta of the liver and after emitting the **right gastric a. (9)** to the pyloric portion of the lesser curvature, and **hepatic rami (5)** to the liver, it continues as the right **gastroduodenal a. (8)**. This divides into the **right gastroepiploic a. (11)** to the greater curvature of the stomach, and the **cranial pancreaticoduodenal a. (10)**. On the greater curvature the right gastroepiploic a. anastomoses with the left artery of like name and the cranial pancreaticoduodenal a. anastomoses with its caudal counterpart, a branch of the cranial mesenteric a., on the duodenum. The **cranial mesenteric a. (32)** emits the **ileocolic a. (36)**, the often duplicated **caudal pancreaticoduodenal a. (17)**, and twelve to fifteen **jejunal aa. (38)**, before terminating as the **ileal a. (20)**. The ileocolic a. branches into the **middle colic a. (33)** going to the transverse colon, the **right colic a. (15)** and the **colic ramus (16)** coursing to the ascending colon. It terminates in the **cecal a. (18)**, the **antimesenteric ileal (21)** and **mesenteric ileal (19) rami**. The mesenteric ileal ramus anastomoses with the ileal a. which has a common stem with the last jejunal a.
The **caudal mesenteric a. (39)** divides into a **cranial rectal a. (40)** running along the rectum, and a **left colic a. (37)** which anastomoses with the middle colic a. on the descending colon.

e) The **PORTAL VEIN (4)** results from the confluence of three main **4** veins, the branches of which, in principle, behave in a similar manner to the accompanying arteries of like name. These three veins are: firstly, the **gastroduodenal v. (8)** which opens into the portal v. from the right at the level of the stomach, and secondly the **splenic v. (27)** which opens into the portal v. from the left, 4 cm caudal to the first vein. Because of its union with the left gastric v., the splenic v. is also known as the gastrosplenic v. The third and largest of the confluent veins is the **common mesenteric v. (7)** resulting from the junction of **cranial (35)** and **caudal mesenteric (34) vv.** A portal blood circulation also occurs at other sites of the body, for example the hypophysis. By definition, a portal system occurs when blood within a vein commences at a capillary bed (in the hepatic portal system this is in the intestinal wall) and further along its course continues to a secondary capillary bed (the liver sinusoids). Blood then flows via collecting veins to the hepatic vv. opening into the caudal vena cava within the liver.

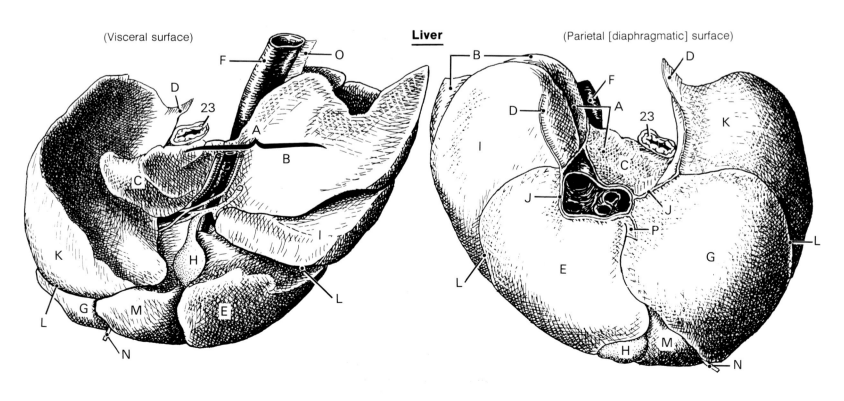

(Visceral surface)   **Liver**   (Parietal [diaphragmatic] surface)

A	Caudate lobe	I	Right lat. lobe
B	Caudate process	J	Coronary ligg.
C	Papillary process	K	Left lat. lobe
D	Triangular ligg.	L	Interlobar notches
E	Right med. lobe	M	Quadrate lobe
F	Caud. vena cava	N	Round lig. of liver
G	Left med. lobe	O	Hepatorenal lig.
H	Gall bladder	P	Falciform lig.

Intestine (ventral view)

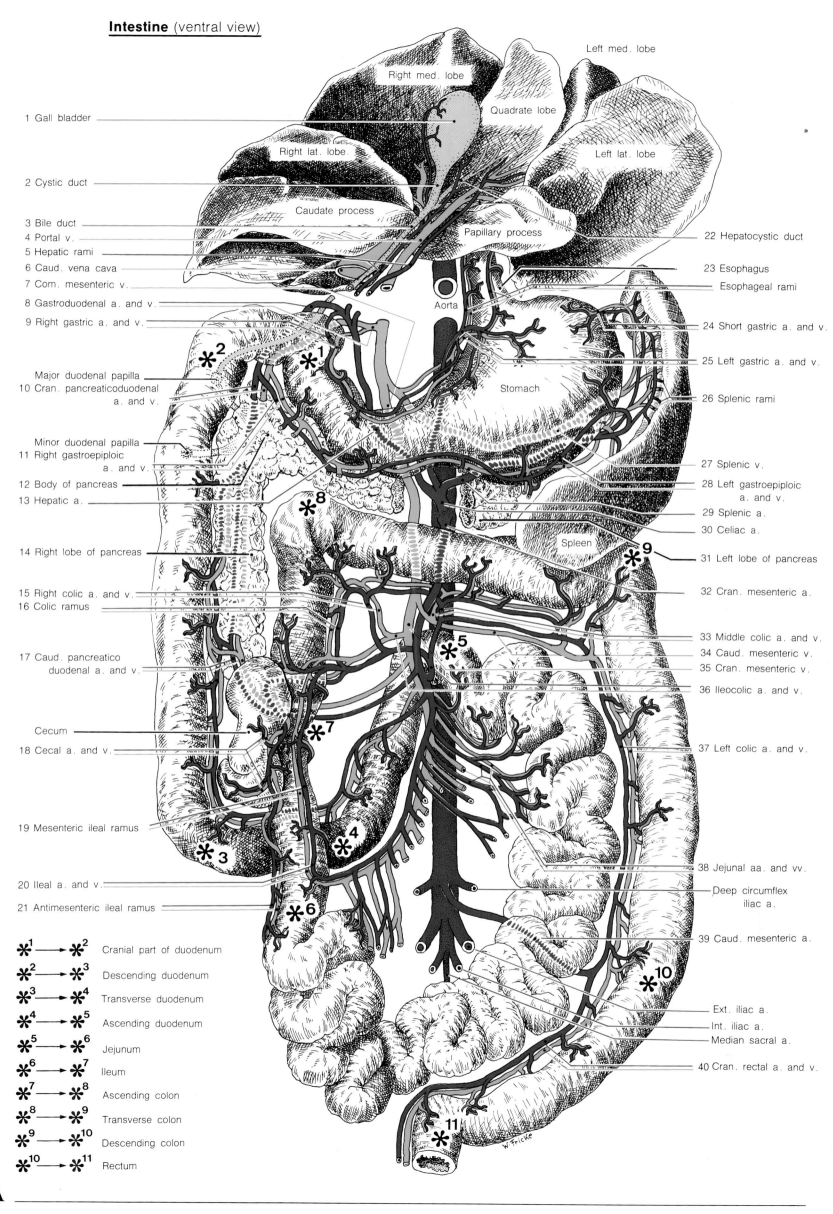

Right med. lobe

Left med. lobe

Quadrate lobe

Right lat. lobe

Left lat. lobe

1 Gall bladder

Caudate process

2 Cystic duct

Papillary process

3 Bile duct
4 Portal v.
5 Hepatic rami
6 Caud. vena cava
7 Com. mesenteric v.
8 Gastroduodenal a. and v.
9 Right gastric a. and v.

Aorta

Major duodenal papilla
10 Cran. pancreaticoduodenal
 a. and v.

Minor duodenal papilla
11 Right gastroepiploic
 a. and v.
12 Body of pancreas
13 Hepatic a.

14 Right lobe of pancreas

15 Right colic a. and v.
16 Colic ramus

17 Caud. pancreatico
 duodenal a. and v.

Cecum
18 Cecal a. and v.

19 Mesenteric ileal ramus

20 Ileal a. and v.
21 Antimesenteric ileal ramus

Stomach

Spleen

22 Hepatocystic duct

23 Esophagus
 Esophageal rami

24 Short gastric a. and v.

25 Left gastric a. and v.

26 Splenic rami

27 Splenic v.
28 Left gastroepiploic
 a. and v.
29 Splenic a.
30 Celiac a.

31 Left lobe of pancreas

32 Cran. mesenteric a.

33 Middle colic a. and v.
34 Caud. mesenteric v.
35 Cran. mesenteric v.
36 Ileocolic a. and v.

37 Left colic a. and v.

38 Jejunal aa. and vv.

Deep circumflex
 iliac a.

39 Caud. mesenteric a.

Ext. iliac a.
Int. iliac a.
Median sacral a.

40 Cran. rectal a. and v.

✳¹ ⟶ ✳² Cranial part of duodenum
✳² ⟶ ✳³ Descending duodenum
✳³ ⟶ ✳⁴ Transverse duodenum
✳⁴ ⟶ ✳⁵ Ascending duodenum
✳⁵ ⟶ ✳⁶ Jejunum
✳⁶ ⟶ ✳⁷ Ileum
✳⁷ ⟶ ✳⁸ Ascending colon
✳⁸ ⟶ ✳⁹ Transverse colon
✳⁹ ⟶ ✳¹⁰ Descending colon
✳¹⁰ ⟶ ✳¹¹ Rectum

W. Fricke

3A

> *After making two ligatures between colon and rectum, the intestine is transected between them, taking care to preserve the caudal mesenteric ganglion. The cranial part of the body is removed including stomach, intestines, liver and pancreas. One can now survey and revise the peritoneal relationships within the pelvic cavity, particularly the topography and caudal extent of the pouches or excavations (see p. 22A).*
> *To study the kidneys serial sections are made in median, paramedian, and sagittal planes and one should refer to demonstration dissections and illustrations such as those on p. 24A.*
> *The urinary bladder is incised and dissected from its ventral aspect.*

a) The **LYMPHATIC SYSTEM OF PELVIC AND LUMBAR REGIONS** is discussed, taking into consideration the natural direction of lymph flow caudocranially. The **sacral lymph nodes (9)** are situated in the angle between the origins of right and left internal iliac aa. Afferent lymph is derived from the proximal parts of rectum and genital organs, the sacrococcygeal part of the vertebral column and its surroundings, and the deep areas of the pelvic limbs. Efferent lymph flows to medial iliac lymph nodes.

The **medial iliac lymph nodes (8)** lie on the aorta at the origin of the deep circumflex iliac a. Afferent lymph flows to them from superficial inguinal lymph nodes by passing through the inguinal space. The medial iliac lymph nodes also receive lymph from the abdominal and pelvic wall, rectum, testis, urinary bladder, ureter and pelvic limb. Lymph from the testis can circumvent the lymph nodes and reach the lumbar aortic lymph nodes directly. For the most part, efferent lymph accumulates in the lumbar trunk which opens cranially into the cisterna chyli. Lateral iliac lymph nodes are absent in the dog.

The **lumbar aortic lymph nodes (6)** are disposed irregularly along the aorta, receiving the smaller portion of their lymph from the medial iliac lymph nodes. They also derive afferent lymph from the caudal thoracic wall, the abdominal wall, the lumbar part of the vertebral column and its adjacency, and the kidneys. Efferent lymph vessels open into the cisterna chyli.

1 b) The **ADRENAL GLAND (1)** is an endocrine gland lying craniomedial to the kidney. The cortex has a yellowish colour due to its content of stored substances similar to fat and used in the synthesis of steroid hormones. The colour is apparent on sectioning the organ. The brownish to gray medulla produces adrenalin and noradrenalin (consult a biochemistry text). In the fresh state the two portions are well demarcated, but definition is lost with the quick onset of postmortem changes. Because of this, adrenal glands may be confused with lymph nodes.

c) Of the **URINARY ORGANS** those discussed here are kidneys, ureters and urinary bladder.

2 I. The **kidneys** are **retroperitoneal (4)**, the left being situated ventral to the first three lumbar vertebrae and the right a half vertebral length more cranially. On a comparative anatomical basis the canine kidney is of the smooth, unipapillary or fused type. Whereas in some marine mammals, for example seals, the **renal lobes (25)** are separated from one another to form an organ resembling a bunch of grapes, in the dog the renal cortex is fused, resulting in a smooth external surface. The **renal medulla (26)** still permits a lobar formation to be recognized in certain planes.

The medullary substance of the renal lobe is shaped like a pyramid and consists of an **external part (28)** or base of the pyramid, and an **internal part (29)** or central renal papilla projecting into the **renal sinus (15)**. In the median plane of the kidney the papillae are fused to form a single **common papilla (13)** or renal crest, hence the term unipapillary kidney. On either side of the median plane between the **pelvic recesses (21)**, the unfused portions of the kidney papillae project as individual **renal pyramids (27)** into the renal sinus (see sagittal and paramedian sections). The kidneys are embedded in an outer **adipose capsule (2)**.

Upon incision the underlying **fibrous capsule (16)** is removed easily from the renal cortex. This is not possible where definite pathological changes occur. One is able to detect pin-point corpuscles in the **renal cortex (20)** which is approximately 2 cm thick. A corpuscle consists of a glomerulus[1] and its surrounding capsule in which the 'primary urine', a blood plasma filtrate, accumulates. In the loop-like kidney tubules which continue on from the corpuscles, a resorption of fluid, glucose, Na and Cl ions occur from the primary filtrate and in a reverse direction, substances of an essential urinary nature are eliminated. Subsequently this 'secondary urine' reaches the papillary ducts. These are the very fine tubular projections of the renal pelvis (see below) into the common renal papilla or renal crest. During ontogenesis they sprout from the primordial bud of the ureter and become connected to the kidney tubules which in turn derive from the metanephros. The very numerous, mainly individual papillary ducts open on the common

papilla or **renal crest (24)**. Thus a close inspection of the crest reveals an **area cribrosa (23)** along its length, at each end of which is situated a slit-like **duct of the common papilla (22)** approximately one to two mm long. Numerous papillary ducts open into these. The organization of the vascular and tubule systems are reflected in the macroscopic appearance of a kidney section. The kidney tubules running in a narrow loop-like fashion together with the papillary ducts, produce a radiating striation in the sectioned renal medulla. As medullary rays these project into the renal cortex (consult histology texts). In the base of the pyramid, that is, in the external part of the renal medulla, the striation is formed by papillary ducts and sectioned large — and small caliber tubules. In the renal papilla, that is, in the internal part of the renal medulla, the striation is formed by papillary ducts and sectioned small caliber tubules. Large caliber tubules are absent.

The **renal pelvis (14)** lies in the renal sinus and is separated by adipose tissue from renal vessels and parenchyma in the region of the renal hilus. At the common renal papilla the pelvis coalesces with the parenchyma by means of the penetrating papillary ducts. Between the unfused portions of the renal papillae, the renal pelvis extends into the recesses of the pelvis.

The vascular system of the kidney produces a complicated topographical relationship to the loop-like renal tubules. The **renal a. (3)** divides into **interlobar aa. (17)** which continue in an arched manner as **arcuate aa. (18)** along the cortico-medullary border. **Interlobular aa. (19)** run peripherally from the arcuate aa. to supply the renal cortex. (Renal lobules are subunits of the renal lobes with no macroscopic boundaries separating them from one another.)

II. Initially, the **ureter (5)** lies retroperitoneally and receives a short 3 mesentery cranial to its opening into the urinary bladder. By that stage the ureter is situated intraperitoneally.

III. The **urinary bladder** (see text illustration and p. 28A) possesses a 4 cranial **apex** or vertex, a body, and a cervix situated caudally. Dorsally at the neck of the bladder converging sections of the ureters pass obliquely through the bladder wall. In so doing the mucous membrane of the organ arches over them as **ureteral columns**, each of which extends to a **ureteral ostium**. The **ureteral folds** or **plicae** continuing caudally, unite at the **internal ostium of the urethra**, the transition between urinary bladder and urethra. The **trigone of the bladder** lies between the ureteral folds. The **median ligament of the bladder (10)** extends ventrally from bladder to linea alba reaching cranially to the umbilicus. During fetal development, the ligament carries the urachus, the fetal urinary duct. The paired **lateral ligaments of bladder (12)** run to the body wall each containing an umbilical a. which, in many cases, is obliterated to give the **round ligament of the bladder (11)**.

d) The **URETHRA** is discussed together with the genital organs on p. 28.

Urinary bladder (ventral view)

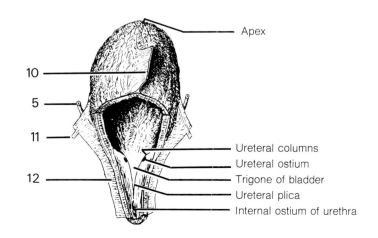

Apex

10

5

11

12

Ureteral columns
Ureteral ostium
Trigone of bladder
Ureteral plica
Internal ostium of urethra

[1] The grammatically correct term is glomerulum.

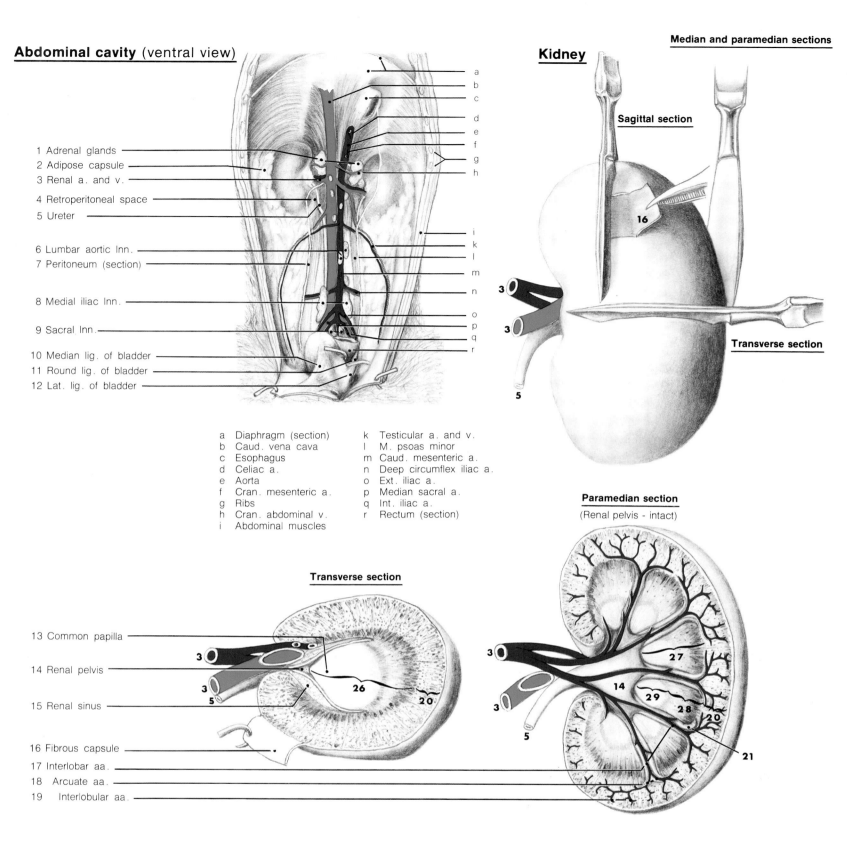

Abdominal cavity (ventral view)

Kidney

Median and paramedian sections

Sagittal section

Transverse section

1 Adrenal glands
2 Adipose capsule
3 Renal a. and v.
4 Retroperitoneal space
5 Ureter
6 Lumbar aortic lnn.
7 Peritoneum (section)
8 Medial iliac lnn.
9 Sacral lnn.
10 Median lig. of bladder
11 Round lig. of bladder
12 Lat. lig. of bladder

a	Diaphragm (section)	k	Testicular a. and v.
b	Caud. vena cava	l	M. psoas minor
c	Esophagus	m	Caud. mesenteric a.
d	Celiac a.	n	Deep circumflex iliac a.
e	Aorta	o	Ext. iliac a.
f	Cran. mesenteric a.	p	Median sacral a.
g	Ribs	q	Int. iliac a.
h	Cran. abdominal v.	r	Rectum (section)
i	Abdominal muscles		

Paramedian section
(Renal pelvis - intact)

Transverse section

13 Common papilla
14 Renal pelvis
15 Renal sinus
16 Fibrous capsule
17 Interlobar aa.
18 Arcuate aa.
19   Interlobular aa.

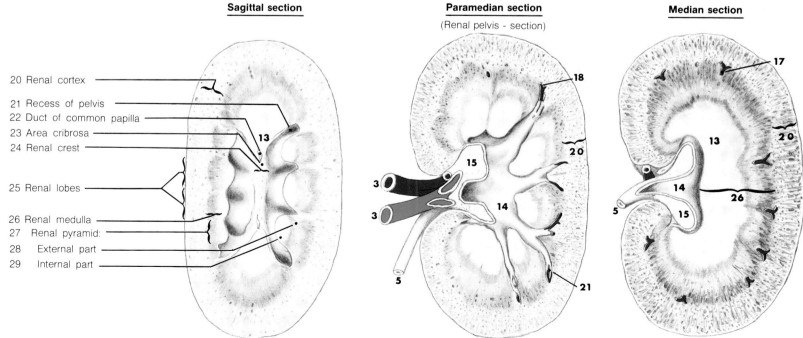

Sagittal section

Paramedian section
(Renal pelvis - section)

Median section

20 Renal cortex
21 Recess of pelvis
22 Duct of common papilla
23 Area cribrosa
24 Renal crest
25 Renal lobes
26 Renal medulla
27   Renal pyramid:
28   External part
29   Internal part

The right pelvic limb and the right half of the osseous pelvis are removed, taking care to preserve blood vessels, nerves, and organs within the pelvic cavity namely rectum, urinary bladder and genitalia. Therefore pubis and ischium are sawn through 2 cm to the right of and parallel to the pelvic symphysis, and the right ilium is separated from the sacrum at the iliosacral articulation. Following this, the right crus of penis or clitoris is separated from the associated section of ischium, and all muscles, nerves and blood vessels between the right femoral region and pelvis are removed. In this way the rectum and internal genital organs are preserved, remaining uninjured within the body of the cadaver.

a) The **TERMINAL DIVISION OF THE AORTA** (21) into an unpaired median sacral a. and paired external and internal iliac aa. takes place at the level of the last lumbar vertebra.

The unpaired **median sacral a.** (19) is a caudal continuation of the aorta ventral to and along the sacrum, and is prolonged caudally as the median coccygeal a.

The **external iliac a.** (28) is accompanied laterally by the genitofemoral n. and it gives off the **deep femoral a.** (11) shortly before its passage through the lacuna vasorum and transition to the femoral a. Before reaching the deep part of the femoral region, the deep femoral a. discharges the **pudendoepigastric trunk** (12). This vascular trunk divides into the **external pudendal a.** (13), passing through the inguinal space, and the **caudal epigastric a.** (29). The latter runs on the deep surface of the m. rectus abdominis and at the level of the umbilicus anastomoses with the cranial artery of like name. The cranial epigastric a. is a branch of the internal thoracic a. At the umbilical level on the external surface of the m. rectus abdominis is a further (superficial) anastomosis between cranial and caudal superficial epigastric aa. They also branch respectively from the internal thoracic and external pudendal aa. At its origin, the **internal iliac a.** (17) emits the umbilical a. running to the urinary bladder in the lateral ligament of the bladder. The vessel is either completely obliterated to give the **round ligament of the bladder** (26) or otherwise supplies the cranial part of the bladder as the cranial vesical a. After a short course and at the ramification of the **caudal gluteal a.** (4), the internal iliac a. continues as the **internal pudendal a.** (10). The caudal gluteal a. discharges the following arteries: **iliolumbar** (16), **cranial gluteal** (3), **lateral coccygeal** (6) and the **dorsal perineal** (8), which, as parietal arteries, supply the wall of the pelvis. Before its entrance into the ischiorectal fossa, the internal pudendal a. provides the visceral arteries of the pelvis, namely the prostatic a. in the male, the vaginal a. in the female, and their respective branches. The middle rectal a. is directed caudally, the uterine a. or a. of ductus deferens depending on sex and the caudal vesical a., cranially. The **prostatic a.** (14) and its counterpart in the female, the **vaginal a.** (9), take a perpendicular course to the prostate gland and vagina respectively. The **middle rectal a.** (7) supplies the ventrolateral region of the ampulla of the rectum. The **uterine a.** (25) runs adjacent to the attachment of the mesometrium and anastomoses cranially with the **uterine r.** (24) of the ovarian a. (23). The latter ramifies directly from the aorta at the level of the fourth lumbar vertebra. The **a. of the ductus deferens** (15) accompanies the deferent duct as far as the epididymis. When the umbilical a. undergoes complete obliteration, the **caudal vesical a.** (27) supplies all of the urinary bladder of its respective side.

b) The **TERMINAL DIVISION OF THE CAUDAL VENA CAVA** (20) occurs at the level of the last lumbar vertebra. The very weak **median sacral v.** (19) continues the caudal direction of its parent. Terminally the caudal vena cava divides into right and left **common iliac**

(22) **vv.**, dividing respectively into an internal and an external iliac v. at the pelvic entrance. In comparison to the like named artery, the **internal iliac v.** (17) is very long and provides the venous visceral stem for either the prostatic or the vaginal v. The concomitant arteries are derived from the internal pudendal a. After the branching of **iliolumbar** (16), **cranial gluteal** (3) and **lateral coccygeal** (6) vv. the internal iliac v. gives origin to the last **parietal vein**, the **caudal gluteal v.** (4). It then continues as the **internal pudendal v.** (10) at the level of the third coccygeal vertebra. The caudal gluteal v. provides the **dorsal perineal v.** (8).

c) The **SACRAL PLEXUS** which is part of the lumbosacral plexus, is formed from the ventral branches of the last two lumbar nerves and the sacral nerves. It may be divided into a strong cranial part, the lumbosacral trunk, and a weaker caudal part.

I. The **lumbosacral trunk** (18) arises from nLv 6 and 7 and nSv 1 and provides the **cranial** (3) and **caudal gluteal** (4) **nn.** which, together with the accompanying vessels of like name, supply gluteal muscles and overlying skin. Then the lumbosacral trunk continues as the **sciatic n.** (2) and passes laterally over the greater ischiatic notch to the thigh.

II. The **caudal part** of the sacral plexus supplies the **pudendal n.** (nSv 1-3 —10) which goes into the ischiorectal fossa in company with the internal pudendal a. and v. (see p. 26A). In most cases, the **caudal cutaneous femoral n.** (5) is derived from nSv 2 and 3, although there are considerable individual variations. At its origin the nerve provides muscle branches to the mm. coccygeus and levator ani and then runs along the sacrotuberous ligament. At the ischiatic tuberosity the nerve supplies the skin of the thigh region. As caudal clunial nn., its branches supply the caudal gluteal region.

The caudal clunial nn. (see p. 26A) also include cutaneous branches of the caudal gluteal n. which reach the skin in company with caudal gluteal a. and v. proximally on the caudal border of the m. gluteus superficialis. The middle clunial nn. (see p. 26A) are formed from fragile dorsal branches of sacral nerves and a subcutaneous branch of the cranial gluteal n. With their accompanying vessels the nerves reach the skin proximally at the cranial border of the m. gluteus superficialis. The cranial clunial nn. are formed from dorsal branches of lumbar cutaneous nerves and supply the skin of the lumbar region (see p. 12).

d) The **SACRAL PORTION OF THE PARASYMPATHETIC DIVISION** includes the **pelvic nn.** (1) running perpendicularly. On the lateral wall of the rectum at the level of prostate or fornix of vagina depending on sex, they meet the hypogastric n. travelling horizontally and form the mixed sympathetic-parasympathetic pelvic plexus with its contained ganglia. From the plexus very fine nerve plexuses accompany blood vessels to the different organs of the pelvic cavity. They are termed rectal, vesical, prostatic, deferent and uterovaginal plexuses depending on the effector organs involved.

ARTERIES, VEINS, AND NERVES OF THE PELVIC CAVITY

Aorta	Caudal vena cava	Sacral plexus
Median sacral a.	Median sacral v.	I. Cranial part (lumbosacral trunk)
Median coccygeal a.	Median coccygeal v.	Cranial gluteal n.
External iliac a.	Common iliac v.	Caudal gluteal n.
Deep femoral a.	External iliac v.	Sciatic n.
Pudendoepigastric trunk	Deep femoral v.	
Internal iliac a.	Pudendoepigastric v.	II. Caudal part
Umbilical a.	Internal iliac v.	Pudendal n.
Caudal gluteal a.	Iliolumbar v.	Caudal cutaneous femoral n.
Iliolumbar a.	Vaginal v./prostatic v.	Caudal clunial nn.
Cranial gluteal a.	Uterine v./v. of ductus deferens	Ramus to m. coccygeus
Lateral coccygeal a.	Caudal vesical v.	Ramus to m. levator ani
Dorsal perineal a.	Middle rectal v.	
Internal pudendal a.	Cranial gluteal v.	
Vaginal a./prostatic a.	Lateral coccygeal v.	
Uterine a./a. of ductus deferens	Caudal gluteal v.	
Caudal vesical a.	Dorsal perineal v.	
Middle rectal a.	Internal pudendal v.	

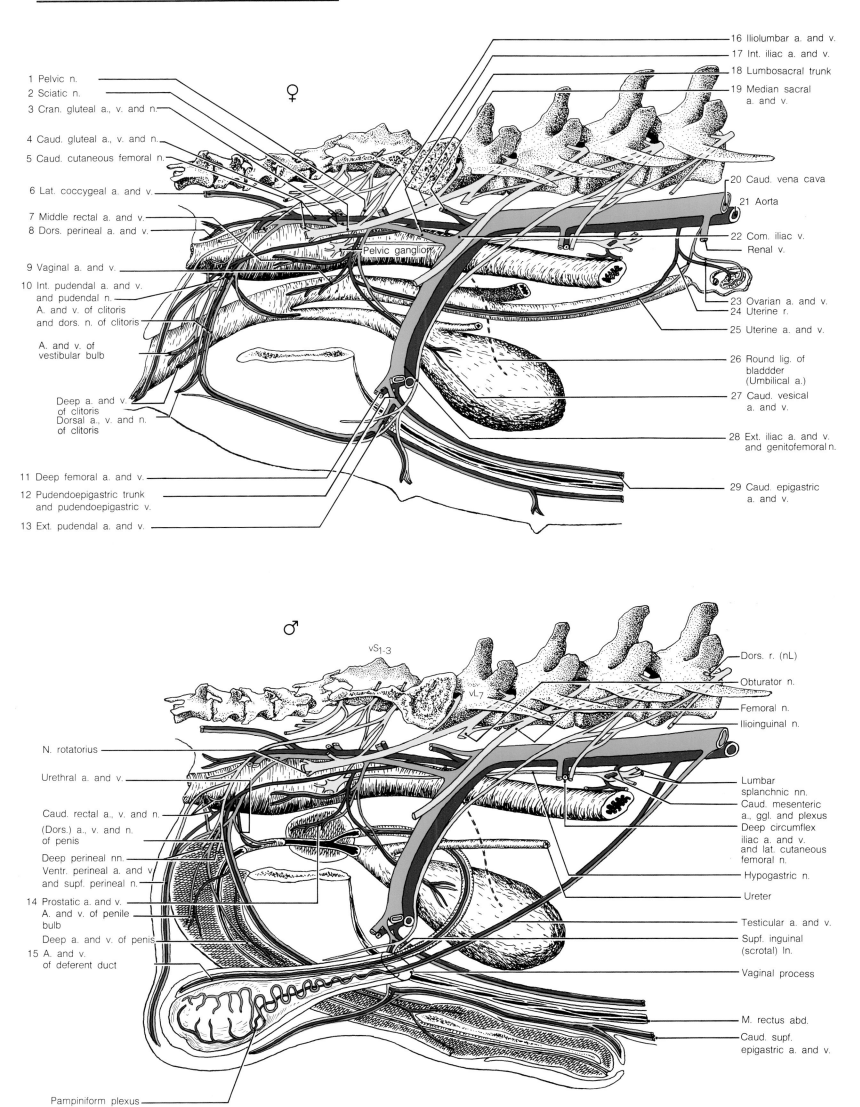

1 Pelvic n.
2 Sciatic n.
3 Cran. gluteal a., v. and n.

4 Caud. gluteal a., v. and n.
5 Caud. cutaneous femoral n.

6 Lat. coccygeal a. and v.

7 Middle rectal a. and v.
8 Dors. perineal a. and v.

9 Vaginal a. and v.
10 Int. pudendal a. and v.
and pudendal n.
A. and v. of clitoris
and dors. n. of clitoris

A. and v. of
vestibular bulb

Deep a. and v.
of clitoris
Dorsal a., v. and n.
of clitoris

11 Deep femoral a. and v.
12 Pudendoepigastric trunk
and pudendoepigastric v.
13 Ext. pudendal a. and v.

♀

Pelvic ganglion

16 Iliolumbar a. and v.
17 Int. iliac a. and v.
18 Lumbosacral trunk
19 Median sacral
a. and v.

20 Caud. vena cava
21 Aorta

22 Com. iliac v.
Renal v.

23 Ovarian a. and v.
24 Uterine r.
25 Uterine a. and v.

26 Round lig. of
bladdder
(Umbilical a.)
27 Caud. vesical
a. and v.

28 Ext. iliac a. and v.
and genitofemoral n.

29 Caud. epigastric
a. and v.

♂

vS1-3

vL7

N. rotatorius

Urethral a. and v.

Caud. rectal a., v. and n.
(Dors.) a., v. and n.
of penis

Deep perineal nn.
Ventr. perineal a. and v
and supf. perineal n.
14 Prostatic a. and v.
A. and v. of penile
bulb
Deep a. and v. of penis
15 A. and v.
of deferent duct

Pampiniform plexus

Dors. r. (nL)
Obturator n.
Femoral n.
Ilioinguinal n.

Lumbar
splanchnic nn.
Caud. mesenteric
a., ggl. and plexus
Deep circumflex
iliac a. and v.
and lat. cutaneous
femoral n.
Hypogastric n.
Ureter

Testicular a. and v.
Supf. inguinal
(scrotal) ln.

Vaginal process

M. rectus abd.
Caud. supf.
epigastric a. and v.

> To obtain the dissection field illustrated on the opposite page, the skin is removed after making incisions around anus and labia of vulva. After docking the tail, its stump is anchored in a dorsal position. Following this, muscles or parts thereof, and blood vessels and nerves, are exposed on the right side of the body. Subsequent to this preliminary exercise, blood vessels, nerves and muscles in that order are demonstrated on the left side of the cadaver.
> Before commencing the dissection it is appropriate to consider the ischiorectal fossa and the pelvic diaphragm.

a) The **ISCHIORECTAL FOSSA** is bounded dorsally by the sacrotuberous ligament and the m. coccygeus and ventrally by the ischiatic arch and the m. obturator internus. It is occupied by adipose tissue, blood vessels and nerves. At the base of the fossa, the mm. coccygeus and levator ani and overlying deep fascia form a 'cutaneous muscular' seal of the caudal aperture of the pelvis. This seal, the pelvic diaphragm, is of prime importance.

b) Within the **PELVIC DIAPHRAGM** (see text illustration, right)

between the right and left mm. levatores ani is a longitudinal lacuna or space. This is referred to as the 'levator entrance' of the pelvic diaphragm and is self-contained. Dorsally it is closed off by the anal diaphragm and the anal canal passing through it, and ventrally by the urogenital diaphragm and the urogenital organs traversing it, as well as perineum (perineal body or corpus perineale in the more restricted sense) lying between the two. Perineum (perineal zone or regio perinealis in the broader sense or in the definition of clinical use) comprises the whole pelvic outlet.

PELVIC DIAPHRAGM AND MM. COCCYGEUS AND LEVATOR ANI

Anal diaphragm	Perineum	Urogenital diaphragm
M. sphincter ani externus	Mm. perinei	M. bulbospongiosus ♀♂
M. sphincter ani internus	M. longitudinalis perinei cutaneus	M. constrictor vestibuli ♀
M. rectococcygeus	M. transversus perinei superficialis	M. constrictor vulvae ♀
	Central tendon of perineum	M. ischiocavernosus ♀♂
	(deep tendinous anchorage	M. retractor penis or clitoridis
	of musculature)	

The **INNERVATION** of the mm. coccygeus and levator ani is derived from the muscle branches of nSv 3. The remaining muscles are supplied by branches of the pudendal n., the muscles of the anal diaphragm by the caudal rectal n., and the muscles of perineum and the urogenital diaphragm by deep perineal nn.
The **m. coccygeus (13)**, situated medial and adjacent to the sacrotuberous ligament, runs between the ischiatic spine and the first four coccygeal vertebrae. The **m. levator ani (12**, see also text illustration right) is situated medial to the m. coccygeus, its fibers directed in a similar manner to those of the m. coccygeus. The linear origin of the m. levator ani lies adjacent to the pelvic symphysis on pubis and ischium and runs dorsally on the medial surface of ilium. The insertion extends from the fourth to the seventh coccygeal vertebra. In contrast to the situation in man, only a few muscle fibers of the m. levator ani are connected with m. sphincter ani externus. In the dog this makes the name of the muscle rather deceptive when referring to its function.
The **m. sphincter ani externus (14)** surrounds the anal canal in common with the m. sphincter ani internus. The latter is the caudal continuation of the smooth circular layer of the rectum. Between the striated m. sphincter ani externus and the smooth m. sphincter ani internus lies the paranal sinus of its respective side. The **m. rectococcygeus (11)** is a prolongation of the longitudinal smooth muscle of the rectum continuing to the ventromedian aspect of the tail (see p. 29A). The **mm. perinei (15)** consist of muscle fiber bundles directed transversely and longitudinally connecting the m. sphincter ani externus and m. bulbospongiosus. Deep to the perineum the central tendon of the perineum forms a median lamina composed of muscle and deep fascia which acts as an anchoring mechanism. In the male dog the **m. bulbospongiosus (21)** covers the bulb of the penis transversely. In the bitch, the muscle consists of the **m. constrictor vestibuli (17)** with its fibers taking an approximately circular course, and the **m. constrictor vulvae (18)** situated slightly caudally, with its fibers running approximately longitudinally. The **m. ischiocavernosus (16)** arises from the ischiatic arch and covers the crus penis or crus clitoridis.
The **m. retractor penis (19)** or m. retractor clitoridis (see p. 29A) is a smooth muscle with only its **penile part (20)** visible on the ventromedian aspect of the penis.

c) At the base of the ischiorectal fossa **ARTERIES, VEINS AND NERVES** pass into the pudendal canal through the pelvic diaphragm. The **pudendal n. (2)** goes into the ischiorectal fossa between the mm. oburator internus and coccygeus. It provides: the **caudal rectal n. (3)** to the anal diaphragm, the elongated **superficial perineal n. (7)** with its branches, the dorsal scrotal or labial nn. to the skin, and the short **deep perineal nn. (6)** to the perineal muscles and urogenital diaphragm. The pudendal n. terminates as the **dorsal n. of the penis (10)** or **clitoris (5)** to the respective parts of those organs.
The **internal pudendal a. (2)** accompanies the pudendal n., and a branch, the urethral a., goes to the pelvic part of the urethra. Then it gives off the **ventral perineal a. (8)** which, in turn, has a branch, the **caudal rectal a. (3)** near its origin, and a dorsal scrotal or labial ramus distally. The internal pudendal a. continues as the **a. of penis (9)** or **clitoris (4)**. These divide respectively into the dorsal aa. of penis or clitoris, a. of bulb of penis or vestibule and the deep a. of penis or clitoris. The **dorsal perineal a. (1)** takes origin from the caudal gluteal a.
The **internal pudendal v.** (see p. 25A —10) has a similar course to the artery with variations in the order of ramification of its branches.

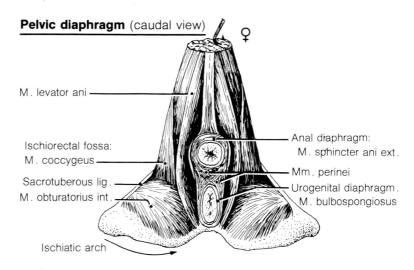

Pelvic diaphragm (caudal view) ♀

M. levator ani

Ischiorectal fossa:
M. coccygeus
Sacrotuberous lig.
M. obturatorius int.

Ischiatic arch

Anal diaphragm:
M. sphincter ani ext.
Mm. perinei
Urogenital diaphragm:
M. bulbospongiosus

ARTERIES, VEINS AND NERVES OF THE ISCHIORECTAL FOSSA (continuation of Table on p. 25)

Internal pudendal a.	Internal pudendal v.	Pudendal n.
Urethral a.	Urethral v.	Caudal rectal n.
Ventral perineal a.	Dorsal v. of penis or clitoris	Superficial perineal n.
Caudal rectal a.	Ventral perineal v.	Dorsal scrotal or dorsal labial nn.
Dorsal scrotal or dorsal labial r.	Caudal rectal v.	Deep perineal nn.
A. of penis or clitoris	Dorsal scrotal or dorsal labial v.	Dorsal n. of penis or clitoris
A. of bulb of penis or vestibule	V. of penis or clitoris	
Deep a. of penis or clitoris	V. of bulb of penis or vestibule	
Dorsal a. of penis or clitoris	Deep v. of penis or clitoris	

Ischiorectal fossa (caudal view)

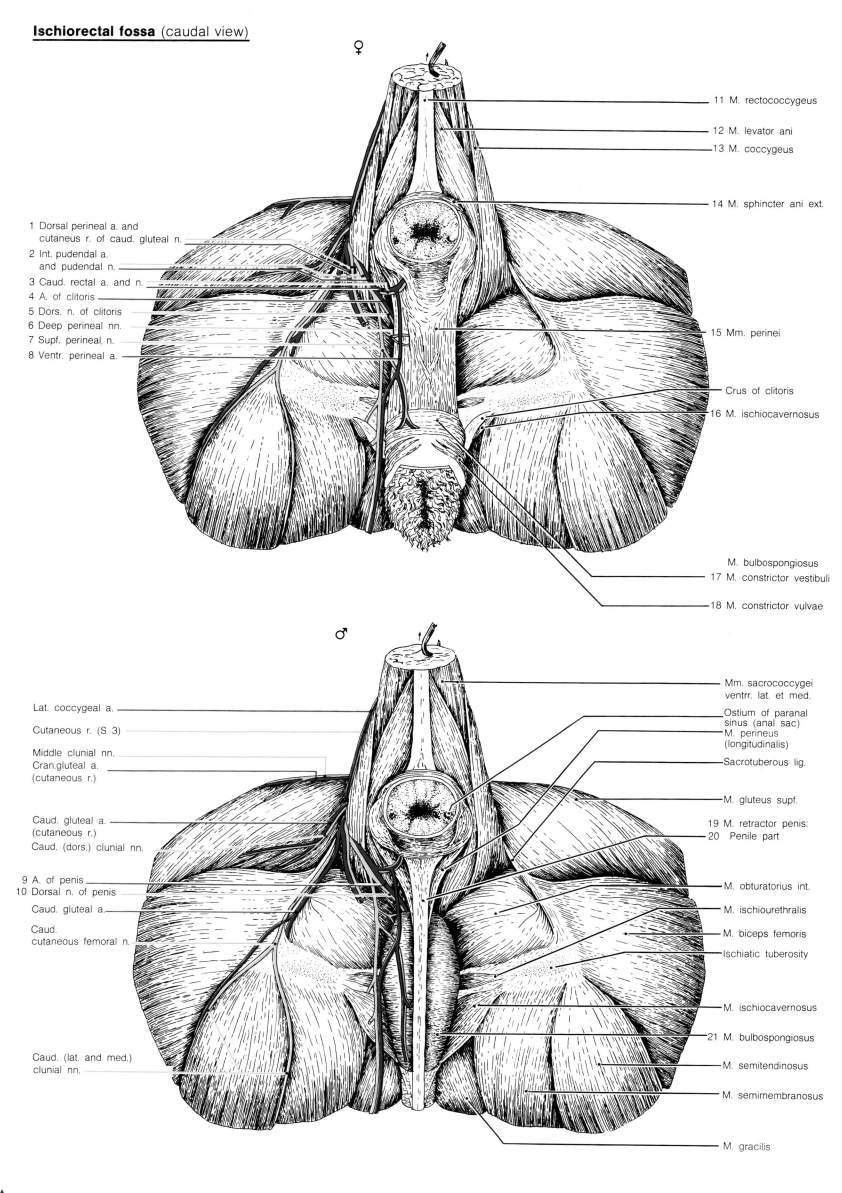

♀

11 M. rectococcygeus

12 M. levator ani

13 M. coccygeus

14 M. sphincter ani ext.

1 Dorsal perineal a. and
cutaneus r. of caud. gluteal n.
2 Int. pudendal a.
and pudendal n.
3 Caud. rectal a. and n.
4 A. of clitoris
5 Dors. n. of clitoris
6 Deep perineal nn.
7 Supf. perineal. n.
8 Ventr. perineal a.

15 Mm. perinei

Crus of clitoris

16 M. ischiocavernosus

M. bulbospongiosus
17 M. constrictor vestibuli

18 M. constrictor vulvae

♂

Mm. sacrococcygei
ventrr. lat. et med.

Ostium of paranal
sinus (anal sac)
M. perineus
(longitudinalis)

Sacrotuberous lig.

M. gluteus supf.

19 M. retractor penis:
20 Penile part

Lat. coccygeal a.

Cutaneous r. (S 3)

Middle clunial nn.
Cran.gluteal a.
(cutaneous r.)

Caud. gluteal a.
(cutaneous r.)
Caud. (dors.) clunial nn.

9 A. of penis
10 Dorsal n. of penis
Caud. gluteal a.

Caud.
cutaneous femoral n.

M. obturatorius int.

M. ischiourethralis

M. biceps femoris

Ischiatic tuberosity

M. ischiocavernosus

21 M. bulbospongiosus

M. semitendinosus

M. semimembranosus

Caud. (lat. and med.)
clunial nn.

M. gracilis

A

In the male dog, the scrotum on the right side of the body is incised parallel to the scrotal raphe, making sure to inspect the tunica dartos during the procedure. The smooth muscle of the tunic extends into the septum of the scrotum which divides the scrotal sac into two. The vaginal process of peritoneum, the m. cremaster (externus) situated laterally, and the internal and external spermatic fascia are all detached from the scrotum. Then they are displaced further by making a longitudinal incision as far as the vaginal ring. In the bitch a similar dissection is made, depending on the development of the vaginal process.

In both sexes the folds of peritoneum associated with the internal genital organs are subdivided into mesenteries and ligaments.

a) The **MESENTERIES OF THE GENITAL ORGANS OF THE BITCH** include the proximal and distal mesovaria, mesosalpinx and mesometrium, all of which are known collectively as the broad ligament of the uterus. They adopt an oblique course to the longitudinal axis of the body and are referred to in German as 'oblique ligaments'. The **mesovarium (17)** conveys the ovarian a. and v. to the ovary. The **mesosalpinx (21)** separates off laterally (see cross-sectional illustration upper right). Thus the parent membrane resolves itself into a long **proximal mesovarium (19)** and a short **distal mesovarium (20)**, the mesothelium of which continues as the superficial epithelium of the ovary. Lateral to the ovary, the thin terminal portion of the uterine tube or oviduct lies between both lamellae of the mesosalpinx, the thicker initial part of the tube being situated medially. Both parts form an arch around the ovary itself. The short distal mesovarium and the long mesosalpinx together form the **ovarian bursa (22)**. This exhibits an ostium of the **ovarian bursa (23)** on the free border of the mesosalpinx medially, and a fat-free site laterally, through which the ovary is glistening. Caudally the mesovarium merges and is continuous with the **mesometrium (14)**. Its attachment to the uterus, the **parametrium (2)**, encloses the uterine a. and v. between its two lamellae, and continues onto the uterine surface as the perimetrium (see p. 27A, cross sectional illustration).

b) Included in the **LIGAMENTS OF THE GONADS OF THE BITCH** are the suspensory ligament of the ovary, the proper ligament of the ovary and the round ligament of the uterus. The **suspensory ligament of the ovary (18)**, regarded as the cranial ligament of the gonad, runs from the diaphragm to the ovary and lateral to the kidney. The attached caudal ligament of the gonad namely the inguinal ligament of the ovary, divides into two parts at the uterine ostium of the uterine tube. The first part, the **proper ligament of the ovary (16)** courses between the ovary and the cranial end of the uterine tube. The second part, the **round ligament of the uterus (15)**, is held to the mesometrium by a long intermediate fold and passes through the **vaginal ring (1)** to the base of the vaginal process of peritoneum. When the vaginal process is absent, and this occurs in approximately 25% of bitches either on one or both sides of the body, the connective tissue portion of the round ligament of the uterus passes through the inguinal space without its peritoneal covering. It may reach the pudendal region.

Since the ovaries are situated near the site of their embryonic formation, that is to say, only an insignificant ovarian descent has occurred, the cranial and caudal peritoneal folds approaching the ovary maintain the same size proportionately during ontogenesis and do not become involuted.

c) Included in the **MESENTERIES OF THE GENITAL ORGANS OF THE MALE DOG** are the proximal and distal mesorchia, the mesoductus deferens, mesofuniculus and mesepididymis. The **proximal mesorchium (3)** and its contained testicular vessels lie in the peritoneal cavity as a vascular fold and enter the vaginal process of peritoneum through the **vaginal ring (4)**. In the vaginal canal the proximal mesorchium and the **mesoductus deferens (10)** are partially fused to the mesofuniculus (see cross section). Comparable to the peritoneal cavity, the vaginal canal and the adjacent vaginal cavity possess a parietal lamina, a folded mesenteric or intermediate lamina, and a visceral lamina. The parietal layer consists of a **perifuniculus (13)** or **periorchium (12)**. The mesenteric or intermediate lamina, as a partial connection from proximal mesorchium and mesoductus deferens, forms the **mesofuniculus (9)**, while the **visceral layer** is the **epiorchium (11)**, a covering of the testis.

The mesofuniculus is the mesentery of the spermatic cord or funiculus, which consists of eight structures, namely the testicular a. and v., the autonomic nerve plexus and lymphatic vessels, the deferent duct and its artery and vein, and the m. cremaster internus. Just proximal to the testis the long proximal mesorchium merges with the short **distal mesorchium (8)** at the lateral separation of the mesepididymis. The **testicular bursa (7)** which is accessible from the lateral side, lies between the mesepididymis and the distal mesorchium (see p. 27A cross-section lower left).

d) The **LIGAMENTS OF THE GONADS OF THE MALE DOG** comprise a cranial and a caudal ligament. The cranial ligament, the suspensory ligament of testis, is more or less completely involuted. At most, it is continued as a prolongation of the vascular fold towards the diaphragm and lateral to the kidney. The caudal ligament, the inguinal ligament of testis, is subdivided by the epididymis into the **proper ligament of the testis (5)** running between testis and epididymis, and the **ligament of the tail of the epididymis (6)** reaching between tail of epididymis and the base of the vaginal process of peritoneum. The vaginal process is anchored to the internal surface of the scrotum by the extraperitoneal section of the ligament of tail of epididymis, also known as the scrotal ligament. Since a distinct testicular descent occurs, the parts of the sections of the ligament initially encountered cranial to the site of gonadal formation involute either completely or almost completely.

PERITONEAL RELATIONSHIPS OF THE GENITAL ORGANS

Mesenteries of Genital Organs			Ligaments of Gonads	
Bitch		**Male dog**	**Bitch**	**Male dog**
Proximal mesovarium		Proximal mesorchium (Vascular fold)	Suspensory lig. of ovary (between ovary and diaphragm)	Suspensory lig. of testis (extensively involuted)
Distal mesovarium	Ovarian or testicular bursa	Distal mesorchium	Proper lig. of ovary (between ovary and termination of uterine tube)	Proper lig. of testis (between testis and tail of epididymis)
Mesosalpinx (lateral separation between proximal and distal mesovarium)		Mesepididymis (lateral separation between proximal and distal mesorchium)	Round lig. of uterus (between termination of uterine tube and base of vaginal process of peritoneum)	Lig. of tail of epididymis (between tail of epididymis and vaginal process of peritoneum)
Mesometrium		Mesoductus deferens		

LAYERS OF ABDOMINAL WALL AND THEIR CONTINUATION ON THE VAGINAL PROCESS OF PERITONEUM

Abdominal wall	Tissue Layers Covering Testis	Vaginal process of peritoneum	
Common integument	Scrotum	Parietal lamina	Perifuniculus
Cutis	Cutis of scrotum		Periorchium
Subcutis	Tunica dartos	Intermediate lamina	Mesofuniculus
External fascia of trunk	External spermatic fascia		Mesoductus deferens
M. obliquus internus abdominis M. transversus abdominis	– – – M. cremaster (externus)		Mesorchium Mesepididymis
Internal fascia of trunk	Internal spermatic fascia	Visceral lamina	Epiorchium
Peritoneum	Vaginal process		

Urogenital ligaments (right side)

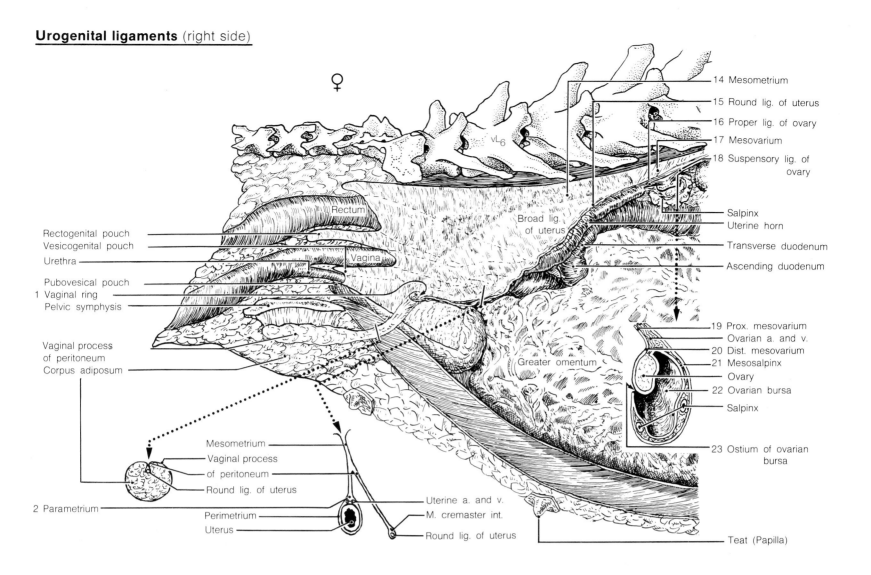

♀

14 Mesometrium
15 Round lig. of uterus
16 Proper lig. of ovary
17 Mesovarium
18 Suspensory lig. of ovary

Rectum

Broad lig. of uterus

vL₆

Salpinx
Uterine horn
Transverse duodenum
Ascending duodenum

Rectogenital pouch
Vesicogenital pouch
Urethra

Pubovesical pouch
1 Vaginal ring
Pelvic symphysis

Vagina

19 Prox. mesovarium
Ovarian a. and v.
20 Dist. mesovarium
21 Mesosalpinx
Ovary
22 Ovarian bursa
Salpinx

Vaginal process of peritoneum
Corpus adiposum

Greater omentum

23 Ostium of ovarian bursa

Mesometrium
Vaginal process of peritoneum
Round lig. of uterus

2 Parametrium

Perimetrium
Uterus

Uterine a. and v.
M. cremaster int.
Round lig. of uterus

Teat (Papilla)

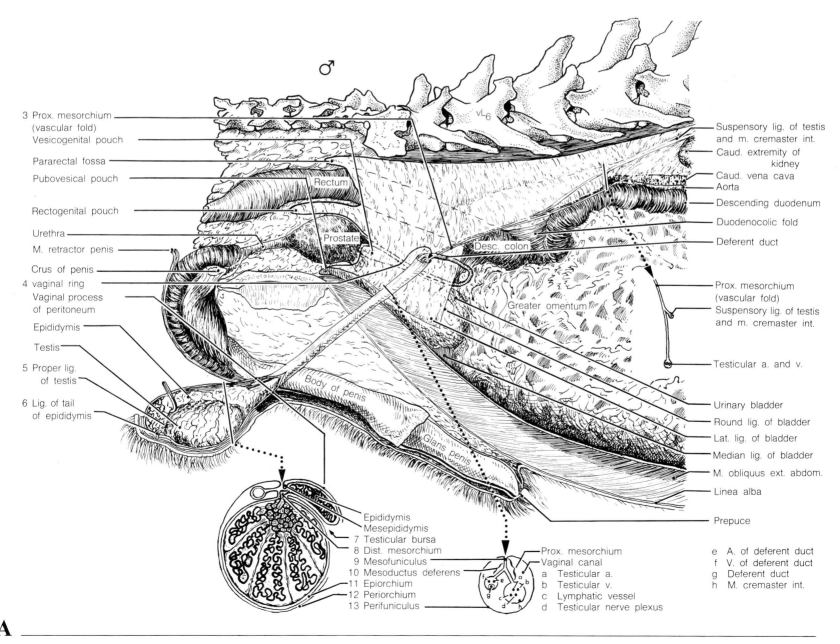

♂

Suspensory lig. of testis and m. cremaster int.
Caud. extremity of kidney
Caud. vena cava
Aorta
Descending duodenum
Duodenocolic fold
Deferent duct

3 Prox. mesorchium (vascular fold)
Vesicogenital pouch
Pararectal fossa
Pubovesical pouch

Rectum

Rectogenital pouch
Urethra
M. retractor penis
Crus of penis
4 vaginal ring
Vaginal process of peritoneum
Epididymis
Testis
5 Proper lig. of testis
6 Lig. of tail of epididymis

Prostate

Desc. colon

Greater omentum

Body of penis

Glans penis

vL₆

Prox. mesorchium (vascular fold)
Suspensory lig. of testis and m. cremaster int.

Testicular a. and v.

Urinary bladder
Round lig. of bladder
Lat. lig. of bladder
Median lig. of bladder
M. obliquus ext. abdom.
Linea alba

Prepuce

Epididymis
Mesepididymis
7 Testicular bursa
8 Dist. mesorchium
9 Mesofuniculus
10 Mesoductus deferens
11 Epiorchium
12 Periorchium
13 Perifuniculus

Prox. mesorchium
Vaginal canal
a Testicular a.
b Testicular v.
c Lymphatic vessel
d Testicular nerve plexus

e A. of deferent duct
f V. of deferent duct
g Deferent duct
h M. cremaster int.

A

> *In the male dog the testis is bisected longitudinally. The urinary bladder and urethra are opened by ventromedian incision and the erectile corpora within the penis are inspected by sectioning the penis transversely at the level of root, body and glans. In the bitch the ovary is exteriorized after making a longitudinal incision to widen the ostium of the ovarian bursa. Then the ovary is bisected sagittally and the uterine tube detached from the mesosalpinx along its whole length. To study the internal relationships of the female genitalia, the entire lengths of uterine horns, body and cervix, and vagina, vestibule and vulva are opened by making a dorsomedian incision.*

In both sexes the genitalia are divided into internal and external genital organs. The accessory glands are included with the internal organs.

a) The **INTERNAL GENITAL ORGANS OF THE MALE DOG** include the organ producing the sexual cells, that is the testis, the organs of transport namely the epididymis and deferent duct (ductus deferens), and the prostate.

1 I. The **testis (8)** is situated in the scrotum. Connective tissue **septula of the testis** radiate from the outer compact **tunica albuginea** and form the **mediastinum testis** in the central part of the testis (see text illustration, left). The interstitial cells of Leydig which produce androgens, and the seminiferous tubules, are situated in the testicular lobules between the septula. The tubules transport semen to straight tubules and these in turn join the rete testis within the mediastinum. The rete is connected to between twelve and sixteen efferent ductules.

Testis (transverse section)

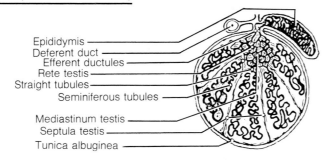

Epididymis
Deferent duct
Efferent ductules
Rete testis
Straight tubules
Seminiferous tubules
Mediastinum testis
Septula testis
Tunica albuginea

2 II. The **epididymis (7)** is arranged into a compact head, formed by the initial part of the epididymal duct and the efferent ductules opening into it, a thin body, and the prominent tail. The body and tail of the epididymis contain the strongly tortuous epididymal duct, some several centimeters in length. On the medial side of the testis the flexuosity decreases to give the deferent duct.

3 III. The **deferent duct (2)** passes through the inguinal space into the abdominal cavity and shortly before reaching the prostate it gives the merest indication of a fusiform increase in cross-section, the **ampulla of the deferent duct (1)**. Accessory glands of the ampulla are present. The narrow end-section of the deferent duct, the **ejaculatory duct (2)**, is surrounded
4 by prostatic parenchyma and opens on the **colliculus seminalis (20)**.

5 IV. The **prostate (19)** is the accessory genital gland in the male dog. Its ductules open adjacent to the colliculus seminalis. The prostate consists of a body or external part with two glandular lobes, and a weak disseminate or internal part. The glandular lobuli of the latter part are situated in the urethral wall, surrounded by the m. urethralis. Domestic mammals other than the dog also have vesicular glands and a bulbourethral gland.

b) Included in the **EXTERNAL GENITAL ORGANS OF THE DOG** are the penis, including the penile part of the urethra and the scrotum.
I. The **penis** consists of a root, body, and glans penis or head, and is formed from two different types of erectile bodies or corpora cavernosa. One is the soft, spongy, unpaired corpus spongiosum penis surrounding the urethra. The other is a hard, paired corpus rigidum penis also termed corpus cavernosum penis[1]. From prepared cross-sections of the penis one can study the tunica albuginea, trabeculae and caver-
6 nae of both types of erectile bodies. The erection of the penis results from a rise in blood pressure in the cavernae followed by an increased blood flow. This is promoted by dilation of the arteries of the penis and constriction of the venous efferent flow. In the region of the **root of the penis (4)**, the **corpus rigidum penis (14')** is fastened to parts of the ischia by means of two **crura penis (5)**. The corpus spongiosum penis lying between the crura, is strongly developed as the **bulb of the penis (6)** and is covered by the m. bulbospongiosus. In the region of the **body of the penis (13)**, both corpora rigida penis join each other midsagittally along the septum penis. In a roof-like manner they overlie the weakly developed corpus spongiosum penis with its contained penile part of the urethra. In the region of the **glans penis (24)** or head of penis, the
7 paired corpora rigida penis fuse to form an **os penis (23)**, grooved lon-

gitudinally. The unpaired **corpus spongiosum penis (14)** and its enclosed penile part of the urethra are situated in this osseous ventromedian groove.

The corpus spongiosum glandis develops from the corpus spongiosum ·penis and at the level of the caudal third of the os penis it produces the conspicuous **bulbus glandis (25)** and cranial to this the **pars longa glandis (26)**.

II. The **male urethra (3)** commences at the **internal urethral ostium (21)** at the neck of the urinary bladder and terminates at the **external urethral ostium (22)** on the tip of the penis. Due to the junction of urethra and deferent ducts at the ostia of deferent ducts, the pelvic urethra is a tube for the combined passage of urine, semen and seminal fluids. Caudal to the prostate, an erectile formation, present in the urethral wall, is surrounded by the circular, striated m. urethralis. The pelvic part of the urethra merges with the penile portion at a caudoconvex arch where a noticeable increase in circumference of the corpus spongiosum occurs.

c) The **INTERNAL GENITAL ORGANS OF THE BITCH** include the organ producing the sexual cells namely the ovary, and the organs of transport, the uterine tube and uterus. During pregnancy the uterus also affords protection to the developing fetus. The vagina and vestibule which are copulatory organs also belong to the internal genital organs.

I. The **ovary (37)** remains in the abdomen after its genesis. Depending 8
on the stage of the sexual cycle, contained **tertiary follicles, preovulatory** or **Graafian follicles**, or **corpora lutea** bulge from the ovarian surface. On sagittal section of the ovary a **medulla** or **vascular zone** and a **cortex** or **parenchymatous zone** are recognizable (see Appendix).

II. The **uterine tube (27)** or **salpinx** is about 10 cm long, commencing at 9
the **abdominal ostium of the uterine tube (30)** within the **infundibulum of the uterine tube (29)**. The fimbriae project, in part, from the ostium of the ovarian bursa. The **ampulla (28)** arches cranially around the ovary and on the lateral side it merges with the narrow, slightly tortuous **isthmus (36)**. This opens at the **uterine papilla (35)**, the uterine opening of the tube on the cranial extremity of the uterine horn.

III. Upon incision of the wall of the **uterus** one is able to define 10;1
peri-(31), **myo- (32)**, and **endometrium (33)** within its wall. The two **uterine horns (34)** unite to form the **body of the uterus (17)** which merges into the cervix of the uterus. At the cervix, the **uterine cavity** 12
(16) is continuous with the narrow **cervical canal (15)**, beginning and terminating at the internal and external uterine ostia respectively. The cervix projects into the cranial part of the vagina by means of the **portio vaginalis uteri (18)**, or vaginal portion of the uterus.

IV. Ventral to the portio, the **vagina** possesses the fornix of the vagina at its cranial end. The vaginal mucous membrane is arranged in longitudinal folds possessing in turn, small transverse folds, the rugae of the vagina. The caudal prolongation of the vagina, the **vestibule (10)**, lies between the lips of vulva and the external ostium of the urethra. At the ostium, an insignificant ring, the hymen, indicates the boundary between vagina and vestibule.

V. The **accessory genital glands**, the **minor vestibular glands** open in two rows adjacent to the external ostium of the urethra.

d) The **EXTERNAL GENITAL ORGANS OF THE BITCH** include the vulva with its labia pudenda or lips, and the clitoris.

I. The **labia pudenda (minora —11)** or lips surround the pudendal rima and with the clitoris form the external vulva.

II. The **clitoris** is the female homologue of the penis and likewise has a 13
root, body and glans (or head) with two types of erectile bodies. The unpaired corpus spongiosum clitoridis forms the glans clitoridis which may project slightly from the fossa clitoridis. The two remaining portions of the unpaired erectile body, the bulb and intermediate part, lie as the **vestibular bulb (9)**, isolated from the clitoris under cutaneous mucous membrane of the vestibule. The paired corpus rigidum of the clitoris is fastened to the ischiatic arch by the crura of the clitoris. It has a similar structure to that of the penis except that an os clitoridis is absent.

[1] In the official nomenclature, the term corpus cavernosum penis is listed instead of corpus rigidum penis.

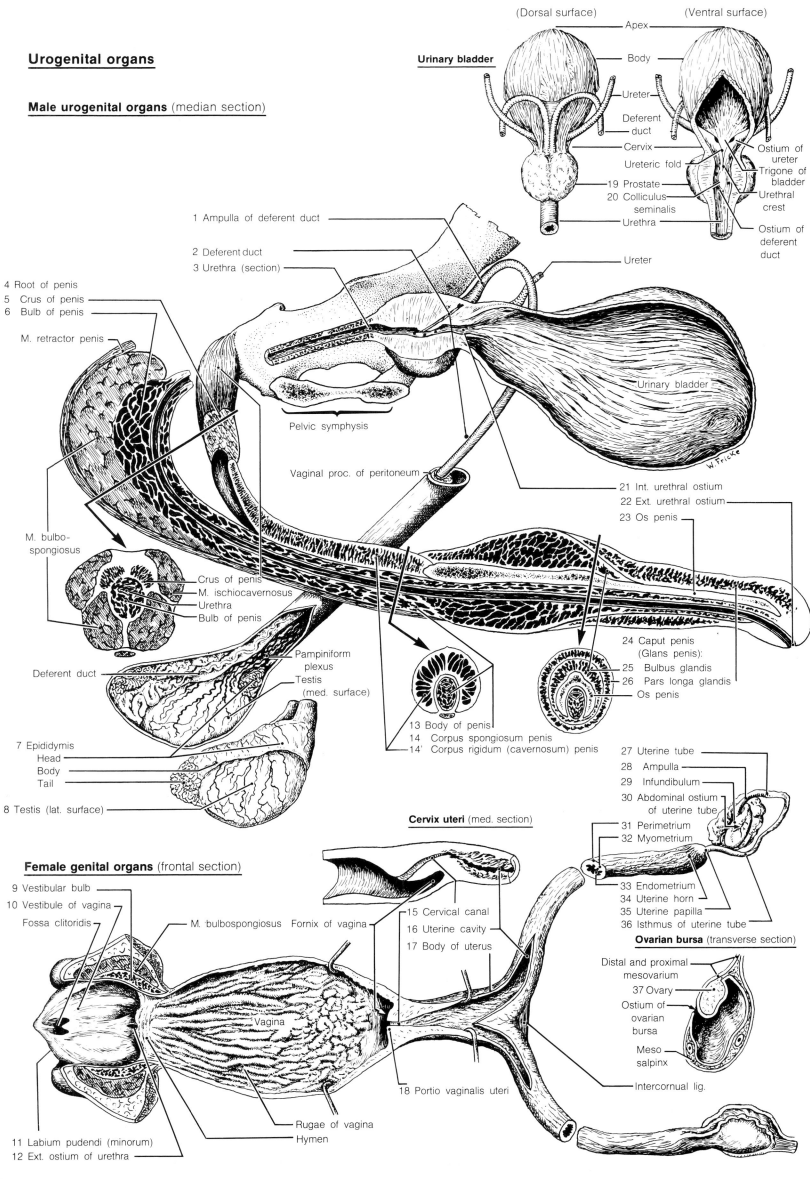

Urogenital organs

Male urogenital organs (median section)

Urinary bladder

(Dorsal surface)   (Ventral surface)

Apex
Body
Ureter
Deferent duct
Cervix
Ureteric fold
19 Prostate
20 Colliculus seminalis
Urethra

Ostium of ureter
Trigone of bladder
Urethral crest
Ostium of deferent duct

1 Ampulla of deferent duct
2 Deferent duct
3 Urethra (section)

4 Root of penis
5 Crus of penis
6 Bulb of penis

M. retractor penis

Pelvic symphysis

Vaginal proc. of peritoneum

Urinary bladder

Ureter

M. bulbo-spongiosus

Crus of penis
M. ischiocavernosus
Urethra
Bulb of penis

Deferent duct

Pampiniform plexus
Testis (med. surface)

21 Int. urethral ostium
22 Ext. urethral ostium
23 Os penis

24 Caput penis (Glans penis):
25 Bulbus glandis
26 Pars longa glandis
Os penis

13 Body of penis
14 Corpus spongiosum penis
14' Corpus rigidum (cavernosum) penis

7 Epididymis
Head
Body
Tail

8 Testis (lat. surface)

27 Uterine tube
28 Ampulla
29 Infundibulum
30 Abdominal ostium of uterine tube
31 Perimetrium
32 Myometrium

Cervix uteri (med. section)

33 Endometrium
34 Uterine horn
35 Uterine papilla
36 Isthmus of uterine tube

Female genital organs (frontal section)

9 Vestibular bulb
10 Vestibule of vagina
Fossa clitoridis

M. bulbospongiosus   Fornix of vagina

15 Cervical canal
16 Uterine cavity
17 Body of uterus

Vagina

Ovarian bursa (transverse section)

Distal and proximal mesovarium
37 Ovary
Ostium of ovarian bursa
Meso salpinx
Intercornual lig.

18 Portio vaginalis uteri

11 Labium pudendi (minorum)
12 Ext. ostium of urethra

Rugae of vagina   Hymen

W. Fricke

The smooth muscles are clearly demonstrated by displacing the free caudal border of the m. levator ani cranially and the anus and its surrounding sphincter muscles caudally.

a) The **SMOOTH MUSCLES OF THE PELVIC DIAPHRAGM** are situated between the pelvic viscera and the skeletal system.

I. The unpaired **m. rectococcygeus** is a concentration of the longitudinal muscle of the rectum ventromedial to the vertebral column and inserts onto the third coccygeal vertebra. It is also known as the caudo-anal ligament.

II. The paired **m. retractor penis** or **clitoridis** depending on the sex, commences deep to the m. levator ani at the first coccygeal vertebra. Caudolaterally on the rectum, it divides into three parts which lie in series and adjacent to one another dorsoventrally. The strong anal part inserts dorsolaterally onto the anus between the anal muscles. The moderately strong penile part meets the corresponding part from the other side caudally and in the midline at the perineum. Then both parts proceed cranially along the penis on its ventromedian aspect. The clitoridean part in the bitch is insignificant. The weak, sometimes absent rectal part curves around the ventral side of the rectum. By contrast, in the horse this part is a very pronounced loop-like suspensory apparatus of the rectum.

b) The **OSSEOUS PELVIC GIRDLE OR CINGULUM** (see also text illustration, left) consists of two ossa coxae, each having three individual bones in the puppy which fuse completely with one another in the adult. In the ventral midline both ossa coxae subscribe to the formation of the pelvic symphysis (1) whereas dorsally they are separated by the sacrum. Together with the sacrum and the first two coccygeal vertebrae, the ossa coxae participate in forming the bones of the pelvis.

Of the three bones of the os coxae which have fused with one another, the pubis and ischium border the obturator foramen (2) and both bones fuse with the third, the ilium, at the acetabulum (3). In the fetus and neonate a minute acetabular bone is detected in the acetabular fossa (4) and this fuses with the remaining three bones at an early stage. The strong ligament of the head of the femur runs between the acetabular fossa and the head of the femur. The acetabular notch (5), bridged across by the transverse ligament of the acetabulum, is a ventromedial interruption to the lunate surface (6) and its covering of articular cartilage. A connective tissue labrum or lip elevates the border of the acetabulum so that the acetabular fossa becomes deeper as a result. Dorsal to the acetabulum in the dog, is a blunt-edged ischiatic spine (7), which, being in the border area between ilium and ischium, is formed from both fused bones. Cranially, the ilium is bounded by the iliopubic eminence (34) on the pubis. The abutting pubes of both sides form the cranial section, and the ischia the caudal section, of the pelvic symphysis.

Os coxae (medial view)

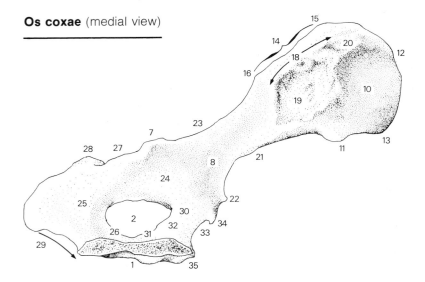

I. The **ilium**, by means of the body (8), participates in the formation of the acetabulum together with the bodies of pubis and ischium. Craniodorsal to the acetabulum is the caud. ventral iliac spine (9), the area of origin of the m. rectus femoris. It merges with the wing of the ilium (10). The alar spine (11), the area of insertion of the m. quadratus lumborum, lies on the cranioventral border of the wing of the ilium, three centimeters ventral from its transition to the iliac crest (12). The crest, situated dorsally, is covered with cartilage. At the angle between spine and crest is the tuber coxae (13) which, in the dog, is identical with the cran. ventral

iliac spine. The tuber sacrale (14) lies on the caudodorsal border of the wing of the ilium and extends from the cran. dorsal iliac spine (15) to the caud. dorsal iliac spine (16). The gluteal surface (17), situated laterally on the wing of the ilium, is the surface of origin of the gluteal muscles.

The indistinct ventral gluteal line (17') and the caudal gluteal line (17") commence at the levels of alar spine and caud. ventral iliac spine respectively. (An additional dorsal gluteal line is present in man.) The sacropelvic surface (18), situated on the medial aspect of the wing of the ilium, is divided into an ear-shaped auricular surface (19) for articulation with the sacrum, and a smooth iliac surface (20) dorsally for the insertion of the mm. longissimus dorsi and iliocostalis. The arcuate line (21) on the cranioventral part of the ilium, in common with the promontory of the sacrum dorsally and the pecten ossis pubis (33) ventrally, takes part in the formation of the terminal line. In the dog the very indistinct tubercle of the m. psoas minor (22) is present on a level approximately halfway along this line. Opposite the tubercle, on the caudodorsal border of the wing of the ilium, is the greater ischiatic notch (23). It extends from the caud. dorsal iliac spine of the tuber sacrale to the ischiatic spine and affords passage to the sciatic n.

II. The **ischium** extends from the acetabulum, where its body (24) is situated, to merge with the smooth ischiatic table (25). The (symphyseal) ramus of the ischium (26) attached medially, participates in forming the caudal part of the pelvic symphysis with its counterpart of the other side. The lesser ischiatic notch (27) faces laterodorsally. At the pronounced ischiatic tuberosity (28) caudolaterally, it is continuous with the concave ischiatic arch (29). The ischiatic arch and tuberosity, the sacrotuberous ligament and the third coccygeal vertebra form the margin of the caudal pelvic aperture.

III. As indicated above, the **pubis** also has its body (30) at the acetabulum. The caudal (or symphyseal) ramus (31) extends caudomedially in the direction of the pelvic symphysis, the cranial ramus (32) craniomedially. At the cranial pelvic aperture the cranial ramus forms the pecten ossis pubis (33) between the iliopubic eminence (34) laterally and the ventral pubic tubercle (35) midsagittally. The obturator groove lies on the inner surface of the pubis suggesting the direction of the obturator n.

c) The **PELVIC DIAMETERS** (see text illustration, right) are significant for the estimation of the natural birth canal in obstetrics. The **vertical diameter** is a vertical line from the cranial end of the pelvic symphysis to the vertebral column. The **true conjugate** measurement extends from the cranial end of the pelvic symphysis to the promontory of sacrum. The **dorsal transverse diameter** joins both ends of the wings of the sacrum, the **middle transverse diameter** both tubercles of m. psoas minor, and the **ventral transverse diameter** both iliopubic eminences.

d) In the dog, the **TERMINAL LINE** lies at the border of the abdomen. In man the line marks the boundary between the 'large' and the 'small' pelvis, a division absent in domestic mammals. In both man and domestic mammals the terminal line is drawn from the promontory of sacrum to the arcuate line of ilium and pecten ossis pubis.

e) The **FLOOR OF THE PELVIS,** in contrast to that in erect man, is formed by pubis and ischium and the contained obturator foramen.

Diameters of pelvis

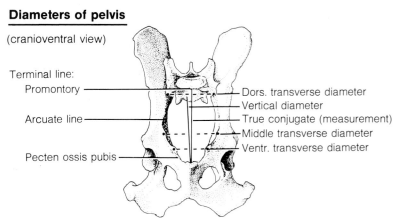

(cranioventral view)

Terminal line:
Promontory ——————— Dors. transverse diameter
——— Vertical diameter
Arcuate line ——————— True conjugate (measurement)
——— Middle transverse diameter
——— Ventr. transverse diameter
Pecten ossis pubis ———————

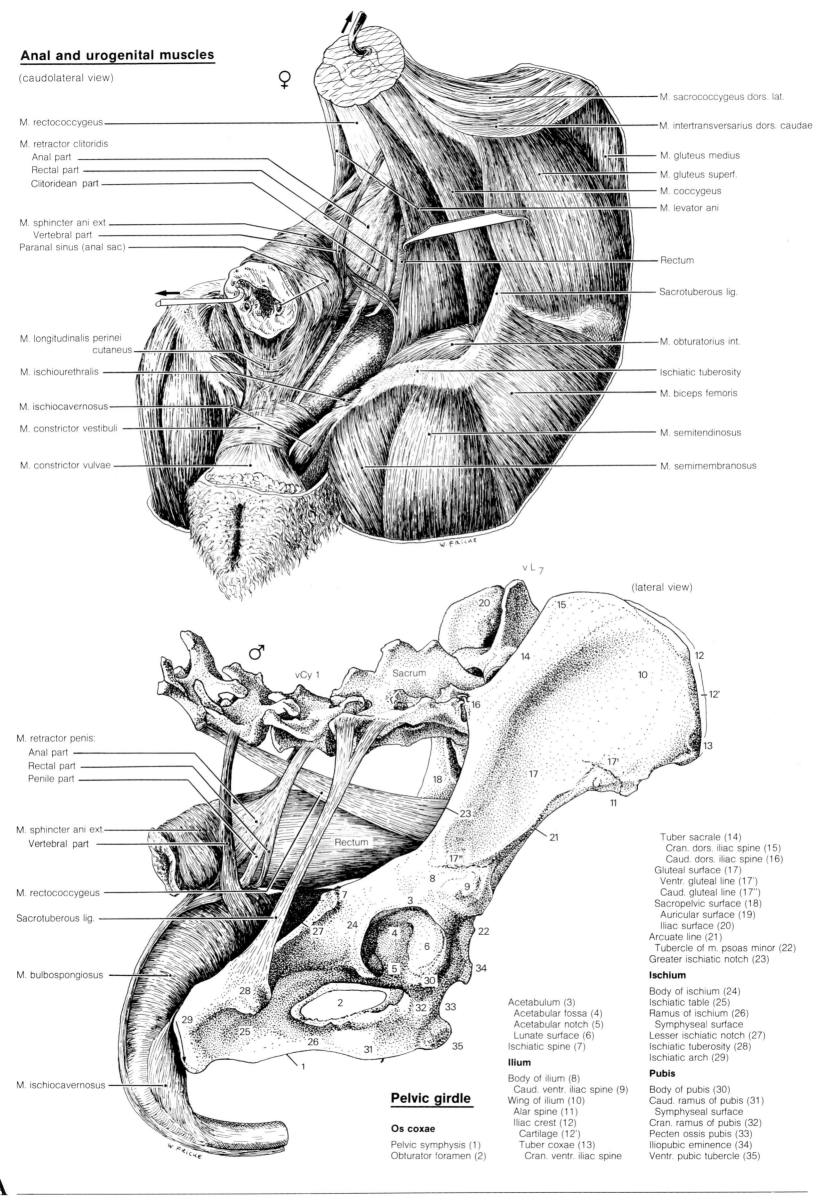

Anal and urogenital muscles

(caudolateral view)

♀

M. rectococcygeus

M. retractor clitoridis
Anal part
Rectal part
Clitoridean part

M. sphincter ani ext
Vertebral part
Paranal sinus (anal sac)

M. longitudinalis perinei cutaneus

M. ischiourethralis

M. ischiocavernosus

M. constrictor vestibuli

M. constrictor vulvae

M. sacrococcygeus dors. lat.

M. intertransversarius dors. caudae

M. gluteus medius

M. gluteus superf.

M. coccygeus

M. levator ani

Rectum

Sacrotuberous lig.

M. obturatorius int.

Ischiatic tuberosity

M. biceps femoris

M. semitendinosus

M. semimembranosus

W. FRICKE

vL7

(lateral view)

♂

vCy 1

Sacrum

M. retractor penis:
Anal part
Rectal part
Penile part

M. sphincter ani ext.
Vertebral part

M. rectococcygeus

Sacrotuberous lig.

Rectum

M. bulbospongiosus

M. ischiocavernosus

W. FRICKE

Pelvic girdle

Os coxae

Pelvic symphysis (1)
Obturator foramen (2)

Acetabulum (3)
Acetabular fossa (4)
Acetabular notch (5)
Lunate surface (6)
Ischiatic spine (7)

Ilium

Body of ilium (8)
Caud. ventr. iliac spine (9)
Wing of ilium (10)
Alar spine (11)
Iliac crest (12)
Cartilage (12')
Tuber coxae (13)
Cran. ventr. iliac spine

Tuber sacrale (14)
Cran. dors. iliac spine (15)
Caud. dors. iliac spine (16)
Gluteal surface (17)
Ventr. gluteal line (17')
Caud. gluteal line (17")
Sacropelvic surface (18)
Auricular surface (19)
Iliac surface (20)
Arcuate line (21)
Tubercle of m. psoas minor (22)
Greater ischiatic notch (23)

Ischium

Body of ischium (24)
Ischiatic table (25)
Ramus of ischium (26)
Symphyseal surface
Lesser ischiatic notch (27)
Ischiatic tuberosity (28)
Ischiatic arch (29)

Pubis

Body of pubis (30)
Caud. ramus of pubis (31)
Symphyseal surface
Cran. ramus of pubis (32)
Pecten ossis pubis (33)
Iliopubic eminence (34)
Ventr. pubic tubercle (35)

C. 7. S. 1: SKELETON OF PELVIC LIMB

> *Like the thoracic limb, the pelvic limb is also connected to a cingulum, namely the pelvic girdle. To study the similarities in construction between the two, the explanations given on p. 7 are to be repeated.*

a) Proximally, the **FEMUR** consists of a head, followed by a thin neck, prominent muscle protuberances, and a body which merges distally with the trochlea of the femur. The head of the femur (1) which is covered by articular cartilage, bears a cartilage-free fovea capitis (2) at its center.

This is the site of insertion of the round ligament of the head of the femur which is anchored in the acetabulum. The neck of the femur (3) is a distinct tapering feature between the head and the muscular protube-rances. Of these, the large greater trochanter (4) laterally, serves as the site of insertion of the gluteal muscles. Medial to the base of the greater trochanter is the trochanteric fossa (5), a hollow for the attachment of the deep muscles of the hip or coxal articulation. Medial to the trochan-teric fossa is the lesser trochanter (6) for the insertion of the m. iliopsoas.

In the dog, the third trochanter (7) is an indistinct projection distal to the greater trochanter. Caudally on the elongated body of the femur (8) is the facies aspera (9), the rough surface of insertion of the mm. adductores magnus and brevis. It is bounded by the lateral (10) and medial (11) lips. Distal and adjacent to the facies aspera, the popliteal surface (12) is bounded laterally by the lateral supracondylar tuberosity (13), a rough area for the origin of the superficial digital flexor muscles.

Caudally, the medial condyle (14) bears the facies articularis sesamoidea medialis (15) for articulation with the medial sesamoid bone of the m. gastrocnemius (64). Proximal to the condyle is the projecting medial epicondyle (16). To this is attached the medial collateral ligament of the femorotibial articulation. The lateral condyle (17), its facies articularis sesamoidea lateralis (18), and the lateral epicondyle (19) present a similar configuration. Between the condyles is a deep intercondylar fossa (20) which is continuous cranially with the trochlea of femur (21). Between the lips, the trochlea provides a gliding surface for the patella (69).

b) The **BONES OF THE LEG** are the tibia and fibula which articulate with each other proximally and distally. In contrast to the femur, the tibia lacks a head and a neck. At the tarsus the distal ends of tibia and fibula project beyond their articular surfaces as the medial and lateral malleoli respectively.

I. At the **tibia**, the menisci of the femorotibial articulation are fastened to the osseous proximal articular surface (22) by ligaments. The medial condyle (23) is separated from the lateral condyle (25) by the intercondy-lar eminence (24) projecting proximomedian, while the lateral condyle of the tibia bears the facies articularis fibularis (26) for articulation with the head of the fibula. Cranial to this is the extensor groove (27) for the passage of the tendon of the m. extensor digitalis longus which arises from the lateral condyle of femur. Proximally at the insertion of the patellar ligament (quadriceps tendon) the body of the tibia (28) posses-ses a tibial tuberosity (29) which merges with the cranial border (29') distally. At the distal end of the tibia the medial malleolus (31) projects beyond the cochlea of the tibia (30) which articulates with the tarsus.

II. Concerning the **fibula**, the head (32) and its articular surface (33) are continuous with the body (34). Distally this is followed by the lateral malleolus (35). Both bones of the leg enclose the crural interosseous space (36).

c) The **BONES OF THE TARSUS** are arranged in three irregular rows. Of the two bones of the proximal row, the talus (37) has a body (38) proximally, a trochlea (39), a neck (40) and a head (41) distally. The calcaneus (42) situated laterally, has a calcaneal tuber (43) projec-ting further proximally. This makes possible its function as an 'attachment-lever' for the extensor muscles of the tarsus. The tendon of the m. flexor digitalis lateralis passes over the sustentaculum tali (44) supporting the talus, and just distally, it unites with the tendon of the m. flexor digitalis medialis to give the deep digital flexor tendon. The central tarsal bone (45) is the only bone of the middle row, and into it 'projects' the particularly large tarsal bone IV. Tarsal bones I to IV (46) form the distal row of the bones of the tarsus.

d) Each of the **METATARSAL BONES I - V** consists, in general, of a base (47) proximally, a body (48), and a head (49) situated distally. A short first metatarsal bone may be absent. Occasionally it is divided into two, the proximal portion fusing with the first tarsal bone.

e) Concerning the **BONES OF THE DIGITS**, the commentary on those of the thoracic limb is generally valid (see p. 7).

Each of the proximal (50) and middle (51) phalanges has a base (52), a body (54) and a head (55) situated distally. The indistinct flexor tubero-sity (53), situated on the proximopalmar surface of the middle phalanx of the third digit is the insertion of the superficial digital flexor tendon of that digit. Each distal phalanx (56) has an articular surface (57) proxi-mally, and an ill-defined extensor process (58) for the attachment of the digital extensor tendon. On its proximoplantar surface, the distal pha-lanx has a flexor tuberosity (59) for the attachment of the deep digital flexor tendon. The sharp-edged unguicular crest (60) projects over the unguicular groove (61). The distal part of the distal phalanx, the ungui-cular process (62), is surrounded by the horn of the claw.

When only the claw of the first digit remains and its supporting phalan-ges are missing, it is referred to as a dewclaw (paraunguicula 63).

f) Of the **SESAMOID BONES** those associated with the origin of the m. gastrocnemius (64) on the femoral condyles have been mentioned. The sesamoid bone of the m. popliteus (65) lies caudolateral to the intercondylar eminence at the transition between the tendon of origin and the muscle proper. Each pair of proximal sesamoid bones (66) lies on the plantar aspect of its respective metacarpophalangeal articulation.

Each distal sesamoid bone (67) is situated on the plantar aspect of the distal phalangeal articulation. Each dorsal sesamoid bone (68) is situated on the dorsal aspect of the metacarpophalangeal articulation. The patel-la (69) is the largest sesamoid bone of the body and lies in the tendon of insertion of the m. quadriceps femoris. It has an articular surface covered with cartilage for articulation with the trochlea of the femur.

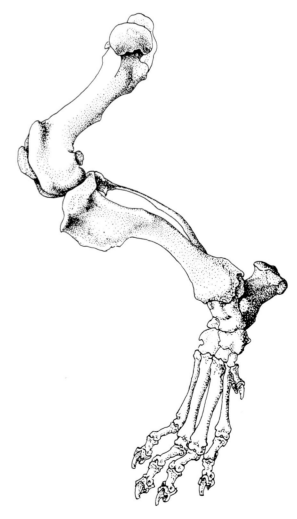

Pelvic limb (medial view)

(Artésien-Normand Basset)

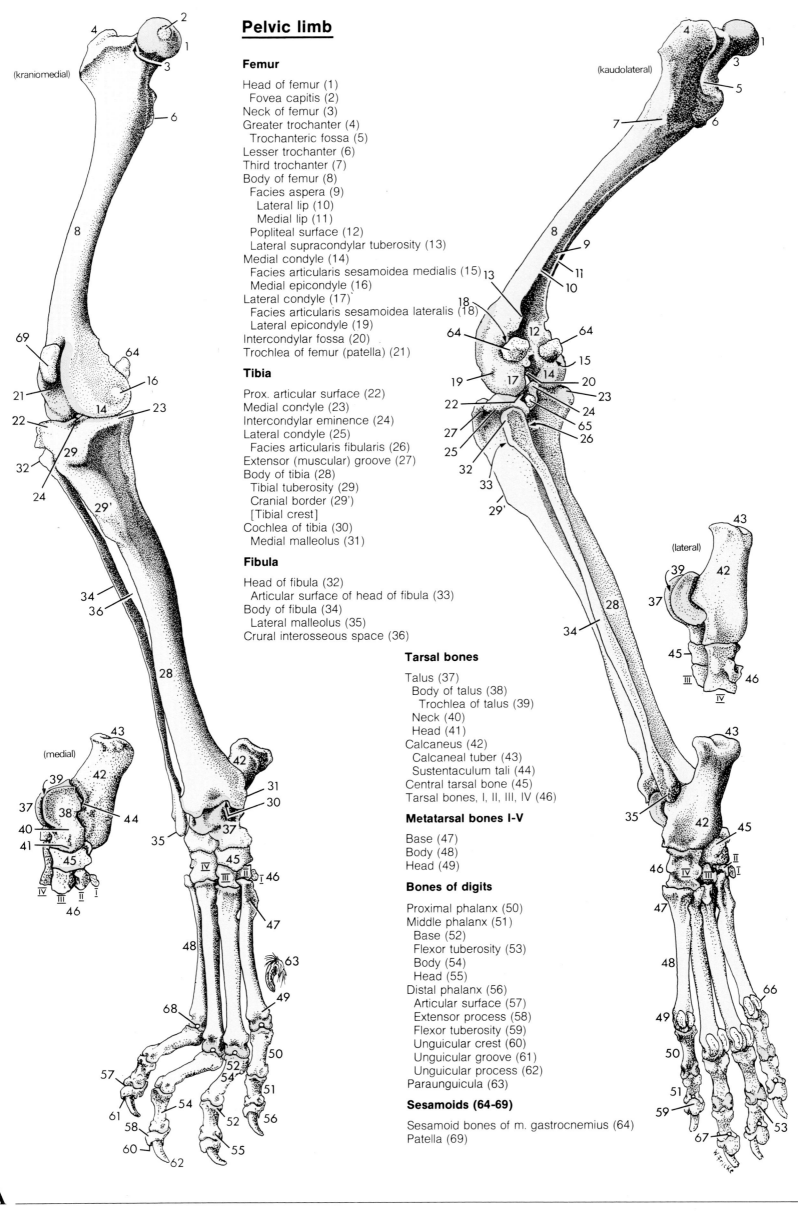

Pelvic limb

Femur

Head of femur (1)
  Fovea capitis (2)
Neck of femur (3)
Greater trochanter (4)
  Trochanteric fossa (5)
Lesser trochanter (6)
Third trochanter (7)
Body of femur (8)
  Facies aspera (9)
    Lateral lip (10)
    Medial lip (11)
  Popliteal surface (12)
  Lateral supracondylar tuberosity (13)
Medial condyle (14)
  Facies articularis sesamoidea medialis (15)
  Medial epicondyle (16)
Lateral condyle (17)
  Facies articularis sesamoidea lateralis (18)
  Lateral epicondyle (19)
Intercondylar fossa (20)
Trochlea of femur (patella) (21)

Tibia

Prox. articular surface (22)
Medial condyle (23)
Intercondylar eminence (24)
Lateral condyle (25)
  Facies articularis fibularis (26)
Extensor (muscular) groove (27)
Body of tibia (28)
  Tibial tuberosity (29)
  Cranial border (29')
  [Tibial crest]
Cochlea of tibia (30)
  Medial malleolus (31)

Fibula

Head of fibula (32)
  Articular surface of head of fibula (33)
Body of fibula (34)
  Lateral malleolus (35)
Crural interosseous space (36)

Tarsal bones

Talus (37)
  Body of talus (38)
    Trochlea of talus (39)
  Neck (40)
  Head (41)
Calcaneus (42)
  Calcaneal tuber (43)
  Sustentaculum tali (44)
Central tarsal bone (45)
Tarsal bones, I, II, III, IV (46)

Metatarsal bones I-V

Base (47)
Body (48)
Head (49)

Bones of digits

Proximal phalanx (50)
Middle phalanx (51)
  Base (52)
  Flexor tuberosity (53)
  Body (54)
  Head (55)
Distal phalanx (56)
  Articular surface (57)
  Extensor process (58)
  Flexor tuberosity (59)
  Unguicular crest (60)
  Unguicular groove (61)
  Unguicular process (62)
Paraunguicula (63)

Sesamoids (64-69)

Sesamoid bones of m. gastrocnemius (64)
Patella (69)

The skin is removed from the pelvic limb as far distally as the tarsus, taking care to preserve the cutaneous veins (see pp. 32A and 33A). Directly after they are exposed, the following muscles are transected at the junctions of their middle and distal thirds: the mm. gluteus superficialis, gluteus medius, and biceps femoris (gluteobiceps). As a variation of the usual procedure, the nerve supply is studied immediately after the muscles are transected and likewise is described with each muscle group. The muscles of the hip include the m. tensor fasciae latae and three groups of muscles namely rump muscles, caudal muscles of thigh and deep muscles of the hip. Each muscle group consists of four muscles.

a) The **SUPERFICIAL LAMINA OF DEEP FASCIA** in the gluteal region, the **gluteal fascia (14)**, is a derivative of the thoracolumbar fascia. It continues towards the thigh as the multilaminate fascia lata which is tensed by the radiating m. tensor fasciae latae.

The **m. tensor fasciae latae (3)** arises on the tuber coxae and its immediate vicinity. The main cranial portion of the muscle radiates in the superficial lamina of the fascia lata, which fuses distally with the patellar ligament (quadriceps tendon). The fan-shaped accessory caudal portion merges with the deep lamina of fascia lata and this ends on the lateral lip of the facies aspera of the femur. **Function:** Tenses the fascia lata and flexes the hip. **Innervation:** Cranial gluteal n.

b) The **RUMP MUSCLES** arise on the gluteal surface of the wing of the ilium (the mm. gluteus medius and profundus) or on sacrum and sacrotuberous ligament (the mm. piriformis and gluteus superficialis). All insert onto the greater trochanter of femur. **Function:** Extension of hip and abduction of limb.

The **m. gluteus superficialis (2)** arises more caudally than the others of the group and inserts further distally at the base of the greater trochanter. This is evident after transection and reflection of the muscle parts.

The **m. gluteus medius (1)** arises dorsally on the gluteal surface of the ilium between the ventral gluteal line and the crest of the ilium.

The **m. piriformis (6[1])** lies deep to the m. gluteus medius and inserts in common with it on the greater trochanter.

The **m. gluteus profundus (7)** has its large area of origin between the ventral and caudal gluteal lines. Its converging fiber bundles insert on the greater trochanter.

Innervation is from cranial and caudal gluteal nn. both of which pass over the greater ischiatic notch. The **caudal gluteal n. (17)** enters the medial surface of the m. gluteus superficialis and exceptionally, the m. piriformis. The **cranial gluteal n. (5)** passes between the mm. gluteus medius and — profundus. It supplies them and generally the m. piriformis before ending in the m. tensor fasciae latae.

c) The **CAUDAL MUSCLES[2] OF THE THIGH** take origin from the ischiatic tuberosity, the sacrotuberous ligament being an additional origin of the m. biceps femoris and the sole origin of the m. abductor cruris caudalis. **Function:** Extension of hip and flexion of knee. In so far as the mm. biceps femoris and semitendinosus insert on the tuber calcanei by means of the tractus calcaneus, they extend the tarsus.

The **m. biceps femoris (4)** terminates by a broad aponeurosis on the patellar ligament (quadriceps tendon), the cranial border of tibia and the crural fascia. In addition, it ends on the calcaneal tuber by means of the tractus calcaneus lateralis. In the distal third of the leg it fuses with the tractus calcaneus medialis associated with m. semitendinosus and m. gracilis. After transecting the muscle it becomes evident that the tractus lateralis receives a fascial reinforcement from the lateral lip of the facies aspera of the femur. (Here the term tract or tractus is used instead of tendon.)

The strip-like **m. abductor cruris caudalis (20)** arises on the distal part of the sacrotuberous ligament under the m. biceps femoris. Together with the caudal border of the m. biceps, its insertion radiates into the crural fascia.

The **m. semitendinosus (16)** situated laterally, inserts onto the proximomedial part of the tibia and continues to the calcaneal tuber in the tractus calcaneus medialis, in common with the m. gracilis.

The **m. semimembranosus (15)** craniomedially has two thick muscle bellies. The cranial inserts onto the medial condyle of femur, the caudal onto the medial condyle of tibia distal to the flexor aspect of the knee. The **innervation** of the muscle group is from the tibial and common fibular nn., arising as sectional branches of the sciatic n. The proximal muscle branches of the **tibial n. (11)** innervate the mm. biceps femoris, semitendinosus and semimembranosus. In addition, a branch of the caudal gluteal n. enters the cranial part of the m. biceps femoris arising from the sacrotuberous ligament. Hence the synonym gluteobiceps is a valid one. Only the m. abductor cruris caudalis is innervated by the **common fibular n. (12)**. The **musculocutaneous r. (13)** of the nerve ends on the skin. Before it divides, the **sciatic n. (18)** passes dorsal to the greater ischiatic notch, lateral to the neck of the femur and its covering

of deep muscles of the hip. The nerve innervates the hip joint and then, between the mm. biceps femoris and adductor magnus, divides into the thicker tibial and the thinner common fibular n. Initially these two nerves are in a common connective tissue sheath, becoming distinct only in the distal half of the thigh.

d) The **DEEP MUSCLES OF THE HIP**, also known as the inner pelvic muscles or 'small pelvic association' (see also text illustration) are the supinators of the thigh.

Innervation comes from the rotatorius n. arising from the sciatic n. at the caudal border of the m. gluteus profundus. The sole exception to this is the m. obturator externus supplied by the obturator n. and not the sciatic n. The **origin** of the muscle group is on the os coxae adjacent to the obturator foramen. The **insertion** is at the trochanteric fossa of the femur. The four muscles lie in a cranial to caudoventral sequence as listed.

The **mm. gemelli (8)** arise in the vicinity of the lesser ischiatic notch bounded cranially by the caudal border of the m. gluteus profundus. The **m. obturator internus (19)** has its area of origin within the pelvis on the medial border of the obturator foramen, its strong tendon running dorsally over the indented midregion of the fan-shaped mm. gemelli.

The **m. obturator externus (9)** has a similar area of origin from the medial border of the obturator foramen, but lies outside the pelvis (see text illustration p. 32). Only its strong deeply situated tendon of insertion is visible at the caudal border of the mm. gemelli.

Innervation: Obturator n.
The **m. quadratus femoris (10)** arises ventromedial to the ischiatic tuberosity and inserts onto the ventral border of the trochanteric fossa as far as the third trochanter.

Deep muscles of hip joint (lateral view)

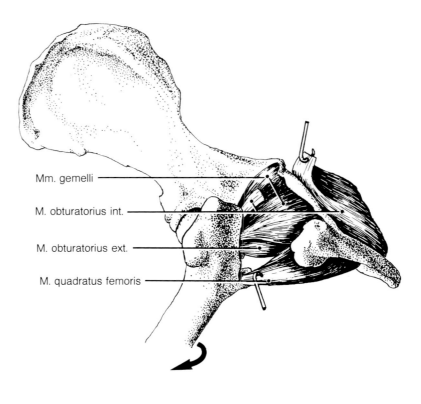

Mm. gemelli
M. obturatorius int.
M. obturatorius ext.
M. quadratus femoris

---

[1]) The m. piriformis was homologized after Henning (1965).
[2]) The caudal muscles are also called the hamstring muscles.

Muscles of the hip joint (coxal articulation)

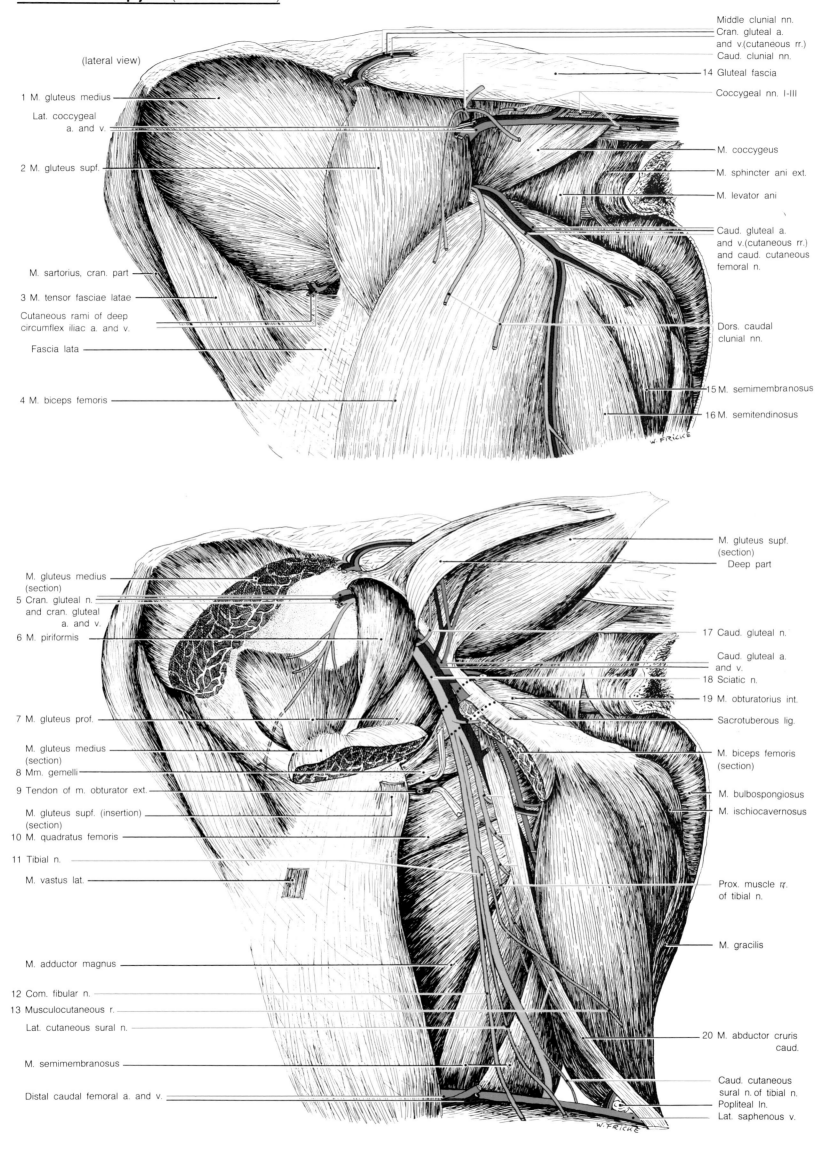

(lateral view)

1 M. gluteus medius

Lat. coccygeal a. and v.

2 M. gluteus supf.

M. sartorius, cran. part

3 M. tensor fasciae latae

Cutaneous rami of deep circumflex iliac a. and v.

Fascia lata

4 M. biceps femoris

Middle clunial nn.
Cran. gluteal a. and v.(cutaneous rr.)
Caud. clunial nn.

14 Gluteal fascia

Coccygeal nn. I–III

M. coccygeus

M. sphincter ani ext.

M. levator ani

Caud. gluteal a. and v.(cutaneous rr.) and caud. cutaneous femoral n.

Dors. caudal clunial nn.

15 M. semimembranosus

16 M. semitendinosus

M. gluteus medius (section)

5 Cran. gluteal n. and cran. gluteal a. and v.

6 M. piriformis

7 M. gluteus prof.

M. gluteus medius (section)

8 Mm. gemelli

9 Tendon of m. obturator ext.

M. gluteus supf. (insertion) (section)

10 M. quadratus femoris

11 Tibial n.

M. vastus lat.

M. adductor magnus

12 Com. fibular n.

13 Musculocutaneous r.

Lat. cutaneous sural n.

M. semimembranosus

Distal caudal femoral a. and v.

M. gluteus supf. (section)
Deep part

17 Caud. gluteal n.

Caud. gluteal a. and v.

18 Sciatic n.

19 M. obturatorius int.

Sacrotuberous lig.

M. biceps femoris (section)

M. bulbospongiosus

M. ischiocavernosus

Prox. muscle rr. of tibial n.

M. gracilis

20 M. abductor cruris caud.

Caud. cutaneous sural n. of tibial n.
Popliteal ln.
Lat. saphenous v.

1A

> *After being exposed, the m. rectus femoris is transected two centimeters distal to its origin in order to obtain a general view of the four heads of the m. quadriceps femoris. In addition, the caudal part of the m. sartorius is either transected or pushed to one side.*

a) The **OBTURATOR N.** (4) runs to the deep surface of the pelvis through a lacuna in the m. levator ani, then in the obturator groove of the pubis to the obturator foramen. It passes through the foramen, innervates the m. obturator externus and provides branches for the adductor muscles of thigh namely the mm. adductor magnus, adductor brevis and pectineus. The nerve then runs to the medial surface of the thigh to innervate the m. gracilis which likewise acts as an adductor. The obturator n. ends in a very weak sensory branch medial to the knee (not illustrated).

The **m. gracilis** (10) arises along the pelvic symphysis, its aponeurosis of origin fusing with that of the other side to give a perpendicular, tendinous 'double lamina', the **symphyseal tendon** (7). This becomes evident after scraping away the fleshy fibers of origin of the m. adductor magnus. At the knee the m. gracilis merges with the crural fascia and participates in the formation of the tractus calcaneus medialis.

The **m. adductor magnus** (8) arises over a wide part of the surface of the symphyseal tendon and from the pubis and the ischium along the pelvic symphysis. It has a broad fleshy insertion onto the facies aspera of the femur.

The **m. adductor brevis** (5) takes origin from the ventral tubercle of the pubis and inserts in common with the m. adductor magnus on the facies aspera. It can be identified on the basis of its typical position at the bifurcation of the obturator n. The nerve branch to m. adductor magnus crosses the m. adductor brevis dorsally, that to m. gracilis ventrally. Proximally the m. adductor brevis is adjacent to the cranial aspect of m. adductor magnus.

1 The **m. pectineus (et adductor longus —6)** must be considered as a double muscle on phylogenetic grounds since it is fused in part to the m. adductor longus. Besides its nerve supply from the obturator n., it often has an additional innervation from the saphenous n. or the femoral ramus of the genitofemoral n. The muscle takes origin from the iliopubic eminence and from the caudal angle of the external inguinal ring. It terminates distally on the medial lip of the facies aspera.

b) In the abdominal region, the **FEMORAL N.** (1) is embedded in the m. iliopsoas and accompanies it through the lacuna musculorum. Then the elongated **saphenous n.** (9) branches from the parent nerve and supplies the knee and the medial side of the pelvic limb as far distally as the first and second digits. After the saphenous n. is given off, the femoral n. supplies m. sartorius and the four heads of the m. quadriceps femoris.

The **m. sartorius** (2) is divided into a cranial and a caudal part, taking origin from the iliac crest and the cranial border of the wing of the ilium respectively. The two parts radiate distally in the crural fascia. **Function:** Adduction of limb and extension of knee.

The **m. quadriceps femoris** (A) consists of the **mm. vastus lateralis (B)**, - **intermedius (C)** and - **medialis (3,D)** arising from the femur, and the **m. rectus femoris (E)** from the ventral caudal iliac spine, craniodorsal to the acetabulum. Taking into consideration its origin from the pelvis, the m. rectus femoris is able to function as an extensor of the knee and a flexor of the hip at the same time. After reflecting the transected parts of the m. rectus femoris, all four heads of m. quadriceps are seen. The strong tendon of the muscle surrounds the patella and terminates 2 in the **patellar lig. (F)** (preferably termed quadriceps tendon). This inserts onto the tibial tuberosity after passing over the distal infrapatellar (subtendineal) bursa. (see p. 3A—7.)

c) The **FEMORAL SPACE, (FEMORAL CANAL[1])** lies in the femoral triangle bounded at its base by the inguinal ligament, cranially by the caudal part of the m. sartorius, and caudally by the m. pectineus (et adductor longus). At a deep level, the grooved or canaliculate femoral space is bounded by the m. iliopsoas, while superficially it is covered by medial femoral fascia. The abdominal access to the space, namely the femoral ring, is closed over solely by peritoneum and transverse fascia.

The space accommodates the femoral a. and v. and the saphenous n. At the apex of the triangle distally, the blood vessels of the thigh proceed deeply from the medial surface of the space into the popliteal region.

The vessels leave the femoral space through a lacuna between the m. adductor magnus and the m. semimembranosus or through the 'adductor cleft' which can be present in the m. adductor magnus.

d) The **MEDIAL SAPHENOUS V.** (11), a large medial cutaneous vein, arises from the femoral v. at the apex of the femoral triangle. It accompanies the saphenous a. and n. to the knee and then the leg where it divides into a **cranial** (12) and a **caudal ramus** (13).

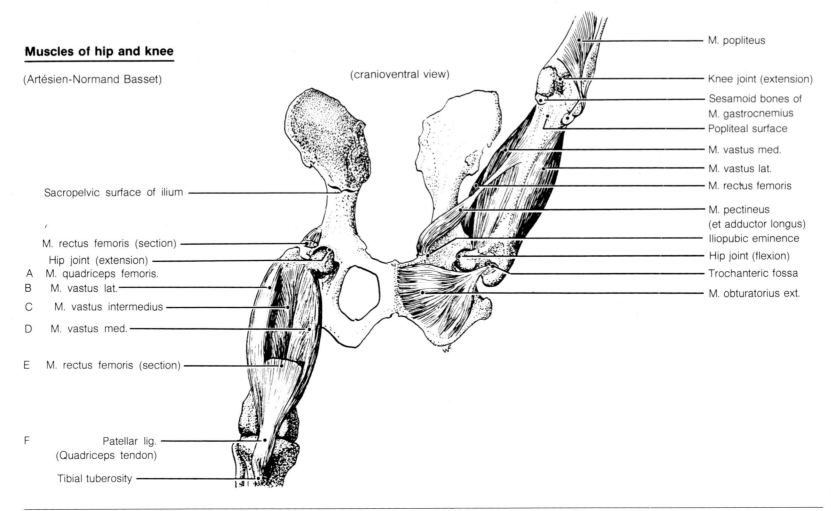

Muscles of hip and knee

(Artésien-Normand Basset)

(cranioventral view)

- M. popliteus
- Knee joint (extension)
- Sesamoid bones of M. gastrocnemius
- Popliteal surface
- M. vastus med.
- M. vastus lat.
- M. rectus femoris
- M. pectineus (et adductor longus)
- Iliopubic eminence
- Hip joint (flexion)
- Trochanteric fossa
- M. obturatorius ext.

Sacropelvic surface of ilium

M. rectus femoris (section)
Hip joint (extension)
A   M. quadriceps femoris.
B   M. vastus lat.
C   M. vastus intermedius
D   M. vastus med.

E   M. rectus femoris (section)

F   Patellar lig. (Quadriceps tendon)

Tibial tuberosity

[1]) Femoral space and femoral canal are perceived by many veterinary anatomists to be synonymous. In human medicine the femoral canal is defined in a completely different manner as the canal through which herniation occurs.

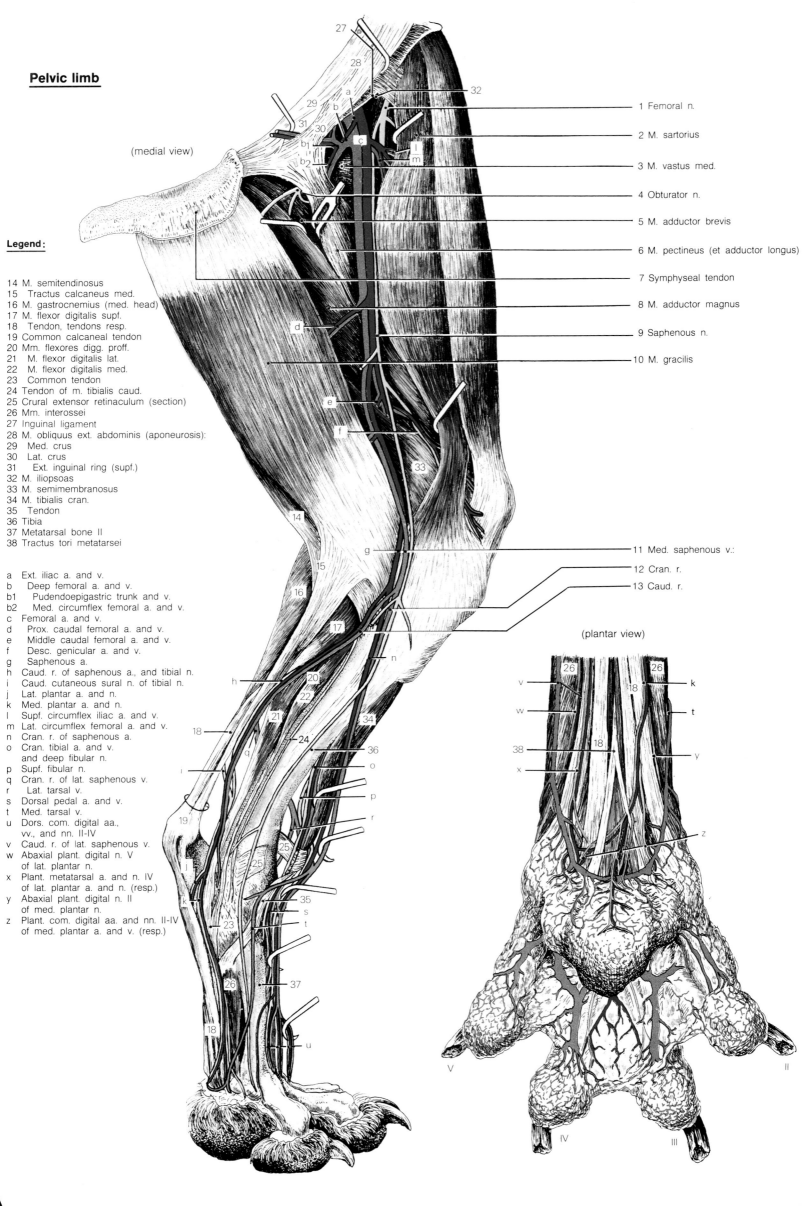

Pelvic limb

(medial view)

Legend:

14 M. semitendinosus
15 Tractus calcaneus med.
16 M. gastrocnemius (med. head)
17 M. flexor digitalis supf.
18 Tendon, tendons resp.
19 Common calcaneal tendon
20 Mm. flexores digg. proff.
21 M. flexor digitalis lat.
22 M. flexor digitalis med.
23 Common tendon
24 Tendon of m. tibialis caud.
25 Crural extensor retinaculum (section)
26 Mm. interossei
27 Inguinal ligament
28 M. obliquus ext. abdominis (aponeurosis):
29 Med. crus
30 Lat. crus
31 Ext. inguinal ring (supf.)
32 M. iliopsoas
33 M. semimembranosus
34 M. tibialis cran.
35 Tendon
36 Tibia
37 Metatarsal bone II
38 Tractus tori metatarsei

a Ext. iliac a. and v.
b Deep femoral a. and v.
b1 Pudendoepigastric trunk and v.
b2 Med. circumflex femoral a. and v.
c Femoral a. and v.
d Prox. caudal femoral a. and v.
e Middle caudal femoral a. and v.
f Desc. genicular a. and v.
g Saphenous a.
h Caud. r. of saphenous a., and tibial n.
i Caud. cutaneous sural n. of tibial n.
j Lat. plantar a. and n.
k Med. plantar a. and n.
l Supf. circumflex iliac a. and v.
m Lat. circumflex femoral a. and v.
n Cran. r. of saphenous a.
o Cran. tibial a. and v.
   and deep fibular n.
p Supf. fibular n.
q Cran. r. of lat. saphenous v.
r Lat. tarsal v.
s Dorsal pedal a. and v.
t Med. tarsal v.
u Dors. com. digital aa.,
   vv., and nn. II-IV
v Caud. r. of lat. saphenous v.
w Abaxial plant. digital n. V
   of lat. plantar n.
x Plant. metatarsal a. and n. IV
   of lat. plantar a. and n. (resp.)
y Abaxial plant. digital n. II
   of med. plantar n.
z Plant. com. digital aa. and nn. II-IV
   of med. plantar a. and v. (resp.)

1 Femoral n.
2 M. sartorius
3 M. vastus med.
4 Obturator n.
5 M. adductor brevis
6 M. pectineus (et adductor longus)
7 Symphyseal tendon
8 M. adductor magnus
9 Saphenous n.
10 M. gracilis
11 Med. saphenous v.:
12 Cran. r.
13 Caud. r.

(plantar view)

C. 7. S. 4: LATERAL SAPHENOUS VEIN, FIBULAR NERVE, FLEXORS OF TARSUS AND EXTENSORS OF DIGITS, TIBIAL NERVE, EXTENSORS OF TARSUS AND FLEXORS OF DIGITS, AND M. POPLITEUS

> *The skin of the limb is removed as far as the distal end of the metatarsus. To demonstrate the insertions of digital extensor and flexor tendons, one also removes skin from the entire second digit. Then the mm. semitendinosus and gracilis are transected 2 cm proximal to their distal myotendinous junctions to display the origins of the leg muscles. After its exposure, the medial head of the m. gastrocnemius is detached from its origin and the m. flexor digitalis superficialis separated from its covering of m. gastrocnemius. To display the common fibular n. and the origins of digital extensors and tarsal flexors, a window is inserted in the broad tendon of insertion of the m. biceps femoris as shown in the illustration opposite. Thus only two two-centimeter-wide strips of tendinous insertion remain, one to the patellar ligament proximally, the other to the crural fascia distally. One preserves the tractus calcaneus lateralis. The leg muscles are displayed and the terminal tendons of digital extensor and flexor muscles are dissected to at least halfway along the metatarsus distally, to distinguish clearly between one group and the other.*

a) The **LATERAL SAPHENOUS V. (27)** arises from the distal caudal femoral v. in the popliteal region. In the distal third of the leg it divides into a **cranial (28)** and a **caudal ramus (29)**. The cranial rami of lateral and medial saphenous vv. anastomose twice on the dorsal aspect of the pes (see p. 34a). The proximal anastomosis occurs on the flexor aspect of tarsus and the distal halfway along the metatarsus. The distal anastomosis is the dorsal superficial arch which provides the dorsal common digital vv. II to IV. The caudal rami of the lateral and medial saphenous vv. also anastomose on the plantar surface at the distal end of the metatarsus. Likewise, this anastomosis is called the plantar superficial arch from which the common plantar digital vv. II to IV are derived.

1 b) The **COMMON FIBULAR N. (23)** is the lateral branch of the sciatic n., the term 'fibular' being synonymous with 'peroneal' in the literature. This is equally true for muscles. The common fibular n. crosses over the lateral head of the m. gastrocnemius. Proximal to this, a branch, the **lateral cutaneous sural n. (24)** penetrates the m. biceps femoris in the popliteal region between its principal and accessory portions, to reach the skin. The common fibular n. enters the musculature of the leg laterally, and divides into a **superficial (10)** and a **deep fibular n. (11)** providing distal muscle branches for the flexors of the tarsus and extensors of digits. Both nerves run distally on the flexor aspect of tarsus, the superficial fibular n. craniolaterally, the deep fibular n. craniomedially through the ring-shaped crural extensor retinaculum. The superficial fibular n. then ramifies to give the common dorsal digital nn. II to IV, the first and second digits also being supplied dorsally by the saphenous n. On the flexor aspect of tarsus the deep fibular n. gives a muscle branch to the **m. extensor digitalis brevis (18)** and then divides into dorsal metatarsal nn. II to IV (see p. 34A).

c) The **TARSAL FLEXOR AND DIGITAL EXTENSOR MUSCLES** lie craniolaterally on the leg. They are described in order craniocaudally.

I. The **m. tibialis cranialis (5)** arises laterally on the cranial border of the tibia. In common with the mm. extensor digitalis longus and extensor digiti I and the deep fibular n., the muscle tendon passes deep to the crural extensor retinaculum and terminates proximally on metatarsal bone I.

II. The **m. extensor digitalis longus (9)** arises on the lateral condyle of femur and after the division of its terminal tendon, ends on the distal phalanges of all digits.

III. The **m. extensor digiti I** is an insignificant muscle situated on the distal two-thirds of the tibia deep to the m. extensor digitalis longus. It arises on the fibula and terminates on the first and second digits (not illustrated).

IV. The **m. fibularis longus (4)** arises on the fibula and proximally on the tibia, and crosses over the tendons of the m. fibularis brevis and the m. extensor digitalis lateralis. It attaches to the plantar aspect of the proximal end of metatarsal bone V and then terminates by a tendon directed medially and transversely to metatarsal bone I.

V. The **m. extensor digitalis lateralis (12)** arises on the proximal third of the fibula and ends on the distal phalanx of digit V.

VI. The **m. fibularis brevis (13)** arises on the distal two-thirds of the fibula and inserts on the proximal end of metatarsal bone V.

1 d) The **TIBIAL N. (25)**, proximal to its passage between the two heads of the m. gastrocnemius, discharges the **caudal cutaneous sural n. (26)**. This cutaneous nerve crosses the lateral head of m. gastrocnemius and either crosses over or pierces the tractus calcaneus lateralis. It innervates the skin of the leg caudally and reunites with the tibial n. two centimeters proximal to the calcaneal tuber. Then the tibial n. runs over the sustentaculum tali where it divides into medial and lateral plantar nn. The **medial plantar n. (21)** branches into the common plantar digital nn. II to IV, while the **lateral plantar n. (20)** gives the plantar metatarsal nn. II to V.

e) The **TARSAL EXTENSOR AND DIGITAL FLEXOR MUSCLES** are situated caudally on the leg.

I. The **m. gastrocnemius (1)** is the extensor muscle of the tarsus. In the tendons of origin of its **lateral (2)** and **medial (3)** heads it contains the lateral and medial sesamoid bones of the m. gastrocnemius articulating with the respective condyles of femur. Due to their moveable articular attachments to underlying bone, these sesamoids are located by joint penetration with a scalpel. The insertion of the m. gastrocnemius approaching the calcaneal tuber is known as the Achilles tendon.

II. The **m. flexor digitalis superficialis (8)** arises from the lateral supracondylar tuberosity of the femur and is covered proximally by the m. gastrocnemius. After transection of the medial head of the m. gastrocnemius, the m. flexor digitalis superficialis is separated from enveloping muscle. Its continuation as the superficial flexor tendon, in common with the tractus calcaneus and the Achilles tendon, forms the **common calcaneal tendon (14)**. The **superficial flexor tendon (17)** winds around the Achilles tendon from its cranial to medial and then caudal aspects. At the calcaneal tuber, the tendon broadens to give the **calcaneal cap or galea (19)** which is attached to the tuber medially and laterally by **retinacula (16)**. An inconstant (acquired) subcutaneous calcaneal bursa overlies the galea while deep to it is the **subtendinous calcaneal bursa (15)**. This may be opened after incising the lateral flexor retinaculum and folding back the calcaneal cap.

III. The **mm. flexores digitales profundi (6**, see also text illustration) include the large **m. flexor digitalis lateralis (7)** and the moderately large m. flexor digitalis medialis[1]. In other domestic mammals, the very weak m. tibialis caudalis which takes origin from the head of the fibula, participates in the formation of the deep digital flexor tendon.

This is not so in the dog, where the very long thin tendon terminates on the central tarsal bone. Both digital flexor muscles take origin from the proximomedial aspect of the tibia. Their tendons go separately over the medial aspect of tarsus, to unite within a short distance in the deep flexor or **common tendon (22)** on the proximal end of the metatarsus. Subsequently this common tendon divides, the tendons going to the distal phalanges of the digits through tunnel-like flexor manicae formed by the superficial flexor tendons.

f) The **SPECIFIC FLEXOR MUSCLE OF THE KNEE** is the **m. popliteus** (see text illustration). Arising on the lateral condyle of the femur, it lies on the flexor aspect of the knee deep to the m. gastrocnemius, and inserts onto the tibia proximomedially. Its contained sesamoid bone is at the transition of the tendon of origin and the muscle belly.

Mm. flexores digitales profundi and m. popliteus (caudal view)

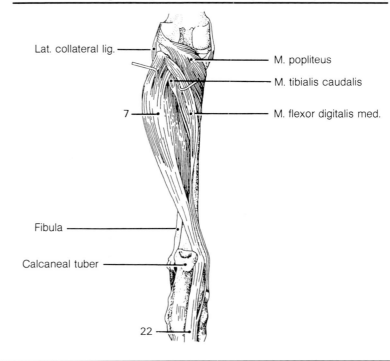

- Lat. collateral lig.
- M. popliteus
- M. tibialis caudalis
- M. flexor digitalis med.
- 7
- Fibula
- Calcaneal tuber
- 22

[1]) M. flexor digitalis medialis is also called m. flexor digitalis longus and the m. flexor digitalis lateralis, m. flexor digiti I.

Pelvic limb

Legend:

30 M. biceps femoris
31 Tractus calcaneus lat.
32 M. abductor digiti V.
33 Popliteal ln.
34 M. semitendinosus
35 Tractus calcaneus med.
36 M. flexor dig. med.
37 Mm. interossei
38 M. sartorius
39 Fascia lata
40 M. adductor magnus
41 M. semimembranosus
42 M. abductor cruris caud.
43 Crural extensor retinaculum
44 Tendon of m. fibularis longus
45 Tarsal extensor retinaculum
46 Tendon of m. fibularis brevis
47 Tendons of m. extensor dig. longus
48 Tendon of m. extensor dig. lat.
49 Calcaneal tuber
50 M. quadratus plantae
51 Mm. interflexorii (section)
52 Mm. lumbricales
53 Tractus tori metatarsei
54 Plantar annular ligg.

a Abaxial plant. digital. n. V of lat. plant. n.
b Plant. metatars. a.
   and n. IV of lat. plant. a. and n.
c Caud. r. of med. saphenous v.
d Caud. r. of saphenous a.
e Med. tarsal v. of med. saphenous v. (cran. r.)
f Abaxial plant. digital n. II of med. plant. n.
g Plant. comm. digital aa. and nn. II-IV of med.
   plant. a. and n. resp.
h Supf. plantar arch
i Plant. com. digital v. III
j Plant. prop. digital aa., vv. and nn.
k Dist. caudal femoral v.
l Lat. tarsal v. of lat. saphenous v. (cran. r.)
m Cran. r. of saphenous a.
n Dors. com. digital a., v., n. IV
o Abaxial dors. digital n. V of supf. fibular n.

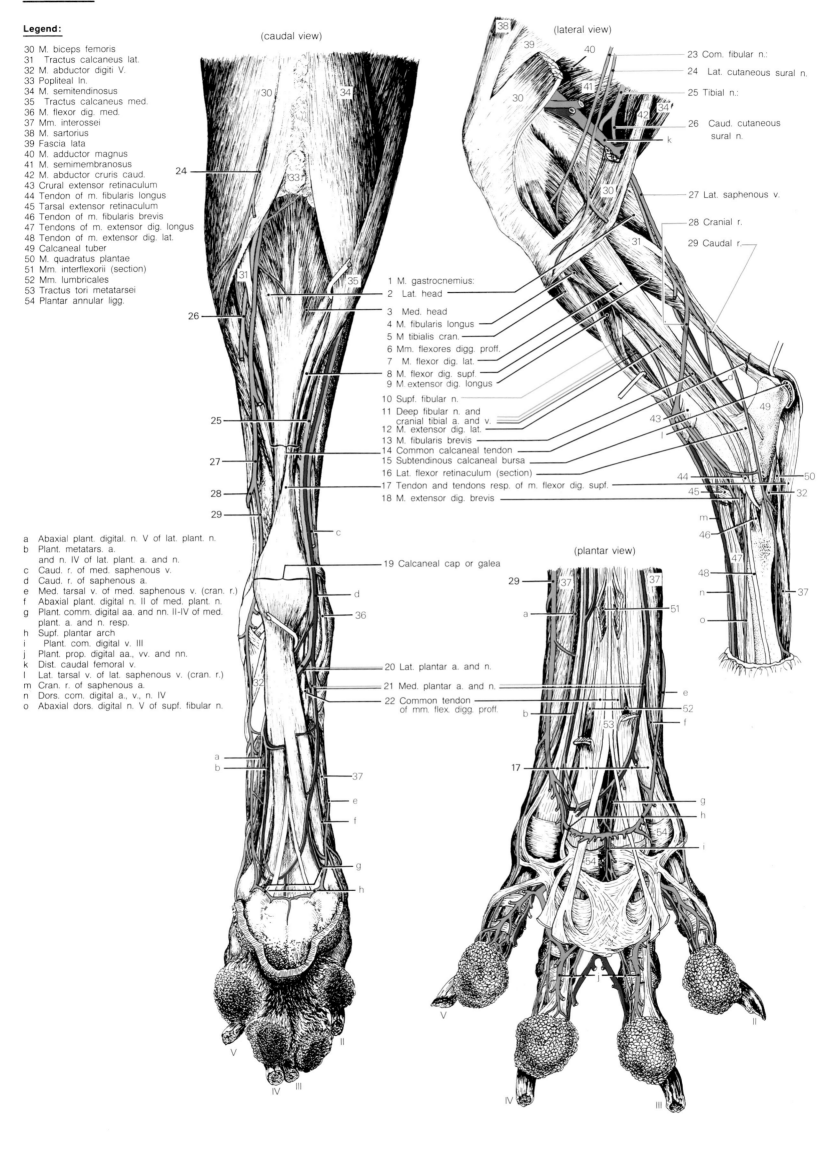

(caudal view)

(lateral view)

(plantar view)

23 Com. fibular n.:
24 Lat. cutaneous sural n.
25 Tibial n.:
26 Caud. cutaneous
    sural n.
27 Lat. saphenous v.
28 Cranial r.
29 Caudal r.

1 M. gastrocnemius:
2 Lat. head
3 Med. head
4 M. fibularis longus
5 M tibialis cran.
6 Mm. flexores digg. proff.
7 M. flexor dig. lat.
8 M. flexor dig. supf.
9 M. extensor dig. longus
10 Supf. fibular n.
11 Deep fibular n. and
   cranial tibial a. and v.
12 M. extensor dig. lat.
13 M. fibularis brevis
14 Common calcaneal tendon
15 Subtendinous calcaneal bursa
16 Lat. flexor retinaculum (section)
17 Tendon and tendons resp. of m. flexor dig. supf.
18 M. extensor dig. brevis

19 Calcaneal cap or galea

20 Lat. plantar a. and n.
21 Med. plantar a. and n.
22 Common tendon
   of mm. flex. digg. proff.

3A

C. 7. S. 5: ARTERIES, CONCOMITANT VEINS, NERVES AND LYMPHATIC VESSELS OF PELVIC LIMB

(see also pp. 31A to 33A)

> *When displaying the arteries, one should also consider the veins, mostly of like name, and revise the nerves. To demonstrate the popliteal a., the m. popliteus is detached at its insertion on the medial border of the tibia.*

a) The **GLUTEAL** and **LATERAL REGION OF THE THIGH** are supplied by parietal branches of the internal iliac a. (see p. 31A). In company with middle clunial nn., the **cranial gluteal a.** (2) and its cutaneous branches reach the body surface proximal to the cranial border of the m. gluteus superficialis. In like manner, deeply situated arterial branches accompanying the **cranial gluteal n.** (1), run between the mm. gluteus medius and profundus, and irrigate them. Proximally, at the caudal border of the m. gluteus superficialis, the large **caudal gluteal a.** (3) also provides cutaneous branches, in company with the ramifying lateral coccygeal a. (see upper p. 31A) and caudal clunial nn. Deep branches of the caudal gluteal a. accompany the sciatic n. caudally and supply the m. gluteus superficialis and origins of the caudal thigh muscles.

b) The **THIGH** and **LEG** are supplied by the distal continuation of the external iliac a. Just proximal to its passage through the lacuna vasorum and its transition to the **femoral a.** (10), the **external iliac a.** (8) discharges the **deep femoral a.** (9). After giving rise to the pudendoepigastric trunk, the deep femoral a. crosses deep or lateral to the origin of the m. pectineus (et adductor longus). Its continuation, the medial circumflex femoral a., goes between the mm. adductor magnus and obturatorius externus and then caudally around the femur to anastomose with the lateral circumflex femoral a.

Two centimeters distal to the lacuna vasorum, the **lateral circumflex femoral a.** (12) and the **superficial circumflex iliac a.** (11) arise either by a common stem or individually from the cranial side of the femoral a. The superficial circumflex iliac a. runs between the mm. sartorius and tensor fasciae latae to the m. rectus femoris. The **proximal caudal femoral a.** (4) arises from the caudal side of the femoral a. and runs under the m. gracilis to the thigh muscles. At the apex of the femoral triangle, the femoral a. discharges the following branches: the **descending genicular a.** (13) craniodistally to the knee, the deeply situated **middle caudal femoral a.** (5) caudodistally to the m. semimembranosus, and the saphenous a. superficially in company with the medial saphenous v. The **saphenous a.** (16) (and its accompanying vein) divides in the proximal third of the leg. Its **cranial ramus** (17) runs distally along the m. tibialis cranialis and gives rise to the common dorsal digital aa. I to IV on the dorsal surface of the pes. The caudal ramus of the saphenous a. runs distally as far as the medial head of the m. gastrocnemius and divides medial to the tarsus. The **medial plantar a.** (19) gives the common plantar digital aa. II to IV and the **lateral plantar a.** (18) contributes to the deep plantar arch. Plantar metatarsal aa. II to V proceed distally from the arch. The **distal caudal femoral a.** (6), the last branch of the femoral a., runs caudodistally. On the flexor aspect of the knee it supplies the caudal muscles of the thigh and the m. gastrocnemius. (The accompanying distal caudal femoral v. gives rise to the lateral saphenous v. which runs to the lateral head of the m. gastrocnemius. It is unaccompanied by an artery. In the distal third of the leg, the lateral saphenous v. divides into a cranial and a caudal ramus.) The **popliteal a.** (14), the continuation of the femoral a., lies deep to the m. popliteus. It gives several branches to the knee and divides into two on the flexor aspect of knee under the detached m. popliteus. Of the two

branches the caudal tibial is short and weak, the cranial tibial elongated and robust. The **cranial tibial a.** (15) passes through the crural interosseous space and proceeds distally on the cranial aspect of the tibia in company with the like named vein and the deep fibular n. Then it passes through the loop-shaped crural extensor retinaculum between the mm. tibialis cranialis and extensor digitalis longus. Its distal continuation on the dorsum of the pes, the **dorsal pedal a.** (7) gives the arcuate a. and its branches, the dorsal metatarsal aa. II to IV

c) The **VASCULAR AND NERVE SUPPLY OF THE PES** arises chiefly from dorsal and plantar common digital aa., vv. and nn. and dorsal and plantar metatarsal aa., vv. and nn. situated at a deeper level.
I. On the **dorsal aspect of the pes**, the dorsal common digital aa. I to IV arise from the cranial ramus of the saphenous a. The dorsal common digital vv. II to IV arise from the dorsal superficial arch, in turn formed by cranial rami of lateral and medial saphenous vv. Dorsal common digital nn. II to IV are derived from the superficial fibular n.
Dorsal metatarsal aa. II to IV arise from the arcuate a. Dorsal metatarsal vv. II to IV arise from the deep dorsal arch which merges with the dorsal pedal v. proximally. Dorsal metatarsal nn. II to IV are branches of the deep fibular n.
II. On the **plantar aspect of the pes** common plantar digital aa. II to IV are branches of the medial plantar a., while plantar metatarsal aa. II to IV arise from the deep plantar arch derived in turn from the lateral plantar a. Common plantar digital vv. II to IV branch from the superficial plantar arch. This is formed from the caudal ramus of the lateral saphenous v. and the medial tarsal v. Plantar metatarsal vv. II to IV arise from the deep plantar arch, formed in turn from the deep ramus of the caudal branch of the lateral saphenous v. Common plantar digital nn. II to IV are branches of the medial plantar n. while plantar metatarsal nn. II to IV stem from the lateral plantar n. The medial and lateral plantar nn. are terminal branches of the tibial n.

d) The **LYMPHATIC VESSELS** commence distally in the region of digits and pads with a well formed lymphatic capillary net that is suited to indirect lymphography. (The injected radiopaque material reaches the lymphatic capillary net indirectly from connective tissue.) From the distal region of the pelvic limb, superficial lymphatic vessels run to the **popliteal lymph node** (not illustrated, see p. 33A —33) in the popliteal region lying between the mm. biceps femoris and semitendinosus. In the main, the superficial vessels then go to the superficial inguinal lymph node and subsequently through the inguinal space to the medial iliac lymph nodes. The deep lymphatic vessels also reach the popliteal lymph node and then the medial iliac lymph nodes, by different routes. In one instance they accompany the tibial and sciatic nn. to the sacral lymph nodes and then to the medial iliac lymph nodes. In another they run with femoral a. and v., through the femoral space and ring, to the inconstant iliofemoral lymph node. From here they proceed to the medial iliac lymph nodes. Lymph flows from these nodes to the cisterna chyli, passing through the aortic lumbar lymph nodes along the route. The cranial continuation of the cisterna chyli is the unpaired thoracic duct which transports lymph to the venous angle between internal and external jugular vv. This has been mentioned several times previously.

Legend p. 34A

20 Lumbosacral tr.	31 Supf. fibular n.	42 Med. tarsal v.
21 Caud. gluteal n.	32 Lat. saphenous v.	43 Abaxial dors. dig. a., v. and n. V
22 Sciatic n.	33 Caud. r.	44 Arcuate a.
23 Rotatorius n.	34 Anastomotic. r. with 16	45 Dors. arch and deep plantar arch resp.
24 Tibial n.	35 Cran. r.	46 Dors. arch and supf. plantar arch resp.
25 Prox. muscular r.	36 Anastomotic r. with 17	47 Dors. and plant. metatarsal aa., vv. and nn. resp.
26 Caud. cutaneous sural n.	37 Caud. r. of saphenous a.	48 Dors. and plant. com. digital aa., vv, and nn. resp.
27 Com. fibular n.	and med. saphenous v.	49 Dors. and plant. proper digital aa., vv., and nn. resp.
28 Musculocutaneous r.	38 Pudendoepigastric tr.	50 Abaxial plant. dig. n. V
29 Lat. cutaneous sural n.	38' Med. circumflex femoral a. and v.	51 Abaxial plant. dig. n. II
30 Deep fibular n.	39 Femoral n.	
	40 Saphenous n.	
	41 Supf. r. of cran. tibial a.	

34

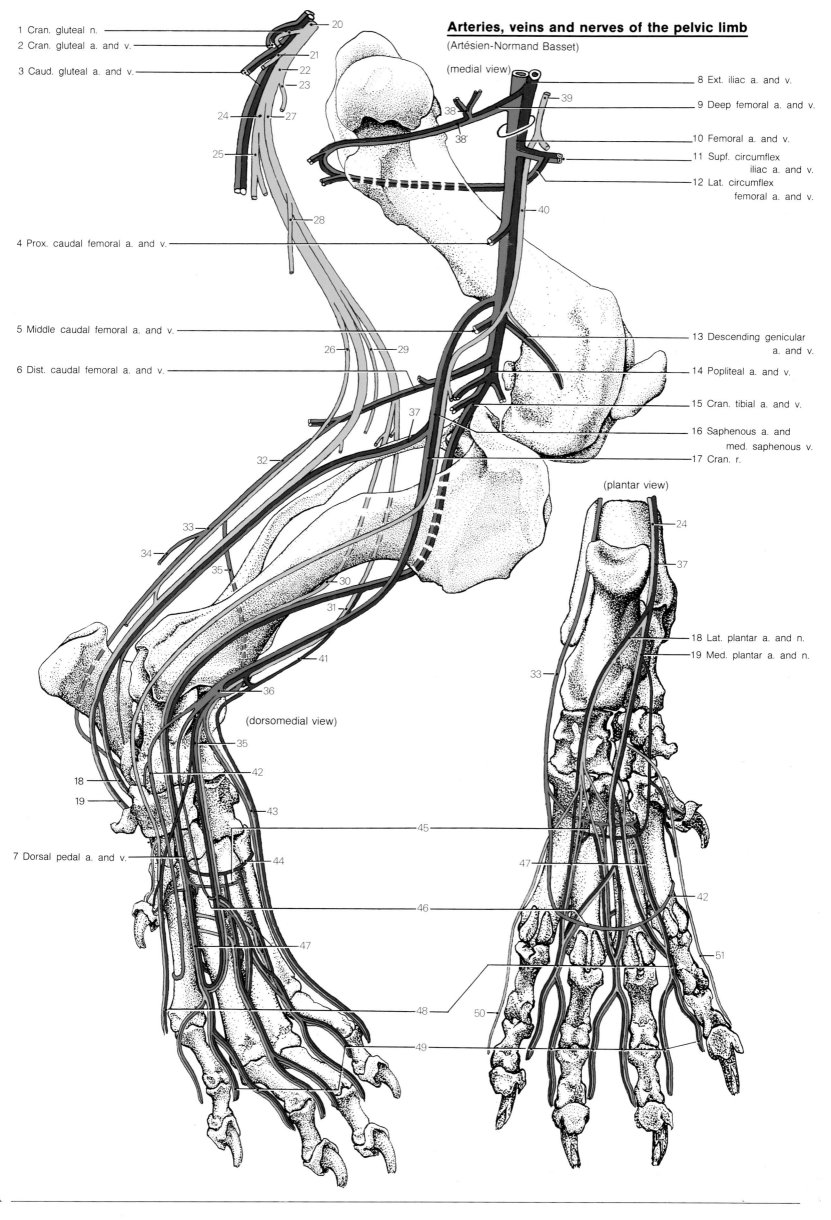

Arteries, veins and nerves of the pelvic limb
(Artésien-Normand Basset)

(medial view)

1 Cran. gluteal n.
2 Cran. gluteal a. and v.
3 Caud. gluteal a. and v.

4 Prox. caudal femoral a. and v.

5 Middle caudal femoral a. and v.
6 Dist. caudal femoral a. and v.

8 Ext. iliac a. and v.
9 Deep femoral a. and v.
10 Femoral a. and v.
11 Supf. circumflex iliac a. and v.
12 Lat. circumflex femoral a. and v.

13 Descending genicular a. and v.
14 Popliteal a. and v.
15 Cran. tibial a. and v.
16 Saphenous a. and med. saphenous v.
17 Cran. r.

(plantar view)

18 Lat. plantar a. and n.
19 Med. plantar a. and n.

(dorsomedial view)

18
19

7 Dorsal pedal a. and v.

20 21 22 23 24 25 27 28 26 29 32 33 34 35 30 31 41 36 35 42 43 44 46 47 48 49 37 38 38' 39 40 45 50 51

4A

CHAPTER 8: HEAD

C. 8. S. 1: SKULL, EXTERNAL SURFACE OF CALVARIA AND BASE OF CRANIUM, FACE, AND HYOID APPARATUS
(see also p. 36A)

The skull is the skeleton of the head, organized into a cranium and a face. Of the seven bones of the (neuro)cranium those numbered I to III and IVc are membranous bones forming the roof of the skull or calvaria. Individual bones numbered IVa and b, V to VII are replacement or primordial bones which replace the cartilaginous primordial skeleton at the base of the skull. **Hints on using the illustration.** *The asterisked (*) skeletal parts and features in the table refer to the upper illustration, those designated with a point or dot (.) refer to the lower. The undesignated bones and features refer to the illustration on p. 36A. The first sectional list (a to g inclusive) on p. 35A is discussed on p. 36.*

1 a) With regard to the **CRANIUM** the calvaria has a temporal fossa (**j**) which one can regard as the site of origin of the m. temporalis and which is bounded by the external frontal crest (**k**), the external sagittal crest (**l**), the nuchal crest (**m**) and the temporal crest (**m'**). The carotid canal housing the internal carotid a. begins at the caudal carotid foramen (**n**) deep to the jugular foramen at the base of the skull. It terminates at the internal carotid foramen (**o**) within the cranial cavity rostrally, and at the external carotid foramen (**p**) on the base of the skull. The jugular foramen (**q**) gives passage to cranial nerves IX, X and XI. The features of the cranial cavity (**r** to **z'**) are described on p. 36.

I. Caudodorsal to the orbit, the **frontal bone** possesses the zygomatic process (**1**) from which the orbital ligament runs to the frontal process of the zygomatic bone. The external ethmoid a. and v. and the ethmoid n. pass through the ethmoidal foramina (**2**). The frontal sinuses (**3a to 3c**) are described on p. 36.

II. The **parietal bone** and

III. the **interparietal bone** project into the cranial cavity by means of their tentorial processes (**4 and 5** on p. 36A).

IV. The **temporal bone** consists of petrosal, tympanic and squamous parts.

a. Of the petrosal part (**6**) only the following features are visible externally: the mastoid process (**7**) for the attachment of sections of the m. sternocleidomastoideus, the stylomastoid foramen (**10**) for the exit of the facial n., the external opening of the canaliculus of the chorda tympani (**11**), and the petrotympanic fissure (**12**). (Nos. **8, 9, 13** and **14** are discussed on p. 36.)

b. The tympanic part (**15**) lies caudal to the temporomandibular articulation on the base of the skull. The external acoustic meatus, commencing at the external acoustic pore (**16**), is separated from the bulla tympanica (**17**) medially by the tympanic membrane. The bulla contains the tympanic cavity of the middle ear into which opens the tympanic ostium of the auditory tube. The other end of the auditory tube, the pharyngeal ostium, opens into the pharynx.

c. The squamous part (**18**) belongs to the roof of the skull, its zygomatic process (**19**) participating in the formation of the zygomatic arch. Caudoventrally, at the base of the arch are located the mandibular fossa (**20**) and its articular surface (**21**), and the definite retroarticular process (**22**) caudal to these.

Temporal bone

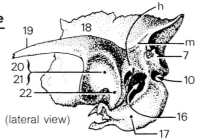

(lateral view)

V. The **ethmoid bone** and its seven features (**23 to 29**), are discussed with the paranasal sinuses and nasal cavity (see p. 36).

VI. The **occipital bone** of the fetus has recognizable sutures between its squama, lateral and basilar parts.
The occipital squama (**30**) has a conspicuous external occipital protuberance (**31**) situated middorsally, which merges with the nuchal crest to either side. The lateral part (**32**) bears an occipital condyle (**33**) on each side of the foramen magnum for participation in the atlantooccipital articulation. Within the foramen magnum and occipital condyle is the condylar canal (**34**) for the passage of emissary vv. from the skull. Rostroventral to this lies the opening at the commencement of the hypoglossal canal (**35**). The external opening of the canal is caudal to the jugular foramen between the occipital condyle and the jugular process (**36**). The basilar part (**37**) borders the foramen magnum (**38**) ventrally.

Midventrally between the jugular foramina it possesses an indistinct, unpaired pharyngeal tubercle (**39**) for the origin of pharyngeal muscles. Adjacent and medial to the tympanic bulla of either side, the basilar part has a muscular tubercle (**40**) for the insertion of the m. longus capitis.

Occipital bone

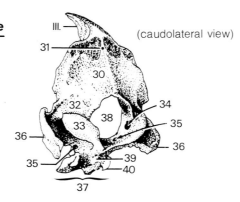

(caudolateral view)

VII. The **sphenoid bone** consists basically of two bones, the basisphenoid and the presphenoid. Each of these has a (horizontal) body situated medially, and a (vertical) wing laterally.

The **basisphenoid bone** bears the sella turcica (**42**) on the inner surface of the body (**41**). Internally, the wing (**43**) carries the foramen rotundum (**44**) for the entrance of the maxillary n. (V2), while externally the foramen ovale (**45**) for the passage of the mandibular n. is seen medial and adjacent to the mandibular fossa. The pterygoid crest (**46**), the osseous ridge of origin of the extrinsic muscles of the ocular bulb, commences ventrally at the alar canal (**47**). The maxillary a. and n. leave the alar canal at the rostral alar foramen (**48**), having entered it at the caudal alar foramen (**49**) and the foramen rotundum respectively.

The body (**50**) of the **presphenoid bone** situated medially, merges with the wing (**51**) laterally. The optic canal (**52**) for the passage of the optic n. (II), is situated cranially on the wing at the base of the orbit. Caudally, the orbital fissure (**53**) affords passage to cranial nerves III, IV, V and VI.

b) Rostroventral to the orbit, the **FACE** houses the pterygopalatine fossa (**A**). The greater palatine canal, containing the greater palatine n., begins in the fossa at the caudal palatine foramen (**B**) and ends at the greater palatine foramen (**C**) on the hard palate. Lesser palatine canals, containing lesser palatine nn., branch from within the greater palatine canal and terminate likewise at lesser palatine foramina (**D**). The sphenopalatine foramen (**E**), lying dorsal to the caudal palatine foramen contains the caudal nasal n. (from V2) going to the nasal cavity. At the choanae (**F**) the nasopharyngeal canal is continuous with the nasopharynx. The orbit (**G**) is closed caudally by the orbital ligament. The paired palatine fissures (**H**), housing the incisive ducts, are located in the hard palate caudal to the upper incisor teeth. Pit-like excavations, the dental alveoli (**J**), accomodate the roots of the teeth thus causing the alveolar juga (**K**) to arch externally. Deep to the alveoli, alveolar canals (**L**) conveying arteries, veins and nerves to the teeth, have their commencement. Interalveolar septa (**M**) are osseous ridges between the alveoli, while a diastema (**N**) or gap lies rostral and caudal to the canine tooth.

VIII. to XVII. are described on p. 36.

XVIII. The **hyoid bone** or **apparatus** has cartilaginous precursors in 2 the branchial arches and is therefore classified as a bone of the skull. The unpaired basihyoid (**90**) element lies transversally at the base of the tongue, which is flanked on both sides by the paired ceratohyoid (**91**) elements. From the basihyoid, the two thyrohyoid (**92**) elements are directed caudodorsally to articulate with the thyroid cartilage of larynx. The epihyoid (**93**) elements extend rostrally from the keratohyoids and these are followed by stylohyoid (**94**) elements directed towards the base of the skull. Connective tissue tympanohyoids (**95**) connect the paired stylohyoids to the mastoid processes of the temporal bones.

35

Cranium

External lamina (a)
Diploë (b)
Internal lamina (c)
Osseous tentorium cerebelli (d)
Temporal meatus (e)
  Canal of transverse sinus (f)
  Groove of transverse sinus (g)
  Retroarticular foramen (h)
Temporal fossa (j) ✽
  External frontal crest (k) ✽
  External sagittal crest (l) ✽ •
  Nuchal crest (m) •
  Temporal crest (m') ✽ •
Carotid canal
  Caudal carotid foramen (n) •
  Internal carotid foramen (o) •
  External carotid foramen (p) •
Jugular foramen (q) •

Cranial cavity
Rostral fossa of cranium (r)
Ethmoidal fossae (s)
Sulcus chiasmatis (t)
Middle fossa of cranium (u)
Hypophyseal fossa (v)
Piriform fossa (w)
Caudal fossa of cranium (x)
Pontine impression (y)
Medullary impression (z)
Petrooccipital fissure (z')

Bones of cranium
I. Frontal bone ✽ •
Zygomatic process (1) ✽ •
Ethmoidal foramina (2) ✽ •
Rostral frontal sinus (3a)
Lateral frontal sinus (3b)
Medial frontal sinus (3c)

II. Parietal bone ✽ •
Tentorial process (4)

III. Interparietal bone ✽ •
Tentorial process (5)

IV. Temporal bone ✽ •
a. Petrosal part (6) ✽ •
  Mastoid process (7) ✽ •
  Internal acoustic meatus
    Internal acoustic pore (8) •
  Facial canal (9)
    Stylomastoid foramen (10) •
  Canaliculus of chorda tympani (11) •
  Petrotympanic fissure (12) •
  Cerebellar (floccular) fossa (13)
  Canal of trigeminal nerve (14)
b. Tympanic part (15) •
  External acoustic meatus
    External acoustic pore (16) ✽
  Bulla tympanica (17) •
  Tympanic ostium of auditory tube (17') •
c. Squamous part (18) ✽ •
  Zygomatic process (19) ✽ •
  Mandibular fossa (20) •
  Articular surface (21) •
  Retroarticular process (22) ✽ •

V. Ethmoidal bone
Lamina cribrosa (23)
Crista galli (24)
Ethmoidal labyrinth (25)
  Ethmoturbinates
  Ectoturbinates (26)
  Endoturbinates (27)
    Dorsal nasal concha (28)
    Middle nasal concha (29)

VI. Occipital bone •
Occipital squama (30) •
  External occipital protuberance (31) •
  Tentorial process (31a) •
Lateral part (32) •
  Occipital condyle (33) ✽ •
  Condylar canal (34) •
  Hypoglossal canal (35) •
  Jugular (paracondylar) process (36) ✽ •
Basilar part (37) •
  Foramen magnum (38) •
  Pharyngeal tubercle (39) •
  Muscular tubercle (40) •

VII. Sphenoid bone •
Basisphenoid bone
Body (41) •
  Sella turcica (42) •
Wing (43) ✽ •
  Foramen rotundum (44) •
  Foramen ovale (45) •
  Pterygoid crest (46) ✽ •
Alar canal (47) •
  Rostral alar foramen (48) ✽ •
  Caudal alar foramen (49) •

Presphenoid bone
Body (50) •
Wing (51) ✽ •
  Optic canal (52) ✽ •
  Orbital fissure (53) ✽ •

Face

Pterygopalatine fossa (A) •
Greater palatine canal •
  Caudal palatine foramen (B) ✽ •
  Greater palatine foramen (C) •
Lesser palatine canals •
  Caudal palatine foramen (B) ✽ •
  Lesser palatine foramina (D) •
Sphenopalatine foramen (E) ✽ •
Choanae (F) •
Orbit (G) ✽ •
Palatine fissure (H) ✽ •
Dental alveoli (J) •
  Alveolar juga (K) •
Alveolar canals (L) •
Interalveolar septa (M) •
Diastema (N) ✽

Bones of face
VIII. Lacrimal bone ✽ •
Fossa of lacrimal sac (54) ✽ •

IX. Zygomatic bone •
Temporal process (55) •
Frontal process (56) •

X. Nasal bone ✽

XI. Maxilla ✽
Body of Maxilla (57) ✽
  Infraorbital canal
    Maxillary foramen (58) ✽ •
    Infraorbital foramen (59) ✽
  Lacrimal canal (60)
  Lacrimal groove (61) •
  Frontal process (62) •
  Zygomatic process (63) ✽ •
  Palatine process (64) •
  Alveolar process (65) •
  Maxillary recess (65') •

XII. Incisive bone ✽ •
Body of incisive bone (66) ✽
  Alveolar process (67) •
  Palatine process (68) •
  Nasal process (69) ✽

XIII. Palatine bone ✽ •
Perpendicular lamina (70) •
Horizontal lamina (71) •

XIV. Pterygoid bone ✽ •
Pterygoid hamulus (72) ✽ •

XV. Vomer •

XVI. Bone of ventral nasal concha
XVII. Mandible
Mandibular canal
  Mandibular foramen (74) •
  Mental foramina (75) •
Body of mandible (76) •
  Ventral border (77) •
  Alveolar border (78) •
  Mylohyoid line (79) •
Ramus of mandible (80) •
  Angle of mandible (81) •

Angular process (82) •
Masseteric fossa (83) •
Pterygoid fossa (84)
Condylar (articular) process (85) •
  Head of mandible (86) •
  Neck of mandible (87) •
  Mandibular notch (88) •
Coronoid (muscular) process (89)

XVIII. Hyoid bone (apparatus) ✽
Basihyoid (90) ✽
Ceratohyoid (91) ✽
Thyrohyoid (92) ✽
Epihyoid (93) ✽
Stylohyoid (94) ✽
Tympanohyoid (95) ✽

$I_1 - I_3$ Incisor teeth
C Canine tooth
$L(P_1)$ Wolf tooth
$P_2 - P_4$ Premolar teeth
$M_1 - M_2$ Molar teeth

(craniodorsal view ✽)

(caudobasal view •)

Thyroid cartilage

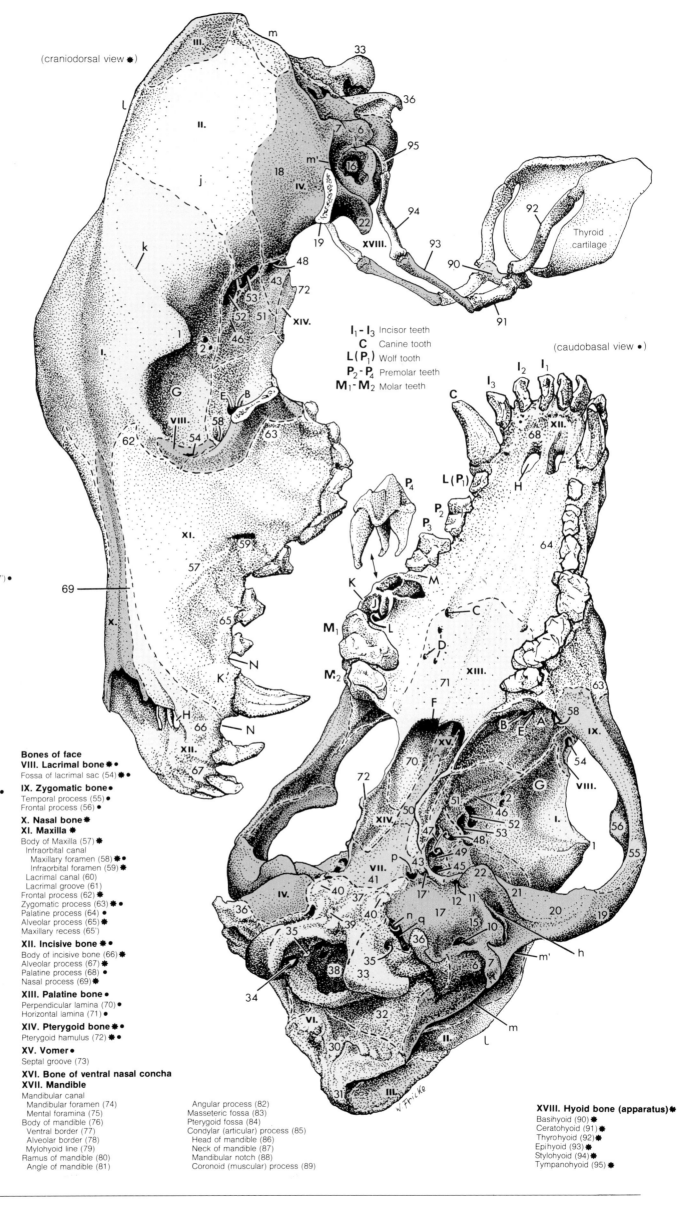

W. Fricke

> *To interpret illustrations on p. 36A, read the instructions on p. 35. (Zero -0- illustration upper left; X - illustration upper right; * - lower illustration; without symbols - see illustrations on p. 35A.)*

1 **a)** The **WALL OF THE SKULL** consists of an outer cortical layer of bone, the external lamina (a), a middle narrow cancellous layer, the diploë (b), and a cortical internal lamina (c). On the dorsal aspect of the frontal bone, the cancellous layer disappears completely and both osseous laminae separate to form the paired frontal sinuses. These are paranasal, in open contact with the nasal cavity by means of apertures.

2 The rostral frontal sinus (3a) lies between the median osseous septum of the frontal sinus and the orbit, while the largest, the lateral frontal sinus (3b) extends into the zygomatic process of the frontal bone. The middle frontal sinus (3c) is very small and lies between the other two. Occasionally it is absent. In some cases a further paranasal sinus, the sphenoidal is produced.

The osseous tentorium cerebelli (d) is formed from the fused tentorial processes (4, 5, and 31a) of parietal, interparietal and occipital bones. It is the bony foundation of the membranous tentorium cerebelli separating cerebrum from cerebellum within the cranial cavity.

The temporal meatus (e) is the outlet of the dorsal part of the vascular system of the brain. As a groove it is partly on the skull wall, as a canal partly within it. As the canal of the transverse sinus (f), the meatus begins dorsally in the midline at the base of the osseous tentorium cerebelli. Laterally at the pyramid-like petrous temporal bone, it continues in the groove of the transverse sinus (g) and terminates externally at the retroarticular foramen (h).

b) The **CRANIAL CAVITY** is bounded by the internal surfaces of the cranial bones. Rostral, middle and caudal fossae of the cranium are present at the base of the cranial cavity. The rostral fossa (r) begins at the paired ethmoidal fossae (s) and ends at the chiasmatic sulcus (t) where both optic nerves cross (the optic chiasm) caudal to the entrances of the **3** optic canals (52). Medially, the middle fossa (u) houses the sella turcica (42) formed by the basisphenoid. Cranial to the dorsum sellae, a median hypophyseal fossa (v) is present for the reception of the hypophysis or pituitary gland. Laterally, the paired piriform fossae (w) accommodate the pear-shaped olfactory lobes. The caudal fossa of the cranium (x) has a shallow pontine impression (y) rostromedially for the pons of the metencephalon, and the medullary impression (z) mid-sagittally for the reception of the medulla oblongata. In the dog the petrooccipital fissure (z') is a very narrow space between the petrous part of the temporal bone and the basioccipital bone.

IV. a. The petrous part of the temporal bone (6) has only the mastoid process (7) on the external surface of the skull while the petrous pyramid limits the cranial cavity basolaterally. In the middle of the medial surface of the petrous pyramid, the internal acoustic pore (8) marks the beginning of the internal acoustic meatus and affords entrance to cranial nerves VII and VIII. The facial canal (9), housing the facial nerve, proceeds deeply from the internal acoustic meatus and terminates externally at the stylomastoid foramen (10). Centrally, within the petrous temporal bone, the canaliculus of the chorda tympani (11) arises from the facial canal and terminates externally at the minute petrotympanic fissure (12). The cerebellar fossa (13), situated dorsocaudally on the petrosal pyramid, contains the ventral paraflocculus of the cerebellum. Rostroventrally, the pyramid is pierced by the canal of the trigeminal nerve (14).

Temporal bone

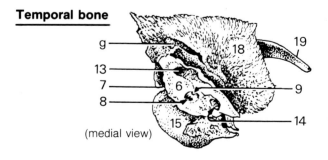

(medial view)

V. The **ethmoidal bone** forms the boundary with the nasal cavity at the lamina cribrosa (23) and the median crista galli (24). The ethmoidal labyrinth (25) and its ethmoturbinates project into the nasal cavity. The smaller ethmoturbinates situated externally are called ectoturbinates (26). The larger, internal endoturbinates (27), situated near the nasal septum, are numbered I to IV in series dorsoventrally. Endoturbinate I forms the osseous base for the dorsal nasal concha (28) while endoturbinate II is similarly associated with the middle nasal concha (29). The bone of the ventral nasal concha (XVI) is the osseous base of the respective nasal concha and belongs properly to the facial bones not the ethmoidal bone.

Ethmoidal bone

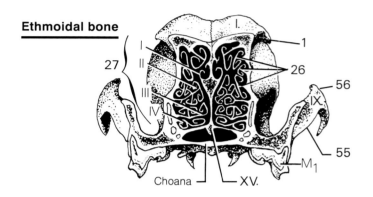

c) The **BONES OF THE FACE (VIII TO XVII)** form the external **4** boundaries of the nose, nasopharyngeal canal and nasopharynx.

VIII. The small **lacrimal bone** has the fossa of the lacrimal sac (54) at its center.

IX. The **zygomatic bone** participates in forming the zygomatic arch with its temporal process (55). Its frontal process (56) is the site of attachment of the orbital ligament.

X. The **nasal bone** lies dorsal to the nose.

XI. The **maxilla** has an infraorbital canal situated centrally in the body of the maxilla (57). The canal is the passage for the infraorbital a. and n. Coming from the pterygopalatine fossa caudally, these structures enter the canal through the maxillary foramen (58) and leave it by the infraorbital foramen (59) rostrolaterally. The lacrimal sac connects with the nasolacrimal duct which passes through the lacrimal canal (60) and the lacrimal groove (61). The frontal (62), zygomatic (63) and palatine (64) processes respectively, border the bones of like name. The alveolar process (65), as the name suggests, bears the alveoli. In contrast to the other domestic mammals, the canine maxilla has no paranasal sinus, but rather a maxillary recess (65') directed laterally.

XII. The **incisive bone** consists of a body (66), and an alveolar (67), a palatine (68), and a nasal (69) process.

XIII. The **palatine bone,** by means of the medial surface of its perpendicular lamina (70), participates in the formation of the nasopharyngeal meatus. Its lateral surface enters into the formation of the orbit. In common with the palatine process of the maxilla, the horizontal lamina (71) forms the osseous base of the hard palate.

XIV. The **pterygoid bone** restricts the nasopharynx laterally and features a hamulus (72) caudoventrally.

XV. The **vomer** has a dorsomedian septal groove (73) for the reception of the nasal septum.

XVI. The **bone of the ventral nasal concha** was mentioned previously as the osseous base of the ventral concha. It is fused to the medial surface of the maxilla.

XVII. The **mandible** consists of a vertical ramus and a horizontal body. The elongated mandibular canal (for the inferior alveolar a., v. and n.) begins caudally at the mandibular foramen (74) and terminates rostrally on the body at the mental foramina (75). The body of the mandible (76) has a ventral border (77), an alveolar border (78), and a very faint mylohyoid line (79) medially for the origin of the m. mylohyoideus. Caudoventrally, the ramus of the mandible (80) has an angle of the mandible (81) featuring an angular process (82). The masseteric fossa (83) is a triangular fossa on the lateral surface of the mandible for the attachment of the m. masseter, while an indistinct pterygoid fossa (84) is present medially, providing insertion to the m. pterygoideus. The condylar process (85) participates with its head of mandible (86) in the temporomandibular articulation, while the neck of mandible (87) merge with the **5** concave mandibular notch (88) situated caudal to the coronoid process (89). The m. temporalis inserts onto this process.

Cranium

- External lamina (a) ○
- Diploë (b) ○
- Internal lamina (c) ○
- Osseous tentorium cerebelli (d) ○
- Temporal meatus (e) ○
  - Canal of transverse sinus (f) ○
  - Groove of transverse sinus (g) ○
  - Retroarticular foramen (h) ○
- Temporal fossa (j)
  - External frontal crest (k)
  - External sagittal crest (l)
  - Nuchal crest (m)
  - Temporal crest (m')
- Carotid canal
  - Caudal carotid foramen (n)
  - Internal carotid foramen (o)
  - External carotid foramen (p)
- Jugular foramen (q) ○

Cranial cavity

- Rostral fossa of cranium (r) ○
- Ethmoidal fossae (s) ○
- Sulcus chiasmatis (t) ○
- Middle fossa of cranium (u) ○
- Hypophyseal fossa (v) ○
- Piriform fossa (w) ○
- Caudal fossa of cranium (x) ○
- Pontine impression (y) ○
- Medullary impression (z) ○
- Petrooccipital fissure (z') ○

Bones of cranium
I. Frontal bone ○
- Zygomatic process (1) ○
- Ethmoidal foramina (2) ○
- Rostral frontal sinus (3a) ○
- Lateral frontal sinus (3b) ○
- Medial frontal sinus (3c) ○

II. Parietal bone ○
- Tentorial process (4) ○

III. Interparietal bone ○
- Tentorial process (5) ○

IV. Temporal bone ○
- a. Petrosal part (6) ○
  - Mastoid process (7)
  - Internal acoustic meatus
    - Internal acoustic pore (8) ○
  - Facial canal (9) ○
    - Stylomastoid foramen (10)
    - Canaliculus of chorda tympani (11)
    - Petrotympanic fissure (12)
  - Cerebellar (floccular) fossa (13) ○
  - Canal of trigeminal nerve (14) ○
- b. Tympanic part (15) ○
  - External acoustic meatus
    - External acoustic pore (16)
  - Bulla tympanica (17)
  - Tympanic ostium of auditory tube (17') ○
- c. Squamous part (18)
  - Zygomatic process (19) ○
  - Mandibular fossa (20)
    - Articular surface (21)
    - Retroarticular process (22)

V. Ethmoidal bone ○
- Lamina cribrosa (23) ○
- Crista galli (24) ○
- Ethmoidal labyrinth (25) ○
  - Ethmoturbinates
  - Ectoturbinates (26) ○
  - Endoturbinates (27) ○
    - Dorsal nasal concha (28) ○
    - Middle nasal concha (29) ○

VI. Occipital bone ○
- Occipital squama (30) ○
  - External occipital protuberance (31) ○
  - Tentorial process (31a) ○
- Lateral part (32) ○
  - Occipital condyle (33) ○
  - Condylar canal (34) ○
  - Hypoglossal canal (35) ○
  - Jugular (paracondylar) process (36)
- Basilar part (37)
  - Foramen magnum (38) ○
  - Pharyngeal tubercle (39)
  - Muscular tubercle (40) ○

VII. Sphenoid bone ○
Basisphenoid bone
- Body (41) ○
  - Sella turcica (42) ○
- Wing (43) ○
  - Foramen rotundum (44) ○
  - Foramen ovale (45) ○
- Pterygoid crest (46)
- Alar canal (47) ○
  - Rostral alar foramen (48)
  - Caudal alar foramen (49)

Presphenoid bone
- Body (50) ○
- Wing (51) ○
  - Optic canal (52) ○
  - Orbital fissure (53) ○

Face

- Pterygopalatine fossa (A)
- Greater palatine canal
  - Caudal palatine foramen (B)
  - Greater palatine foramen (C)
- Lesser palatine canals
  - Caudal palatine foramen (B)
  - Lesser palatine foramina (D)
- Sphenopalatine foramen (E)
- Choanae (F) ○
- Orbit (G)
- Palatine fissure (H) ○
- Dental alveoli (J)
  - Alveolar juga (K)
  - Alveolar canals (L)
- Interalveolar septa (M)
- Diastema (N)

Bones of face ○
VIII. Lacrimal bone ○
- Fossa of lacrimal sac (54) ○

IX. Zygomatic bone ○
- Temporal process (55) ○
- Frontal process (56) ○

X. Nasal bone ○
XI. Maxilla x
- Body of Maxilla (57) X
  - Infraorbital canal
    - Maxillary foramen (58) X
    - Infraorbital foramen (59)
  - Lacrimal canal (60) X
  - Lacrimal groove (61) X
- Frontal process (62) X
- Zygomatic process (63) X
- Palatine process (64) X
- Alveolar process (65) X
- Maxillary recess (65') X

XII. Incisive bone ○
- Body of incisive bone (66) ○
- Alveolar process (67) ○
- Palatine process (68) ○
- Nasal process (69) ○

XIII. Palatine bone ○
- Perpendicular lamina (70) ○
- Horizontal lamina (71) ○

XIV. Pterygoid bone ○
- Pterygoid hamulus (72) ○

XV. Vomer ○
- Septal groove (73) ○

XVI. Bone of ventral nasal concha ○
XVII. Mandible *
- Mandibular canal
  - Mandibular foramen (74) *
  - Mental foramina (75) *
- Body of mandible (76) *
  - Ventral border (77) *
  - Alveolar border (78) *
  - Mylohyoid line (79) *
- Ramus of mandible (80) *
  - Angle of mandible (81) *
  - Angular process (82) *
  - Masseteric fossa (83) *
  - Pterygoid fossa (84) *
  - Condylar (articular) process (85) *
    - Head of mandible (86) *
    - Neck of mandible (87) *
    - Mandibular notch (88) *
  - Coronoid (muscular) process (89) *

XVIII. Hyoid bone (apparatus)
- Basihyoid (90)
- Ceratohyoid (91)
- Thyrohyoid (92)
- Epihyoid (93)
- Stylohyoid (94)
- Tympanohyoid (95)

(paramedian section, o)

(XI. Maxilla, x)

(XVII. Mandible *)

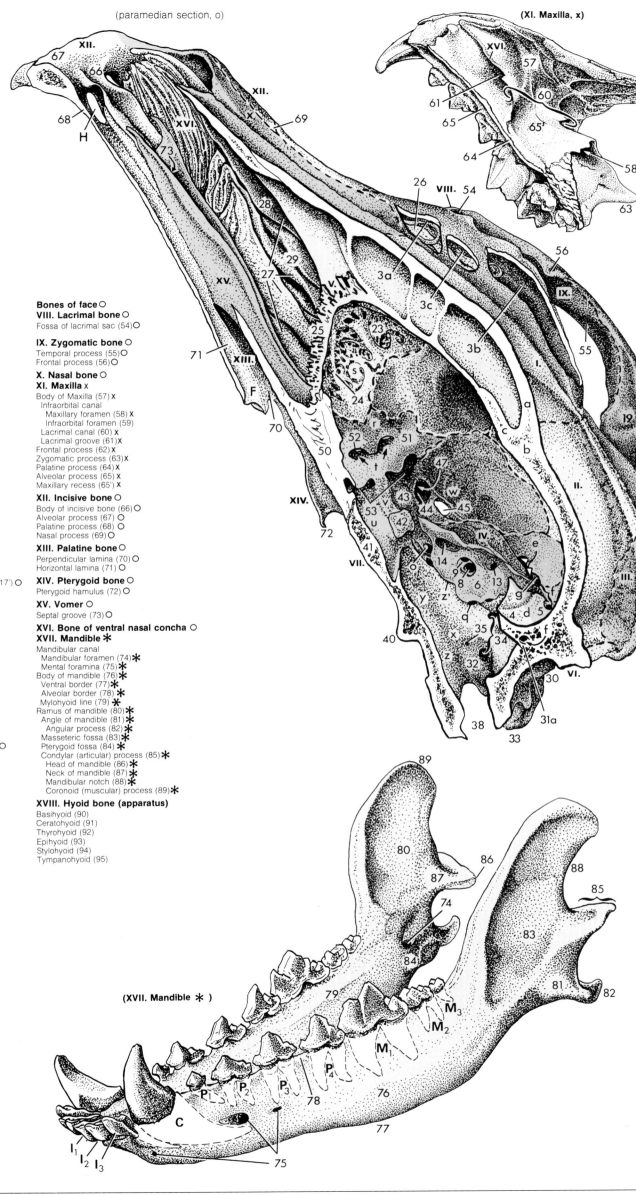

The skin is removed from the left side of the head taking care to preserve superficial arteries, veins and nerves, and cutaneous muscles. The external ear is severed at its base and removed. Following this, the m. cutaneous faciei, a part of the platysma, is exposed, detached from underlying structures and reflected rostrally, keeping intact the mm. malaris, zygomaticus and parotidoauricularis. To display arteries, veins and nerves, each of the facial muscles is either retracted or its coarse muscle bundles are pushed aside serially along the course of the veins and nerves. After its exposure, the m. levator nasolabialis is transected and both parts reflected. The maxillary v. and facial n. are displayed by removing all the parotid gland except for a small nut-sized remnant at the beginning of the parotid duct.

a) The **LYMPHATIC SYSTEM** of the head (see also text illustration) includes the palpable parotid and mandibular lymph nodes situated superficially, and the deeper medial retropharyngeal and inconstant lateral retropharyngeal lymph nodes.

The **parotid lymph node (23)** lies at the level of the temporomandibular articulation at and deep to the rostral border of the parotid gland. Afferent lymphatic vessels come from the superficial head regions dorsal to a line joining eye and base of ear. Efferent vessels go to the medial retropharyngeal lymph node.

The **mandibular lymph nodes (30)** lie rostroventral to the mandibular gland. Numbers of afferent lymphatic vessels come from the deeper regions of the head and a superficial field ventral to the line joining eye and base of ear. Likewise efferent vessels reach the medial retropharyngeal lymph node.

The **medial retropharyngeal lymph node (29)** lies caudal to pharynx and mandibular gland and deep to the latter. Lymph is derived from the deeper regions of head, and from mandibular and parotid lymph nodes. Efferent vessels unite with the jugular trunk (tracheal duct). The medial retropharyngeal lymph node may be connected to an inconstant lateral retropharyngeal lymph node and also receives a part of the lymph from the parotid node. When present, the lateral retropharyngeal node lies at the caudal border of the parotid gland, on a level with the wing of the atlas. Its efferent vessels run to the medial retropharyngeal node.

Lymph nodes of head (lateral view)

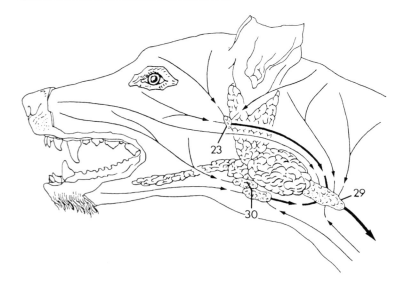

b) The **SUPERFICIAL VEINS** arise from the **external jugular v. (15)** which, caudal to the mandibular gland, bifurcates into a maxillary v. dorsally and a linguofacial v. ventrally. (The superficial arteries of external ear and face stem from the external carotid and the facial a. respectively, and are considered when the veins are exposed.)

The **maxillary v. (14)** collects blood from the external ear and in particular from the deeper head regions such as palate, cranial cavity, eye and mandible. Of its branches, the **caudal auricular v. (13)** passes within the (removed) parotid gland to the caudal aspect of ear, and three centimeters rostrally the curved **superficial temporal v. (11)** runs over the temporal region. The **rostral auricular v. (12)** branches from the latter to reach the cranial border of the external ear. The maxillary v. continues to run deeper and medial to the temporomandibular articulation. After forming the pterygoid and palatine plexuses and the ophthalmic plexus, the maxillary v. anastomoses at the medial angle of eye with a terminal branch of the facial v. and deep v. of the face (see below). The **linguofacial v. (16)** provides the outflow of blood from tongue and face. Ventral to the mandibular gland it bifurcates into the lingual v. directed ventrorostrally, and the facial v. dorsorostrally. Just rostral to its origin, the **lingual v. (18)** discharges the **hyoid venous arch (17)**

transversely to anastomose with the lingual v. of the other side. Subsequently, the lingual v. crosses the caudal border of the m. mylohyoideus dorsally, the **sublingual v. (7)** branching at a distance of 2 cm along its length. This branch is seen by moving the caudal border of the m. stylohyoideus rostrally.

Initially, the **facial v. (10)** runs ventrolaterally along the mandible where it gives off the **submental v. (8)**. This is directed ventrally over the insertion of the m. digastricus and continues ventromedial to the body of the mandible as far as the chin. The facial vein, after providing the **inferior labial v. (6)** to the lower lip, and then the **angular v. of the mouth (5)** to the oral commissure, continues obliquely, curving over the face to the medial angle of the eye. On the caudal aspect of the facial v. at the level of the angle of mouth, the **deep facial v. (9)** runs deeply under the zygomatic arch to the orbit. From the rostral aspect of the facial v. the **superior labial v. (4)** branches to the upper lip and the weak **lateral nasal v. (3)** is directed laterally on the nose. The facial v. also anastomoses with the infraorbital v. This vein takes origin from the termination of the deep facial v. and runs rostrally through the infraorbital canal. (The **facial a.** — 10 ends there by anastomosing with the infraorbital a.) Finally, the facial v. is distributed on the dorsum of nose in the form of a T. The **dorsal nasal v. (2)** runs rostrally in the direction of the tip of the nose, while the **angular v. of the eye (1)** goes in the opposite direction, to the medial angle of eye, where it anastomoses with branches of the deep facial and maxillary vv.

c) The **FACIAL NERVE (VII — 26)** consists of two parts, the facial n. proper which is motor, and the intermediate n. with its sensory, special sensory and parasympathetic fiber types. The individual facial n. enters the facial canal at the internal acoustic pore of the petrous part of the temporal bone. At the geniculate ganglion it discharges the intermediate n. and from this the chorda tympani runs through the petrotympanic fissure to join the lingual n. (from V 3). The greater petrosal, a parasympathetic nerve, also arises from the intermediate n. After uniting with the deep petrosal, a sympathetic nerve, the greater petrosal n. continues as the n. of the pterygoid canal to the pterygopalatine ganglion (see p. 48A). Within the facial canal the (proper) facial n. gives the stapedius n. to the m. stapedius. The facial n. is solely motor with the exception of its predominantly sensory internal auricular ramus. The ramal sensory fibers are conveyed to the facial n. in part by the vagus n. (X) and partly by the intermediate n. After the facial n. emerges from the stylomastoid foramen, the internal auricular ramus leaves the facial n. and subsequently pierces the auricular cartilage to supply the external acoustic meatus as far as the tympanic membrane. The facial n. continues by running around the osseous external acoustic meatus in a ventral convex arch within the parotid gland. From the convexity of the facial n. terminal branches arise in a caudorostral sequence and are described below. These are combined with sensory portions of the fifth cranial and second cervical nn. and innervate deep and superficial muscles of the face, including platysma and the caudal belly of the m. digastricus.

(For the subsequent demonstration of nerves, the parotid gland is removed.)

The **caudal auricular n. (24)** innervates the caudal auricular muscles and provides a **platysmal ramus (20)** for the platysma of the neck. Consecutive digastric and stylohyoid rami supply their respective muscles. From part of the convexity of the facial n., the ramus colli and ventral buccolabial ramus arise from a common origin. The **ramus colli (27)**, directed caudally, innervates the m. parotidoauricularis and is joined by the transverse cervical n. (n. transversus colli), a motor part of the second cervical n. The **ventral buccolabial ramus (28)**, accompanying the facial v. rostrally, communicates with the dorsal buccolabial ramus at the level of the orbit and both innervate cheek, lips and nose. The **dorsal buccolabial ramus (25)** which runs two centimeters dorsoparallel to the parotid duct, arises from the facial n. rostral and deep to the parotid lymph node and gland. The **auriculopalpebral n. (22)** is the terminal branch of the facial n. Its **rostral auricular rami (19)** supply the rostral and dorsal auricular muscles and its **palpebral ramus (21)** forms a broadly ramifying plexus caudal to its entry into the muscles of eyelid and nose.

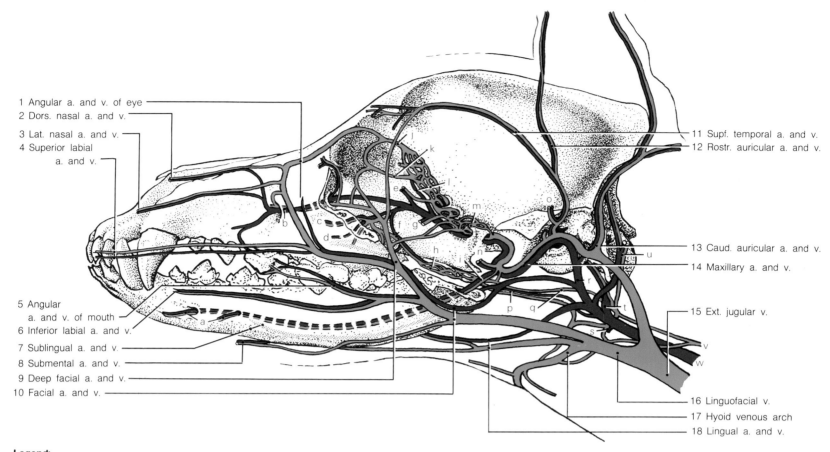

1 Angular a. and v. of eye
2 Dors. nasal a. and v.
3 Lat. nasal a. and v.
4 Superior labial a. and v.
5 Angular a. and v. of mouth
6 Inferior labial a. and v.
7 Sublingual a. and v.
8 Submental a. and v.
9 Deep facial a. and v.
10 Facial a. and v.

11 Supf. temporal a. and v.
12 Rostr. auricular a. and v.
13 Caud. auricular a. and v.
14 Maxillary a. and v.
15 Ext. jugular v.
16 Linguofacial v.
17 Hyoid venous arch
18 Lingual a. and v.

Legend:

a Mental rr.
b Infraorbital a. and v.
c Sphenopalatine a. and v.
d Greater palatine a. and v.
e Rostr. deep temporal a. and v.

f Lesser palatine a. and v.
g Buccal a. and v.
h Palatine plexus
i Inferior alveolar a. and v.
j Dors. ext. ophthalmic v.

k Ventr. ext. ophthalmic v.
l Ophthalmic plexus and ext. ophthalmic a.
m Pterygoid plexus
n Deep temporal a. and v.

o Transverse facial a. and v.
p Ascending pharyngeal a. and v.
q Pharyngeal v.
r Ext. carotid a.
s Cran. thyroid a. and v.

t Int. carotid a. and carotid sinus
u Occipital a. and v.
v Int jugular v.
w Com. carotid a.

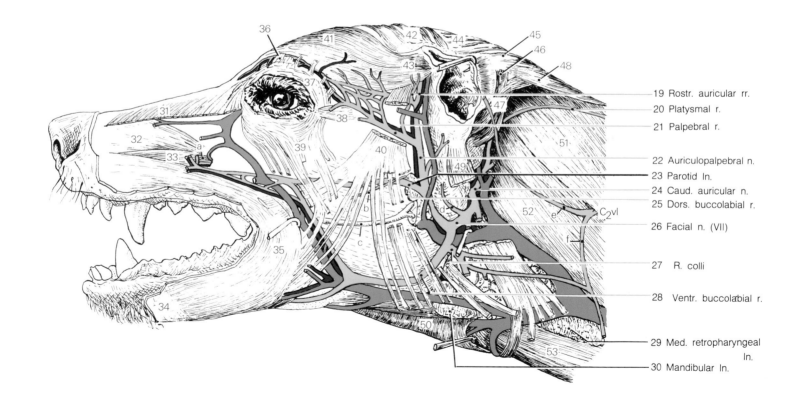

19 Rostr. auricular rr.
20 Platysmal r.
21 Palpebral r.
22 Auriculopalpebral n.
23 Parotid ln.
24 Caud. auricular n.
25 Dors. buccolabial r.
26 Facial n. (VII)
27 R. colli
28 Ventr. buccolabial r.
29 Med. retropharyngeal ln.
30 Mandibular ln.

Legend:

31 M. levator nasolabialis (section)
32 M. levator labii superioris
33 M. caninus
34 M. mentalis
35 M. orbicularis oris
36 M. levator anguli oculi med.

37 M. orbicularis oculi
38 M. retractor anguli oculi lat.
39 M. malaris
40 M. zygomaticus (section)
41 M. frontoscutularis
42 M. interscutularis

43 M. scutuloauricularis supf.
44 M. occipitalis
45 M. cervicoauricularis
46 M. cercicoauricularis medius
47 M. cervicoauricularis prof.
48 M. cervicoauricularis supf.

49 M. parotidoauricularis (section)
50 M. mylohyoideus
51 M. cleidocervicalis
52 M. sternocleidomastoideus
53 M. sternohyoideus

a Infraorbital n.
b Transverse facial r.
c Parotid duct
d Int. auricular r.
e Great auricular n.
f Transverse cervical n.

The muscles of the face and those associated with the mandible are two of the six muscle groups of the head. The division into muscle groups results from a consideration of their genesis and innervation. The facial muscles are supplied by the facial n. (VII), and the muscles of the mandibular space and those of mastication by the mandibular n. (V 3), a major branch of the trigeminal n. (V). The muscles of the ocular bulb are innervated by the oculomotor (III), trochlear (IV) and abducens (VI) nn. Pharyngeal muscles are supplied by glossopharyngeal (XI) and vagus (X) nn., and the muscles of larynx by the vagus n. (X). The hypoglossal n. (XII) supplies the muscles of the tongue.

a) The **MUSCLES OF THE FACE**, innervated by the facial n. and hence known as **FACIAL MUSCLES**, are divided into a superficial and a deep group. The deep facial muscles (the mm. stapedius, occipitohyoideus and stylohyoideus) are either not demonstrated or are discussed with those of the mandibular space (m. digastricus). In the main, the superficial facial muscles have the characteristics of cutaneous muscles. Arising chiefly on smooth bony areas or from fascia, they either radiate in the skin or are arranged in a sphincter-like manner around facial openings by a looping course of their fibers. Facial muscles are not antagonistic to one another and do not act on articulations. On the contrary, they determine facial expression by their disposition and are therefore called mimetic muscles. In more primitive animals the facial muscles are stratified into three more or less uniform layers, one upon the other; in phylogenetically advanced development the individual facial muscles are derived and become differentiated from these layers.

The transverse muscle fibers of the very weak m. sphincter colli superficialis lie in the ventral neck region and are still distinct in the laryngeal region. No facial muscles are derived from this muscle.

The platysma with its muscle fibers running longitudinally, divides the auricular muscles caudally. By means of the m. cutaneus faciei the platysma radiates into the lips.

Originally, in a phylogenetic sense, the m. sphincter colli profundus consisted of a transverse cutaneous muscle layer which, even yet, is evident in the direction of the muscle fibers of the mm. malaris, zygomaticus and parotidoauricularis. In the course of phylogenesis, the original uniform muscle coverage seen for example in the mole, has divided into individual ear muscles and the muscles of eyelids, nose, lips and cheek.

I. Of the four groups of **auricular muscles**, caudal, rostral, ventral and dorsal, two of the most superficial muscles of each are described. **Functionally** the auricular muscles are classified into tensors of the scutiform cartilage, outward movers of the conchal fissure, levators and depressors of the external ear, and inward movers of the conchal fissure. Each muscle derives its name from the sites of its origin and insertion. **Innervation:** Auricular and cervical branches of facial n. The **scutiform cartilage (11)** lies rostrodorsal to the external ear between individual ear muscles, and is a moveable site of attachment for several muscles. Of the caudal auricular muscles, the **m. cervicoauricularis superficialis (8)** and its rostral continuation, the **m. cervicoscutularis (9)** are fused at their origin in the nuchal midline, separating only at their insertions to auricular and scutiform cartilages respectively. Both muscles elevate the ear, the latter also tensing the scutiform cartilage. (The underlying mm. cervicoauricularis medius and cervicoauricularis profundus move the conchal fissure outwards; see page 37A, 46 and 47).

Of the dorsal auricular muscles the weak **m. occipitalis (10)** runs rostrally in an arch from the external sagittal crest of the parietal bone, contacting the scutiform cartilage on its lateral side and thus tensing it.

By means of its transverse muscle bundles the **m. interscutularis (13)** lying rostrally, joins right and left scutiform cartilages and tenses them. Of the rostral auricular muscles, the **m. frontoscutularis (1)**, which is continuous with the previous muscle, runs in a rostroconvex arch across the midline, thus connecting both scutiform cartilages and tensing them. Rostrally, muscle bundles shear off the **m. frontoscutularis**, and without attaining the midline, radiate into the upper eyelid. The **m. scutuloauricularis superficialis (12)** runs from the scutiform cartilage to the rostral border of the auricular cartilage, thus moving the conchal fissure inwards.

Of the ventral auricular muscles, the **m. parotidoauricularis (15)**, a depressor, runs from the laryngeal region to the base of the ear superficial to the parotid gland. The variable m. mandibuloauricularis, a depressor also, runs deeply between mandible and base of ear and can be either of some size or absent (see p. 38A, 27).

II. The **muscles of eyelid and nose** are supplied by the auriculopalpebral n., a branch of the facial n.

The **m. orbicularis oculi (3)** is the annular muscle that closes the eyelids.

The **m. retractor anguli oculi lateralis (2)** radiates into the lateral angle of the eye.

The **m. levator anguli oculi medialis (14)** runs near the medial angle of the eye in the upper eyelid.

The **m. levator nasolabialis (4)** begins at the medial angle of the eye and widens out over the nose to radiate into the upper lip.

The **m. malaris (5)** is interwoven with muscle fibers of platysma, its very delicate fibers running from cheek into lower eyelid.

III. The **muscles of the lips and cheek** are innervated by dorsal and ventral buccolabial rami of facial n.

The **m. orbicularis oris (6)** surrounds the oral cleft in a circular manner with a median interruption in both the upper and the lower lip.

Rostrally the **m. buccinator (20)** is situated deep to the m. orbicularis oris and is interwoven with it. It radiates from the upper lip, over the cheek, to the alveolar border of the mandible. In the cheek an intercrossing of muscle fibers occurs to form a longitudinal muscle raphe (see p. 38A lower right).

The **m. zygomaticus (7)** courses between the angle of the mouth and the scutiform cartilage. From its caudal border transverse muscle bundles shear away towards the mandibular space.

The **m. caninus** (see p. 37A, 33) arises ventral to the infraorbital foramen and radiates into the upper lip at the level of the canine tooth.

The **m. levator labii superioris** (see p. 37A, 32) arises rostral to the infraorbital foramen and proceeds into the upper lip dorsoparallel to the m. caninus.

The mandibular muscles are demonstrated on the right side of the head by removing the skin and the superficial facial muscles overlying the muscles of mastication and the temporal region. Innervation from the mandibular n. (V 3) is demonstrated after the removal of the mandible (see p. 39A).

b) The **MUSCLES ASSOCIATED WITH THE MANDIBLE** are subdivided into superficial muscles of the intermandibular region and external and internal muscles of mastication, in accordance with function and position on mandible. Each group consists of two muscles. As constrictors of the oral cleft, the muscles of mastication, particularly in carnivors, are very strong with tendinous tissue permeating through them. They also possess distinct osseous fossae at origin and/or insertion such as the temporal, masseteric and pterygoid fossae for reception of the respective muscles of like name. Of the muscles of the intermandibular region, only the digastric opens the mouth, and as a levator of the tongue the m. mylohyoideus promotes the movement of food between the cheek teeth, thus assisting in mastication.

I. Of the **superficial muscles of the intermandibular region**, the **m. digastricus (18)** runs from the jugular process to the ventral border of the body of the mandible. The two muscle bellies are separated only by a weak, indistinct tendon. Only the rostral belly is innervated by the

mandibular n., the caudal one being supplied by the facial n. The **m. mylohyoideus (19)** has a linear origin from the very weak mylohyoid line on the medial side of the body of mandible, and its transverse muscle fibers run to a median muscle raphe. The hammock-like muscle elevates and supports the tongue and floor of the mouth.

II. Of the **external muscles of mastication**, the **m. temporalis (16)** runs from the base and around the border of the temporal fossa to the coronoid process of the mandible. An accessory portion with a completely divergent muscle fiber direction commences at the caudal end of the zygomatic bone and courses dorsoparallel (to the bone) as far as the rostral border of the coronoid process. The varying fiber direction is recognizable after the removal of the superficial tendinous aponeurosis. The **m. masseter (17)** consists of a superficial and a deep portion taking origin from the lateral and medial aspects of the zygomatic bone. The two portions insert respectively into the masseteric fossa and the border surrounding it.

38

Muscles of head

(Regions of face)

1 M. frontoscutularis

2 M. retractor anguli oculi lat.

3 M. orbicularis oculi

4 M. levator nasolabialis

5 M. malaris

6 M. orbicularis oris

7 M. zygomaticus

15 M. parotido-auricularis

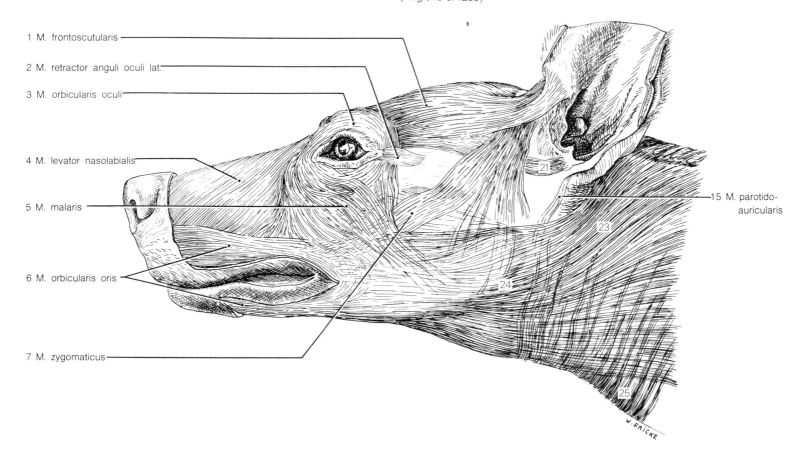

(Frontal region)

(Masseteric region; Temporal region)

8 M. cervicoauricularis supf.

9 M. cervicoscutularis

10 M. occipitalis

11 Scutiform cartilage

12 M. scutulo-auricularis supf.

13 M. interscutularis

14 M. levator anguli oculi med.

16 M. temporalis:
Superficial tendon (section)
Principal part
Accessory part

17 M. masseter (section)

18 M. digastricus

19 M. mylohyoideus

20 M. buccinator

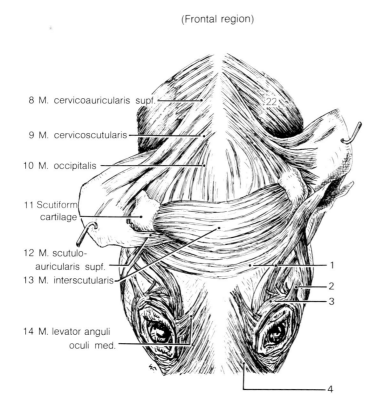

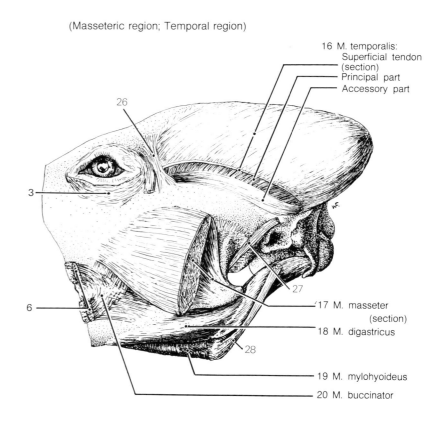

Legend:

21 M. zygomaticoauricularis
22 Platysma:
23   M. cutaneus colli
24   M. cutaneus faciei

25 M. sphincter colli supf.
26 Orbital lig.
27 M. mandibuloauricularis (section)
28 M. stylohyoideus

C. 8. S. 5: DEEP MUSCLES OF MASTICATION, TRIGEMINAL NERVE (V), MAXILLARY NERVE (V 2) AND MANDIBULAR NERVE (V 3)

*The head is bisected midsagittally. The right half is used for the present dissection and involves the disarticulation of the right side of the mandible and subsequent demonstration of internal muscles of mastication, and the mandibular and maxillary nn. The masseter muscle is removed in layers so that one is able to observe the different fiber directions of the superficial and deep parts of the muscle and its strong central tendons. Two centimeters rostral to the temporomandibular articulation, the terminal branches of the **masseteric n.** (3) are then demonstrated on the cut surface of the muscle after the nerve courses laterally over the mandibular notch. The zygomatic arch is sawn through at the temporomandibular articulation and rostral to the attachment of the orbital ligament. The isolated piece of arch is then removed. The m. digastricus is detached at its insertion onto the ventral border of the body of the mandible, taking care to preserve the branch of the mylohyoid n. innervating it. Subsequently the m. mylohyoideus is incised at its origin along the mylohyoid line in common with the oral mucous membrane lying deeper. With a strong sideways movement of the mandible it is possible to transect: a) the insertion of the m. temporalis onto the medial and lateral surfaces of the coronoid process of mandible; and b) the mm. pterygoideus medialis and lateralis at the pterygoid fossa. Following this, the inferior alveolar a., v., and n. are cut through at the mandibular foramen, the entrance to the mandibular canal. The preparation is completed by disarticulating the mandible after transecting associated ligaments. The articular cavity is subdivided into two 'storeys' by the articular disc. To widen the field of dissection, the m. temporalis is removed in part as far as the dorsal contour of the periorbita. The **deep temporal nn.** (2) within the m. temporalis are preserved and followed retrogressively to their ramification from the masticatorius n. in common with the masseteric n. To demonstrate the branches of the maxillary n., the zygomatic gland is retracted from the pterygopalatine fossa.*

a) The **INTERNAL MUSCLES OF MASTICATION** include the strong m. pterygoideus medialis and the weak m. pterygoideus lateralis. They run from the pterygoid and adjacent bones, to the indistinct pterygoid fossa on the medial surface of the ramus of the mandible. The mm. pterygoideus medialis and pterygoideus lateralis are innervated by the nerves of like name situated deeply. These leave the mandibular n. (V 3) directly after its emergence from the foramen ovale.

The **m. pterygoideus medialis (8)** is covered by a conspicuous tendinous aponeurosis on its external surface and crossed superficially by the mandibular n.

The **m. pterygoideus lateralis (18)** is essentially smaller, lying in the bifurcation between the buccal n. and the parent mandibular n.

b) The **TRIGEMINAL N.** (V, see also text illustration) has a large sensory root and a smaller motor root. After leaving the brain but before passing through the skull, the sensory fibers of the large root are associated with the trigeminal ganglion (see p. 48A). At the ganglion, the peripheral fibers, the continuations of the neurons form the three branches of the trigeminal nn. The first branch, the ophthalmic n., containing sensory fibers, passes through the orbital fissure. The second branch, the maxillary n., also sensory, passes from the alar canal to ramify in the pterygopalatine fossa. The third branch, the mandibular n. (V 3), unites a sensory portion with the motor root before passing through the foramen ovale.

1 c) The **MANDIBULAR N.** (V 3 —5) arises at the foramen ovale medial to the temporomandibular articulation and discharges the nerve branches (nerve fiber types - see p. 48A) listed.

The **masticatorius n. (4,** see above) supplies purely motor branches, the deep temporal n. and the masseteric n., to the corresponding external muscles of mastication.

The **buccal n. (1)** in common with the masticatorius n., passes dorsolaterally over the m. pterygoideus lateralis. Its sensory fibers supply the oral mucous membrane and its autonomic fibers the zygomatic and buccal glands.

The **lingual n. (10)** is the direct continuation of the mandibular n. Ventrocaudally, the chorda tympani of the facial n. joins the commencement of the lingual n. at an acute angle. The special sensory fibers of the chorda tympani supply the taste buds of the rostral two-thirds of the tongue. Its autonomic contribution reaches the mandibular gland and rostrally, the sublingual gland. Sensory fibers of the lingual n. supply the tongue and mucous membrane of the floor of mouth. This is also supplied by sensory fibers of the **sublingual n. (11).**

The **auriculotemporal n. (7)** branches near the emergence of the mandibular n. from the foramen ovale and runs caudally around the temporomandibular articulation in a convex arch. Its autonomic section supplies the parotid gland, and a sensory branch, the n. of the external acoustic meatus, supplies the meatus as far as the tympanic membrane.

Sensory fibers in the rostral auricular rami and the transverse ramus of face go to the rostral border of external ear and the face respectively. The **mylohyoid n. (9)** innervates the rostral belly of the m. digastricus, providing it with motor fibers, while its submental rami contribute sensory supply to the chin region.

The **inferior alveolar n. (6)** enters the mandibular canal at the mandibular foramen (where it was previously transected). It provides sensory 2 fibers to the teeth of the mandible and the skin of the chin via mental rami emerging through mental foramina.

d) The **MAXILLARY N.** (V 2 —16) enters the alar canal through the foramen rotundum and leaves it through the rostral alar foramen. This purely sensory nerve gives three main branches.

The **zygomatic n. (14)** proceeds to the ocular bulb and bifurcates within the periorbita into a ventral **zygomaticofacial ramus (13)** and a dorsal **zygomaticotemporal ramus (12).** The latter contains autonomic fibers from the pterygopalatine ganglion (see below) and conveys these to the lacrimal gland. After leaving the orbit both rami supply the skin of face dorsal to the eye.

The **infraorbital n. (15),** a sensory nerve, continuing the direction of the maxillary n., passes through the maxillary foramen in the pterygopalatine fossa to enter the infraorbital canal. Within the canal it provides branches to the teeth, and, leaving the infraorbital foramen, branches to nose and upper lip. The **pterygopalatine n. (17)** leaves the maxillary n. ventrorostrally. At its dorsal border, medial and deep to the infraorbital n., the pterygopalatine n. exhibits the brownish pterygopalatine ganglion. This contains both parasympathetic and sympathetic fibers. These fibers arrive at the ganglion via the thin nerve of the pterygoid canal from facial n. (VII), and the sympathetic trunk respectively. After some of the contained fibers synapse with postganglionic neurons, the pterygopalatine n. continues as far as its end branches. These provide sensory fibers to mucous membrane and autonomic fibers to the glands of palate and nasal cavity. In dorsoventral series the three end branches are in order: The **caudal nasal n. (21),** passing through the sphenopalatine foramen to the lateral nasal gland and conchae; the **greater palatine n. (20)** reaching the hard palate through the foramen of like name; and the **lesser palatine n. (19)** extending mainly to the soft palate through the minor palatine foramen.

Trigeminal n. (V)

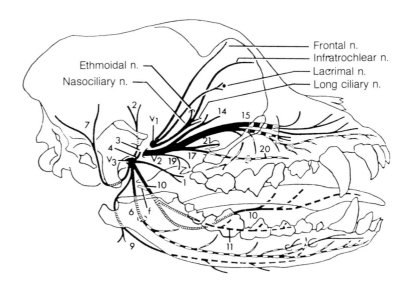

Ethmoidal n.
Nasociliary n.
Frontal n.
Infratrochlear n.
Lacrimal n.
Long ciliary n.

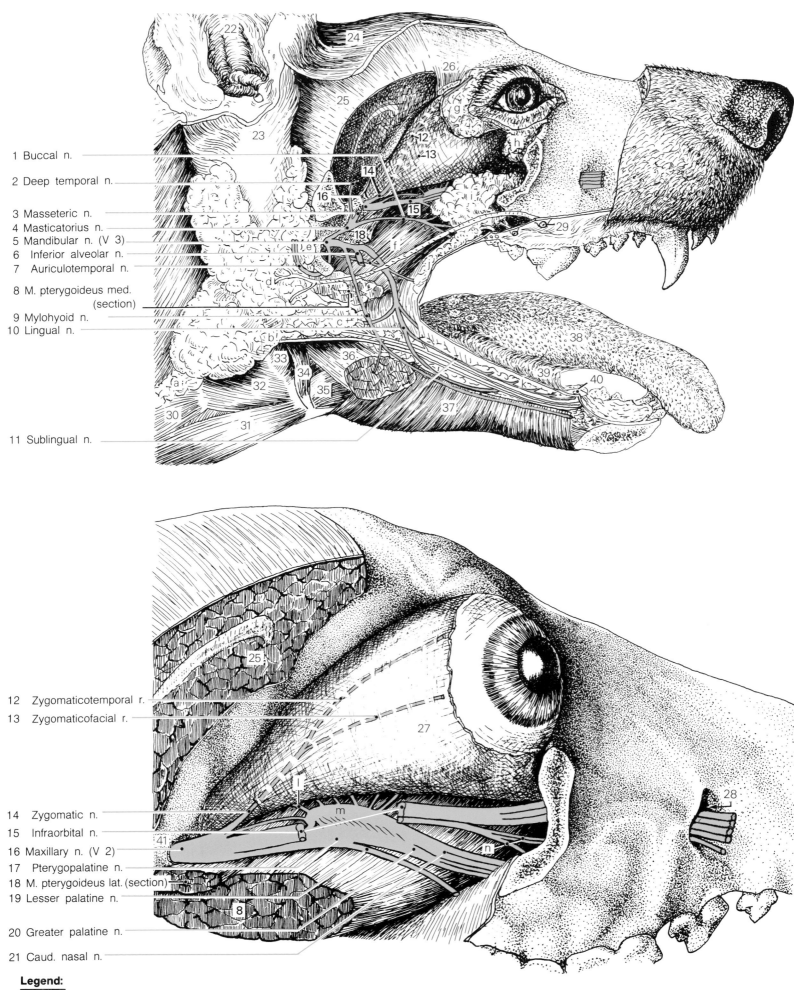

1 Buccal n.

2 Deep temporal n.

3 Masseteric n.
4 Masticatorius n.
5 Mandibular n. (V 3)
6 Inferior alveolar n.
7 Auriculotemporal n.

8 M. pterygoideus med.
         (section)
9 Mylohyoid n.
10 Lingual n.

11 Sublingual n.

12 Zygomaticotemporal r.

13 Zygomaticofacial r.

14 Zygomatic n.
15 Infraorbital n.
16 Maxillary n. (V 2)
17 Pterygopalatine n.
18 M. pterygoideus lat. (section)
19 Lesser palatine n.

20 Greater palatine n.

21 Caud. nasal n.

Legend:

22 Auricle (Scapha)	30 M. sternothyroideus	38 Tongue	a Med. retropharyngeal ln.	h Supf. gl. of third eyelid
23 Auricular cartilage	31 M. sternohyoideus	39 Frenulum	b Mandibular gl. and duct	i Zygomatic gl. and duct
24 M. frontoscutularis (section)	32 M. thyrohyoideus	40 Sublingual caruncle	c Monostomatic sublingual	j Buccal gll.
25 M. temporalis (section)	33 M. hyopharyngeus	41 Rostr. alar foramen	gl. and duct	k Polystomatic sublingual gl.
26 Orbital lig. (section)	34 M. stylohyoideus		d Parotid gland and duct	l N. of pterygoid canal
27 Periorbita	35 M. hyoglossus		e Parotid ln.	m Pterygopalatine ggl.
28 Infraorbital for.	36 M. digastricus (section)		f Chorda tympani	and orbital rami
29 Parotid papilla	37 M. mylohyoideus		g Lacrimal gl.	n Superior alveolar rami

1
2
With the separation of upper and lower eyelids, parts of the lacrimal apparatus are exposed as a result. Following this, the apex of the triangular cartilage supporting the third eyelid is freed from its site of attachment in the orbital cavity. On the lateral surface of the cartilage, the cutaneous covering is removed to expose the accessory lacrimal glands, namely the superficial gland of the third eyelid, while numerous lymph nodules are observed on the medial surface of the third eyelid. To expose and dissect the ocular bulb, more of the m. temporalis is ablated (see p. 40A, middle illustration), remaining periorbita removed, and the cone of eye muscles retracted ventrolaterally away from the osseous part of the orbital cavity. At this stage, one sees the trochlea (16) dorsomedial to the ocular bulb and should detach it from its site of attachment to the orbit. The optic n. lies centrally in the cone of extrinsic eye muscles and is surveyed by retracting them.

a) The **LACRIMAL APPARATUS** (see also text illustration) includes
3 the lacrimal glands and their system of excretory ducts. The **lacrimal gland (11)** lying deep to the orbital ligament and the **superficial gland of the third eyelid (15)** or accessory lacrimal gland, secrete lacrimal fluid. This flows from the glands through narrow ductules into the **superior conjunctival fornix (A)** of the conjunctival sac. With closure of the eyelids, the cornea is moistened by a film of lacrimal fluid. This collects deep to the medial angle of the eye at the **lacrimal lake (B)** in the middle of which is the **lacrimal caruncle (C)** projecting from surrounding fluid. A few millimeters away from the medial angle of the
4 eye, the **lacrimal puncta (12)** lie on the pigmented margins of the reflected upper and lower eyelids. Lacrimal fluid flows through these small openings to superior and inferior **lacrimal canaliculi (13)** which, in turn, unite at the **lacrimal sac (14)**. The nasolacrimal duct begins there, lies at first in the osseous lacrimal canal and then more rostrally in the lacrimal groove of the maxilla. The duct conveys the lacrimal fluid to the **nasolacrimal ostium,** a millimeter-sized opening in the nasal vestibule about one centimeter caudal to the ventral angle of the naris. The opening lies facing the nasal septum at the beginning of the ventral nasal meatus where a distinct pigmented border is visible (see p. 41A —73).

Lacrimal apparatus (lateral view)

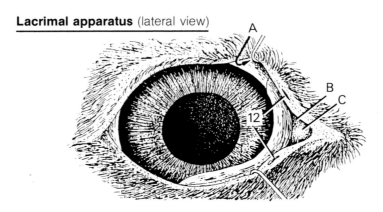

b) The **OPTIC N. (II —8)** consists of a bundle of nerve fibers beginning at the retina. Shortly after leaving the ocular bulb the nerve receives a myelin sheath and goes to the optic chiasm. Its continuation then proceeds to the diencephalon in the optic tract. Due to its developmental history, the optic nerve is regarded as a part of the brain. It is surrounded by a continuation of the three meninges and the myelin sheath is formed from glial cells (oligodendroglia) of the central nervous system.

5 **c)** The **OPHTHALMIC N. (V 1,** see text illustration p. 39) is the first branch of the trigeminal n. (V). Lacrimal, frontal and nasociliary nn. branch from it.

The **lacrimal n. (22)** is thread-like and accompanies the artery of like name and the zygomaticotemporal ramus of the maxillary n. Its autonomic fibers from the pterygopalatine ganglion supply the lacrimal gland, and its sensory part, the upper eyelid.

The **frontal n. (19)** leaves the orbit lateral to the trochlea and adjacent to the angular v. of the eye. It supplies sensory innervation to the skin of the frontal region.

The **nasociliary n. (5)** branches into the infratrochlear, ethmoidal and long ciliary nn. which are followed to it retrogressively (see lower p. 40A) during the dissection. The **infratrochlear n. (17)** leaves the orbit ventromedial to the trochlea and supplies the skin at the medial angle of the eye. The **ethmoidal n. (3)**, containing both sensory and autonomic fibers, goes through the ethmoid foramen (site of identification) and then through the lamina cribrosa to the fundus of the nasal cavity. One is able to follow it deeper between the m. rectus medialis and the m. obliquus dorsalis to its origin from the nasociliary n. Likewise, the **long ciliary nn. (4)** branch from the nasociliary n. and accompany the optic n. laterally. Their sympathetic fibers innervate the m. dilator pupillae and provide sensory fibers to the conjunctiva.

d) The **NERVES** and **MUSCLES OF THE EYE** are discussed together. In the ongoing dissection, the nerves to the extrinsic eye muscles (III, IV and VI) are used to identify the muscles themselves and the nerves are also followed retrogressively to their origins. The muscles of the ocular bulb include the mm. rectus dorsalis, — medialis, — ventralis, and — lateralis, the mm. obliquus dorsalis and — ventralis, the m. levator palpebrae superioris, and the m. retractor bulbi enveloping the optic n. In the main, the recti muscles move the ocular bulb medially towards the nose or laterally towards the temporal region, as well as dorsally and ventrally. The mm. obliquus dorsalis and ventralis produce inward and outward rotation of the bulb respectively.

I. The **oculomotor n. (III)** innervates all extrinsic muscles of the eye except the mm. obliquus dorsalis and rectus lateralis.

II. The **trochlear n. (IV)** supplies only the m. obliquus dorsalis.

III. The **abducens n. (VI)** innervates the m. rectus lateralis and the lateral portion of the m. retractor bulbi.

The **oculomotor n. (6)** terminates with its **ventral ramus (7)** passing into the **m. obliquus ventralis (23)** distally. The ramus is followed retrogressively between the mm. recti lateralis and ventralis and finally through the m. retractor bulbi to the lateral aspect of the optic nn. The **ciliary ganglion (10)** is detected there as a brownish body, the size of a millet seed. Cobweb-like short ciliary nn. from the ganglion accompany the optic n. and penetrate the sclera. They supply parasympathetic fibers to the mm. ciliaris and sphincter pupillae and sympathetic fibers to the dilator pupillae. The dorsal branch of the oculomotor n. innervates the **m. rectus dorsalis (21)** and the more superficial **m. levator palpebrae superioris (20),** both of which are crossed in succession by the frontal n. (from V 1). The **trochlear n. (1)** enters the **m. obliquus dorsalis (18)** at the junction of its proximal and middle thirds. The tendon of insertion of the muscle turns around the cartilaginous trochlea, changing from a longitudinal to a transverse direction. It terminates deep to the insertion of the m. rectus dorsalis. The adjacent **m. rectus medialis (9)** ventromedially, is supplied by the oculomotor n. The **m. rectus lateralis (24)** (an abductor or outward drawer of the bulb) is penetrated on its dorsal aspect of the **abducens n. (2).**

e) The **EXTERNAL NOSE** (see text illustration) extends from its base or **root (D)** to the **dorsum (E)** and then the **apex (G).** There each of the

External nose (lateral view)

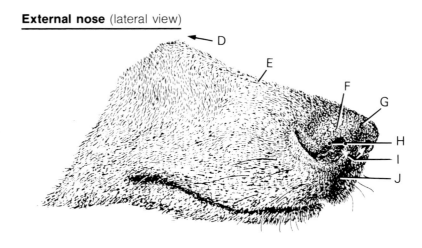

nares (I) is bounded by a medial and a lateral **ala of the nose (H)** and the **planum nasale (F).** These rostral features are supported by the cartilages of nose and septum. The planum nasale is formed by bare, modified skin in the region of the nares, and has a median **philtrum (J).** The nasal septum is membranous in the region of nares, followed by cartilage in its rostral two-thirds and bone in its caudal third.

40

Lacrimal apparatus, Accessory organs of eye, and Cranial nn. II, III, IV, V 1, V 2, and VI

Legend:

25 M. frontoscutularis (section)
26 M. temporalis (section)
27 Orbital lig. (section)
28 Periorbita (section)
29 Maxilla (section)
30 Straight fold
31 Alar fold
32 Rostr. alar for.
33 M. pterygoideus lat. (section)
34 M. pterygoideus med. (section)
35 M. retractor bulbi
36 M. rectus ventr.
37 Zygomatic bone (section)

a   Maxillary n. (V 2)
b   Zygomatic n.
b'   Zygomaticotemporal r.
b'' Zygomaticofacial r.
c   Nasolacrimal duct
d   Infraorbital n.
e   N. of pterygoid canal
f   Pterygopalatine n.
g   Pterygopalatine ggl.
     and orbital rr.
h   Superior alveolar rr.
i   Caud. nasal n.
j   Greater palatine n.
k   Lesser palatine n.
l   Short ciliary nn.

1 Trochlear n. (IV)

2 Abducens n. (VI)

3 Ethmoidal n.

4 Long ciliary nn.

5 Nasociliary n.

6 Oculomotor n. (III)
  Dors. r.
7 Ventr. r.

8 Optic n. (II)
9 M. rectus med.
10 Ciliary gl.

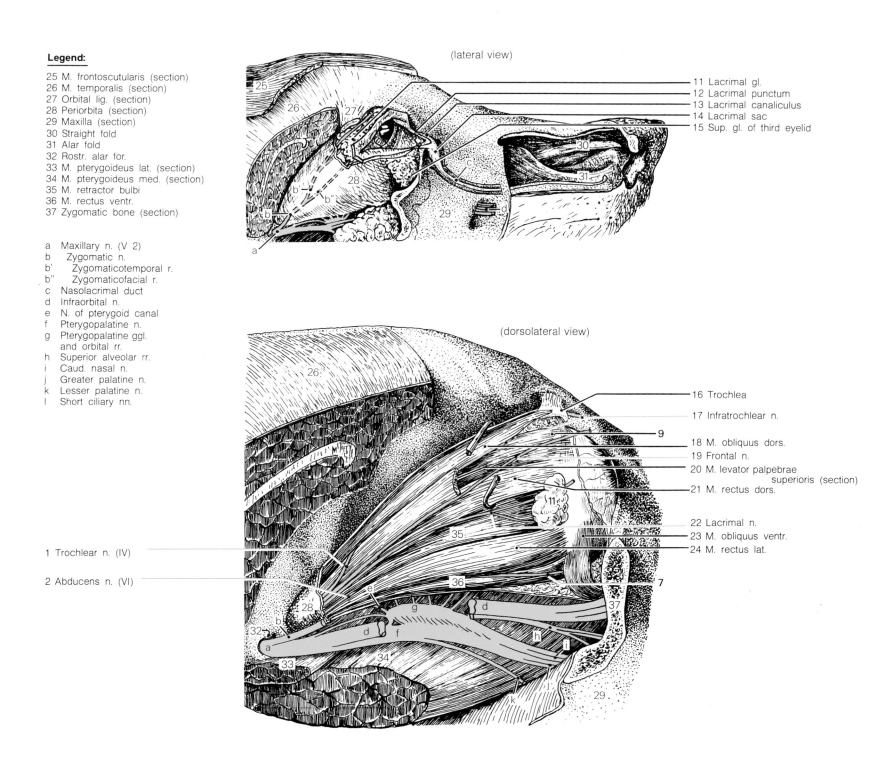

(lateral view)

11 Lacrimal gl.
12 Lacrimal punctum
13 Lacrimal canaliculus
14 Lacrimal sac
15 Sup. gl. of third eyelid

(dorsolateral view)

16 Trochlea
17 Infratrochlear n.
18 M. obliquus dors.
19 Frontal n.
20 M. levator palpebrae
     superioris (section)
21 M. rectus dors.
22 Lacrimal n.
23 M. obliquus ventr.
24 M. rectus lat.

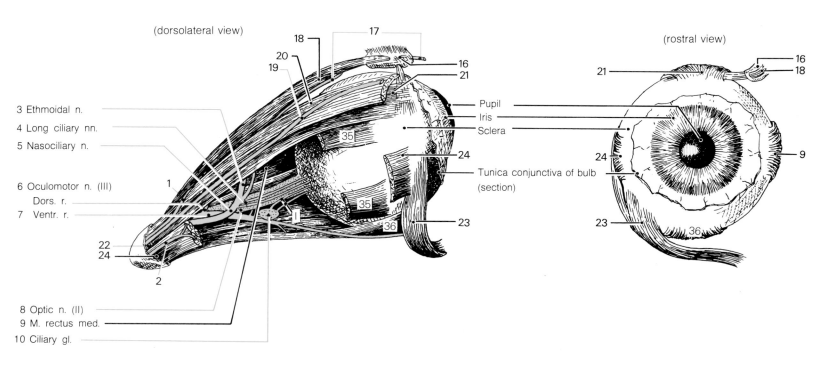

(dorsolateral view)

(rostral view)

Pupil
Iris
Sclera
Tunica conjunctiva of bulb
(section)

1 **a) NOSE:**

2 I. The **nasal cavity** begins at the naris with the vestibule of the nose (33) which is covered by a pigmented cutaneous mucous membrane. The vestibule houses: the **straight fold (39)** dorsally, a process of the dorsal nasal concha; the **alar fold (37)** just ventral to this, an offshoot of the ventral nasal concha; and the indistinct **basal fold (35)** ventral to this again. The nasal cavity proper is covered with respiratory mucous membrane and contains the dorsal and ventral nasal conchae (see p. 36). The middle nasal concha coming from the caudal part of the nasal cavity inserts between them. The lateral nasal gland lies in the maxillary recess of the maxilla, and like the nasolacrimal duct opens into the nasal vestibule. Both the secretion from the gland and the lacrimal fluid provide the moisture on the planum nasale. The **fundus of the nose (43)** accommodates the ethmoidal labyrinth and its covering of olfactory mucous membrane.

II. Four **nasal meatuses** pass through the nasal cavity. The **dorsal nasal meatus (40)** between the dorsal nasal concha and the nasal bone is directed to the olfactory organ and hence is an olfactory meatus. Initially, the **middle nasal meatus (38)** lies between the dorsal and ventral nasal conchae and then divides at the middle nasal concha to proceed to the paranasal sinus. Hence it is known as a sinuosal meatus. The **ventral nasal meatus (36)**, known also as the respiratory meatus, lies between the ventral nasal concha and the hard palate. It is directed caudally to the fundus of the nose and is continued by the nasopharyngeal meatus as far as the choana. Olfactory, sinuosal and respiratory meatuses merge medially into the **common nasal meatus (41)** extending over the entire dorsoventral extent of the nasal cavity along the nasal septum.

III. The **olfactory organ** lies in the fundus of the nose, its olfactory mucous membrane covering the ethmoidal labyrinth (see p. 50).

IV. The **vomeronasal organ** (see p. 50) likewise has an olfactory mucosa. It lies on the floor of the nasal vestibule at the transition to the nasal cavity proper and is in direct association with the cartilaginous nasal septum. It communicates with the roof of the oral cavity through the incisive (or nasopalatine) duct. Functionally it is useful as an oral olfactory organ for the reception of odors or pheromones.

V. The **olfactory nn. (I)** consist of fibers of olfactory cells running chiefly from the olfactory organ in the fundus of the nose and also from the vomeronasal organ in the nasal vestibule, via the vomeronasal n. The nerves pass through the lamina of the ethmoidal bone to the olfactory bulb of the telencephalon (see p. 48A).

b) The LARYNX is palpable ventral to the boundary between head and neck, forming part of the respiratory apparatus and serving in phonation. Its entrance can be sealed off from pharynx and from the path of food as occurs during swallowing. From within outwards the wall of larynx consists of mucous membrane, cartilage and laryngeal muscles.

The **innervation** of the laryngeal muscles and mucous membrane is provided by the vagus n. Its caudal laryngeal n. innervates all laryngeal muscles except for the most caudal and external of them, namely the m. cricothyroideus. This is supplied by the cranial laryngeal n. also a branch of the vagus (see p. 24A —15). Cranial to the rima glottidis sensory innervation is provided by the internal ramus of the cranial laryngeal n., and caudal to it by the caudal laryngeal n.

3 I. The **laryngeal mucosa** lines the lumen of the larynx. At the entrance of larynx, on the margins of the vocal folds, and in the lateral ventricles it is a cutaneous squamous epithelium whereas in the remaining areas a ciliated respiratory mucosa is present.

II. The **cavity of the larynx** (see p. 41A) includes the vestibule of larynx, the rima glottidis and glottis, and the infraglottic cavity.

The **vestibule of the larynx (E)** extends from the border of the epiglottis to the **vestibular folds (F")**. The **glottis (F)** is formed from both 4 vocal cords whose free borders, the **vocal folds (F)**, limit the intermembranous part of the rima glottidis. The intercartilaginous part of the 5 rima is formed by the arytenoid cartilage of its respective side and its mucous membrane lining. The vocal ligament lies deep to the vocal cord and the m. vocalis limits it laterally. Between the vestibular and the vocal 6 fold lies the entrance to the **lateral ventricle of the larynx (F')**. The infraglottic cavity extends from the vocal cords to the first tracheal ring.

III. Of the four **laryngeal cartilages,** portions of the arytenoid cartilages and the **epiglottic cartilage (22)** consist of elastic cartilage, the 7 remaining cartilaginous tissue being hyaline. The laminae of the unpaired **thyroid cartilage (27)** leave the thyroid cartilage open dorsally and enclose and protect the remaining cartilages to an extensive degree. The rostral cornu of the thyroid lamina is attached to the thyroid element of the hyoid apparatus, the caudal cornu to the cricoid cartilage. The **arytenoid cartilage (28)** is paired. It has a muscular process caudally, a vocal process ventrally for attachment of the vocal muscle and ligaments, a cuneiform process rostrodorsally, and a corniculate process caudodorsally. The unpaired **cricoid cartilage (29)** limits the laryngeal lumen dorsally between the laminae of the thyroid cartilage.

IV. Of the **laryngeal muscles** the m. cricoarytenoideus dorsalis is the 8 only one which widens the rima glottidis functionally, and it is important clinically. It runs from the cricoid cartilage to the muscular process of the arytenoid cartilage and by a lever action tenses the vocal ligament attached to the vocal process of the arytenoid cartilage. (Muscles narrowing the rima glottidis see p. 65).

c) The ORAL CAVITY includes the vestibule of the mouth (32) 9 between cheeks, lips and dental arches, and the **oral cavity proper (31)** between the dental arches of either side. It merges with the oropharynx caudal to the last molar teeth. The roof of the oral cavity is formed by the hard palate. The transverse **palatine rugae (42)** are bisected by a 10 midsagittal **palatine raphe** (not illustrated). Caudal to the upper central incisor teeth is the **incisive papilla (34)** situated in the midline, on which open the two incisive ducts from the nasal cavity. Each passes through a palatine fissure and ventral to its nasal opening is connected to the ipsilateral vomeronasal organ. The floor of the oral cavity proper bears the tongue and the frenulum.

d) The PHARYNX surrounds the pharyngeal cavity (see p. 41A). Within it, an intrapharyngeal ostium marks the intersection between the nasopharynx associated with respiration, and a pars digestoria associated with the passage of food material.

I. The **pars digestoria (B-D)** consists of an oral part, the oropharynx, also known as the isthmus faucium, and a laryngeal part, the laryngopharynx. The **oropharynx (B)** extends between the last molar tooth of either side and the base of epiglottis. The floor of the oropharynx is formed by the base of tongue, its side by the palatoglossal arch, which houses the palatine tonsil within the tonsillar sinus, and its roof by the soft palate. The **laryngopharynx (C)** commences at the intrapharyngeal ostium into which the rostral part of the larynx projects. The caudal part of the laryngopharynx is the esophageal vestibule. This merges with the esophagus over a distinct mound, the **limen pharyngoesophageum (30)**.

II. The **nasopharynx (A)** extends from the choanae to the **intrapharyngeal ostium (24)** bounded by the free border of the soft palate, the **arch of the soft palate (25)** and the **palatopharyngeal arch (26)**. Approximately halfway along the nasopharynx open the paired **pharyngeal ostia of the auditory tubes (44)**. The auditory tube opens into 11 the middle ear at the tympanic ostium.

PHARYNX

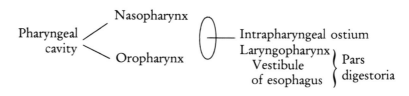

The lymphatic 'pharyngeal ring' is formed by pharyngeal tonsil on the roof of the nasopharynx, tonsil of the soft palate on the roof of the oropharynx, **palatine tonsil (23)** on the sides of the oropharynx, and 12 lingual tonsil on its floor.

Laryngeal cartilages (craniolateral view)

22 Epiglottic cartilage

Arytenoid cartilage

54

49 Thyroid cartilage

53

50

58

51

52

Median section of head

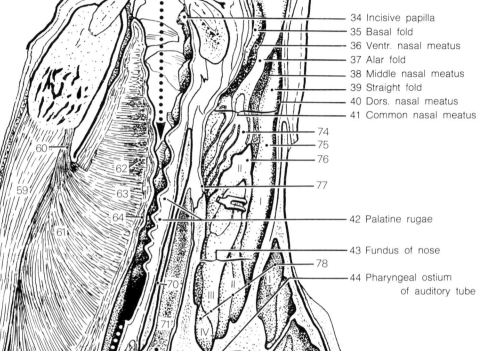

31 Oral cavity proper
32 Vestibule of mouth
33 Vestibule of nose
34 Incisive papilla
35 Basal fold
36 Ventr. nasal meatus
37 Alar fold
38 Middle nasal meatus
39 Straight fold
40 Dors. nasal meatus
41 Common nasal meatus
42 Palatine rugae
43 Fundus of nose
44 Pharyngeal ostium
   of auditory tube

23 Palatine tonsil

24 Intrapharyngeal ostium:
25 Arch of soft palate
26 Palatopharyngeal arch

27 Thyroid cartilage

28 Arytenoid
   cartilage

29 Cricoid cartilage
30 Limen pharyngo-
   esophageum

Cavity of larynx

A Nasopharynx
B Pars digestoria:
B' Oropharynx
C Laryngopharynx
D Vestibule of esophagus

Cavity of pharynx

E Vestibule of larynx
F Glottis and vocal fold
F' Lateral ventricle of larynx
F" Vestibular fold
G Infraglottic cavity

The dissection is performed simultaneously on medial and lateral sides of the specimen. The respiratory mucosa of the nasopharynx, from the pharyngeal ostium of the auditory tube as far as the intrapharyngeal ostium, is dissected away from underlying tissue in order to display muscles of soft palate, the rostral constrictor muscles of the pharynx and cranial nerves IX—XI. The m. longus capitis is also detached from its insertion on the muscular tubercle of the occipital bone (see p. 41A).

To demonstrate the external acoustic meatus a wedge-shaped piece of tissue is excised from the auricular cartilage and removed, beginning cranial and caudal to the quadrangular tragus (see lower p. 42A —76). This is similar to the technique used in the surgical treatment of otitis externa. For this purpose one begins the rostral incision at the tragohelicine or pretragic incisure and the caudal incision at the intertragic incisure. Both incisions converge on the osseous external acoustic meatus.

1 a) The **PHARYNGEAL MUSCLES** include dilators of the pharynx (I), muscles of the soft palate (II), and rostral, middle and caudal constrictors of pharynx (III-V).

The **innervation** of the pharyngeal muscles is derived from pharyngeal rami of the glossopharyngeal and vagus nn. coming from the pharyngeal plexus. This lies dorsolaterally on the pharynx. The m. tensor veli palatini supplied by the mandibular n. (V 3), and the m. stylopharyngeus caudalis innervated solely by the glossopharyngeal n. are the two exceptions.

I. The **m. stylopharyngeus caudalis (14)** is the only **dilator of the pharynx**. It extends from the caudal aspect of the stylohyoid element to 2 the dorsolateral wall of pharynx. II. Both **muscles of the soft palate** arise rostral to the tympanic bulla at the base of the skull. (Dissect from the medial side.) The **m. tensor veli palatini (2)** proceeds around the hamulus of the pterygoid bone to the soft palate. The **m. levator veli palatini (7)** runs perpendicularly into the soft palate. III. Of the two **rostral constrictors of the pharynx**, the **m. pterygopharyngeus (5)** arises at the hamulus and crosses lateral to the m. levator veli palatini before attaching to the wall of the pharynx. The **m. palatopharyngeus (6)** runs between the aponeurosis of the soft palate and the wall of the pharynx. IV. The **middle constrictor of the pharynx**, the **m. hyopharyngeus (16)**, runs to the pharynx dorsally from the thyrohyoid element. (Dissect from the lateral side.) V. Of the two **caudal constrictors of the pharynx**, the **m. thyropharyngeus (19)** arises from the thyroid cartilage and the ill-defined **m. cricopharyngeus (17)** from the cricoid cartilage. Both insert into the dorsal wall of pharynx.

b) The **CRANIAL NERVES OF THE VAGUS GROUP** all pass through the jugular foramen and are crossed laterally by the hypoglossal n. (XII).

The **glossopharyngeal n. (IX —9)**, shortly after its emergence from the jugular foramen, gives off the (parasympathetic) tympanic n. which continues as the minor petrosal n. through the otic ganglion to reach the mandibular n. (V 3) (see p. 48A). Then the glossopharyngeal n. innervates the m. stylopharyngeus caudalis and divides into lingual and pharyngeal rami medial to the tympanic bulla. The **lingual rami (13)** cross the stylohyoid element and the m. styloglossus arising from it, medially. They provide sensory innervation to the base of tongue and special sensory innervation to vallate and foliate papillae on its surface. Together with the nerve branch of like name from the vagus n., and sympathetic fibers from the cervical ganglion, the **pharyngeal ramus (10)** of the glossopharyngeal n. forms the pharyngeal plexus dorsolaterally on the pharynx. The plexus controls the motor and sensory supply to the pharynx.

The **vagus n. (X —8)**, after passing through the jugular foramen gives a sensory portion to the facial n., namely the internal auricular ramus, to supply the external acoustic meatus (see p. 37A). The **pharyngeal ramus (10)** communicates with the branch of like name from the glossopharyngeal n. and proceeds to the pharyngeal plexus. The vagus n. bends around caudally, exhibiting the distinct distal or nodose ganglion, the commencement of the **cranial laryngeal n. (15)**. The external ramus of the latter innervates the m. cricothyroideus while its internal ramus passes through the thyroid foramen to give sensory innervation to the laryngeal mucosa cranial to the rima glottidis. The vagus n. runs in the vagosympathetic trunk dorsoparallel to the common carotid a., the recurrent laryngeal n. being given off within the thoracic cavity (see p. 19A). In turn, this provides tracheal and esophageal rami and its continuation, the **caudal laryngeal n. (20)**, proceeds to the larynx deep to the m. cricothyroideus. With the exception of the latter muscle the caudal laryngeal n. innervates all laryngeal muscles and the laryngeal mucosa caudal to the rima glottidis.

The **accessory n. (XI—11)** runs dorsolateral to the distal or nodose ganglion (X) to innervate the mm. trapezius and sternocleidomastoideus.

c) The **AUTONOMIC NERVOUS SYSTEM OF THE HEAD** (see p. 48A) includes: the parasympathetic division, the nerve cells of which lie in rhombencephalon and midbrain; and the sympathetic division arising from the spinal cord of the thoracic region.

3 I. The **sympathetic trunk** (see page 19A) containing preganglionic fibers runs to the junction of head and neck in the **vagosympathetic** trunk (18). There the vagosympathetic trunk bifurcates, the vagus n. lying dorsally, the sympathetic trunk and its cranial cervical ganglion ventrally. After synapsing in the **cranial cervical ganglion (12)** unmyelinated postganglionic sympathetic fibers run to their effector organs or fields of innervation (for example glands, mucosae and intrinsic muscles of eye). They generally accompany the larger blood vessels of the head in the adventitia as reticula (for example the internal and external carotid plexuses). II. Preganglionic fibers of the **parasympathetic division**, 4 after emerging from the brain, follow cranial nerves III, VII, IX, X for the first part of their course. Within a short distance they ramify to their respective parasympathetic ganglia where synapsing occurs with postganglionic neurons. These reach their fields of innervation in common with branches of other cranial nerves. The vagus, cranial nerve X, behaves completely differently to the others in that it supplies its parasympathetic section to the body cavities and none to other cranial nerves.

d) The **ARTERIES OF THE HEAD** stem from the **common carotid a. (33)** which bifurcates (the carotid bifurcation) at the boundary of head and neck to give a large external carotid and a smaller internal carotid a. Caudal to the bifurcation, the common carotid a. discharges the **cranial thyroid a. (34)** to the thyroid gland. The external carotid a. continues the direction of the common carotid cranially. At its origin, the **internal carotid a. (27)** has a convex widening, the **carotid sinus (27)**, containing pressor receptors. At the carotid bifurcation is the glomus caroticum containing chemoreceptors. The internal carotid a. passes through the carotid canal in an atypical loop-like arrangement, and several cerebral arteries are given off within the cranial cavity. The **external carotid a. (29)** reaches the temporomandibular articulation, to continue as the maxillary a. At the beginning of the external carotid a., the **occipital a. (28)** branches to the occipital region, parallel, initially, to the internal carotid a. At approximately the same level, the **cranial laryngeal a. (31)** leaves the ventral aspect of the external carotid a. to enter the thyroid foramen along with the nerve of like name. Likewise, the lingual and the facial a. arise 2 cm along the external carotid a. on its ventral aspect. The **lingual a. (32)** goes between the mm. styloglossus and hyoglossus to tongue and tonsils. At first, the **facial a. (30)** travels medial to and along the mandible, there giving rise to the **sublingual a. (41)** (the accompanying sublingual v. arises from the lingual v.). Then the facial a. runs between the insertion of the m. digastricus and the ventral border of the mandible, around it, and onto the lateral surface of mandible. Its further course in the face is discussed with the facial v. (see p. 37). The **caudal auricular a. (23)** and the **superficial temporal a. (22)**, with its branch, the **rostral auricular a. (21)**, arise from the dorsal convexity of the terminal arch of the external carotid a. They supply external ear and temporal region. As the rostral continuation of the external carotid a., the **maxillary a. (24)** runs ventrally around the temporomandibular articulation where the **inferior alveolar a. (26)** goes to the mandibular canal and the **caudal deep temporal a. (25)** to the m. temporalis. After its passage through the alar canal, the maxillary a. gives the **external ophthalmic a. (35)** to the ocular bulb and the **rostral deep temporal a. (36)** to the m. temporalis. The terminal branches of the maxillary a. **(infraorbital —37, greater palatine —39 and lesser palatine —40, and sphenopalatine —38)** run with like-named nerves through appropriately denoted osseous canals or foramina, to the areas they irrigate. (The sphenopalatine a. traverses the sphenopalatine foramen with the caudal nasal n.)

e) The **EXTERNAL ACOUSTIC MEATUS (3)** is supported proximally by an **auricular cartilage (1)** resembling a curled piece of paper, 5 followed by an annular cartilage and finally by the osseous meatus. The external meatus is arranged as a long perpendicular part and a horizontal tympanic part both of which one is able to survey once they are opened. Between the two parts is an approximate right-angled bend (about 100 degrees in breeds with erect ears and more than 110 degrees in breeds with pendant ears). The **annular cartilage (4)** and the osseous tube which extends only a few millimeters to the terminating tympanic membrane, form the horizontal part of the external acoustic meatus. In so doing, they participate in the prolongation of the auricular cartilage. For clinical inspection, the external acoustic meatus must be stretched 6 at its bend by pulling the pinna caudolaterally.

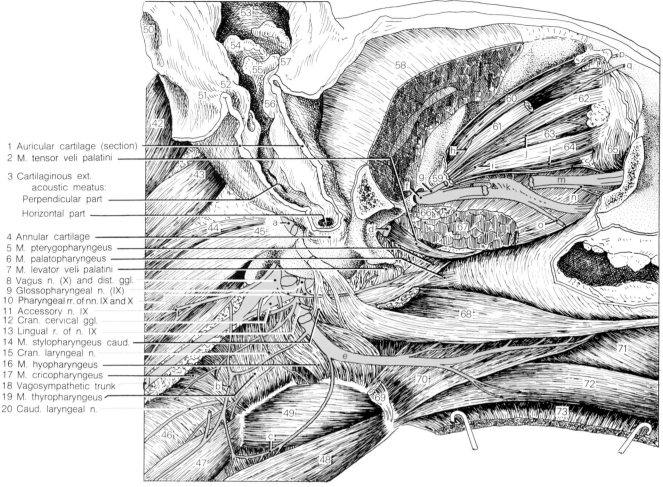

Legend:

42 M. cleidocervicalis
   M. sternocleidomastoideus
43 M. sternooccipitalis
44 M. sternomastoideus
45 M. cleidomastoideus
46 Esophagus
47 M. sternothyroideus
48 M. sternohyoideus
49 M. thyrohyoideus
50 Cutaneous marginal sac
51 Lat. process of antitragus
52 Med. process of antitragus
53 Scapha
54 Anthelix
55 Med. crus of helix
56 Lat. crus of helix
57 Spine of helix
58 M. temporalis (section)
59 Periorbita (section)
60 M. obliquus dors.
61 M. levator palpebrae
   superioris (section)
62 M. rectus dors.
63 M. retractor bulbi
64 M. rectus lat.
65 M. obliquus ventr.
66 M. pterygoideus lat.
   (section)
67 M. pterygoideus med.
   (section)
68 M. styloglossus (section)
69 M. stylohyoideus (section)
70 M. hyoglossus
71 M. genioglossus
72 M. geniohyoideus
73 M. mylohyoideus (section)
74 Antitragus
75 Helix
76 Tragus
77 M. digastricus (section)

1 Auricular cartilage (section)
2 M. tensor veli palatini

3 Cartilaginous ext.
   acoustic meatus:
   Perpendicular part
   Horizontal part

4 Annular cartilage
5 M. pterygopharyngeus
6 M. palatopharyngeus
7 M. levator veli palatini
8 Vagus n. (X) and dist. ggl.
9 Glossopharyngeal n. (IX)
10 Pharyngeal rr. of nn. IX and X
11 Accessory n. IX
12 Cran. cervical ggl.
13 Lingual r. of n. IX
14 M. stylopharyngeus caud.
15 Cran. laryngeal n.
16 M. hyopharyngeus
17 M. cricopharyngeus
18 Vagosympathetic trunk
19 M. thyropharyngeus
20 Caud. laryngeal n.

Legend:

a Facial n. (VII)
b Cervical n. (C 1)
c Ansa cervicalis
d Mandibular n. (V 3)
e Hypoglossal n. (XII)

f Maxillary n. (V 2)
g Zygomatic n.
h Trochlear n. (IV)
i Abducens n. (VI)
j Lacrimal n.

k Pterygopalatine n.
   and ggl.
l Ventr. r. of oculomotor n. (III)
m Infraorbital n.
n Superior alveolar rr.

o Lesser palatine n.
p Infratrochlear n.
q Frontal n.

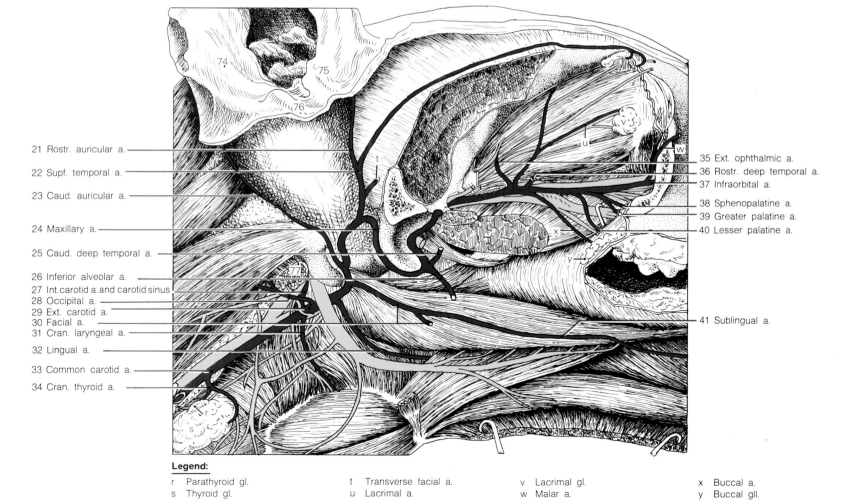

21 Rostr. auricular a.

22 Supf. temporal a.

23 Caud. auricular a.

24 Maxillary a.

25 Caud. deep temporal a.

26 Inferior alveolar a.
27 Int. carotid a. and carotid sinus
28 Occipital a.
29 Ext. carotid a.
30 Facial a.
31 Cran. laryngeal a.

32 Lingual a.

33 Common carotid a.

34 Cran. thyroid a.

35 Ext. ophthalmic a.
36 Rostr. deep temporal a.
37 Infraorbital a.

38 Sphenopalatine a.
39 Greater palatine a.
40 Lesser palatine a.

41 Sublingual a.

Legend:

r Parathyroid gl.
s Thyroid gl.

t Transverse facial a.
u Lacrimal a.

v Lacrimal gl.
w Malar a.

x Buccal a.
y Buccal gll.

C. 8. S. 9: TONGUE AND LINGUAL MUSCLES, HYPOGLOSSAL NERVE (XII), SALIVARY GLANDS, AND DENTITION

a) The **TONGUE** and its **frenulum (16)** are situated on the floor of the mouth. In its root at the oropharynx, the tongue contains the basihyoid element and there it is flanked on both sides by the two keratohyoid elements (see p. 35A). Both body and apex of tongue project well up into the oral cavity. The mucous membrane on the dorsum of tongue bears mechanical papillae, the whole dorsum being covered by fine thread-like **filiform papillae (13)**, while conspicuous, large, separate **conical papillae (11)** are present on the root of the tongue. The taste or gustatory papillae, containing taste buds innervated by special sensory fibers, include vallate, foliate and some of the fungiform papillae. The four to six **vallate papillae (12)** lie on the root of the tongue, each being surrounded by a wall and a fossa. At the same level indistinct leaf-shaped **foliate papillae (14)** are present on both sides of the tongue, while mushroom-shaped **fungiform papillae (15)** are present over the whole dorsum. Marginal papillae, present only in new-born pups, form a fimbriated hem of mechanical papillae particularly on the rostral half of the tongue border. Individual marginal papillae are conical and 2-4 mm long. They function in sucking and disappear by the change from a liquid to a solid diet. The mucous membrane of the tongue is supplied by the lingual n. and sensory lingual rami of the glossopharyngeal n. The chorda tympani from the facial n. (VII) supplies the fungiform papillae in the rostral two-thirds of the tongue, while the glossopharyngeal n. innervates the vallate and foliate papillae with special sensory fibers in the caudal one-third. The parasympathetic part of the chorda tympani innervates lingual glands including the gustatory glands of the vallate papillae. As a morphological specialization in carnivores the lyssa is situated ventrosagittally at the apex of the tongue, deep to the mucous membrane. It is a rod-sphaped formation, of maximum length 4 cm, covered by connective tissue and containing striated muscle cells, fat, and cartilaginous tissue.

b) The **TONGUE MUSCLES** include the proper or intrinsic muscle of the tongue, the extrinsic tongue muscles coming from outside the organ and radiating within it, and the hyoid muscles inserting into the basihyoid element.
Innervation is derived from the hypoglossal n. (XII) except for the hyoid muscles which are supplied predominantly through the cervical ansa connecting hypoglossal and first cervical n. After its emergence from the hypoglossal canal, the **hypoglossal n. (XII —4)** crosses the sympathetic trunk and the glossopharyngeal n. laterally, and communicates with the first cervical n. via the cervical ansa, before reaching the lingual muscles. Although the vagus n. participates in the innervation of the canine tongue its role is not absolutely clear. From its developmental history as a nerve of the branchial arches, it participates in the genesis of tongue. In man its innervation of base of tongue near the base of the epiglottis is proven.

I. The **m. lingualis proprius**, the proper muscle of the tongue, contains longitudinal, transverse and perpendicular fibers (see p. 41A). II. The extrinsic tongue muscles radiate from the stylohyoid element (**m. styloglossus —5**), the basihyoid (**m. hyoglossus —7**), and chin (**m. genioglossus —19**). III. Of the hyoid muscles, the **m. geniohyoideus (20)** runs from chin to basihyoid, and the **m. thyrohyoideus (6)**, as a cranial continuation of the m. sternothyroideus, from thyroid cartilage to basihyoid. The mm. sternohyoideus and sternothyroideus are regarded as 'long hyoid' muscles (see p. 6). The mm. stylohyoideus, occipitohyoideus and keratohyoideus are not considered in detail.

c) The **SALIVARY GLANDS** of the oral cavity are divided into those lying in the surrounding wall which have many short ducts, and appendicular glands more or less removed from the oral cavity, connected to it by a relatively few, long ducts. The appendicular salivary glands opening into the oral vestibule are the parotid and zygomatic glands, those opening into the oral cavity proper being the mandibular and monostomatic sublingual glands. The **parotid gland (1)** is tri-lobed and situated at the base of the ear, the parotid duct terminating on the **parotid papilla (10)** dorsal to the upper sectorial tooth (P4). The **zygomatic gland (9)** lies ventrolaterally in the orbit under the rostral end of the zygomatic arch. It opens by one large duct and several small secretory ducts caudal to the upper sectorial tooth. The **mandibular gland (3)** and the **monostomatic sublingual gland (2)** open by ducts directed rostrally and parallel to each other onto the **sublingual caruncle (18)**. The caruncle lies on the floor of the oral cavity rostral to the lingual frenulum. At the level of the frenulum the **polystomatic sublingual gland (17)** opens into the oral cavity by many short ducts and is thus a gland of the oral wall. This is true also for buccal and palatine glands. The **parasympathetic nerve supply** (see p. 48A) of the palatine glands is provided by the nerve of the pterygoid canal the fibers of which synapse with postganglionic neurones in the pterygopalatine ganglion. The parotid, zygomatic and buccal glands are innervated by the glosso-

pharyngeal n. (IX) via the minor petrosal n., and the otic ganglion. The mandibular and sublingual glands are innervated by the facial n. through the chorda tympani.

d) Regarding **DENTITION**, the individual teeth of the upper and lower dental arches are identified on the basis of a dental formula. In mammals this is deduced from a complete dentition numbering forty-four teeth. Of the domestic mammals only the pig has such a dentition.

Temporary dentition $\frac{iiicopppoo O}{iiicopppoo o}$ = (7 + 7) x 2 = max. 28 milk teeth
(Deciduous teeth)

Permanent dentition $\frac{IIICLPPPMMO}{IIICLPPPMMM}$ = (10 + 11) x 2 = 42 perm. teeth
(Permanent teeth)

To interpret the dental formula, the numerator represents teeth of the upper arch, the denominator those of the lower arch. Symbols in lower case represent deciduous teeth, symbols in upper case permanent teeth. Symbol (o) is the site at which a tooth will be present in the permanent dentition. Symbol (O) represents a total absence of a tooth in canine dentition. The permanent teeth (see p. 43A) either replace the deciduous teeth or develop without having temporary precursors, as for example the first premolar or wolf tooth and all molars. The **incisor teeth (I)** of the upper arch are larger than those of the lower. On the upper arch a distinct **diastema (8)** or space is present between the lateral incisor and the **canine tooth (C)** whereas in the lower arch it occurs between the canine and the first premolar. The first **premolar (P1)** of each arch, sometimes referred to as a **wolf tooth (dens lupinus —L)**, has no temporary precursor, and like the incisor and canine teeth, has only one root. The remaining three **premolar teeth (P2-4)** of each arch are cutting teeth, each having two roots. An exception to this is P4 of the upper arch, which has three roots, is the strongest premolar, and is also a sectorial tooth. The lower sectorial is the first molar (M1). The **molar teeth (M)** have surfaces used for crushing, the upper molars having three roots, the lowers two. In animals, growth and wear of teeth permit one to make an approximate estimation of age. Both criteria however are related to breed and behavior. Regarding tooth growth the following approximate data are worth noting:

up to the third week (of age): edentulous
from the sixth week: complete deciduous dentition
from the third month: commencement of growth of
permanent incisors
from the sixth month: complete permanent dentition

Three hard substances, enamel, cement and dentin, participate in tooth structure. **Enamel (23)** covers the free portion of the tooth namely the crown. It is a very hard, conspicuously white layer approximately one millimeter thick. **Cement (29)** consists of interwoven osseous tissue, approximately one millimeter thick, surrounding the whole **root of the tooth (28)** and extending to the **neck (25)** at the beginning of the enamel layer. **Dentin (21)**, a tissue resembling bone, consists of a calcified ground substance and collagen fiber bundles. Within its covering of enamel and cement, it forms the bulk of the tooth surrounding the **pulp cavity (27)**. This contains the dental pulp, nerves, blood and lymph vessels. These enter the cavity through the **apical foramen (31)** and the **root canal (30)**. On the **crown of the tooth (22)**, one differentiates five contact surfaces (see p. 43A, 36-41) namely the occlusal or masticatory surface, the vestibular surface, the lingual surface, and the mesial (rostral) and distal (caudal) contact surfaces between adjacent teeth. From the anatomical viewpoint, the neck of the tooth is regarded as the site of contact between enamel and cement, whereas clinically the term encompasses the region between alveolus and gingiva. Thus the dental crown projects beyond the gingiva and the clinical root is that portion of the tooth fixed in the alveolus and covered by gingiva. (Dental anesthesia: see Appendix.)

Teeth are implanted or 'wedged' into the osseous alveoli by means of **parodontium (26)** which produces a springy fibrous union, the gomphosis. The term parodontium embraces alveolus, periodontium and cement. Collagen fibers of the periodontium (desmodontium) run in different directions between cement and alveolus. In the main, fibers descend steeply towards the apex of the tooth. This permits the pressure of mastication to be transformed into tensile stress in the periodontium and results in tooth movement. Approaching the neck of the tooth, the fibers gradually become horizontal and finally reverse direction of climb steeply.

Tongue, Lingual mm., Hypoglossal n. (XII), and Salivary Glands (lateral view)

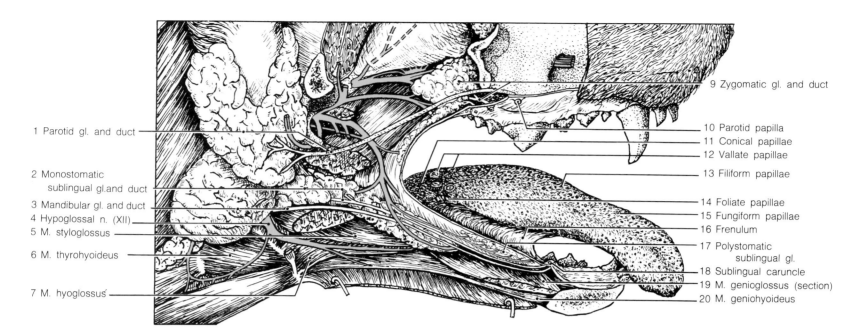

1 Parotid gl. and duct
2 Monostomatic sublingual gl.and duct
3 Mandibular gl. and duct
4 Hypoglossal n. (XII)
5 M. styloglossus
6 M. thyrohyoideus
7 M. hyoglossus

9 Zygomatic gl. and duct
10 Parotid papilla
11 Conical papillae
12 Vallate papillae
13 Filiform papillae
14 Foliate papillae
15 Fungiform papillae
16 Frenulum
17 Polystomatic sublingual gl.
18 Sublingual caruncle
19 M. genioglossus (section)
20 M. geniohyoideus

Permanent teeth (lateral view)

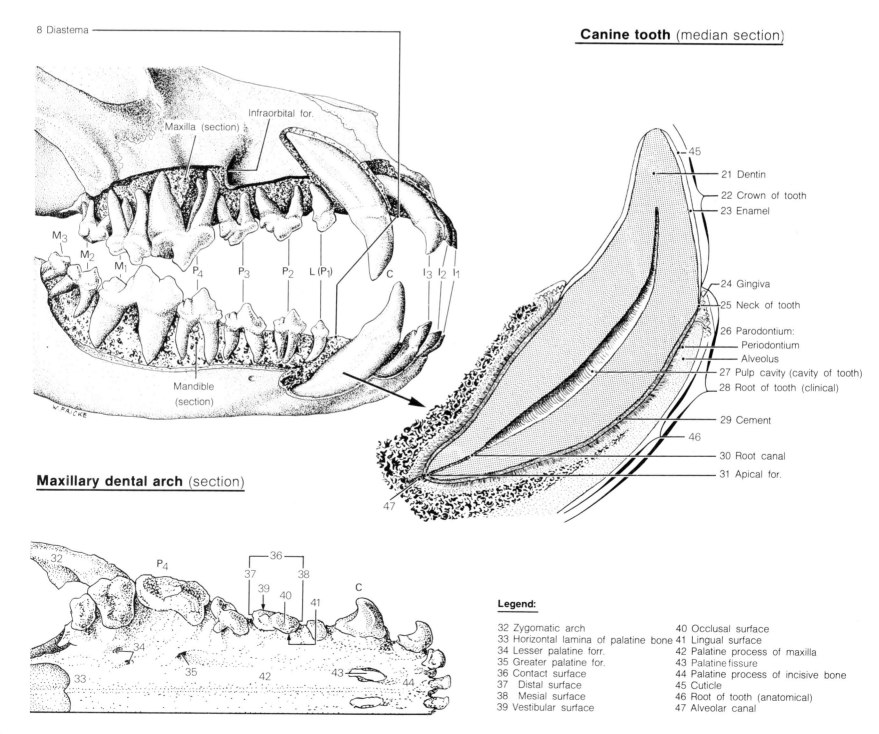

8 Diastema

Maxilla (section)
Infraorbital for.

M₃ M₂ M₁ P₄ P₃ P₂ L (P₁) C I₃ I₂ I₁

Mandible (section)

W. FRICKE

Maxillary dental arch (section)

Canine tooth (median section)

45
21 Dentin
22 Crown of tooth
23 Enamel
24 Gingiva
25 Neck of tooth
26 Parodontium:
   – Periodontium
   – Alveolus
27 Pulp cavity (cavity of tooth)
28 Root of tooth (clinical)
29 Cement
46
30 Root canal
31 Apical for.
47

32 P₄ 36 37 38 C
39 40 41
34
33 35 42 43 44

Legend:

32 Zygomatic arch
33 Horizontal lamina of palatine bone
34 Lesser palatine forr.
35 Greater palatine for.
36 Contact surface
37 Distal surface
38 Mesial surface
39 Vestibular surface
40 Occlusal surface
41 Lingual surface
42 Palatine process of maxilla
43 Palatine fissure
44 Palatine process of incisive bone
45 Cuticle
46 Root of tooth (anatomical)
47 Alveolar canal

C. 9. S. 1: SPINAL CORD AND MENINGES

a) The **THREE MENINGES OF THE SPINAL CORD** arise from the meninges of the brain caudal to the foramen magnum. The ectomeninx lying externally, surrounds the cobweb-like arachnoid and the soft pia mater, the latter lying tightly adjacent to the spinal cord. Arachnoid and pia mater are collectively known as the endomeninx (leptomeninx) (for corresponding relationships of cerebral meninges see p. 45).

The **ectomeninx (1)** consists of two laminae. The external lamina is identical with the periosteal lining (**periosteum —2**) of the vertebral canal. The internal stout lamina is the **dura mater (3)**. It envelopes the spinal cord, the roots of the spinal nerves and ganglia, and forms the initial external covering of epineurium on the spinal nerves. Cranial to the fourth cervical vertebra the dorsal and ventral spinal nerves have a common dural envelope (see p. 44A, upper right). Caudal to this each spinal nerve root has an individual dural sheath (middle illustration).

1 The **epidural space (4)** containing fat, lymph vessels and large venous plexuses, lies between both laminae of the ectomeninx. Fat upholsters the spinal cord which is deformed passively by the curvature of the vertebral column.·

The **arachnoid (6)** lies deep to the subdural **space (5)** which is normally very narrow and artificially wider with fixation. Superficially, this cobweb-like membrane consists of a layer of epithelioid cells continuing to the peripheral nerves (not illustrated) as a deep neurothelial lining of the internal perineurium. From the underlying connective tissue of the arachnoid with its sparse number of blood vessels, a loose system of trabeculae radiates deeper. In its meshes the arachnoid accommodates the **subarachnoid space (8)** and its contained cerebrospinal fluid. (For the corresponding relationships of brain, see p. 45A, middle illustration.) The subarachnoid space extends to the spinal ganglia and, between the cerebellum and beginning of the spinal cord, widens out as the **cerebellomedullary cistern (24)**. Around the conus medullaris and the filum terminale (see below), the subarachnoid cavity is relatively wide and therefore suitable for lumbar puncture when obtaining cerebrospinal fluid.

The **pia mater (32)** with its rich endowment of blood vessels, is tightly pressed against the surface of the spinal cord. The **denticulate ligaments (28)** are attached by their wide bases onto the outer surface of the pia mater between the roots of the spinal nerves. The ligaments penetrate the arachnoid, and their apices taper to end on the deeper surface of the dura mater. Thus, they anchor the spinal cord to the surrounding meninges.
The arterial supply of the spinal cord is mainly from the **ventral spinal a. (36)** extending along the whole length of the spinal cord in the **median fissure (13)**.

The venous **ventral internal vertebral plexus (52)** is embedded in epidural fat within the vertebral canal on its ventral aspect. (The **ventral external vertebral plexus (51)** runs externally along the ventral aspect of the vertebral column.)

2;3 **b)** The **SPINAL CORD** arises from the brain at the level of the first cervical vertebra. The **cervical intumescence (7)**, the site of emergence of the brachial plexus, lies at the junction between neck and thorax. Likewise, the **lumbar intumescence (21)** is situated in the lumbar region at the egress of the lumbosacral plexus. At the level of the fifth lumbar vertebra the cord tapers conically at the **conus medullaris (22)** and this merges into the **filum terminale (43)** at the level of the sixth lumbar vertebra. The **cauda equina (44)** is formed within the vertebral canal by the filum terminale, the roots of the last lumbar nerves and the sacral nerves accompanying it lengthwise. In the cervical and thoracic regions the roots of spinal nerves run approximately transversely and leave the vertebral canal by direct routes through the intervertebral foramina. In the lumbar region and even more distinctly in the sacral, they leave the vertebral canal after running caudally and almost parallel to one another within the canal. Along the entire spinal cord of the embryo the origin of nerves and their course through the intervertebral foramina

lie at approximately the same levels as the foramina. Due to a retardation of growth of the spinal cord with increasing age, in contrast to that of the vertebral column, a difference in the level of emergence of spinal nerves from the cord, and their exit from the vertebral column results. This is known erroneously as the 'ascent of the spinal cord'. In the cervical part of the cord, the spinal roots of the **accessory n. (23)** arise and take a longitudinal course cranially between the dorsal and ventral roots of spinal nerves. After their passage through the foramen magnum the spinal roots of the accessory n. connect with cranial roots derived from the brain. The accessory n. then leaves the cranial cavity through the jugular foramen. Superficially and on cross-sectioning the spinal cord, one can determine the **median sulcus (47)** and the **dorsolateral sulcus (46)** where dorsal roots enter the spinal cord. Both sulci limit the **dorsal funiculus (45)** longitudinally. Only on the cervical part of the cord is it possible to identify a **dorsal intermediate sulcus (27)** between them. The **fasciculus gracilis (25)** and **fasciculus cuneatus (26)** lie medial and lateral respectively to the dorsal intermediate sulcus. The **lateral funiculus (50)** lies between the dorsolateral sulcus and the very indistinct or absent **ventrolateral sulcus (53)**. Its position is defined by the outflow of the ventral roots of spinal nerves. The **ventral funiculus (49)** is situated ventromedially bounded by the median fissure. It contains numerous unnamed tracts connecting comparative parts of brain and spinal cord, which are unable to be defined exactly at a macroscopic level.

c) In a **CROSS SECTION OF THE SPINAL CORD** one can differentiate **white matter (34)** with its myelinated nerve tracts, from **gray matter (35)** shaped like a butterfly. In addition, the gray matter has a **dorsal horn (9)**, a **ventral horn (12)**, and a **lateral horn (10)** in thoracic and lumbar parts of the cord. Taking spatial factors into consideration, the lateral horn should be called preferably a column. Lying between dorsal and ventral horns in the cervical part of the cord is a scattered group of ganglion cells, the **reticular formation (48)** which continues into the brain. The **central canal (11)** of the spinal cord runs in the central region of the gray matter.

d) The **ROOTS OF THE SPINAL NERVES** consist of a dorsal and a ventral series. The **ventral root (31)** provides egression for **motor fibers (41)** beginning at large perikarya or nerve cell bodies in the ventral horn. **Sympathetic fibers (39)** come from the lateral horns of the thoracic and lumbar parts of the spinal cord and also exit via the ventral root. (Parasympathetic fibers begin in the intermediocentral column of the sacral part of the cord, also known by several authors as the lateral horn or column.) The **dorsal root (29)** of the spinal nerve is predominantly afferent, containing **sensory fibers (37)** coming for example from skin. Their nerve cell bodies lie in the **spinal ganglion (30)** in the vicinity of the intervertebral foramen. Central fibers proceed to the spinal cord.

e) SYMPATHETIC FIBERS have their nerve cell bodies in the lateral horn of the spinal cord and exit via the ventral root. After leaving this they reach the sympathetic trunk, particularly the **ganglia (33)** thereof, in the **white rami communicantes (14)**. After synapsing with a second neuron in the ganglia of the sympathetic trunk a portion of the nerve fibers return to segmental (somatic) nerves in **gray rami communicantes (15)**. Of the myelinated fibers which do not synapse in the sympathetic trunk, some reach the succeeding paravertebral ganglion, for example the **celiac ganglion (17)**, in the **lesser splanchnic n. (16)** and some in the lumbar splanchnic nn. At the paravertebral ganglia the myelinated fibers synapse with unmyelinated or poorly myelinated postganglionic neurons. Postganglionic fibers then proceed as periarterial plexuses in the adventitia of the visceral arteries to their effector organs, for example the **small intestine (42)**. They merge with intramural plexuses, known, according to their site as **subserosal (18)**, **myenteric (19)** and **submucosal (20)** plexuses. Ganglion cells situated in the plexuses are predominantly parasympathetic. **Parasympathetic fibers (38)** run from the **dorsal vagal trunk (40)** through the solar plexus to the respective parts of the intestine, in company with intestinal vessels.

Spinal cord

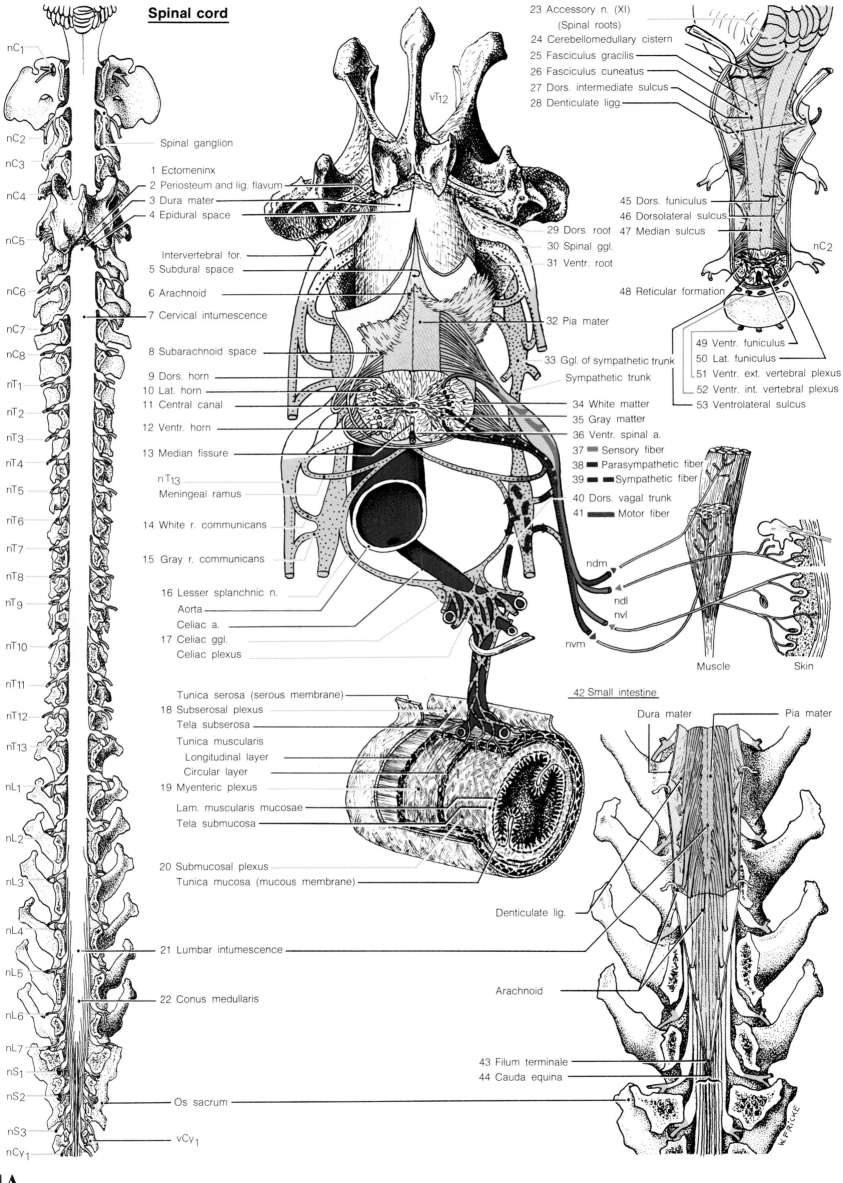

nC1
nC2
nC3
nC4
nC5
nC6
nC7
nC8
nT1
nT2
nT3
nT4
nT5
nT6
nT7
nT8
nT9
nT10
nT11
nT12
nT13
nL1
nL2
nL3
nL4
nL5
nL6
nL7
nS1
nS2
nS3
nCy1

Spinal ganglion

1 Ectomeninx
2 Periosteum and lig. flavum
3 Dura mater
4 Epidural space
Intervertebral for.
5 Subdural space
6 Arachnoid
7 Cervical intumescence
8 Subarachnoid space
9 Dors. horn
10 Lat. horn
11 Central canal
12 Ventr. horn
13 Median fissure
n T13
Meningeal ramus
14 White r. communicans
15 Gray r. communicans
16 Lesser splanchnic n.
Aorta
Celiac a.
17 Celiac ggl.
Celiac plexus
Tunica serosa (serous membrane)
18 Subserosal plexus
Tela subserosa
Tunica muscularis
Longitudinal layer
Circular layer
19 Myenteric plexus
Lam. muscularis mucosae
Tela submucosa
20 Submucosal plexus
Tunica mucosa (mucous membrane)
21 Lumbar intumescence
22 Conus medullaris
Os sacrum
vCy1

vT12

23 Accessory n. (XI)
   (Spinal roots)
24 Cerebellomedullary cistern
25 Fasciculus gracilis
26 Fasciculus cuneatus
27 Dors. intermediate sulcus
28 Denticulate ligg.

29 Dors. root
30 Spinal ggl.
31 Ventr. root

32 Pia mater

33 Ggl. of sympathetic trunk
Sympathetic trunk

34 White matter
35 Gray matter
36 Ventr. spinal a.
37 ▭ Sensory fiber
38 ▬ Parasympathetic fiber
39 ▬▬ Sympathetic fiber
40 Dors. vagal trunk
41 ▬ Motor fiber

45 Dors. funiculus
46 Dorsolateral sulcus
47 Median sulcus

nC2

48 Reticular formation

49 Ventr. funiculus
50 Lat. funiculus
51 Ventr. ext. vertebral plexus
52 Ventr. int. vertebral plexus
53 Ventrolateral sulcus

ndm
ndl
nvl
nvm

Muscle   Skin

42 Small intestine

Dura mater   Pia mater

Denticulate lig.

Arachnoid

43 Filum terminale
44 Cauda equina

W.FRICKE

4A

C. 9. S. 2: BRAIN (ENCEPHALON)

1 a) The **MENINGES OF THE BRAIN** are a continuation of those of the spinal cord. The **ectomeninx (23)** or pachymeninx consists of the internal covering of **periosteum (24)** of the cranium, namely endocranium and tough dura mater. Both laminae which are separated from each other in the spinal cord by the epidural space, fuse in the cranial cavity to form a combined ectomeninx. They separate from each other, however, at two locations. The first results in the **diaphragma sellae (33)** of the hypophysis, the second is along the course of the **sinuses of the dura mater (10, 26)**. In the first instance the internal periosteal covering of the sella turcica coats over and surrounds the projecting hypophysis or pituitary gland. The dura mater lifts away from the border of the sella turcica and covers the gland dorsally as the diaphragma sellae. This has a central opening for that part of the hypophysis projecting through it. Usually when removing the brain, one tears away the portion of the hypophysis covered by the diaphragm. The ectomeninx covers the entire internal surface of the cranial cavity.

In the dorsal midline, the **dura mater (25)** detaches itself from the periosteum between the two cerebral hemispheres and at the boundary between cerebrum and cerebellum, and projects into the cranial cavity as the falx cerebri and the membranous tentorium cerebelli respectively. Extending from the ethmoidal to the occipital bone as a dorsomedian limiting wall, the **falx cerebri (27)** separates the two cerebral hemispheres. Likewise, the **membranous tentorium cerebelli (28)** is formed from dura mater supported by the processes of the osseous tentorium cerebelli emanating from parietal and interparietal bones. It projects over the cerebellum in a tent-like manner, separating it from the cerebrum.

The **arachnoid (11, 30)** and its neurothelium lie on the deep surface of the dura mater. It stretches across the **subarachnoid space (12)** by means of a net of connective tissue trabeculae. The space contains cerebrospinal fluid and surrounds the brain, which as it were, swims in a fluid envelope. Depending on the different portions of the brain it is associated with, the subarachnoid space varies widely. On the superficial convexity of the cerebral gyri, dura and pia mater lie close together and the subarachnoid space is very narrow. In contrast, the subarachnoid space overlying the cerebral sulci is very wide since the pia plunges deeper, giving a greater clearance between it and the dura. The arachnoid occupies this space completely. The wider sections of the arachnoid, the cisterns, have specific names. The **cerebellomedullary cistern (31)** lies dorsolaterally between cerebellum and spinal cord, while the **intercrural cistern (34)** is situated ventrally and in the midline between the cerebral peduncles. By means of **arachnoid granulations (10)** the arachnoid 'proliferates' into the sinuses of the dura mater and at other sites within the bones of the cranium. Previously the granulations were indicated as sites of resorption of cerebrospinal fluid. On the basis of recent investigations however, this is to be doubted.

The **pia mater (13)** or leptomeninx adheres to the cerebral surface accompanying it into the depths of the sulci.

2;3;4 b) The **STRUCTURE OF THE INDIVIDUAL SUBDIVISIONS OF THE BRAIN** should be dealt with in a caudorostral sequence, with the brain bisected midsagittally. The divisions of the brain develop from the embryonic neural tube and the three cerebral vesicles (other text classifications give two). In turn these differentiate into the five subdivisions of the brain and the cerebellum as follows:

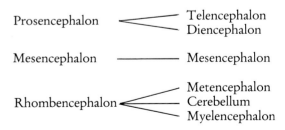

Prosencephalon ———— Telencephalon
                    Diencephalon

Mesencephalon ———————— Mesencephalon

Rhombencephalon ———— Metencephalon
                     Cerebellum
                     Myelencephalon

I. In its basic structural plan, the **rhombencephalon (47)** and its contained **fourth ventricle (41)** resemble the spinal cord and are yet distinct from it. With reservations this is also valid for the mesencephalon and diencephalon. The nucleus and sites of emergence of the hypoglossal n. (XII), a motor nerve, lie ventromedially (see p. 46A) and this is comparable relatively to the motor ventral horn of the spinal cord. The exit sites of sensory or special sensory nerves, for example the vestibulocochlear n. (VIII), lie dorsolaterally, an analogous relationship to the sensory dorsal horn of the spinal cord. Nuclei and sites of egression of nerves containing predominantly parasympathetic fibers, for example the vagus n. (X), are situated in between. This site is analogous in position to the lateral horn. Cranial nerves V to XII arise from the rhombencephalon. (In addition, cranial nerve XI has a root from the spinal cord, and cranial nerve V has sensory nuclei in mesencephalon and spinal cord.)

5 The **reticular formation (43)** is present in the rhombencephalon as well as the adjacent mesencephalon and spinal cord. It consists of scattered collections of nerve cells and fibers and, as a superimposed autonomic center, coordinates sensory, motor and autonomic function (for example circulatory and respiratory centers). Important nerve tracts pass through the ventral part of the rhombencephalon. Included in these is the **pyramidal tract (44)** ventromedially, which decussates at the transition of the spinal cord.

The rhombencephalon is divided into metencephalon, myelencephalon, and cerebellum. The **metencephalon (45)**, the rostral part of the rhombencephalon, has the distinct **pons (46)** situated ventrally. It consists of transverse fiber tracts which go to the cerebellar peduncle (see p. 46A). The central region of the pons is traversed by longitudinal nerve tracts, among others, for example, the pyramidal tract connecting brain and spinal cord. The **medulla oblongata (42)** or myelencephalon extends from pons to spinal roots of the first cervical n. The **cerebellum (29)** 6 superimposes dorsally and is connected with the remaining central nervous system by the peduncles (see p. 46A). The rostral cerebellar peduncle is connected to the mesencephalon, the middle peduncle to the pons of the metencephalon, and the caudal peduncle to the medulla oblongata.

7 II. Three main parts of the **mesencephalon (19)** should be discussed. The **tectum (36)** and the **tectal lamina (37)** overlie the **mesencephalic aqueduct (40)** dorsally, while the **rostral colliculus (38)** connects with the optic tract and the **caudal colliculus (39)** with acoustic pathways in the lateral lemniscus. The **tegmentum (20)** is situated ventral to the mesencephalic aqueduct. It contains motor nuclei of nerves (III and IV) innervating muscles of the ocular bulb, and a part of the reticular formation as well as the red nucleus and the substantia nigra. The paired **crus cerebri (21)** and the **intercrural fossa (22)** lie ventrally.

8 III. Three parts of the **diencephalon (2)** should be mentioned, namely the thalamus, epithalamus and hypothalamus. The **thalamus (3)** limits the loop-shaped **third ventricle (35)** dorsolaterally. The paired sections contact each other in the **interthalamic adhesion[1] (4)** midsagittally, and exhibit the geniculate bodies laterally. The **lateral geniculate body** (see p. 46A —34) lies at the end of the optic tract and is also connected to the rostral colliculus of the tectal lamina. The **medial geniculate body** (see p. 46A —33) is connected to the caudal colliculus of the tectal lamina. Functionally the thalamus is the last site of synapsing of all ascending tracts to the cerebrum, with the exception of the olfactory tract. Thus the thalamus is known as the 'gate to consciousness' because in the cerebrum one is made aware of sense impressions. The **epiphysis or pineal body (1)** belongs to the epithalamus. This appendage of the 9 third ventricle, shaped like a pine cone, is situated rostral to the tectal lamina and is the homologue of the third eye of lower vertebrates. The **hypothalamus (5)** lies ventrolateral to the third ventricle and has a 10 predominantly autonomic function. The **optic n. (6,** cranial nerve II see p. 46) arises at the optic chiasm which continues caudally as the optic tract. Its fibers traverse the hypothalamus to the region of the lateral geniculate body and the rostral colliculus. The **hypophysis (7)** attached 11 ventrally, has a **neurohypophysis (9)** or posterior lobe developed from the hypothalamus, and an **adenohypophysis (8)** or anterior lobe derived from the roof of the pharynx. The **mamillary body (32)** is situated on the floor of the diencephalon caudal to the hypophysis, at the border of the mesencephalon.

IV. The **telencephalon (14)** comprises the **cerebrum (15)** and the **rhinencephalon (16)**. The falx cerebri of the dura mater divides the cerebrum into two cerebral hemispheres. Nevertheless the central commissural pathways in the **corpus callosum (17)** and **rostral commissure (18)** remain undivided. The surface of the cerebrum is increased considerably by cerebral sulci and between them, the cerebral gyri which are convex across their widths. Deep prominent furrows separate the lobes of the cerebrum one from the other.

[1]) Instead of an adhesion (contact) some authors recognize a fusion of the paired parts and speak of an intermediate mass (massa intermedia).

Brain and Meninges

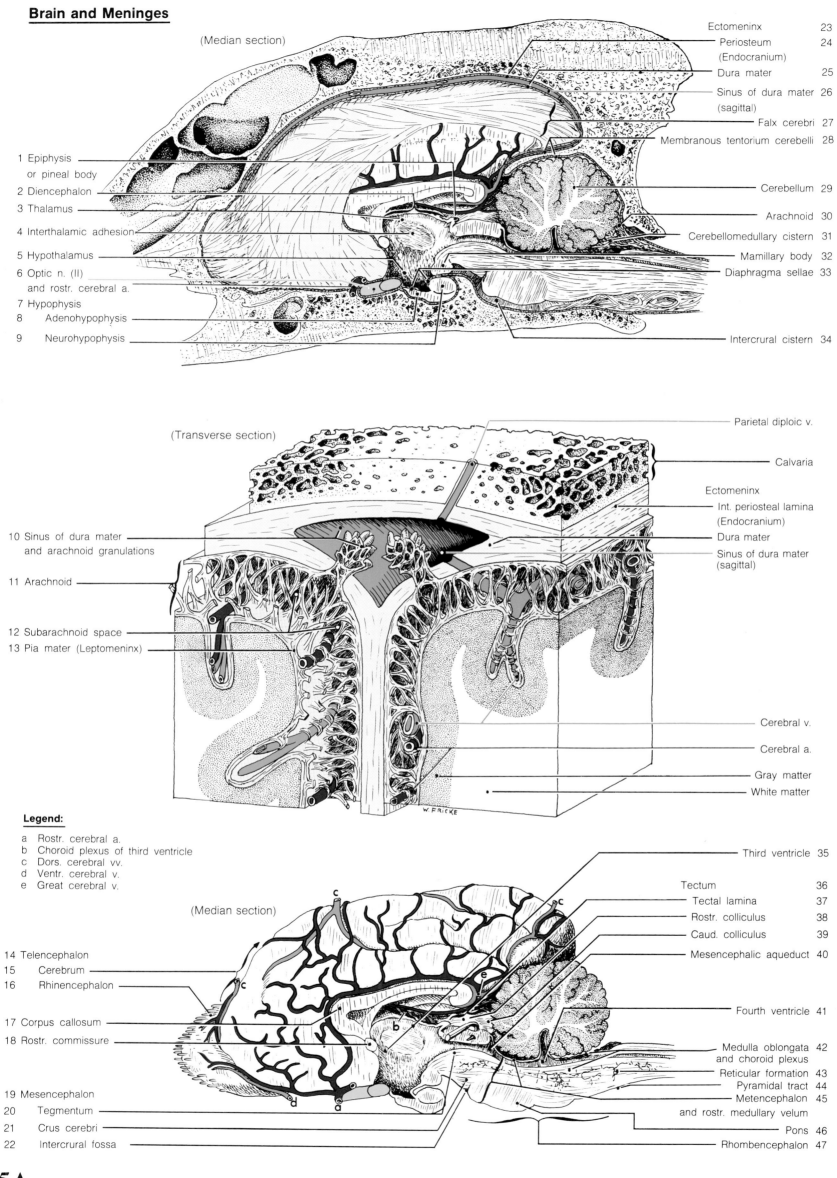

(Median section)

Ectomeninx 23
Periosteum 24
(Endocranium)
Dura mater 25
Sinus of dura mater 26
(sagittal)
Falx cerebri 27
Membranous tentorium cerebelli 28

1 Epiphysis
or pineal body
2 Diencephalon
3 Thalamus
4 Interthalamic adhesion
5 Hypothalamus
6 Optic n. (II)
and rostr. cerebral a.
7 Hypophysis
8 Adenohypophysis
9 Neurohypophysis

Cerebellum 29
Arachnoid 30
Cerebellomedullary cistern 31
Mamillary body 32
Diaphragma sellae 33

Intercrural cistern 34

(Transverse section)

Parietal diploic v.

Calvaria

Ectomeninx
Int. periosteal lamina
(Endocranium)
Dura mater
Sinus of dura mater
(sagittal)

10 Sinus of dura mater
and arachnoid granulations
11 Arachnoid

12 Subarachnoid space
13 Pia mater (Leptomeninx)

Cerebral v.

Cerebral a.

Gray matter
White matter

W. FRICKE

Legend:

a Rostr. cerebral a.
b Choroid plexus of third ventricle
c Dors. cerebral vv.
d Ventr. cerebral v.
e Great cerebral v.

(Median section)

Third ventricle 35
Tectum 36
Tectal lamina 37
Rostr. colliculus 38
Caud. colliculus 39
Mesencephalic aqueduct 40

14 Telencephalon
15 Cerebrum
16 Rhinencephalon

17 Corpus callosum
18 Rostr. commissure

Fourth ventricle 41

Medulla oblongata 42
and choroid plexus
Reticular formation 43
Pyramidal tract 44
Metencephalon 45
and rostr. medullary velum

19 Mesencephalon
20 Tegmentum
21 Crus cerebri
22 Intercrural fossa

Pons 46
Rhombencephalon 47

With the advice of tutorial staff, cerebral hemispheres and cerebellum are sectioned horizontally through the corpus callosum to open the lateral ventricles of the brain (see lower page 46A). Then one half of the brain is sectioned transversely to study basal ganglia and internal capsule at the level of the hypophysis (see text illustration). As indicated on p. 46A lower right, the hippocampus is sectioned in part, to demonstrate the field of the geniculate bodies and the optic tract.

1 a) During ontogenesis the **CEREBRAL HEMISPHERES (22, 31)** cover the remaining parts of the brain with the exception of cerebellum and base of the brain. This developmental process, known as rotation, occurs rostrally, dorsally and caudoventrally and involves the irregular 'curling up' of portions of the cerebral hemispheres including lateral ventricles. This is evident from the lateral view and also from preparations injected to show the ventricles of the brain (see lower p. 47A; and text illustration p. 47).
The **cortex (23)** or pallium comprises the peripheral grey matter of the brain. During ontogenesis, the neuroblasts of the vesicle of the telencephalon migrate to the periphery, resulting in peripheral **grey matter (23)** and central **white matter (24)**. Thus, in comparison with the spinal cord, white and grey matter have changed positions relative to each other.

2 b) The **CORPUS STRIATUM** and its **caudate nucleus (26)** lie on the floor of the **lateral ventricle (28)**. The ganglion cells of the corpus do not migrate peripherally, but remain as subcortical basal ganglia.

3 c) The **INTERNAL CAPSULE (A)** consists of 'projection fibers' connecting brain and other deeper or caudal parts of the central nervous system. These fibers come from the motocortex of the cerebral cortex and penetrate the corpus striatum caudoventrally in a converging fan-shaped manner. Thus the corpus striatum is divided into the mediodorsal **caudate nucleus (B)** and the **lentiform nucleus (D)**. This lentiform nucleus consists of the lateral **putamen (C)** and the medial **globus pallidus (E)**.

d) The **COMMISSURES** run chiefly in the **corpus callosum (25)**, see also p. 45A —17), and also in the **rostral commissure** (see p. 45A —18) to form the transverse attachments of both cerebral hemispheres.

the optic canal (see also p. 47A, a) to the **optic chiasm (7)**. There the nerve fibers originating from the medial half of the retina cross the midline, while those from the lateral half proceed further without decussating. From the optic chiasm the optic n. continues proximocaudally by means of the **optic tract (8, 32)** and ends in the **lateral geniculate body (33)**. (The neighbouring **medial geniculate body (34)** is concerned with the acoustic pathway.)

III. The **oculomotor n.** arises from the mesencephalon caudolateral to the mamillary body.

IV. The **trochlear n.** arises from the mesencephalon on the caudodorsal aspect of the caudal colliculus (see lower p. 46A) and runs in an arch between cerebrum and cerebellum. It then passes around the mesencephalon to the base of the brain and through the orbital fissure (see also p. 47A, b) to the cone of extrinsic muscles of the eye.

V. The **trigeminal n.** arises caudally from the rhombencephalon between **pons (13)** and **cerebellum (15)**. It has a thick sensory root and a thin motor root. The nerve cell bodies of the sensory (afferent) fibers are situated in the large conspicuous trigeminal ganglion. This lies deep to the dura, between the rostral end of the petrous temporal bone and the transit passages in the skull for the three trigeminal branches.

VI. The **abducens n.** arises from the rhombencephalon at the caudal border of the pons in the angle between **pyramid (16)** and **trapezoid body (14)**.

VII. The **facial n.** emerges from the rhombencephalon between the cerebellum and the rostral border of the trapezoid body. It passes through the internal acoustic pore into the facial canal.

VIII. The **vestibulocochlear n.** proceeds from the rhombencephalon within the lateral prolongation of the trapezoid body to the border of the cerebellum. It then reaches the internal acoustic pore.

Transverse section through the brain at the level of the optic chiasm

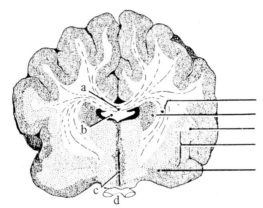

a	Corpus callosum
b	Lateral ventricle
c	Third ventricle
d	Optic chiasm

A	Internal capsule
B	Caudate nucleus
C	Putamen
D	Lentiform nucleus
E	Globus pallidus

e) The multifunctional **LIMBIC SYSTEM** is not the exact marginal zone (limbus = margin) delineated between cerebrum and diencephalon. It includes a portion of the rhinencephalon previously known as the limbic part of the rhinencephalon. Of the several other parts of the system only five need be named here, the **hippocampus (30)**, the **fornix (29)**, the septal nuclei of the **telencephalic septum (27)**, previously known as the septum pellucidum, and sections of the lentiform nucleus.

f) The **RHINENCEPHALON**, specifically its **basal part (2)**, begins at the **olfactory bulb (3)** where the **olfactory filaments (1)** of the olfactory nn. terminate. The nerve cell bodies and dendrites with their contained olfactory receptors are situated in the olfactory epithelium in the fundus of nose and in the vomeronasal organ. A second series of neurons synapse in the olfactory bulb. Initially their fibers travel in the short **olfactory peduncle (4)** and, following its division, partly in the **lateral olfactory tract (5)** and partly in the **medial olfactory tract (6)**. After synapsing again with a further neuron series, fibers from the lateral olfactory tract proceed to the primary cortex of the **piriform lobe (9)**. Subsequently they reach the limbic system in particular the amygdaloid body. After some synapsing occurs, fibers from the medial olfactory tract proceed to the septal part of the rhinencephalon and along other sites to septal nuclei of the telencephalic septum (septum pellucidum).

g) SITES OF EMERGENCE OF CRANIAL NERVES

I. The **olfactory nn.** are attached to the rhinencephalon.

II. The **optic n.** and its contained fibers extend from the retina, through

IX. The **glossopharyngeal n.** leaves the medulla oblongata in the caudal angle between trapezoid body and cerebellum and proceeds to the jugular foramen (see also p. 47A, h).

X. The **vagus n.** proceeds likewise to the jugular foramen caudoparallel to the glossopharyngeal n.

XI. The **accessory n.** has a thick spinal section and a thinner cranial section. The **spinal roots (19)** pass between the dorsal and ventral roots of cervical nerves on the lateral aspect of the spinal cord and through the foramen magnum. The **cranial roots (18)** arise from the medulla oblongata caudal to the vagus n. Then both sections pass in common through the jugular foramen, the cranial section being included in the vagus as the internal ramus thereof. The spinal section becomes the external ramus which innervates the mm. trapezius, sternocleidomastoideus and cleidocervicalis.

XII. The **hypoglossal n.** emerges by means of numerous nerve threads from the caudal end of the medulla oblongata between the pyramid and the **olive (20)**. After the union of these, the hypoglossal n. leaves the skull by the hypoglossal canal and extends to the tongue muscles.

h) The **ARTERIAL SUPPLY TO THE BRAIN** comes from the 4 **arterial circle of the brain (11)** supplied in turn by the **internal carotid a. (10)** laterally, and the unpaired **basilar a. (12)** caudally. The latter arises from the confluence of the left and the right **vertebral a. (21)** at the level of nC 1.

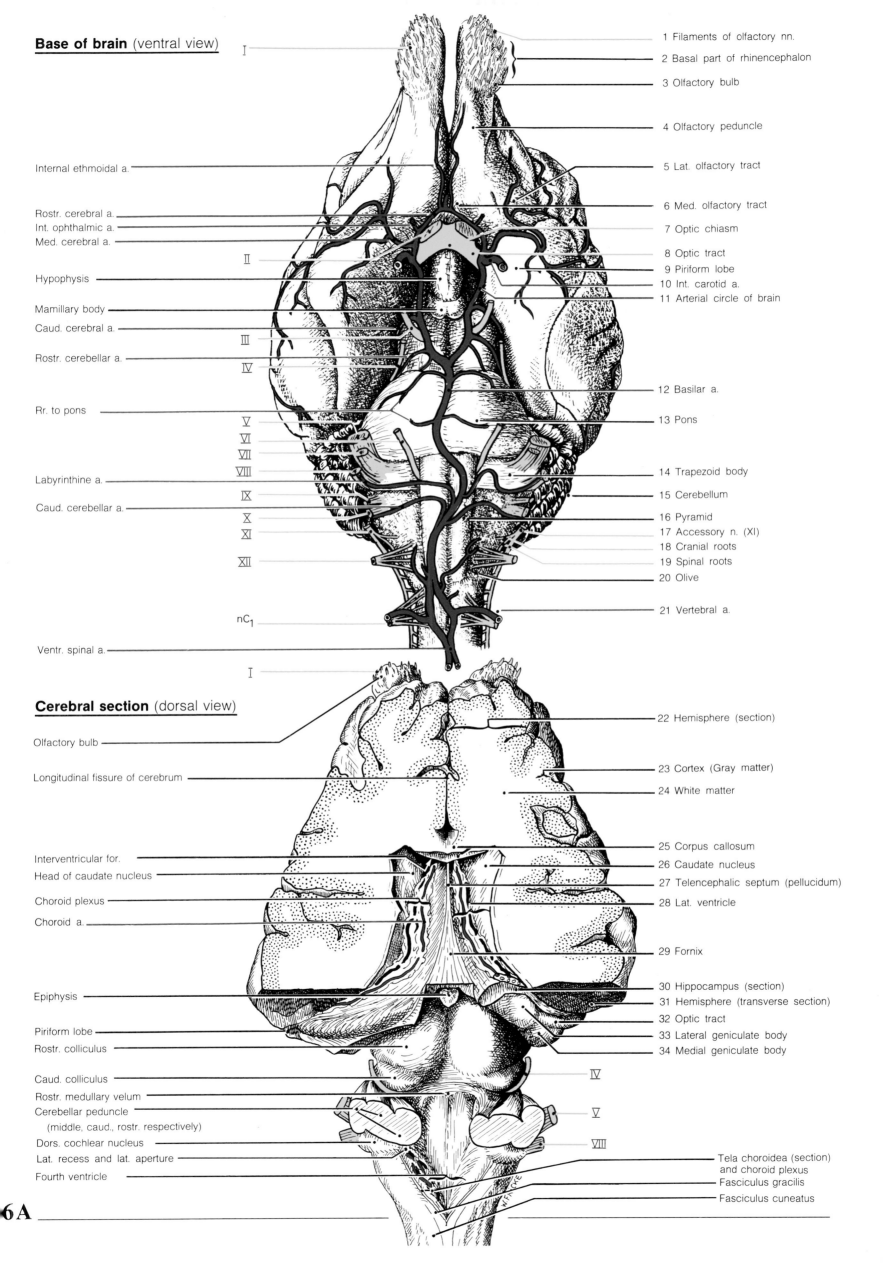

Base of brain (ventral view)

I

Internal ethmoidal a.

Rostr. cerebral a.
Int. ophthalmic a.
Med. cerebral a.

Hypophysis

Mamillary body
Caud. cerebral a.

Rostr. cerebellar a.

II

III

IV

Rr. to pons

V
VI
VII
VIII

Labyrinthine a.

Caud. cerebellar a.

IX

X
XI

XII

nC₁

Ventr. spinal a.

1 Filaments of olfactory nn.

2 Basal part of rhinencephalon

3 Olfactory bulb

4 Olfactory peduncle

5 Lat. olfactory tract

6 Med. olfactory tract

7 Optic chiasm

8 Optic tract
9 Piriform lobe
10 Int. carotid a.
11 Arterial circle of brain

12 Basilar a.

13 Pons

14 Trapezoid body

15 Cerebellum

16 Pyramid
17 Accessory n. (XI)
18 Cranial roots
19 Spinal roots
20 Olive

21 Vertebral a.

I

Cerebral section (dorsal view)

Olfactory bulb

Longitudinal fissure of cerebrum

Interventricular for.
Head of caudate nucleus

Choroid plexus

Choroid a.

Epiphysis

Piriform lobe
Rostr. colliculus

Caud. colliculus
Rostr. medullary velum
Cerebellar peduncle
  (middle, caud., rostr. respectively)
Dors. cochlear nucleus
Lat. recess and lat. aperture
Fourth ventricle

22 Hemisphere (section)

23 Cortex (Gray matter)

24 White matter

25 Corpus callosum
26 Caudate nucleus
27 Telencephalic septum (pellucidum)
28 Lat. ventricle

29 Fornix

30 Hippocampus (section)
31 Hemisphere (transverse section)
32 Optic tract
33 Lateral geniculate body
34 Medial geniculate body

IV

V

VIII

Tela choroidea (section)
  and choroid plexus
Fasciculus gracilis

Fasciculus cuneatus

6A

C. 9. S. 4: CEREBRAL VEINS AND SINUSES OF DURA MATER, VENTRICLES OF BRAIN AND CEREBROSPINAL FLUID, BRAIN STEM, AND SURFACE CONFIGURATION OF BRAIN (see also p. 45A)

a) The **VEINS OF THE BRAIN** open into the sinuses of the dura mater unaccompanied by arteries, and valves and a muscular tunica media are absent. The veins are classified into a superficial and a deep group.

Of the superficial veins two to four **dorsal cerebral vv. (1)** open into the dorsal sagittal sinus. The ventral cerebral vv. (A and B), lying superficially on the base of the brain and lateral to the temporal lobe, open at different sites into the system of ventral venous sinuses.

The deep cerebral veins come from, among other locations, the corpus callosum, corpus striatum and choroid plexus. At the mesencephalon dorsally, they unite with the **great cerebral v. (6)** which continues caudally in the tentorium cerebelli between cerebrum and cerebellum, as the straight sinus.

1 b) The **SINUSES OF THE DURA MATER** are veins of the ectomeninx (see p. 45A) modified for discharging blood, and they run chiefly between periosteum (endocranium) and the dura mater. Due to the position of venous sinuses in either ectomeninx or osseous canals, and the absence of muscular walls, their diameters remain constant at all times. Absence of venous valves makes blood flow in different directions possible. The sinuses receive the above-named cerebral veins and the veins of the wall of the cranium, namely the **diploic vv. (2).** Then the sinuses take blood by way of emissary veins, to the veins of the head and subsequently to the **internal jugular v. (15).** Corresponding to their locations on the brain, one differentiates a dorsal from a ventral system of venous sinuses.

The ventral system commences with a right and a left **cavernous sinus (8)** each of which is connected rostrally with the ipsilateral venous **ophthalmic plexus (17)** of the orbit. The cavernous sinuses are also connected to each other by **intercavernous sinuses (14)** to either side of the hypophysis, the rostral sinus being inconstant, the caudal constant. Caudally each cavernous sinus continues as the **ventral petrosal sinus (12)** within the petrooccipital canal. After discharging an emissary v. through the jugular foramen to join the internal jugular v., and after anastomosing with the **sigmoid sinus (11),** the ventral petrosal sinus joins the **basilar sinus (10).** Following its passage through the condylar canal and the foramen magnum, the basilar sinus opens into the **ventral internal vertebral plexus (venous) (9).** The **dorsal petrosal sinus (7)** runs in the tentorium cerebelli to the transverse sinus.

The dorsal system of the venous sinuses commences with the **dorsal sagittal sinus (3)** which receives venous branches from the nasal cavity. After having received the **straight sinus (4),** a continuation of the great cerebral v., the dorsal sagittal sinus bifurcates caudally at the osseous tentorium cerebelli into a right and a left **transverse sinus (5).** Each sinus runs in the temporal meatus, initially in the osseous canal, and more ventrally in the transverse sinuosal groove. There it gives off the sigmoid sinus to anastomose with the ventral system and continues into the **temporal sinus (13).** This, in turn, passes through the retroarticular foramen (j) as the emissary vein of the retroarticular foramen to open into the **maxillary v. (16).**

c) The **VENTRICLES OF THE BRAIN** (see text illustration) deve- **2** lop from the lumen of the neural tube. The **right** and **left lateral ventricles (S),** regarded also as the first and second ventricles, are situated in the right and left cerebral hemispheres respectively (see p. 46A). The **interventricular foramen (W)** connects the two lateral ventricles to the **third ventricle (Y).** In the diencephalon, this ventricle encircles the interthalamic adhesion centrally and protrudes ventrally as a **recess (Z).** The narrow **mesencephalic aqueduct (X)** arises from the ring-shaped third ventricle at the transition of diencephalon to mesencephalon. The aqueduct merges with the ample **fourth ventricle (V)** at the transition from mesencephalon to rhombencephalon. More caudally, this continues into the narrow **central canal of the spinal cord (U).** On inspection, the floor of the fourth ventricle (see p. 46A) appears rhomboid-shaped, hence the reason for so-naming the surrounding part of the brain. Rostrally in the metencephalon, the roof of the fourth ventricle is formed from very thin rostral medullary velum, and caudally in the myelencephalon, from caudal medullary velum and the tela choroidea. (Following the embalming procedure, the rostral medullary velum becomes glued to the floor of the fourth ventricle to interrupt its connection with the aqueduct.) The very thin caudal medullary velum is united to the cerebellum and to the cerebellar peduncle by a narrow margin. It bears the tela choroidea (see lower p. 46) and the choroid plexus of blood vessels. In the **lateral recess of the fourth ventricle (T)** are the lateral apertures of the fourth ventricle opening into the subarachnoid space.

d) The **CEREBROSPINAL FLUID** fills the ventricles of the brain, the **3** aqueduct and the central canal of the spinal cord. In the fourth ventricle, these internal fluid spaces are connected with the external fluid space, namely the subarachnoid space (see p. 45A). The actual connections or openings are the median aperture, and the paired lateral apertures situated at the lateral ends of the lateral recesses of the fourth ventricle (see lower p. 46) and lead to the cerebellomedullary cistern (see p. 45A —31).

e) The **CHOROID PLEXUSES** (see p. 46A) lying in all four ventricles (but not in the mesencephalic aqueduct) produce cerebrospinal fluid. In the formation of the choroid plexus, pia mater containing garlanded blood vessels, projects into the ventricular lumen of its respective side. There it is covered with a modified cuboidal ependyma which also lines the walls of the ventricle. The cerebrospinal fluid originates from the blood which retains blood corpuscles and the protein fraction. The fluid selectively penetrates the pores of the capillary endothelium, the basal membrane and the modified ependymal cells, to reach the ventricle. The blood-liquor barrier modifies the resulting cerebrospinal fluid as opposed to blood serum, by retaining different blood fractions. The secretory capacity of the ependymal cells also contributes to this.

f) The **BRAIN STEM** is defined differently in different texts. Undisputedly, the rhombencephalon and mesencephalon belong to the brain stem but some workers also include the diencephalon and the basal ganglia of the cerebrum which are then known as stem ganglia.

g) The **CONFIGURATION OF THE CEREBRAL SURFACES** is indicated in part, in the accompanying illustration, showing lobes of the cerebrum and their limiting gyri and sulci.

Ventricles of brain

(Cast preparation by Prof. Böhme)

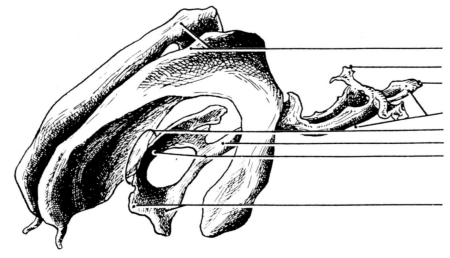

S Lateral ventricles

T Lateral recess of fourth ventricle

U Central canal of spinal cord

V Fourth ventricle

W Interventricular foramen

X Mesencephalic aqueduct

Y Third ventricle

Z Optic and neurohypophyseal recesses (respectively)

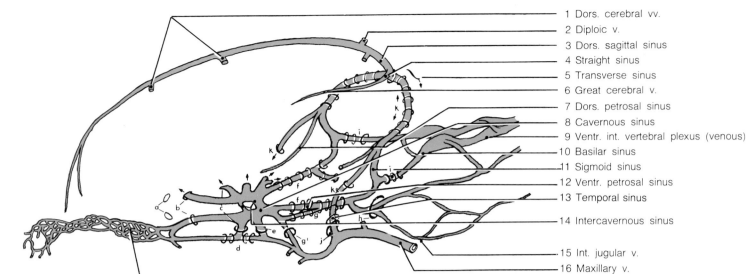

Sinuses of dura mater

1 Dors. cerebral vv.
2 Diploic v.
3 Dors. sagittal sinus
4 Straight sinus
5 Transverse sinus
6 Great cerebral v.
7 Dors. petrosal sinus
8 Cavernous sinus
9 Ventr. int. vertebral plexus (venous)
10 Basilar sinus
11 Sigmoid sinus
12 Ventr. petrosal sinus
13 Temporal sinus
14 Intercavernous sinus
15 Int. jugular v.
16 Maxillary v.
17 Ophthalmic plexus

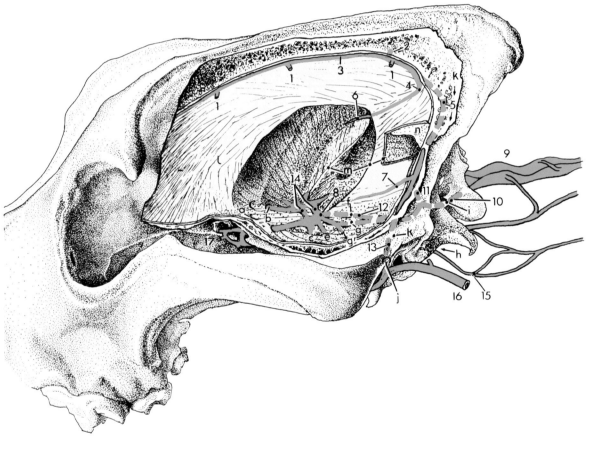

Cranial cavity and sinuses of dura mater

a Optic canal
b Orbital fissure
c For. rotundum
d Alar canal
e For. ovale
f Petrooccipital canal
g Carotid canal
g' External carotid for.
h Jugular for.
i Condylar canal
j Retroarticular for.
k Temporal meatus
l Falx cerebri
m Membranous tentorium cerebelli
n Tentorial process   (section)

Lateral surface of brain

1 Dors. cerebral vv.

A Ventr. cerebral vv. (rostr.)
B Ventr. cerebral vv. (caud.)
C Dors. cerebellar vv.
D Int. carotid a.
E Rostr. cerebellar a.
F Labyrinthine a.
G Caud. cerebellar a.
H Presylvian sulcus
J Lat. rhinal sulcus
K Cruciate sulcus
L Marginal sulcus
L' Marginal gyrus
M Middle suprasylvian sulcus
M' Middle suprasylvian gyrus
N Middle ectosylvian sulcus
N' Middle ectosylvian gyrus
O Pseudosylvian fissure
P Rostr. sylvian gyrus
Q Caud. sylvian gyrus
R Cerebellum

CLASSIFICATION OF CRANIAL NERVES I TO XII: NERVES OF SPECIAL SENSES (I, II, VIII), NERVES OF EYE MUSCLES (III, IV, VI), NERVES OF BRANCHIAL ARCHES (V, VII, IX, X, XI), HYPOGLOSSAL NERVE XII. NERVES IX, X AND XI BELONG TO THE VAGUS GROUP

Nerve/s	Classification	Origin	Route through skull	Fiber type	Region of innervation	Comments
I Olfactory nn.	Nerves of special sense	Telencephalon	Lamina cribrosa	Special sensory	Olfactory region in fundus of nose and vomeronasal organ	Receptor cells are primary sense cells
II Optic n.	Nerve of special sense	Diencephalon	Optic canal	Special sensory	Retina	Extroversion of diencephalon enveloped by 3 meninges
III Oculomotor n. Motor root	Nerve to eye muscles	Mesencephalon	Orbital fissure	Motor	Extrinsic muscles of eye, except mm. obliquus dors., rectus lat. and lateral part of m. retractor bulbi	Preganglionic parasympathetic fibers synapse in the ciliary ganglion. Sympathetic fibers to m. dilator pupillae via short ciliary nn.
Parasympathetic root				Parasympathetic	Mm. sphincter pupillae and ciliaris	
IV Trochlear n.	Nerve to eye muscles	Mesencephalon	Orbital fissure	Motor	M. obliquus dors.	Smallest cranial nerve
V Trigeminal n. Ophthalmic n. VI	Nerve of branchial arch I	Mesencephalon	Orbital fissure	Sensory	Dorsum of nose, ethmoidal bone, lacrimal gland, eyelid	Trigeminal terminal branches receive parasympathetic fibers through: the pterygopalatine ggl. from VII, otic ggl. from IX, mandibular and sublingual ggll. from chorda tympani (VII) which supply salivary gl., lacrimal and nasal glands. Clinical significance: Trigeminal neuralgia.
Maxillary n. V 2		Rhombencephalon	For rotundum and alar canal	Sensory	Face, nasal cavity, teeth, upper dental arch, temporal region	
Mandibular n. V 3 Sensory root		Rhombencephalon	For ovale	Sensory	Oral cavity, temporal and auricular region, teeth, lower dental arch, skin of chin	
Motor root				Motor	Muscles of mastication, muscles of intermandibular space; mm. tensor tympani, tensor veli palatini	
VI N. abducens	Nerve to eye muscles	Rhombencephalon	Orbital fissure	Motor	M. rectus lat. and part of m. retractor bulbi	With III and VI contained in common sheath of dura
VII Facial n. (Intermediofacial n.) Facial part Intermedius part	Nerve of branchial arch II	Rhombencephalon	Facial canal from internal acoustic pore to stylomastoid for.	Motor	Muscles of face, mm. stapedius, occipitohyoideus, stylohyoideus	Clinical significance: facial paralysis. The n. intermedius may be regarded as cranial nerve XIII.
Internal auricular r.			Stylomastoid foramen	Sensory	External acoustic meatus and tympanic membrane	The internal acoustic r. receives a sensory part from vagus n. and n. intermedius.
Chorda tympani			Petrotympanic fissure	Special sensory and parasympathetic	Rostral 2/3 of tongue, sublingual and mandibular gll.	Unites with lingual n. from V 3
Greater petrosal n.			Petrosal canal, pterygoid canal	Parasympathetic	Glands of nose and palate, lacrimal gland	Unites with sympathetic fibers of deep petrosal n. to give n. of pterygoid canal
VIII Vestibulocochlear n. (Statoacoustic n.)	Nerve of special sense	Rhombencephalon	Internal acoustic pore	Special sensory	Macula of utricle, maculae of saccule, crista ampullaris, cochlea	Connected to the nerves of the eye within C.N.S.
IX Glossopharyngeal n. Pharyngeal r.	Nerve of branchial arch III and Nerve of vagus group	Rhombencephalon	Jugular for.	Sensory and motor	Part of pharyngeal musculature and mucosa	Ramus to the carotid sinus mediates impulses from sinuosal baroceptors to circulatory center
Lingual rr.				Sensory / Special sensory	Pharynx, tongue, caudal 1/3 of tongue	Communicate with lingual n. from V 3
Tympanic n.			Tympanic canaliculus	Parasympathetic, sensory	Zygomatic, buccal, parotid gll.; mucosa of tympanic cavity	Continues as lesser petrosal n. to synapse in the otic ganglion
X Vagus n.	Nerve of branchial arch IV and Nerve of vagus group	Rhombencephalon	Jugular for.	Motor / Sensory / Parasympathetic	Muscles of larynx and part of pharyngeal musculature / Pharynx, larynx / Thoracic and abdominal viscera	The 'wandering' nerve. Nerve cell bodies of the distal ganglion (nodose) are cultured in the diagnosis of rabies.
XI Accessory n. Cranial root	Nerve of branchial arch IV and Nerve of vagus group	Rhombencephalon	Jugular for.	Motor	Mm. sternocleidomastoideus, trapezius, cleidocervicalis	Phylogenetically a separation from vagus n.
Spinal root		Cervical spinal cord	For. magnum			
XII Hypoglossal n.	Also a cervical nerve	Rhombencephalon	Hypoglossal canal	Motor	Muscles of tongue and hyoid muscles	Connected to C 1 by the ansa cervicalis (cervical loop)

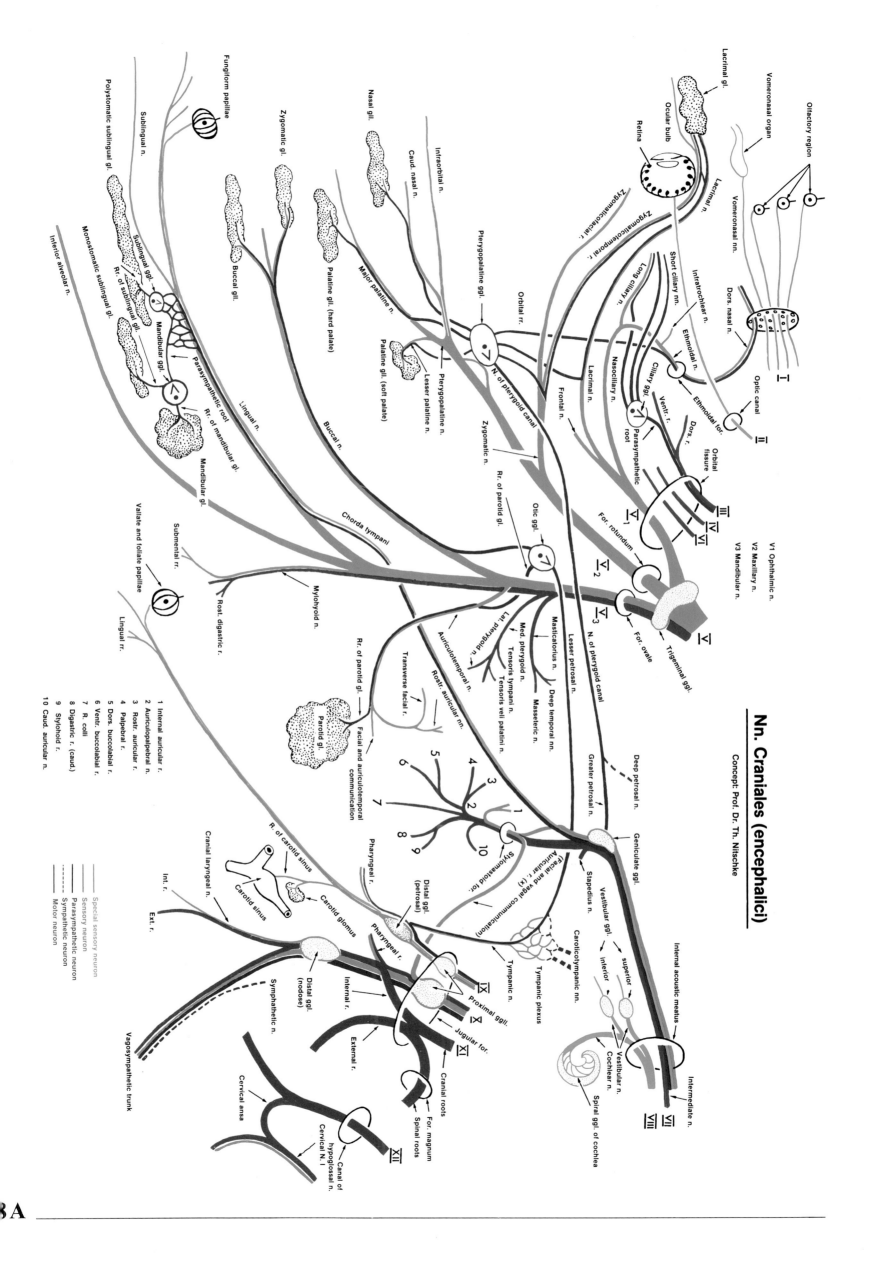

Nn. Craniales (encephalici)

Concept: Prof. Dr. Th. Nitschke

Olfactory region

Vomeronasal organ

Lacrimal gl.

Ocular bulb

Retina

Olfactory bulb

Vomeronasal nn.

Lacrimal n.

V1 Ophthalmic n.
V2 Maxillary n.
V3 Mandibular n.

Zygomaticofacial r.

Zygomaticotemporal r.

Long ciliary n.

Short ciliary nn.

Infratrochlear n.

Dors. nasal n.

Optic canal

Lacrimal n.

Nasociliary n.

Ethmoidal n.

Ethmoidal for.

Ciliary ggl.

Ventr. r.

Dors. r.

Parasympathetic root

Orbital fissure

II

Frontal n.

Pterygopalatine ggl.

Orbital rr.

7

N. of pterygoid canal

Pterygopalatine n.

Lesser palatine n.

Zygomatic n.

Rr. of parotid gl.

For. rotundum

III

IV

VI

V₁

Nasal gll.

Caud. nasal n.

Infraorbital n.

Zygomatic gl.

Major palatine n.

Palatine gll. (hard palate)

Palatine gll. (soft palate)

Buccal gl.

Buccal n.

V₂

For. ovale

N. of pterygoid canal

V₃

V

Trigeminal ggl.

Fungiform papillae

Polystomatic sublingual gl.

Sublingual n.

Sublingual ggl.

Monostomatic sublingual gl.

Rr. of sublingual gl.

Mandibular ggl.

Parasympathetic root

Lingual n.

Rr. of mandibular gl.

Mandibular gl.

Inferior alveolar n.

Chorda tympani

Otic ggl.

Rr. of parotid gl.

7

Lat. pterygoid n.

Med. pterygoid n.

Tensoris tympani n.

Tensoris veli palatini n.

Masticatorius n.

Deep temporal nn.

Masseteric n.

Lesser petrosal n.

Deep petrosal n.

Greater petrosal n.

Geniculate ggl.

Valliate and foliate papillae

Submental r.

Lingual rr.

Rost. digastric r.

Mylohyoid n.

Rost. auricular nn.

Auriculotemporal n.

Transverse facial r.

Parotid gl.

Facial and auriculotemporal communication

Stapedius n.

Vestibular ggl.

Caroticotympanic nn.

Tympanic plexus

Tympanic n.

Internal acoustic meatus

Intermediate n.

superior

inferior

Vestibular n.

Cochlear n.

Spiral ggl. of cochlea

VII

VIII

1 Internal auricular r.
2 Auriculopalpebral n.
3 Rostr. auricular r.
4 Palpebral r.
5 Dors. buccolabial r.
6 Ventr. buccolabial r.
7 R. colli
8 Digastric r. (caud.)
9 Stylohoid r.
10 Caud. auricular n.

R. of carotid sinus

Cranial laryngeal n.

Int. r.

Ext. r.

Carotid sinus

Carotid glomus

Pharyngeal r.

(Facial and vagal communication)

Styliomastoid for.

(Facial and vagal communication)

Distal ggl. (petrosal)

Pharyngeal r.

Distal ggl. (nodose)

Internal r.

Sympathetic n.

External r.

Proximal ggll.

Jugular for.

Cranial roots

Spinal roots

IX

X

XI

For. magnum

Cervical ansa

Canal of hypoglossal n.

Cervical N. 1

XII

Vagosympathetic trunk

Special sensory neuron
Sensory neuron
Parasympathetic neuron
Sympathetic neuron
Motor neuron

3A

The **ENDOCRINE SYSTEM** includes the endocrine glands and those organs which, besides their essential function, liberate hormones. For example, the stomach and intestines are functioning endocrine organs due to the presence of gastrointestinal endocrine cells. Hormones belong to different chemical groups such as steroids, peptides and amines, and act in very small concentrations. The endocrine glands, compared to their output, are remarkably small, and contrary to earlier opinion, an individual cell type is able to form several different hormones. The secreted hormones are messenger substances influencing the development of specific organs such as genital organs and coordinating different body functions with the cooperation of the autonomic nervous system. The endocrine cells secrete hormones mostly into the intercellular space and its direct vicinity. The secretion can then have a direct effect on neighbouring cells or tissues, as for example the androgens synthesized in the testicular interstitial cells affecting neighbouring seminiferous tubules. The hormones, however, can also have an indirect effect, by being absorbed into the vascular or lymphatic systems and then transported to more remote target cells, such as those of the prostate, and their receptors. To facilitate hormone absorption, blood capillaries exhibit a special endothelial structure involving numerous fenestrations. Gonadal capillaries are an exception, having a continuous endothelial lining which exhibits a blood-testis (gonad) barrier. Presumably this provides a functional security for the organs involved.

1 The **HYPOTHALAMUS (1, g)** is the control center of the hypophysis or pituitary gland and is very closely connected to it, topographically, developmentally and functionally.

The **HYPOTHALAMOHYPOPHYSEAL SYSTEM** is the overall central regulatory mechanism of the endocrine system. By means of glandotrophic or regulatory hormones, this regulatory system influences the peripheral endocrine glands (for example the adrenal cortex, but not, however the adrenal medulla). Thus, to be more specific, regulatory hormones of the hypothalamohypophyseal system work indirectly on peripheral target cells or organs such as prostate, through peripheral endocrine secreting cells, such as testicular interstitial cells. In addition, hypophyseal effector hormones are formed which work directly on the peripheral cells or organs, such as prolactin and its direct action on the mammary gland.

The **HYPOPHYSIS** (see p. 45A) or pituitary gland has an anterior lobe, the **adenohypophysis (1, a-c)** which develops from the ectoderm of the roof of the pharynx and migrates towards the sphenoidal bone. There it attaches to a posterior lobe, the **neurohypophysis (1, d, e)** arising from the diencephalon.

a) The **neurohypophysis** receives its effector hormones, oxytocin and vasopressin (adiuretin) from the hypothalamus, where they are formed in the supraoptic and paraventricular nuclei respectively. The efficiency, however, of the blood-brain barrier at the formation site prevents the absorption of the hormones into the bloodstream. Therefore, by the principle of neurosecretion, they are conducted from the site of synthesis to the neurohypophysis through the fibers of the supraopticohypophyseal and paraventriculohypophyseal tracts (see text illustration a and b). A blood-brain barrier is absent at the neurohypophysis, so that the hormones are absorbed through fenestrated blood capillaries and transported further by the vascular system. Vasopressin (adiuretin) promotes the resorption of water from renal canaliculi and raises blood pressure. At the periphery, oxytocin acts on the myoepithelium of the appropriate parts of sweat and mammary glands as well as the smooth muscle of uterus, where it causes labour during parturition.

b) The **adenohypophysis** is acted upon by releasing hormone and release-inhibiting hormone, two regulatory hormones formed in the hypothalamus. As the names suggest, these influence the release of hormones from the anterior lobe of the hypophysis. The hypothalamic hormones are formed by the nerve cells of the tuberal region of the hypothalamus and are transported through the fibers of tuberoinfundibular tract (text illustration c) to the median eminence (text illustration j). This feature projects into the infundibulum of the third ventricle as a crescent-shaped arch of the tuber cinereum.

The **portal system** (text illustration d-h) of the hypothalamohypophyseal system makes the transport of releasing hormones for example from the median eminence to the adenohypophysis. To permit this, the **rostral hypophyseal a. (d)** breaks up into a **capillary net I (e)** in the **median eminence (j)** and this receives the releasing hormones from the hypothalamus. The hormones then proceed to the adenohypophysis via the **hypophyseal portal v. (f)** where they are released into a **capillary net II (q)**. There, both regulatory and effector hormones are received into the bloodstream.

The **THYROID GLAND (2a**, see p. 6), under the impact of thyroid stimulating hormone (TSH) of the adenohypophysis, stimulates cell metabolism by producing thyroxin (T3) and tri-iodo-thyronine (T4), and thus regulates body growth. The hormones are bound to a thyroglobulin and stored in the follicles of the thyroid gland which is therefore referred to as a storage gland. The parafollicular C-cells influence calcium metabolism and are antagonists of the **parathyroid glands (2b**, see also p. 6).

The **ADRENAL GLAND (3**, see p. 24) is organized into a **cortex (a)** and a **medulla (b)**. The three zones of the adrenal cortex synthesize steroid hormones. Mineralocorticoids (aldosterone), which influence salt and water balance, are produced in the subcapsular zona glomerulosa (arcuata), while glucocorticoids influencing carbohydrate metabolism are formed in the adjacent central zona fasciculata (a bundle or column layer). Representative of this hormone group are cortisone and hydrocortisone, anti-inflammatory agents of considerable medical significance. The innermost layer of the adrenal cortex, the reticular zone, produces androgens.

The **GONADS (TESTIS —4A AND OVARY —4B**, see p. 28) possess **testicular interstitial cells (4A, a)**, (the interstitial cells of Leydig), between the testicular tubules, **ovarian interstitial cells (4B, a)**, (cells of the theca interna), and the **corpus luteum (4B, b)**. These are target cells for luteinizing hormone (LH) in the female and interstitial cell stimulating hormone (ICSH) in the male, both emanating from the adenohypophysis. The target cells synthesize androgens, with testosterone and hydrotestosterone as the most important representatives in the male. In the ovary, androgens are aromatized into estrogens in the epithelium of the follicle. Luteinizing hormone also promotes the genesis of the corpus luteum. In the ovary, the follicle stimulating hormone (FSH) of the adenohypophysis stimulates the growth of secondary and tertiary follicles as well as (together with LH) estrogen and androgen synthesis. In the testis, FSH influences the cells of Sertoli by stimulating the production of the androgen-binding protein. Further FSH promotes the genesis of LH-receptors of the cells of Leydig.

The **PLACENTA (4C)** takes over the formation of progesterone and estrogen, previously synthesized in the corpus luteum, during the second half of pregnancy. At that stage, progesterone synthesis in the corpus luteum ceases and the placenta alone assumes the 'endocrinal security' of pregnancy.

The **PINEAL GLAND (5)** or epiphysis of cerebrum synthesizes melatonin and antigonadotrophin (see 45.9.).

The **KIDNEYS** produce renin in the modified **myoid cells (6, a)** of the afferent glomerular arterioles of the **juxtaglomerular complex (6)**, and 2 in this way regulate their own blood flow and the blood pressure of the whole body.

In man the **ENDOCRINE PART OF THE PANCREAS** consists of 1 to 2 million **pancreatic islets (7a)**. By synthesizing insulin their B-cells promote glycogen synthesis from glucose and as a result reduce the blood sugar level. The A-cells synthesize glucagon which raises this level. Furthermore endocrine substances are formed in the D-cells which belong to the group of gastrointestinal hormones.

The **GASTROINTESTINAL ENDOCRINE CELLS (8a)** are scattered among the mucosal epithelial cells of stomach and intestine.

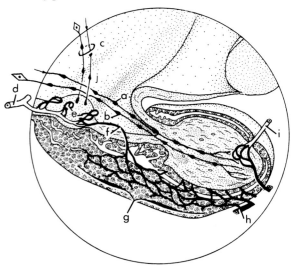

a Supraopticohypophyseal tract
b Paraventriculohypophyseal tract
c Tuberoinfundibular tract
d Rostr. hypophyseal a.
e Capillary net I
f Hypophyseal portal vein
g Capillary net II
h Hypophyseal v.
i Caudal hypophyseal a.
j Median eminence

Endocrinology

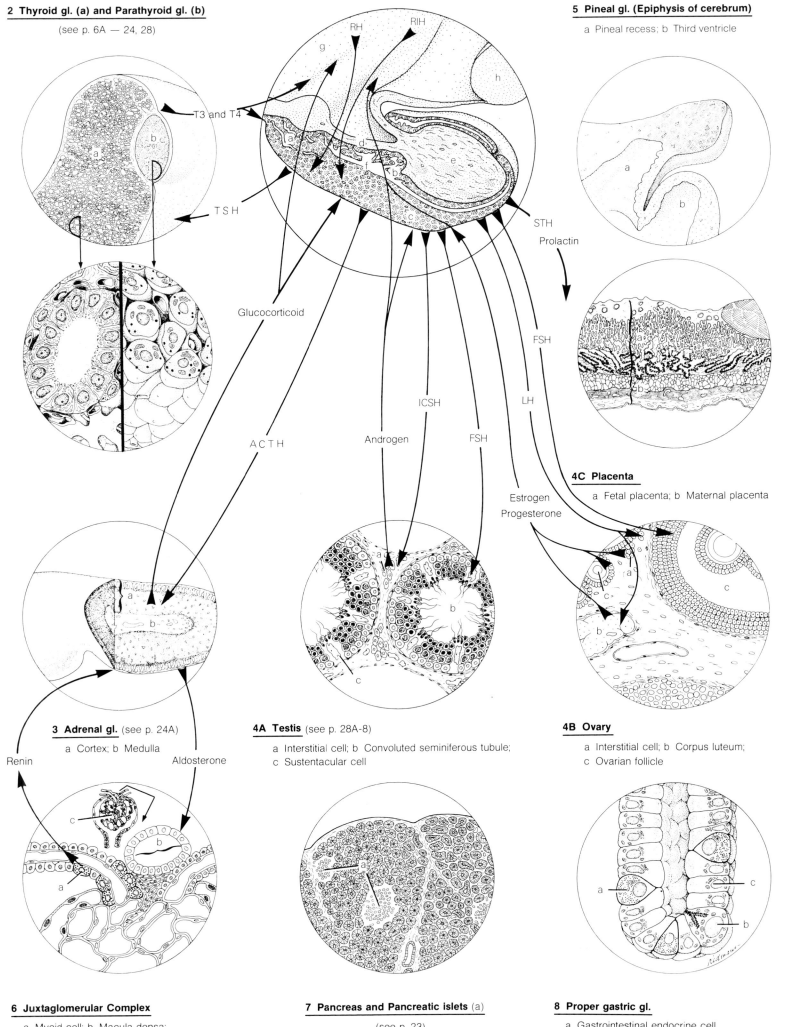

1 Hypothalamus — Hypophysis (see p. 45A)

Adenohypophysis: a Pars tuberalis; b Pars intermedia; c Pars distalis

Neurohypophysis: d Infundibulum; e Lobus nervosus; f Hypophyseal cavity;
g Hypothalamus; h Mamillary body

2 Thyroid gl. (a) and Parathyroid gl. (b)

(see p. 6A — 24, 28)

5 Pineal gl. (Epiphysis of cerebrum)

a Pineal recess; b Third ventricle

RH   RIH

T3 and T4

TSH

STH
Prolactin

Glucocorticoid

FSH

ICSH   LH

ACTH   Androgen   FSH

4C Placenta

a Fetal placenta; b Maternal placenta

Estrogen
Progesterone

Renin   Aldosterone

3 Adrenal gl. (see p. 24A)

a Cortex; b Medulla

4A Testis (see p. 28A-8)

a Interstitial cell; b Convoluted seminiferous tubule;
c Sustentacular cell

4B Ovary

a Interstitial cell; b Corpus luteum;
c Ovarian follicle

6 Juxtaglomerular Complex

a Myoid cell; b Macula densa;
c Glomerulus

7 Pancreas and Pancreatic islets (a)

(see p. 23)

8 Proper gastric gl.

a Gastrointestinal endocrine cell
b Parietal exocrine cell; c Principal exocrine cell

9A

The **FIVE SENSES** should be mentioned here, purely with reference to their histology and physiology.

Peripheral receptor cells, chiefly in association with neighbouring supporting cells, form the receptor organ which receives adequate stimuli of a mechanical or chemical nature. The sense cells are divided into primary and secondary types depending on their morphology. Primary sense cells are modified nerve cells with short dendritic receptors and long axons or fibers leading to the central nervous system, for example olfactory cells. Secondary sense cells are receptor cells of non-neural origin, for example taste cells and those of the vestibulocochlear organ. The impulses of these receptors are conducted further by means of synapses to myelinated nerve fibers which, in turn, proceed to the central nervous system.

The **ORGAN OF OLFACTION** is particularly well developed in the dog. The fundus of the nasal cavity covered with **olfactory mucosa** is known as the area of olfaction. It is lined with **olfactory epithelium** in which the olfactory cells (primary sense cells or olfactory neurosensory epithelial cells) are flanked by supporting cells. Together with the olfactory glands, these lend a yellow-brownish colour to the whole olfactory area. The dog as a macrosmatic animal, possesses an olfactory area of considerable size, some fifteen times larger in an average-sized dog than in man.

The **vomeronasal organ** is also known as the organ of Jacobson after one of the first workers to describe it. It is a blind tube a few millimeters thick, clothed on its inner surface with olfactory epithelium and supported externally by **vomeronasal cartilage (2)**. The organ is situated directly at the cartilaginous nasal septum, caudal to the **incisive papilla (5)**. At its open end, the organ is connected to the roof of the oral cavity by the **vomeronasal duct (4)** and the **incisive duct (3)**. The function of the organ is not clarified absolutely. It appears to be an accessory olfactory organ, its principal role being the reception of olfactory stimulants from food and the perception of pheromones, hence its influence on breeding. The special sensory innervation is supplied by the **vomeronasal n. (1**, a part of the first cranial nerve). In man and many mammals the organ regresses early in ontogenesis.

The **ORGAN OF SIGHT**, the eye, is an evagination of brain and meninges. The fibrous tunic of the bulb, situated externally, can be interpreted as a continuation of the compact dura mater while the middle tunic, namely the choroid or vascular tunic, may be regarded as a continuation of the soft leptomeninx (arachnoid and pia mater). The internal tunic, the retina, is a continuation of the cerebral surface with its concentration of nerve cells.

Functionally, the eye with its five chief constituent parts can be likened to a photographic camera.

1. The **eyelids, upper (6)** and **lower (12)** close the eye and correspond to a camera shutter.

2. The **fibrous tunic of the eye** consists of the **sclera (24)** caudally and the **cornea (11)** rostrally. The former provides the shape of the ocular bulb and corresponds to the camera casing, the latter to the first element of the camera lens.

The transparent cornea retains its transparency due to a precise state of hydration and swelling, maintained externally by lacrimal fluid and internally by aqueous humour (see below). Light refracts on the external surface of the cornea (not on its internal surface), and, as stated above, this is equated with the inflexible first element of the objective lens of the camera.

The tensile fibers in the white sclera maintain mechanical stresses and strains, particularly those caused by the considerable internal ocular pressure and the traction of the extrinsic muscles of the eye. The sclera is modified at the optic disc to form a lamina cribrosa, similar in principal to that of the ethmoidal bone, for the exit of optic nerve fibers. The scleral pigment cells, positioned predominantly at the border with choroid, serve as 'darkening' for the camera casing.

The **middle tunic, the vascular tunic of the bulb,** possesses a **choroid (21)**, heavily pigmented relative to the sclera, and a limiting membrane (of Bruch) between the choroid and the internal tunic. Between the choroid and the limiting membrane is a well developed vascular net. In domestic mammals, one of the layers of choroid, the **tapetum lucidum (22)**, is a particular apparatus for the utilization of incidental light by the eye. This is achieved by photoreceptor cells of the internal tunic of retina. In carnivors, special tapetal cells contain guanine crystals which reflect and scatter light. The tapeta of the other domestic mammals contain special connective tissue fibers. Due to its reflection, incidental

light can act almost twice on the receptor cells of the retina, because the outer pigment layer of the retina is either free of, or poorly endowed with pigment in the region of the tapetum. Depending on breed and size of dog the tapetum occupies an approximately triangular area dorsal to the optic disc.

3. The **iris (13)** corresponds to the diaphragm of a camera. It is part of the middle or vascular tunic of the bulb and lies on the rostral side of the lens. It surrounds the **pupil (9)** freely and regulates the passage of light by means of the **m. dilator pupillae (15)** equipped with radiating fibers, and the **m. sphincter pupillae (14)** with its fibers running in a circular manner. On its caudal aspect, the iris is covered with heavily pigmented epithelium of the pars caeca retinae (see below). Pigments may also be present in the stroma of the iris. These determine eye color which is genetically fixed.

4. The **ciliary body** is also a constituent part of the vascular or middle tunic and forms a circular elevation around the lens. Contraction of the smooth **m. ciliaris (17)** of the ciliary body changes the tension of the **zonular fibers (16)** attaching to the lens. This in turn changes its shape during accommodation. Hence lens and ciliary body, due to fibers emanating on the latter and radiating onto the former, can be likened to the regulating second element of the objective lens of a camera. This makes possible the sharp focussing of the image on the retina, in other words, the film. When observed in greater detail, the surface of the ciliary body is increased by the presence of approximately eighty folds attached to the lens by the zonular fibers. When the m. ciliaris contracts, the folds of the body approach the lens and the fibers relax. As a result, particularly in the young individual, the lens increases its convexity due to intrinsic elasticity. The accompanying increase in light refraction in the course of accommodation causes closely situated objects to be sharply 'drawn' on the retina. In the resting state, the lens is maintained in an oblate or less convex form due to traction on the zonular fibers by the relaxed m. ciliaris.

The **lens (10)**, since it develops from ectoderm and not from the brain, is not a part of the middle or vascular tunic of eye. It is located deep to pupil and iris. The lens is covered by the lens capsule, a cuticular product of its epithelial cells, into which the zonular fibers are anchored. The anterior epithelium of the lens remains a single cell layer during ontogenesis, while the cells of the posterior epithelium elongate, lose their nuclei and become lens fibers. These can attain a length of a centimeter. The lens fibers which have almost a circular course, emanate from two lens sutures or stars which have a Y-shaped configuration.

5. The **internal tunic of the bulb, the retina**, consists of ten layers (refer to an histology text) and in the ninth layer numbered from within, bears the photoreceptors, namely the rods and cones. As stated previously, the retina is therefore the equivalent of a photographic film.

The light-sensitive **pars optica retinae (20)** extends from the ocular fundus to the base of the iris. There it merges with the receptor-free, heavily pigmented **pars caeca retinae (19)** at the **ora serrata (18)**. The pars caeca retinae covers the inner surface of the ciliary body and the iris.

The photoreceptors comprise 95% rods and only 5% cones. The light sensitivity of the retina (regulated in the photographic camera by different light sensitive film = DIN.No) is regulated by the tenth histological layer of the retina, the pigment epithelium. During intense light, the cell processes of the pigment epithelium penetrate the pars optica and regularly envelop the receptors. With darkness, the pigment epithelium retracts from the field of the rods and cones.

Photoreceptors are absent from the **optic disc (25)**. The disc is the site of passage of the **optic n. (26)**, and **retinal blood vessels (27)** appear at its periphery. These are visible on ophthalmological examination. The macula, the site of sharpest vision, is situated a few millimeters dorsolaterally. In the dog, as compared with primates, it is underdeveloped and scarcely detectable, the cone population being only slightly increased. In man the analogous macula, the macula lutae, has a yellowish colour.

Within the eye and anterior to the lens, the **anterior (8)** and the **posterior chamber (7)** are located anterior and posterior to the iris respectively.

The **vitreous chamber (23)** is situated posterior to the lens. The aqueous humour of the anterior and posterior chambers is clear and devoid of blood vessels. To be resorbed, the aqueous humour flows into the lacunae of a spongy system of trabeculae, the space of the iridocorneal angle, at the angle of like name. It then flows to the venous sinuses of the sclera (the canals of Schlemm in man) where it is received by ciliary vv. With damage to the drainage system, internal ocular pressure increases and results in glaucoma.

Organic of Olfaction and Organ of Sight

Vomeronasal organ (medial view)

Olfactory epithelium

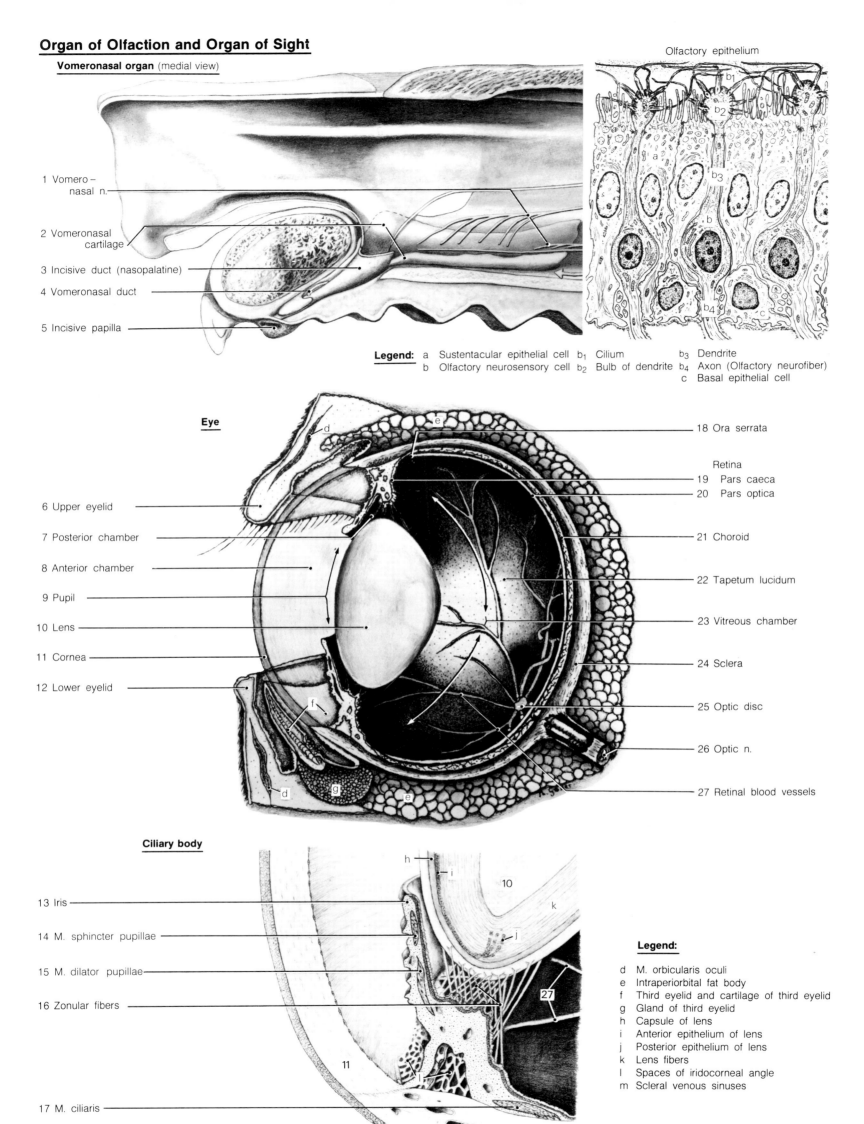

1 Vomero –
  nasal n.

2 Vomeronasal
  cartilage

3 Incisive duct (nasopalatine)

4 Vomeronasal duct

5 Incisive papilla

Legend: a   Sustentacular epithelial cell   b1   Cilium   b3   Dendrite
          b   Olfactory neurosensory cell   b2   Bulb of dendrite   b4   Axon (Olfactory neurofiber)
                                                     c   Basal epithelial cell

Eye

6 Upper eyelid

7 Posterior chamber

8 Anterior chamber

9 Pupil

10 Lens

11 Cornea

12 Lower eyelid

18 Ora serrata

Retina

19   Pars caeca

20   Pars optica

21 Choroid

22 Tapetum lucidum

23 Vitreous chamber

24 Sclera

25 Optic disc

26 Optic n.

27 Retinal blood vessels

Ciliary body

13 Iris

14 M. sphincter pupillae

15 M. dilator pupillae

16 Zonular fibers

17 M. ciliaris

Legend:

d   M. orbicularis oculi
e   Intraperiorbital fat body
f   Third eyelid and cartilage of third eyelid
g   Gland of third eyelid
h   Capsule of lens
i   Anterior epithelium of lens
j   Posterior epithelium of lens
k   Lens fibers
l   Spaces of iridocorneal angle
m   Scleral venous sinuses

Th. Goller

The **VESTIBULOCOCHLEAR ORGAN** has its receptors in the **internal ear**, situated in the **petrosal part of the petrous temporal bone (1)**. The organ of hearing also includes: a) the external ear (see p. 42) with an auricular cartilage moved by muscles and including the **external acoustic meatus (2)**; and b) the middle ear. The **middle ear (3)** is located in the tympanic cavity and is separated from the external ear by the **tympanic membrane (4)**. This is covered by modified skin on its external surface, and a mucous membrane on its internal surface. The membrane forms an air-tight seal between the external and the middle ear. The middle ear, however, is ventilated through the **auditory tube (5**, see 41.13) leading from the pharynx. The **auditory ossicles** of the middle ear, namely the **malleus (6)**, **incus (7)** and **stapes (8)** and the tympanic membrane, function as sound amplifiers. (A deficient apparatus signifies a hearing difficulty not deafness.) The base of the stapes projects into the (oval) **vestibular window (9)**.

The **internal ear** is housed in a complicated labyrinthine system, the **osseous labyrinth (10)** enveloping and protecting the **membranous labyrinth (11)**

The **perilymphatic space (12)** lies between the osseous and membranous labyrinths and in the scala tympani and the scala vestibuli of the cochlea (see below). It is filled with perilymph. Endolymph fills the endolymphatic space in the vestibule, semicircular ducts, and the **cochlear duct (13)**. Endo- and perilymph from endo- and perilymphatic spaces are conducted to the meningeal covering of the petrous part of the temporal bone by way of the **endolymphatic (14)** and the **perilymphatic duct (15)** respectively.

Sound waves are conducted from the middle ear through the vestibular window, to the **vestibule (16)** and **cochlea (17)** where they stimulate the organ of hearing. Sound waves reach the apex of the cochlea by the **scala vestibuli (18)**. There it connects with the **scala tympani (19)** through the helicotrema. An equalisation of pressure between scala tympani and middle ear is produced across the membranous (round) **cochlear window (20)**.

The three principal sections of the osseous or membranous labyrinth, as the case may be, are:

1. The cochlea and the **auditory receptor organ** namely the **spiral organ (21) (of Corti)**. The osseous cochlea consists of three gyri becoming noticeably narrower towards the apex and positioned around a central osseous axis, the **modiolus (22)**. From this on osseous spiral **lamina (23)** with its membranous covering projects out into the cochlea and subdivides it into an overlying scala vestibuli and a underlying scala tympani lying ventrally. Both scalae contain perilymph. Inserted between the two and attached to the modiolus is the cochlear duct containing endolymph. The spiral organ (of Corti), associated with the sense of hearing, lies on the duct floor of the basal lamina. Its secondary sense cells are surrounded by dendrites which then course in the modiolus to bipolar ganglion cells contained in the **spiral ganglion (24**, see p. 48). The central processes of these bipolar ganglion cells from the **cochlear n. (25)**. Peripheral processes of bipolar ganglion cells associated with equilibrium (see below) reach the superior and inferior parts of the **vestibular ganglion (26)** while the central processes form the **vestibular n. (27)**. This unites with the cochlear n. to give the **vestibulocochlear n. (28)** which passes out of the **internal acoustic meatus (29)** to the brain.

2. In the static maculae of the **vestibule** and its **utricle (30)** and **saccule (31)**, straight line movements are registered in association with the sense of equilibrium or balance.

3. The three **osseous semicircular canals** contain within them the membranous semicircular ducts. Within these, angular accelerations are registered by pectinate ampullary crests also associated with the sense of balance (for details consult histology texts).

The **ORGAN OF TASTE**, endowed with special **taste receptors (a)**, perceives substances to be tasted. In common with **supporting or sustentacular cells (b)**, and **basal cells (c)**, receptors are integrated into a barrel-shaped **taste-bud (d)** located in the surface epithelium of the tongue. Taste receptors are secondary sense cells, from which stimuli are conducted via synapses to special sensory nerve fibres. These go to the brain (see p. 43A, 11 and 12; gustatory glands —x).

Superficial and visceral **SENSORY INNERVATION** mediate sensations particularly temperature, taste and pain, by free and encapsulated nerve endings. A special form is deep sensation where information is perceived from so-called proprioceptors particularly associated with muscle and tendon.

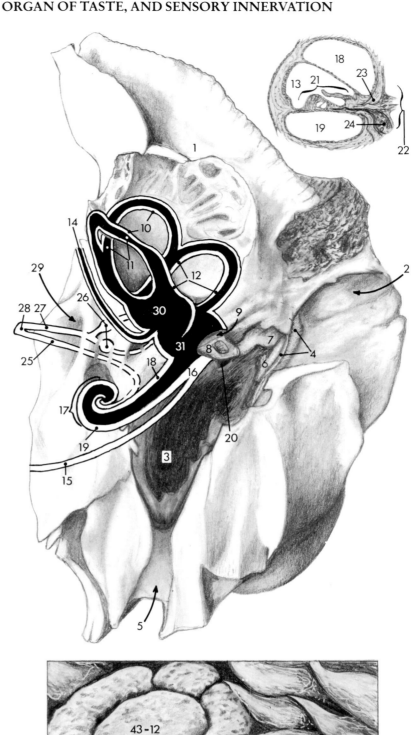

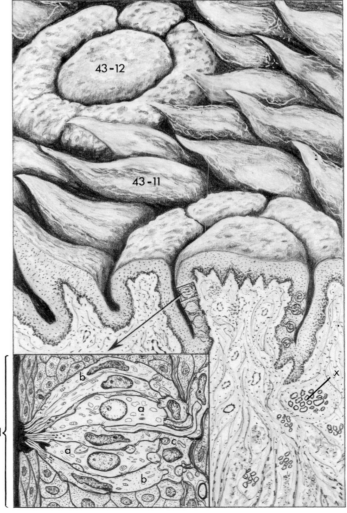

APPENDIX OF SPECIFIC ANATOMY:

ARTHROLOGY: ARTICULATIONS OF TRUNK (TURN TO P. III)

NAME	BONES INVOLVED	SHAPE	FUNCTION	COMMENTS
1. **Atlantooccipital art.** (illustrated) (referred to as first head articulation)	Occipital condyle and articular fovea of atlas	Ellipsoid art. Simple art.	Flexion; nodding of the head	The first two articulations are stabilized by the dens of the axis which is anchored to the floor of the vertebral canal by the ligaments of the apex of the dens of the axis and the transverse ligament of the atlas
2. **Atlantoaxial art.** (illustrated) (referred to as second head articulation)	Dens and articular surface of axis, and fovea of dens and articular surface of atlas	Trochoid art. Simple art.	Rotation; shaking of the head	
3. **Intervertebral symphysis**	Bodies of adjacent vertebrae	Intervertebral discs between vertebrae	Slight possibility of movement	Dorsal and ventral longitudinal ligaments are situated on the respective aspects of the vertebral bodies
4. **Art. of the articular processes**	Articular processes of adjacent vertebrae	Plane art.	Arthrodial joint	In the cervical and lumbar regions slight margin for movement is present; in the thoracic region there is almost none
5. **Art. of head of rib**	Articular surface of head of rib and caudal costal fovea and cranial costal fovea respectively of the contiguous vertebrae	Spheroid art. Composite art.	Ginglymus (hinge joint)	Both articulations make possible the fluctuation of thoracic volume during respiration. The intercapital ligament (see III 9) connects the heads of the ribs of like numeration of both sides and lies dorsal to the intervertebral disc. The intercapital ligament holds the head of each rib in its costal fovea and is absent on the first pair and last two pairs of ribs
6. **Costotransverse art.**	Articular surface of the tubercle of the rib and the costal fovea of the transverse process of the appropriate vertebra	Condylar art. Simple art.	Ginglymus (hinge joint)	
7. **Sternocostal artt.**	Ends of costal cartilages 1 to 8 and sternum	Condylar art. Simple art.	Ginglymus (hinge joint)	The ninth rib is not connected to the sternum by an articulation

Atlantooccipital and atlantoaxial articulations

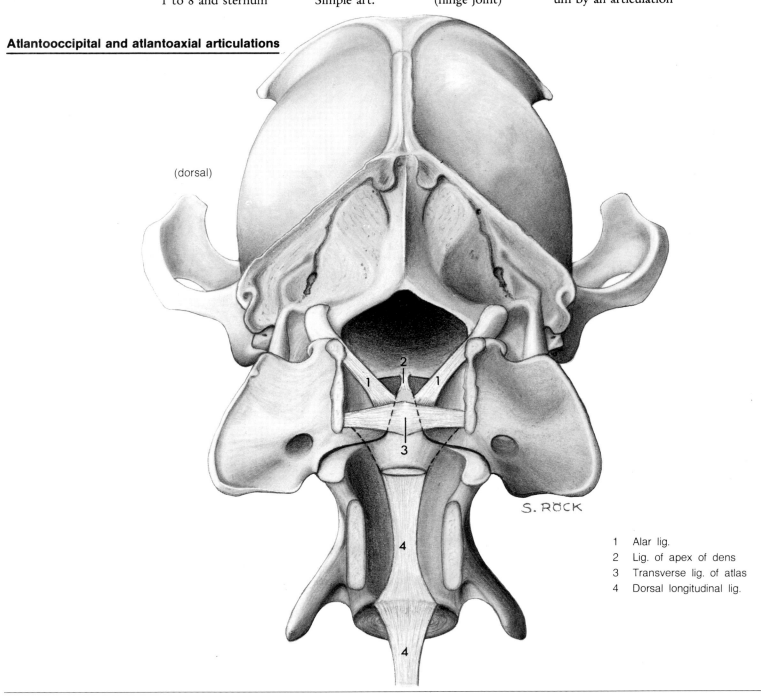

(dorsal)

S. RÖCK

1 Alar lig.
2 Lig. of apex of dens
3 Transverse lig. of atlas
4 Dorsal longitudinal lig.

NAME	BONES INVOLVED	SHAPE	FUNCTION	COMMENTS
1. Art. of humerus (illustrated) (Shoulder joint)	Glenoid cavity of scapula and head of humerus	Spheroid art. Simple art.	Ginglymus	Typical ligaments are absent; their function is taken over by contractile tension 'bands' within the ligaments of subscapular and infraspinous muscles.
2. Cubital art. (Elbow joint) (illustrated) Humeroulnar art. Humeroradial art. Proximal radioulnar art.	Capitulum and trochlea of humerus, head of radius, articular circumference of radius, and radial notch of ulna; ulna	Ginglymus Ginglymus Trochoid art. Composite art.	Ginglymus, (snatching) Trochoid (Pivot)	Surgical intervention with rupture of the medial or lateral collateral ligaments, isolation of the anconeal process, fracture of the bones forming the articulations, and luxation which occurs only with flexion of the articulation
3. Distal radioulnar art.	Ulnar notch of radius Articular circumference of ulna	Condylar art. Simple art.	Trochoid art.	Surgical intervention in fractures of the lateral styloid process and growth disorders (e. g. radius curvus). The articulation communicates with the antebrachiocarpal art.
Artt. of manus: 4. Antebrachiocarpal art.	Trochlea of radius, and head of ulna and carpal bones	Condylar art. Composite art.	Almost a free articulation	Surgical intervention in instability and trauma; feasibility of treatment: arthrodesis
5. Metacarpophalangeal art.	Metacarpal bone Proximal phalanx Sesamoid bones	Ginglymus Composite art.	Ginglymus, uniaxial	Surgical intervention in fractures of sesamoid bones and phalanges; panaritium; feasibility of treatment: digital amputation, prophylactic amputation of dewclaws to prevent injuries
6. Proximal interphalangeal art. of manus	Proximal phalanx Middle phalanx	Sellar art. Simple art.	Ginglymus, uniaxial	
7. Distal interphalangeal art. of manus	Middle phalanx Distal phalanx sesamoid bone	Sellar art. Composite art.	Ginglymus, biaxial	The dewclaws may be duplicated in certain dog breeds for example the St. Bernard

8. The **mediocarpal art.** is situated between the proximal and distal rows of carpal bones.
9. The **carpometacarpal art.** lies between the distal row of carpal bones and the metacarpal bones.
10. The **intercarpal artt.** are situated in the perpendicular articular spaces between the carpal bones.

Humeral and cubital articulations (shoulder and elbow joints)

Humeral art. (medial)

Cubital art. (medial)

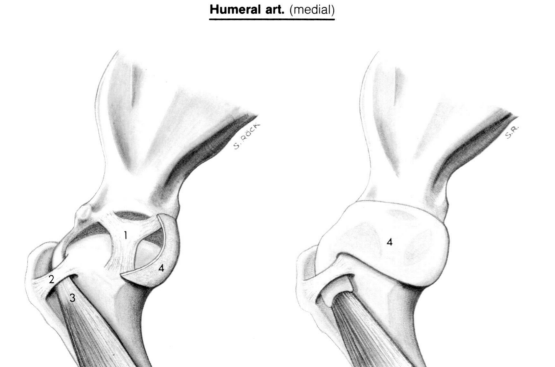

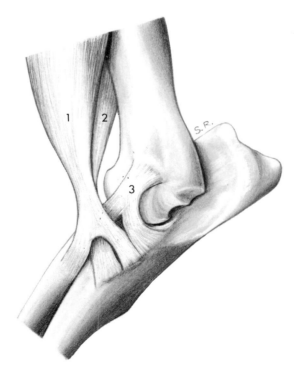

1 Medial glenohumeral lig.
2 Transverse ligament (retinaculum)
3 M. biceps brachii
4 Articular capsule

1 M. biceps brachii
2 M. brachialis
3 Medial collateral lig. of elbow

(see p. 8)

ARTICULATIONS OF THE PELVIC LIMB (TURN TO P. III)

NAME	BONES INVOLVED	SHAPE	FUNCTION	COMMENTS
1. **Sacroiliac art.** (illustrated) (Joint of sacrum)	Auricular surfaces of sacrum and ilium	Plane art. Simple art.	Amphiarthrosis	The strong articular capsule and the ligaments (sacrotuberal, dorsal and ventral iliac ligg.) relax to widen the pelvis at parturition
2. **Coxal art.** (illustrated) (Hip joint)	Ilium, ischium and pubis at the acetabulum, and the head of the femur	Spheroid art. Composite art.	Free joint	The ligament of the head of the femur anchors the head of the femur to the acetabulum and, at least in young dogs, brings blood vessels to the head of femur
3. **Articulation of knee** (stifle joint): Femorotibial art.	Lateral and medial condyles of tibia and femur, sesamoid bones of m. gastrocnemius and sesamoid bone of m. popliteus	Spiral art. Composite art.	Ginglymus with a brake function, uniaxial	To compensate for the incongruency of the articular surfaces of femur and tibia, two fibrocartilaginous menisci lie between them. The menisci also serve as a shock absorber mechanism and are anchored to each other and the tibia by meniscal ligaments The meniscofemoral lig. provides attachment to the femur
Femoropatellar art.	Trochlea of femur Patella	Sliding art. Simple art.	Sliding	The patella is a sesamoid bone situated in the terminal tendon of the quadriceps muscle. The tendon part associated with the insertion, the patellar ligament, anchors the patella to the tibial tuberosity

The **articulation of the knee (stifle joint)** is composed of the femorotibial and the femoropatellar articulation. The femorotibial articulation possesses two synovial cavities in open communication with each other and these in turn communicate with the femoropatellar articulation, the proximal tibiofibular articulation, and the articulation of the sesamoid bones of the m. gastrocnemius with the femur. The capsule of the stifle joint also surrounds the tendons of origin of the mm. extensor digitalis longus and popliteus as capsular tendon sheaths. The joint capsule is attached to the outer convexitiy of each meniscus. The cruciate ligaments of the articulation are covered by synovial membrane in a positional relationship similar to a portion of intestine and its associated peritoneum. An extensive body of adipose tissue, the fat body, lies distal to the patella between the synovial and fibrous strata of the articular capsule.

4. The **proximal and distal tibiofibular articulations** are simple, tight synovial articulations with limited synovial cavities. The proximal communicates with the synovial cavity of the femorotibial articulation and the distal with that of the tarsal articulation (see p. 55). The crural interosseous membrane could be perceived as the ligament between these articulations.

Sacroliliac and coxal articulations (ventral, see p. 29)

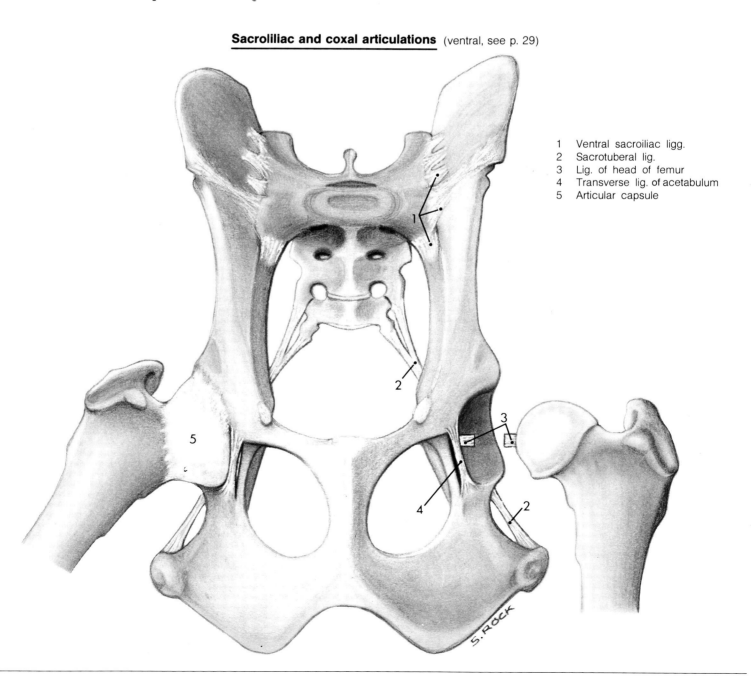

1 Ventral sacroiliac ligg.
2 Sacrotuberal lig.
3 Lig. of head of femur
4 Transverse lig. of acetabulum
5 Articular capsule

NAME	BONES INVOLVED	SHAPE	FUNCTION	COMMENTS
Articulations of foot:				
5. Tarsocrural art.	Cochlea of tibia with lateral malleolus of fibula and trochlea of talus	Cochlear art. Composite art.	Ginglymus, uniaxial	Surgical interference is necessary among other things, in fractures of the malleolus and tarsal bones. In complicated splinter fractures arthrodesis is necessary.

6. The **tarsal articulation** actually comprises four horizontal articulations, each with its own synovial cavity, and also the intertarsal articulations (see below). Only the proximal two, the tarsocrural and the proximal intertarsal articulation, are ginglymus and trochoid articulations respectively. Both distal articulations (the distal intertarsal and tarsometatarsal articulations) are tight joints permitting almost no scope for movement.

7. The **intertarsal articulations** are situated perpendicularly between the tarsal bones as tight joints. (The digital articulations are similar to those of the thoracic limb.)

Tarsal articulation (medial)

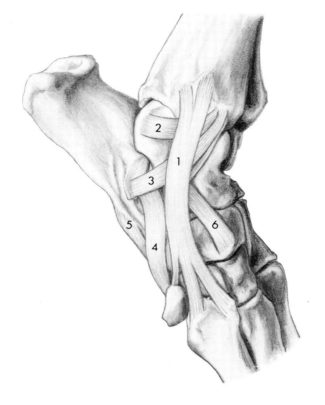

	Medial collateral ligament of tarsus
1	Long medial collateral ligament of tarsus
	Short collateral ligament of tarsus
2	Tibiotalar part
3	Tibiocalcaneal part.
4	Calcaneometatarsal part
5	Long plantar ligament
	Dorsal tarsal ligament
6	Talocentral ligament

ARTICULATIONS BETWEEN BONES OF SKULL

1. Temporomandibular art. (illustrated)	Condylar process of mandible and mandibular fossa of temporal bone	Condylar art. Simple art.	Ginglymus	The fibrocartilaginous articular disc divides the articular cavity into two levels

2. The **intermandibular suture** connects both halves of the mandible between the central incisor teeth of the lower arch. The attachment is partly by connective tissue and partly by cartilage, and has a tendency to fracture. The suture ossifies as the animal ages.

3. The **hyoid bone** is attached to the temporal bone by the tympanohyoid element consisting of connective tissue, while the osseous thyrohyoid element is attached to the thyroid cartilage by an articulation. The connections between the individual elements of the hyoid bone are due in part to synovial articulations, in part to connective tissue.

Temporomandibular articulation

(ventrolateral)

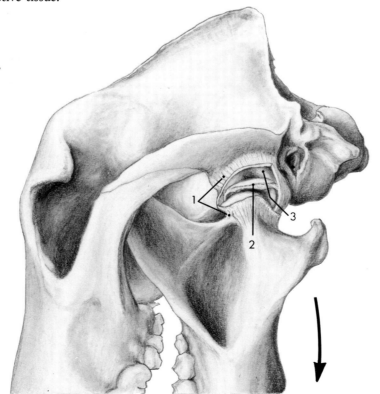

1	Articular capsule
2	Articular disc
3	Articular cavity

MYOLOGY

DORSAL MUSCULATURE CONNECTING TRUNK AND THORACIC LIMB — SUSPENSION OF THORACIC LIMB (TURN TO P. 5)

MUSCLE	ORIGIN	INSERTION	INNERVATION	FUNCTION	COMMENTS
M. trapezius	A dorsomedian position in region from 3rd cervical to 9th thoracic vertebra overlying spinous processes	**Thoracic part:** Dorsal third of spine of scapula **Cervical part:** Dorsal two thirds of spine of scapula	Dorsal ramus of accessory n. (XI)	Protractor of thoracic limb	Together, muscles of right and left side form a trapezium
M. cleidocervicalis	A dorsomedian position in the neck region	Clavicular intersection	Dorsal ramus of accessory n. (XI)	Protractor of thoracic limb	Regarded by some workers as part of m. trapezius
M. omotransversarius	Acromion of scapula	Transverse process of atlas	nC 4 vm	Protractor of thoracic limb; bends neck to one side	Covers superficial cervical lymph nodes
M. rhomboideus: — capitis — cervicis — thoracis	Nuchal crest A dorsomedian position from second cervical to sixth thoracic vertebra	Scapular cartilage	nCvm nCvm nCvm	Fixation of thoracic limb; elevates and retracts thoracic limb; raises neck	Muscles of right and left side form a rhomboid; covered by m. trapezius
M. latissimus dorsi	Thoracolumbar fascia	Fascia of brachium; crests of greater and lesser tubercles of humerus	Thoracodorsal n.	Retractor of thoracic limb; flexor of shoulder joint; progression of trunk with fixation of limb	Tendon of insertion forms the arch of axilla

VENTRAL MUSCULATURE CONNECTING TRUNK AND THORACIC LIMB—SUSPENSION OF TRUNK (TURN TO P. 6)

MUSCLE	ORIGIN	INSERTION	INNERVATION	FUNCTION	COMMENTS
Mm. pectorales superficialis: M. pectoralis transversus M. pectoralis descendens	Manubrium and cranial part of body of sternum Manubrium of sternum	Crest of greater tubercle of humerus	Cranial and caudal pectoral nn.	Trunk-limb attachment; adductor, protractor and retractor of thoracic limb	Forms the lateral pectoral groove with the clavicular part of m. deltoideus
M. pectoralis profundus:	Manubrium and body of sternum		Cranial and caudal pectoral nn.	Supports trunk; retractor of thoracic limb; fixes shoulder joint	Is divided into a principal and an accessory part
Principal part Accessory part		Greater and lesser tubercles of humerus; Fascia of brachium			
M. serratus ventralis: — cervicis —thoracis	Transverse processes from 2nd to 7th cervical vertebrae First seven to ten ribs	Facies serrata of scapula	nCvm Long thoracic n.	Most important muscle supporting trunk; raises neck; when thoracic limb is fixed, auxillary inspiratory muscle	A dentate or serrated muscle. Dorsal scapular a. courses along boundary between both parts
M. sternocleidomastoideus: M. cleidomastoideus M. sternomastoideus M. sternooccipitalis	Clavicular intersection Manubrium of sternum Manubrium of sternum	Mastoid process of temporal bone Mastoid process of temporal bone Nuchal crest	Ventral ramus of accessory n. (XI)	Protractor of thoracic limb; depresses head; rotation of head and neck laterally with the contraction of the ipsilateral muscle	The jugular groove is situated on the lateral surface of the muscle
Deltoideus m.	See lateral muscles of shoulder and muscles of brachium (p. 57)				

LONG HYOID MUSCLES (TURN TO P. 6)

MUSCLE	ORIGIN	INSERTION	INNERVATION	FUNCTION	COMMENTS
M. sternohyoideus	Sternum: Internal surface of manubrium and first costal cartilage	Basihyoid element	nC 1 vm	Retractor of hyoid and tongue	In the ventral midline of the neck (linea alba colli) muscles of right and left side are attached to each other
M. sternothyroideus	Sternum: Internal surface of manubrium and first costal cartilage	Thyroid cartilage of larynx	nC 1 vm	Synergist of the m. sternohyoideus and retractor of the thyroid cartilage	At its origin, connected to m. sternohyoideus

MEDIAL MUSCLES OF SHOULDER AND MUSCLES OF BRACHIUM (TURN TO P. 8)

MUSCLE	ORIGIN	INSERTION	INNERVATION	FUNCTION	COMMENTS
M. teres major	Caudal border of scapula	Teres tuberosity of humerus	Axillary n.	Flexor of shoulder joint	Not a round muscle in the dog
M. subscapularis	Subscapular fossa of scapula	Lesser tubercle of humerus	Subscapular and axillary nn.	Predominantly an extensor of the shoulder joint	Functions as a medial contractile 'ligament' of the shoulder joint
M. coracobrachialis	Coracoid process of scapula	Crest of lesser tubercle of humerus	Musculocutaneous n.	Extensor and adductor of shoulder joint	Its tendon of origin is surrounded by a tendon sheath
M. biceps brachii	Supraglenoid tubercle of scapula	Radial tuberosity and proximomedial aspect of ulna	Musculocutaneous n.	Flexor of elbow joint	Two heads of origin of muscle are present in man
M. brachialis	Caudodistal to neck of humerus	Proximomedial aspect of ulna	Musculocutaneous n.	Flexor of elbow joint	Muscle has long fibers and a considerable up and down stroke
M. tensor fasciae antebrachii	M. latissimus dorsi	Antebrachial fascia	Radial n.	Tensor of the antebrachial fascia	Only moderately pronounced in dog

LATERAL MUSCLES OF THE SHOULDER AND MUSCLES OF BRACHIUM (TURN TO PP 6 AND 9)

MUSCLE	ORIGIN	INSERTION	INNERVATION	FUNCTION	COMMENTS
M. deltoideus:					
Clavicular part (m. cleidobrachialis)	Clavicular intersection	Crest of humerus	Accessory axillary n.	Clavicular part: protractor of thoracic limb	To perform surgery on the head of the humerus access is gained between acromial part of m. deltoideus and m. supraspinatus
Scapular part	Aponeurosis at spine of scapula	Deltoid tuberosity of humerus	Axillary n.	Flexor of shoulder joint	
Acromial part	Acromion	Deltoid tuberosity of humerus	Axillary n.		
M. teres minor	Infraglenoid tubercle and distal third of caudal border of scapula	Deltoid tuberosity	Axillary n.		Lies deep to the scapular part of m. deltoideus
M. supraspinatus	Supraspinous fossa and spine of scapula	Greater tubercle of humerus	Suprascapular n.	Extensor of shoulder joint; fixes or stabilises the articulation	Lies cranial to acromion approaching its insertion
M. infraspinatus	Infraspinous fossa and spine of scapula	Greater tubercle of humerus	Suprascapular n.	Flexor of shoulder; functions as a lateral contractile 'ligament' of the articulation	Strongly pennate; located deep to the scapular part of m. deltoideus
M. triceps brachii		Common tendon of all heads inserts at olecranon with a synovial bursa underlying it	Radial n.	Extension of elbow joint; long head also a flexor of the shoulder joint	The muscle occupies completely the triangle formed by scapula, humerus and olecranon. Its caudal contour forms the visible tricipital border
Long head	Caudal border of scapula				
Lateral head	Laterally on humerus				
Medial head	Medially on humerus				
Accessory head	Caudally on the neck of the humerus				
M. anconeus	Around the border of the olecranon fossa	A fleshy insertion together with the m. triceps brachii laterally on the olecranon	Radial n.	Extensor of elbow joint	Can be included as the fifth head of the m. triceps brachii

CAUDOMEDIAL MUSCLES OF THE ANTEBRACHIUM: FLEXOR MUSCLES ARISING FROM MEDIAL EPICONDYLE OF HUMERUS (TURN TO P. 10)

MUSCLE	ORIGIN	INSERTION	INNERVATION	FUNCTION	COMMENTS
M. flexor dig. superficialis	Medial epicondyle of humerus	Flexor tuberosity of middle phalanx of each of digits 2-5	Median n.	Flexor of digits and joints of carpus	Continuing as superficial digital flexor tendon
M. flexor dig. profundus	Three heads:	Flexor tuberosity of distal phalanx of each digit		Flexor of joints of carpus and the distal interphalangeal joints	The muscle heads unite in the distal third of the antebrachium to give the deep digital flexor tendon which passes through the flexor manica
Humeral head	Medial epicondyle of humerus		Ulnar and median nn.		
Radial head	Radius		Median n.		
Ulnar head	Ulna		Ulnar n.		
M. flexor carpi ulnaris	Two heads:	Accessory carpal bone	Ulnar n.	Flexor and supinator of joints of carpus	Two heads are separate throughout their entire length
Humeral head	Medial epicondyle of humerus				
Ulnar head	Olecranon				
M. flexor carpi radialis	Medial epicondyle of humerus	McII and McIII	Median n.	Flexor of joints of carpus	A bifurcated terminal tendon
M. pronator teres	Medial epicondyle of humerus	Craniomedial aspect of radius	Median n.	Pronation (rotation of antebrachium inwards)	Lies medial to elbow joint
M. pronator quadratus		Bridges the interosseous space	Median n.	Pronation (rotation of antebrachium inwards)	Constant only in carnivors

CRANIOLATERAL MUSCLES OF ANTEBRACHIUM: EXTENSORS ARISING FROM LATERAL EPICONDYLE OF HUMERUS (TURN TO P. 10)

MUSCLE	ORIGIN	INSERTION	INNERVATION	FUNCTION	COMMENTS
M. extensor digitalis com.	Lateral epicondyle of humerus	Distal phalanx of each digit 2 to 5	Radial n.	Extensor of digits and joints of carpus	Between the terminal tendons of m. ext. dig. lat. and m. ext. carpi ulnaris one has access to the radioulnar ligament
M. extensor digitalis lat.	Lateral epicondyle of humerus	Distal phalanx of each digit 3 to 5	Radial n.	Extensors of digits	
M. abductor pollicis longus	A thin triangular muscle arising craniolaterally on the antebrachium	Mc I	Radial n.	Extensor and abductor of digit I	A sesamoid bone lies on the tendon of insertion of the muscle
M. extensor pollicis	Craniolateral middle third of ulna	Digits I and II	Radial n.	Extensor of digits I and II	Constant only in carnivors
M. extensor carpi radialis	Lateral supracondylar crest	Mc II and Mc III	Radial n.	Most important extensor; fixes the joints of carpus	Bi- or trifurcated terminal tendon
M. extensor carpi ulnaris [M. ulnaris lateralis]	Lateral epicondyle of humerus	Mc V and accessory carpal bone	Radial n.	Extensor of joints of carpus	Due to attachment to the accessory carpal bone, the muscle can function as a flexor and abductor of the joints of carpus
M. supinator	Lateral supracondylar crest; lateral collateral lig.; annular lig. of radius	Proximocranially on radius	Radial n.	Supination of antebrachium and manus	Lies deep to m. ext. dig. com.
M. brachioradialis (supinator longus)	Lateral supracondylar crest	Mediodistally on radius and antebrachial fascia	Radial n.	Supination of the elbow joint	May be absent in the dog

VERTEBRAL MUSCLES: A. DORSAL (AUTOCHTHONOUS) MUSCULATURE (TURN TO P. 12)

MUSCLE	ORIGIN	INSERTION	INNERVATION	FUNCTION	COMMENTS
M. splenius	Spinous processes of first 3 thoracic vertebrae	Nuchal crest	Dorsal rami of corresponding spinal nerves	Extension and lateral flexion of head and neck	Muscle shape resembles a spleen or strap
M. iliocostalis — lumborum — thoracis	Crest of ilium	Transverse processes of lumbar vertebrae, angles of ribs, transverse processes of last two cervical vertebrae	Dorsal rami of corresponding spinal nerves	Fixation of lumbar region and ribs; erection and lateral flexion of vertebral column	The m. iliocostalis thoracis arises without transition from the m. iliocostalis lumborum
M. longissimus — lumborum — thoracis — cervicis — capitis	Spinous processes of sacrum, lumbar and thoracic vertebrae; wing of ilium; transverse processes of thoracic and cervical vertebrae	Transverse and accessory processes of the spinal column; tubercles of ribs; wing of atlas and mastoid process of temporal bone	Dorsal rami of corresponding spinal nerves	Fixation and extension of the vertebral column; erection of upper body; elevation of head and neck; lateral flexion of neck by contraction of muscle of one side	Lies directly under the thoracolumbar fascia medial to the adjacent m. iliocostalis; the longest muscle of the vertebral column
M. spinalis et semispinalis thoracis et cervicis	Spinous processes from cervical vertebra 2nd (axis) to 11th thoracic vertebra		Dorsal rami of corresponding spinal nerves	Fixation of dorsum and neck	Situated medial and adjacent to the m. longissimus
M. semispinalis capitis M. biventer cervicis M. complexus	Transverse processes from 4th to 2nd thoracic vertebrae; caudal articular processes from 1st thoracic vertebra to 3rd cervical vertebra	Squamous occipital bone	Dorsal rami of corresponding spinal nerves	Elevation and lateral flexion of head	Dorsal portion: m. biventer cervicis has tendinous intersections directed transversally. Ventral portion: m. complexus
Mm. multifidi — multifid. cervicis — multifid. thoracis — multifid. lumb.	Articular and mamillary processes from 2nd cervical vertebra (axis) to sacrum	Spinous processes of respective preceding vertebrae	Dorsal rami of corresponding spinal nerves	Fixation and rotation of the spinal column; elevation of neck	Repeated pennate muscle series, each muscle taking origin from 1 to 4 vertebrae
Mm. interspinales	Run between the spinous processes of lumbar, thoracic and cervical vertebrae		Dorsal rami of corresponding spinal nerves	Fixation and lateral flexion of vertebral column with unilateral action of muscles	Each muscle spans adjacent spinous processes
Mm. intertransversarii	Extend between transverse, mammillary, articular and accessory processes of vertebrae		Dorsal rami of corresponding spinal nerves	Fixation and lateral flexion of the vertebral column with unilateral action of muscles	In the neck and tail, the muscles are subdivided into a dorsal and a ventral series
M. sacrococcygeus dors. med.	Mammillary processes from 7th lumbar vertebra to the last coccygeal vertebra	Spinous processes of the respective preceding vertebrae	Dorsal rami of the corresponding spinal nerves	Elevation and lateral flexion of the tail	Caudal continuation of the mm. multifidi
M. sacrococcygeus dors. lat.	Aponeurosis of m. longissimus, and mammillary processes from the 4th lumbar vertebra caudally	Slender tendinous attachments to the respective preceding vertebrae	Dorsal rami of the corresponding spinal nerves	Elevation and lateral flexion of tail	Caudomedial continuation of the m. longissimus

MUSCLES OF VERTEBRAL COLUMN: B. VENTRAL MUSCLES OF VERTEBRAL COLUMN (TURN TO P. 12)

MUSCLE	ORIGIN	INSERTION	INNERVATION	COMMENTS	FUNCTION
Mm. scaleni M. scalenus dors. M. scalenus med. (of first rib)	Transverse processes of 4th or 5th cervical vertebra	8th rib 1st rib	Ventral rami of spinal nn.	Elevation of ribs (auxilliary inspiratory muscles); depression and lateral flexion of neck	A m. scalenus ventralis is absent in the dog
M. longus capitis	Ventral to the transverse processes of cervical vertebrae 1-6	Basilar part of occipital bone	Ventral rami of spinal nn.	Flexion and lateral inclination of head and neck	Surgical access to vertebral discs afforded between muscles of right and left sides
M. longus colli	Transverse processes of 5th and 6th thoracic vertebrae	Transverse process of 1st cervical vertebra (atlas)	Ventral rami of spinal nn.	Flexor of neck	Muscle has a plaited appearance

MUSCLES OF VERTEBRAL COLUMN: C. DORSAL MUSCLES OF HEAD (MOVERS OF HEAD) (TURN TO P. 12)

MUSCLE	ORIGIN	INSERTION	INNERVATION	COMMENTS	FUNCTION
M. rectus capitis dorsalis major	Spinous process of axis	Nuchal crest	Dorsal ramus of 1st cervical n.	Elevation of head	Direct cranial continuation of the ligamentum nuchae
M. rectus capitis dorsalis minor	Atlas	Occipital bone	Dorsal ramus of 1st cervical n.	Elevation of head	Situated deep to the m. rectus capitis dorsalis major
M. obliquus capitis caudalis	Spinous process of axis	Wing of atlas	Dorsal ramus of 2nd cervical n.	Rotation of atlas around the dens of axis	Oblique cranial continuation of lig. nuchae
M. obliquus capitis cranialis	Wing of atlas	Nuchal crest	Dorsal ramus of 1st cervical n.	Extension and rotation of head	Occupies atlanto-occipital space

RESPIRATORY MUSCLES (TURN TO P. 13): EXSPIRATORY MUSCLES — drawing the ribs caudomedially = narrowing the thorax

MUSCLE	ORIGIN	INSERTION	INNERVATION	COMMENTS	FUNCTION
M. serratus dorsalis caudalis	Thoracolumbar fascia	Caudal borders of last 3 ribs	Intercostal nn.	Exspiration	The serrated termination lies deep to the mm. intercostales externi
Mm. intercostales interni	Occupy the intercostal spaces		Intercostal nn.	Exspiration	Lie deep to the mm. intercostales externi
Mm. subcostales	Situated medial to the mm. intercostales interni, each m. subcostalis is directed across one rib to insert onto the next but one rib		Intercostal nn.	Exspiration	These muscles belong to the system of mm. intercostales interni
M. retractor costae	Transverse process of the first lumbar vertbra	Caudal border of last rib	Intercostal nn.	Exspiration	
M. transversus thoracis	Medially on sternum	Medially near the costochondral junctions of ribs 2 to 8	Intercostal nn.	Exspiration	Cranial continuation of the m. transversus abdominis

RESPIRATORY MUSCLES: INSPIRATORY MUSCLES — drawing the ribs craniolaterally = widening the thorax

MUSCLE	ORIGIN	INSERTION	INNERVATION	COMMENTS	FUNCTION
M. serratus dorsalis cranialis	Supraspinous lig. of first 8 thoracic vertebrae	Cranial borders of ribs 3 to 10	Intercostal nn.	Inspiration	Has a wide surface as aponeurosis of origin
M. rectus thoracis	Rib 1	2nd to 4th costal cartilages; deep fascia of trunk in the region of origin of m. rectus abdominis	Intercostal nn.	Inspiration	Cranial continuation of m. rectus abdominis
Mm. intercostales externi	Occupy the intercostal spaces		Intercostal nn.	Inspiration	Situated superficial to the mm. intercostales interni crossing them almost at right angles
Mm. levatores costarum	Transverse processes of thoracic vertebrae	Cranial borders of sequential ribs caudally	Intercostal nn.	Inspiration	Vertebral portions of mm. intercostales interni
Diaphragm Sternal part Costal part Lumbar part	Central tendon	Sternum 9th to 13th rib lumbar vertebrae 3rd and 4th	Phrenic n.	Inspiration	An arched septum cranially between the thoracic and the abdominal cavity. Three openings: - Aortic hiatus - Oesophageal hiatus - For. of vena cava

ABDOMINAL MUSCULATURE: Muscles have aponeuroses of origin and insertion with large surface areas (turn to p. 14)

MUSCLE	ORIGIN	INSERTION	INNERVATION	FUNCTION	COMMENTS
M. obliquus ext. abdominis	Costal part: from 4th and 5th ribs Lumbar part: thoracolumbar fascia	Abdominal tendon: linea alba Pelvic tendon: at the pecten ossis pubis with inguinal ligament	Ventral branches of corresponding intercostal and lumbar nerves	In general: A contractile corsetlike mechanism with an adaptability to the volume and change of volume of the abdominal organs.	Inguinal space: Abdominal and pelvic tendons form the external inguinal ring; Sheath of rectus: regularly forms the external lamina
M. obliquus int. abdominis	Inguinal part: tuber coxae and proximally at inguinal ligament. Lumbar part: thoracolumbar fascia	Linea alba and costal arch	Ventral branches of corresponding intercostal and lumbar nerves	Ventral abdominal pressure in urination, defecation and parturition, assisted by the position of the diaphragm at inspiration; flexion of the vertebral column by contraction of m. rectus abdominis, auxiliary respiratory musculature assisting diaphragm during exspiration;	Inguinal space: formation of internal inguinal ring in common with the inguinal ligament and m. rectus abdominis; contributes to separation of the m. cremaster (externus); sheath of rectus: cranially: internal lamina; umbilical region: external and internal laminae; caudally: external lamina.
M. transversus abdominis	Costal part: sternal ribs Lumbar part: thoracolumbar fascia and lumbar transverse processes	Linea alba	Ventral branches of corresponding intercostal and lumbar nerves	direct compression: mm. rectus abdominis and transversus abdominis; oblique compression: mm. obliquus int. and ext. abdominis;	Contributes to: separation of m. cremaster (externus); sheath of rectus: fundamentally as with the m. obliquus int. abdominis except that its two laminae are shifted 2 vertebrae caudad.
M. rectus abdominis	Costal cartilages of sternal ribs and sternum	Pecten ossis pubis (brim of pelvis)	Ventral branches of corresponding intercostal and lumbar nerves		Attachment between thorax and pelvis; segmentation of muscle by tendinous intersections

DEEP OR SUBLUMBAR MUSCLES: ADJACENT AND VENTROLATERAL TO THE LUMBAR COLUMN (TURN TO P. 20)

MUSCLE	ORIGIN	INSERTION	INNERVATION	FUNCTION	COMMENTS
M. quadratus lumborum	Last 3 thoracic vertebrae and transverse processes of lumbar vertebrae	Alar spine to the auricular surface of the ilium	Ventral branches of lumbar nerves	Reinforcement, arching or dorsal flexion of lumbar vertebral column	Thick muscle, projects over the m. psoas major cranially and laterally
M. psoas major	Transverse processes of 2nd and 3rd lumbar vertebrae; vertebral ends of last two ribs	As m. iliopsoas combined muscle has a common insertion on trochanter minor of femur	Ventral branches of lumbar nerves	Protractor of pelvic limb; flexor and supinator of hip joint; stabilises the vertebral column when pelvic limb is fixed	Passes through the lacuna musculorum to the femur. The mm. psoas major and iliacus form the m. iliopsoas
M. iliacus	Sacropelvic surface of ilium		Ventral branches of lumbar nerves		
M. psoas minor	Last 3 thoracic vertebrae and first 4 lumbar vertebrae, tendinous fascia of m. quadratus lumborum	Tubercle of the m. psoas minor on the ilium	Ventral branches of the lumbar nerves	With the vertebral column fixed, increases the vertical position of pelvis; with the pelvis fixed, muscle fixes the lumbar vertebral column or arches it dorsally	The wide terminal tendon is covered by iliac fascia

DIAPHRAGM OF PELVIS, MM. COCCYGEUS AND LEVATOR ANI (TURN TO P. 26)

MUSCLE	ORIGIN	INSERTION	INNERVATION	FUNCTION	COMMENTS
M. coccygeus	Ischiatic spine	First 4 coccygeal vertebrae	nSv 3	Unilateral contraction: lateral movement of tail	'Clamping' of tail
M. levator ani	Medially on ischium, pubis and ilium	Fourth to 7th coccygeal vertebrae (hemal processes)	nSv 3	Bilateral contraction: lowering of tail	Muscular basis of diaphragm of pelvis
M. sphincter ani externus	A muscular loop coming from the coccygeal vertebrae, it surrounds the ostium of anus		Caudal rectal n., a branch of the pudendal n.	Closure of ostium of anus	A voluntary muscle; on either side, between the sphincter muscles of anus, lies the paranal sinus
M. sphincter ani internus	Concentrated circular muscle of the caudal part of rectum		Caudal rectal n.	Closure of ostium of anus	Smooth muscle
M. rectococcygeus	External longitudinal muscle of rectum	Midsagittally on the ventral aspect of 1st coccygeal vertebra	Caudal rectal n., a branch of the pudendal n.	Stabilisation of anal canal and rectum	Smooth muscle
M. perinei	Connects anal and genital muscles		Deep perineal nn., branches of pudendal n.	Muscular fastening between anus and urogenital region	Anchored deeply in the tendinous centre of the perineum
M. bulbospongiosus (Bitch: M. constrictor vestibuli M. constrictor vulvae)	M. urethralis (Vestibule of vagina) (Labia of vulvae)	Root of penis	Deep perineal nn., branches of pudendal n.	Ejection of contents from urethra	Transversely directed over the bulb of penis

MUSCLE	ORIGIN	INSERTION	INNERVATION	FUNCTION	COMMENTS
M. ischiocavernosus	Ischiatic arch	Corpus cavernosum penis or clitoridis	Deep perineal nn., branches of pudendal n.	Elevation of the penis during erection	Covers the crus of penis or clitoris
M. retractor penis - clitoridis	Sacrum or 1st coccygeal vertebra	Anus; rectum; penis or clitoris	Deep perineal nn., branches of pudendal n.	Retractor of penis or clitoris	Smooth muscle with rectal, anal and penile or clitoric parts

MUSCLES OF THE HIP JOINT (TURN TO P. 31): Innervation from cranial or caudal gluteal nerves

MUSCLE	ORIGIN	INSERTION	INNERVATION	FUNCTION	COMMENTS
M. tensor fasciae latae	Tuber coxae	Fascia lata	Cranial gluteal n.	Flexor of hip joint; protractor of pelvic limb; extensor of knee (stifle); tensor of fascia lata	Divided into a principal and an accessory portion

Rump or croup muscles: Insertion on greater trochanter of femur

MUSCLE	ORIGIN	INSERTION	INNERVATION	FUNCTION	COMMENTS
M. gluteus superficialis	Gluteal fascia; sacrum, 1st coccygeal vertebra; sacrotuberous ligament	On distal part of greater trochanter	Caudal gluteal n.	Extensor of hip; retractor of limb	Present only in carnivors as an independent muscle, in other domestic species fused to the m. gluteobiceps
M. gluteus medius	Gluteal surface of ilium; on wing of ilium dorsally	Greater trochanter	Cranial gluteal n.	Extensor of hip; retractor and abductor of pelvic limb	Largest of the gluteal muscles of the dog
M. piriformis	Ventral and lateral aspects of sacrum; sacrotuberous lig.	Greater trochanter of femur	Cranial gluteal (or caudal gluteal) n.	Extension of hip; retractor and abductor of pelvic limb	A 'pear shaped' muscle between the mm. gluteus medius and profundus
M. gluteus profundus	Gluteal surface of ilium	Greater trochanter	Cranial gluteal n.	To assist the abduction effect of the m. gluteus medius	Lies directly on the hip joint deep to the m. gluteus medius

Caudal muscles of thigh: Innervated by the tibial nerve with the exception of m. abductor cruris caudalis

MUSCLE	ORIGIN	INSERTION	INNERVATION	FUNCTION	COMMENTS
M. biceps femoris (M. gluteobiceps)	Ischiatic tuber; Sacrotuberal lig.	Crural fascia; patellar lig.; cranial border of tibia; calcaneus by means of the lateral calcaneal tract	Cranial portion: caudal gluteal n. Caudal portion: tibial nerve	Extension of hip and stifle; caudal portion of muscle flexes the knee, abducts the pelvic limb and extends tarsus	Homologue of the long head of the m. biceps femoris of man
M. abductor cruris caudalis	Sacrotuberal lig.	Crural fascia	**Fibular n.!**	Abductor of limb (assisting the caudal portion of m. biceps femoris)	Homologue of the short head of the m. biceps femoris of man
M. semitendinosus	Ischiatic tuber	Tibia; calcaneal tuber by means of the medial calcaneal tract together with the tendon of m. gracilis	Tibial n.	Weight bearing: extensor of hip, stifle and tarsus; Non-weight bearing: Flexor of stifle, adductor and retractor of pelvic limb	In man it is semitendinous; has an oblique tendinous intersection proximally
M. semimembranosus	Ischiatic tuber	Medial condyles of femur and tibia	Tibial n.	Weight bearing: extensor of hip and knee (stifle); Non-weight bearing: retraction, abduction and inward rotation of pelvic limb	Semimembranous in man; the two muscle bellies end resp. proximal and distal to the plantar aspect of knee

The inner pelvic or deep muscles of the hip: Serve more acute adaptability during the course of movement

MUSCLE	ORIGIN	INSERTION	INNERVATION	FUNCTION	COMMENTS
M. gemelli	Lesser sciatic notch or incisure	Trochanteric fossa of femur	Rotator n., a branch of the sciatic n.	Supinator, rotating the femur outwards; auxiliary extensor of hip joint	In man muscle is divided in two, resulting in twin muscles
M. obturatorius internus	Inner surface of pelvis at the obturator foramen	Trochanteric fossa of femur	Rotator n., a branch of the sciatic n.	Supinator, rotating the femur outwards; auxiliary extensor of hip joint	Closes the obturator foramen internally on the floor of the pelvis
M. obturator externus	External surface of pelvis at obturator foramen, lies deep to adductor muscles	Trochanteric fossa of femur	Obturtor n.	Supinator of femur, rotating it outwards; adductor of the pelvic limb	Closes the obturator foramen external to the floor of pelvis
M. quadratus femoris	Ventromedial to the ischiatic tuber	Distal to the trochanteric fossa at the 3rd trochanter of femur	Rotator n., a branch of the sciatic n.	Supinator of femur, rotating it outwards; auxiliary extensor of the hip joint	Adjacent caudolaterally to the mm. gemelli

MEDIAL MUSCLES OF THIGH: Adductors of hip (turn to p. 32)

MUSCLE	ORIGIN	INSERTION	INNERVATION	FUNCTION	COMMENTS
M. gracilis	An aponeurotic origin with the symphyseal tendon at the pubic symphysis	Crural fascia; medial calcaneal tract	Obturator n.	Adductor (perhaps an extensor of the knee via the crural fascia); extensor of tarsus	A thin muscle; in the racing Greyhound, a site of muscle rupture after inactivity
M. adductor magnus	Fleshy origin at the symphyseal tendon and pubic symphysis	Facies aspera of the femur	Obturator n.	Adductor and retractor of the limb	The cut surface has a sex-specific shape
M. adductor brevis	Ventral pubic tubercle	Proximal to the facies aspera of femur	Obturator n.	Adductor	Situated at a bifurcation of the obturator n.
M. pectineus (and adductor longus)	Iliopubic eminence os pubis	Medial lip of facies aspera	Obturator n. (in addition saphenous n. or femoral ramus of genitofemoral n.)	Adductor of limb; flexor of hip joint	Due to amalgamation regarded as a 'double muscle'

MUSCLES OF THIGH: Extensors of knee (stifle)

MUSCLE	ORIGIN	INSERTION	INNERVATION	FUNCTION	COMMENTS
M. sartorius: cranial part caudal part	 Tuber coxae iliac crest	 Crural fascia Cranial border of tibia	Femoral n.	Flexor of hip joint; protractor and adductor of limb; extensor of knee (crural fascia)	In man known as the tailor's muscle; in dog contributes to the craniomedial contour of the thigh
M. quadriceps femoris - vastus lateralis - vastus medialis - vastus intermed. - rectus femoris	 Femur: craniolateral Femur: craniomedial Femur: craniomedial Ilium: caudal ventral iliac spine	By the quadriceps tendon (patellar lig.) to the tibial tuberosity; distal infrapatellar bursa situated deep to tendon	Femoral n.	Largest of extensors of knee; fixation of limb; flexor of hip via the m. rectus femoris	A thick muscle mass, it lies on the cranial aspect of the thigh, covered by m. tensor fasciae latae, fascia lata and m. sartorius

Specific flexor of knee (stifle): Situated at the plantar aspect of joint (turn to p. 33)

MUSCLE	ORIGIN	INSERTION	INNERVATION	FUNCTION	COMMENTS
M. popliteus	Lateral condyle of femur	Tibia, proximomedial aspect	Tibial n.	Flexor of knee; pronator of tibia	The sesamoid bone of the m. popliteus

FLEXORS OF TARSUS (HOCK) AND EXTENSORS OF DIGITS: Situated craniolaterally on tibia; innervation: Fibular n. (turn to p.33)

MUSCLE	ORIGIN	INSERTION	INNERVATION	FUNCTION	COMMENTS
M. tibialis cranialis	Lateral to cranial border of tibia	Base of Mt1	Fibular n.	Flexor of tarsus; supinator of pes	Passes through the loop-like opening of crural extensor retinaculum
M. extensor digitorum brevis	Distal part of calcaneus	With m. extensor digitorum longus to digits 2 to 4	Fibular n.	Extensor of digits	Extension is its sole function
M. extensor digitorum longus	Lateral condyle of femur	Phalanx III of each digit	Fibular n.	Extensor of digits; protraction of pes	Passes through the loop-like openings of crural and tarsal extensor retinacula
M. extensor digiti I	Fibula deep to the m. cranialis tibialis	Phalanx I of 1st and 2nd digit	Fibular n.	Extensor of corresponding digits	Passes through the crural extensor retinaculum
M. fibularis longus	Lateral condyle of tibia and head of fibula	Tendon of insertion directed mediotransversely to Mt1	Fibular n.	Auxiliary flexor of tarsus; pronator of pes	Transverse tendon of insertion forms a stirrup' with the m. tibialis cranialis
M. extensor digitalis lateralis	Proximal part of fibula; lateral collateral ligament of knee joint	Phalanx III of 5th digit	Fibular n.	Extensor of 5th digit	Crosses deep to the tendon of m. fibularis longus
M. fibularis brevis	Distal two thirds of fibula	MtV	Fibular n.	Flexor of tarsus	Well developed only in man and carnivors

EXTENSORS OF TARSUS AND FLEXORS OF DIGITS: Situated on plantar aspect of tibia: Innervation: Tibial n. (turn to p. 33)

MUSCLE	ORIGIN	INSERTION	INNERVATION	FUNCTION	COMMENTS
M. gastrocnemius	Distal part of femur (the medial and the lateral head)	By means of the calcaneal tendon to the calcaneal tuber	Tibial n.	Extensor of tarsus; flexor of knee	Sesamoid bone lies in each head (medial and lateral) of origin of the muscle
M. flexor digitorum superficialis	Lateral supracondylar tuberosity of femur, between the two heads of m. gastrocnemius	Phalanx II of each digit	Tibial n.	Flexor of corresponding digits; extensor of the tarsus; accessory flexor of knee	Tendon forms a cap covering the calcaneal tuber; at digital level, each tendon forms a flexor manica
Mm. flexores digitorum profundi: - flexor digit. lat. - flexor digit. med.	Tibia and fibula proximocaudally	Phalanx III of each digit	Tibial n.	Flexor of digits; auxiliary extensor of tarsus	Distal to the sustentaculum tali the tendons merge to form the deep flexor tendon
M. tibialis caudalis	Proximal part of fibula	Central tarsal bone	Tibial n.	Extensor of tarsus	In the other domestic mammals, muscle participates in formation of deep flexor tendon

FACIAL MUSCULATURE: 1. M. sphincter colli superficialis; 2. Platysma; 3. M. sphincter colli profundus (turn to p. 38)

MUSCLE	ORIGIN	INSERTION	INNERVATION	FUNCTION	COMMENTS
M. sphincter colli superficialis	Situated ventrally on the neck, the individual muscle fibers, which are not continuous with one another, are interwoven in the cervical superficial fascia		Ramus colli of facial n.	Tenses and moves the skin on the ventral and lateral aspects of the neck	Muscle fibers very sparse
M. platysma	Covers the cervical region with muscle fibers directed longitudinally; rostrally, is continuous with cutaneous muscle of face and lips which radiates into m. orbicularis oris of upper and lower lip		Platysmal ramus of caudal auricular n., rami colli (VII), dorsal and ventral buccolabial rami	Tenses and moves skin of cervical and mandibular regions; retractor of commissure of mouth; tenses skin of lip region	Two muscle layers present in neck region

M. sphincter colli profundus:

AURICULAR MUSCLES

MUSCLE	ORIGIN	INSERTION	INNERVATION	FUNCTION	COMMENTS
M. cervicoauricularis superficialis	Midsagittally, dorsum of neck	Dorsal surface of auricular cartilage	Caudal auricular n., a branch of facial n.	Long levator of auricular cartilage	M. cervicoauricularis superficialis and m. cervicoscutularis fused at origin
M. cervicoscutularis	Midsagittally, dorsum of neck	Caudomedial aspect of scutiform cartilage	Caudal auricular n.	Elevates auricular cartilage and tenses the scutiform cartilage	
Mm. cervicoauricularis profundus and medius	External sagittal crest	Lateral border of auricular cartilage	Caudal auricular n.	Move auricular cartilage outwards	Covered by m. cervicoauricularis superficialis
M. occipitalis	External sagittal crest	Superficial fascia of head	Rostral auricular rami of auriculopalpebral n.	Tenses scutiform cartilage	Shaped like arch, very delicate muscle fibers
M. interscutularis	Runs transversely between right and left scutiform cartilages		Rostral auricular rami of auriculopalpebral n.	Tenses scutiform cartilage	
M. frontoscutularis	Rostral continuation of m. interscutularis		Rostral auricular rami of auriculopalpebral n.	Tenses scutiform cartilage	
M. scutuloauricularis superficialis	Scutiform cartilage	Rostral border of auricular cartilage	Rostral auricular rami of auriculopalpebral n.	Elevates and erects auricular cartilage	
M. parotidoauricularis	Parotid and cranial neck regions	Ventrolateral part of base of ear	Ramus colli	Depresses auricular cartilage and lays it back	
M. mandibuloauricularis	Between angular and articular processes of mandible	Ventral part of base of ear	Caudal auricular n., a branch of facial n.	So-called 'muscle of the auditory canal'	May be absent

MUSCLES OF LIPS AND CHEEK

MUSCLE	ORIGIN	INSERTION	INNERVATION	FUNCTION	COMMENTS
M. orbicularis oris	Muscle encircles the oral rima; no connection to bones of face			Closes the oral rima tightly	
M. buccinator	Runs between lips and alveolar processes of maxilla and mandible; is interwoven with m. orbicularis oris encircling the oral rima			Lateral boundary of oral cavity; transport of food from buccal part of vestibule to oral cavity proper	
M. zygomaticus	Scutiform cartilage	Radiates into the m. orbicularis oris	Dorsal and ventral buccolabial rami of facial n.	Retractor of angle of mouth; protractor of scutiform cartilage	
M. caninus	Ventral to infraorbital foramen	Upper lip		Retractor of upper lip	
M. levator labii maxillaris	Rostral to infraorbital foramen	Tendons diverge in region of naris; upper lip		Elevation and retraction of upper lip and nasal plate	

MUSCLES OF EYELID AND NOSE

MUSCLE	ORIGIN	INSERTION	INNERVATION	FUNCTION	COMMENTS
M. orbicularis oculi	Ring of muscle surrounding the eye and almost completely enclosing it		Auriculopalpebral n., a branch of facial n.	Narrowing and closing of ocular rima	
M. retractor anguli oculi lateralis	Deep temporal fascia	Lateral angle of eye	Auriculopalpebral n., a branch of facial n.	Retractor of the lateral angle of eye	
M. levator anguli oculi medialis	Frontal fascia	Medial to upper eyelid	Auriculopalpebral n., a branch of facial n.	Raises medial portion of upper eyelid; erection of tactile hairs	
M. levator nasolabialis	Maxilla in region of medial angle of eye; frontal fascia	Upper lip and lateral to the naris	Auriculopalpebral n., a branch of facial n.	Raises the upper lip; dilation of naris	
M. malaris	Deep fascia of face in buccal region	Lower eyelid	Auriculopalpebral n., a branch of facial n.	Depressor of the lower eyelid	

MUSCLE	ORIGIN	INSERTION	INNERVATION	FUNCTION	COMMENTS
SUPERFICIAL MUSCLES OF THE MANDIBULAR REGION					
M. digastricus	Jugular process of occipital bone	Ventral border of mandible	Caudal belly: digastric ramus of facial n.; Rostral belly: mylohyoid n., a branch of mandibular n.	Opening mouth	Weak tendon joins the two bellies
M. mylohyoideus	Medial side of mandible (mylohyoid line)	Basihyoid element; median muscle raphe	Mylohyoid n., a branch of mandibular n.	Lifts the tongue to press it against palate	Hammock shaped; supports floor of oral cavity
EXTERNAL MUSCLES OF MASTICATION					
M. temporalis	Temporal fossa	Coranoid process of mandible	Masticatory n., a branch of mandibular n.	Muscle of mastication; raises mandible and presses it against maxilla	Accessory portion runs dorsoparallel to the zygomatic arch
M. masseter	Zygomatic arch	Masseteric fossa	Masticatory n., a branch of mandibular n.		Arranged as a superficial and a deep part
INTERNAL MUSCLES OF MASTICATION					
M. pterygoideus - medialis -lateralis	Pterygoid bone and vicinity	Medial pterygoid fossa on mandible	Pterygoid nn., branches of mandibular n.	Synergist of m. masseter; unilateral contraction draws mandible to side	Mandibular n. is situated between both muscles
(extrinsic) EYE MUSCLES (turn to p. 40)		Nerves: III, IV, VI			
M. obliquus dorsalis	Medial border of optic foramen	Dorsally on the ocular bulb under the insertion of the m. rectus dorsalis	Trochlear n. (IV)	Rotates the ocular bulb inwards (medially)	A U-shaped course around trochlea
M. obliquus ventralis	Ventral to the fossa of the lacrimal sac	Laterally on the ocular bulb under the insertion of the m. rectus lateralis	Oculomotor n. (III)	Rotates the ocular bulb outwards (laterally)	
M. rectus ventralis	A common tendinous ring on the pterygoid crest and around the optic foramen provides origin for the recti muscles of the eye	Rostroventral on ocular bulb	Oculomotor n. (III)	Rotation of ocular bulb ventrally around a horizontal axis	
M. rectus dorsalis		Rostrodorsal on ocular bulb	Oculomotor n. (III)	Rotation of ocular bulb dorsally around a horizontal axis	
M. rectus medialis		Rostromedial on ocular bulb	Oculomotor n. (III)	Rotation of ocular bulb medially around a perpendicular axis	
M. rectus lateralis		Rostrolateral on ocular bulb	Abducens n. (VI)	Rotation of ocular bulb laterally around a a perpendicular axis	
M. retractor bulbi	Between optic foramen and orbital fissure	On ocular bulb; covered by recti muscles	Oculomotor n. (lateral portion also from abducens n.)	Retractor of ocular bulb	Surrounds the optic n. (II) like a sleeve
M. levator palpebrae superioris	Dorsal to the optic foramen	By a broad tendon to the upper eyelid	Oculomotor n.	Raises the upper eyelid	Lies on the m. rectus dorsalis
(intrinsic) SMOOTH MUSCLES OF EYE					
M. ciliaris	Scleral ring	By fibers of zonule to the lens capsule	Parasympathetic fibers from the short ciliary nn. (from III)	Accommodation: approximation of ciliary body to lens	Has circular and meridional fibers
M. sphincter pupillae	A circular course around the border of the pupil		Parasympathetic fibers from the oculomotor n. (III)	Constriction of pupil	Smooth muscle of ectodermal origin
M. dilator pupillae	A radial course around the border of the pupil		Sympathetic fibers	Dilation of pupil	

PHARYNGEAL MUSCLES: Attached by means of the dorsomedian raphe of the pharynx to the pharyngeal tubercle of the occipital bone (turn to p. 42)

MUSCLE	ORIGIN	INSERTION	INNERVATION	FUNCTION	COMMENTS
M. stylopharyngeus caudalis	Caudal aspect of stylohyoid element	Dorsolateral wall of pharynx	Glossopharyngeal n. (IX)	Dilates pharynx	The only dilator of pharynx

MUSCLES OF THE SOFT PALATE

MUSCLE	ORIGIN	INSERTION	INNERVATION	FUNCTION	COMMENTS
M. tensor veli palatini	Rostral to bulla tympanica	Pterygoid hamulus and tendinous insertion of the soft palate	Mandibular n. (V3)	Tensor of the soft palate	During swallowing both muscles are constricted so that the auditory tube opens and its wall becomes tense (equalisation of pressure)
M. levator veli palatini	Rostral to bulla tympanica	Soft palate	Pharyngeal plexus (IX and X)	Levator of soft palate	

ROSTRAL CONSTRICTORS OF PHARYNX

MUSCLE	ORIGIN	INSERTION	INNERVATION	FUNCTION	COMMENTS
M. pterygopharyngeus	Pterygoid hamulus	Pharyngeal raphe	Pharyngeal plexus (IX and X)	Constrictor and protractor of pharynx	Rostral to its insertion it crosses the m. levator veli palatini larally
M. palatopharyngeus	Aponeurosis of palate	Dorsolateral wall of pharynx	Pharyngeal plexus (IX and X)	Constrictor and protractor of pharynx; depressor of palate	Situated medial to the m. levator veli palatini

MIDDLE CONSTRICTOR OF PHARYNX

MUSCLE	ORIGIN	INSERTION	INNERVATION	FUNCTION	COMMENTS
M. hyopharyngeus	Thyrohyoid element	Pharyngeal raphe	Pharyngeal plexus (IX and X)	Constrictor of pharynx	Frequently divided in two

CAUDAL CONSTRICTOR OF PHARYNX

MUSCLE	ORIGIN	INSERTION	INNERVATION	FUNCTION	COMMENTS
M. thyropharyngeus	Thyroid cartilage	Pharyngeal raphe	Pharyngeal plexus (IX and X)	Constrictor of pharynx	A transition occurs between the muscles of both sides due to an absence of the raphe
M. cricopharyngeus	Cricoid cartilage	Pharyngeal raphe	Pharyngeal plexus (IX and X)	Constrictor of pharynx	

MUSCLES OF TONGUE AND HYOID APPARATUS: Radiating respectively in the tongue and onto the hyoid elements (turn to p. 43)

MUSCLE	ORIGIN	INSERTION	INNERVATION	FUNCTION	COMMENTS
M. lingualis proprius	Muscle within tongue		Hypoglossal n. (XII)	Intrinsic movement of tongue	Longitudinal, transverse and perpendicular fibers within tongue substance

EXTRINSIC TONGUE MUSCLES

MUSCLE	ORIGIN	INSERTION	INNERVATION	FUNCTION	COMMENTS
M. styloglossus	Stylohyoid element	Tongue (muscle fibers radiating laterally)	Hypoglossal n. (XII)	Draws tongue caudodorsally	Covers the stylohyoid element laterally
M. hyoglossus	Basihyoid element	Tongue (muscle fibers radiating laterally and caudally)	Hypoglossal n. (XII)	Draws tongue caudoventrally	Situated at root of tongue
M. genioglossus	Chin (medial surface of mandible)	Tongue	Hypoglossal n. (XII)	Draws tongue rostrally and ventrally	Radiates in a fan-like manner into tongue

MUSCLES OF HYOID APPARATUS

MUSCLE	ORIGIN	INSERTION	INNERVATION	FUNCTION	COMMENTS
M. geniohyoideus	Chin (symphysis of mandible)	Basihyoid element	Ansa cervicalis (XII and C1)	Draws tongue rostrally	Muscular base of floor of oral cavity
M. thyrohyoideus	Thyroid cartilage	Basihyoid element	Ansa cervicalis (XII and C1)	Connects basihyoid element to larynx	Cranial continuation of m. sternothyroideus.

LARYNGEAL MUSCLES (INTRINSIC MUSCULATURE OF LARYNX) (TURN TO P. 41)

MUSCLE	ORIGIN	INSERTION	INNERVATION	FUNCTION	COMMENTS
M. cricothyroideus	Ventrolateral part of cricoid cartilage	Caudal part of thyroid cartilage	Cranial laryngeal n. a branch of vagus n.	Constrictor of rima vocalis; tensor of vocal ligaments	
M. cricoarytenoideus dorsalis	Dorsolateral part of cricoid cartilage	Muscular process of arytenoid cartilage		Dilator of rima vocalis	
M. cricoarytenoideus lateralis	Craniolateral part of cricoid cartilage	Muscular process of arytenoid cartilage		Constrictor of rima vocalis	
M. arytenoideus transversus	Muscular process of arytenoid cartilage	Median tendinous connection with analogous muscle of opposite side	Caudal laryngeal n. derived from vagus n.	Constrictor of rima vocalis	
M. thyroarytenoideus - M. ventricularis - M. vocalis	Ventromedial part of the thyroid cartilage	Muscular process of arytenoid cartilage; Vocal process of arytenoid cartilage		Constrictor of rima vocalis, dilator of lateral ventricle of larynx	

LYMPHATIC SYSTEM

LYMPHOCENTER LYMPH NODE	LOCATION	AREA/ORGANS DRAINED BY AFF. LYMPHATICS	EFFERENT LYMPHATICS	COMMENTS	PAGE/ILL.
1. **Parotid ln.**	On or in the parotid gland at level of temporomandibular articulation	Chiefly superficial head regions dorsal to line joining eye and ear	Medial retropharyngeal ln.	Palpable; deep nodes lie in parotid gl.	37/23
2. **Mandibular lnn.**	In the mandibular region rostroventral to the mandibular gl.	Superficial and deep parts of face region of skull; muscles; salivary glands ventral to line joining eye and ear	Medial retropharyngeal ln.	Palpable; lymph nodes of mandibular space	37/30
Retropharyngeal lc.					
3. Lateral retropharyngeal ln.	Caudal to parotid gland at the level of wing of atlas	Superficial and deep regions of the head; tonsillar tissue of lymphatic pharyng. ring; cervical musculature adjacent to head	Medial retropharyngeal ln.	Inconstant	37/-
4. Medial retropharyngeal ln.	Dorsomedial to pharynx	Afferent lymphatic vessels from parotid and mandibular lnn.	Jugular trunk (tracheal duct)	behind the pharynx	37/29
5. **Superficial cervical lnn.**	Cranial to shoulder deep to the m. omotransversarius	Superficial parts of neck, head, abdominal wall and thoracic limb	In angle between external and internal jugular veins or caudal end of jugular trunk	Palpable	6/31
Deep cervical lc.					
6. Cranial deep cervical ln.	Craniodorsal to thyroid gland	Deep parts of neck; cervical viscera	In one of the lymph collecting ducts or the mediastinal lc.	Inconstant; danger of confusion with parathyroid glands	6/-
7. Middle deep cervical ln.	In the middle third of the neck	Cervical viscera			
8. Caudal deep cervical ln.	On the trachea, cranial to 1st rib	Shoulder and brachial region			
Mediastinal lc.					
9. Cranial mediastinal lnn.	In precardial mediastinum	Deep parts of thorax; also parts of neck, shoulder, pleura, thoracic viscera, lymph from 7, 8, 10, 11, 12 to 14	End of jugular trunk and/or thoracic duct	Middle and caudal mediastinal lymph nodes are absent in dog	17/1
Ventral thoracic lc.					
10. Cranial sternal ln.	On the sternum, at cranial border of m. transversus thoracis	Thoracic wall; pectoral girdle; diaphragm; mediastinum; cranial mammae; ventral abdominal wall	Cranial mediastinal lnn.	Caudal sternal lnn and cranial epigastric ln are absent in dog	17/2
Dorsal thoracic lc.					
11. Intercostal ln.	Proximally in 5th or 6th intercostal space	Deep sections of dorsum, shoulder; abdominal muscles; aorta	Cranial mediastinal lnn.	Very often absent	17/-
Bronchial lc.					
12. Right tracheobronchial ln.	Cranially on the right principal bronchus			Cranial tracheobronchial lnn. are absent in dog	17/3
13. Middle tracheobronchial ln.	Dorsal to bifurcation of trachea	Lung and remaining thoracic viscera	Cranial mediastinal lnn.	Pulmonary lnn. are situated at hilus of lung	17/F
14. Left tracheobronchial ln.	Dorsal to left principal bronchus				17/G

Legend

Tj = Jugular trunk (Tracheal Duct)
Vw = Venous angle
Dt = Thoracic duct
Cc = Cisterna chyli
Tv = Visceral trunk
Tl = Lumbar trunk
● = deeply situated
⊗ = inconstant
○ = superficial

LYMPHOCENTER LYMPHNODE	LOCATION	AREA/ORGANS DRAINED BY AFF. LYMPHATICS	EFFERENT LYMPHATICS	COMMENTS	PAGE/ ILL.
Axillary lc.					
15. Axillary ln.	Caudal to shoulder joint on the lateral thoracic v.	Superficial parts of thoracic wall and ventral abdominal wall; thoracic and abdominal mammae	In angle between external and internal jugular veins or end of jugular trunk (tracheal duct)	Palpable; cubital ln. and axillary ln. of first rib are absent in dog	8/21
16. Accessory axillary ln.	2 cm caudal to 15.	As in 15.	As in 15.	Inconstant	8/3
Celiac lc.					
17. Hepatic (portal) lnn.	On the portal vein	The whole area supplied by the celiac artery and its branches	Visceral trunk	Lymph nodes of liver	22/21
18. Gastric ln.	Adjacent to pylorus on the lesser curvature of stomach	Stomach	Hepatic lnn. or splenic lnn.	Inconstant	
19. Splenic lnn.	At the ramifications of splenic artery and vein in greater omentum	Spleen and afferent vessels from gastric ln.	Visceral trunk	Group of small lymph nodes	22/-
20. Pancreaticoduodenal lnn.	Between cranial part of duodenum and pancreas	Duodenum, pancreas, stomach, greater omentum	Hepatic lnn.	Inconstant	22/-
Cranial mesenteric lc.					
21. Jejunal lnn.	Proximally in mesojejunum	From organs supplied by the cranial mesenteric artery	Visceral trunk	Very large lymph node conglomerate	22/8
22. Colic lnn.	In ascending and transverse mesocolons	Lymph vessels from caudal mesenteric lnn.	Visceral trunk	Iliocolic and cecal lnn. are absent in dog	22/9
Caudal mesenteric lc.					
23. Caudal mesenteric lnn.	In descending mesocolon	From organs supplied by caudal mesenteric artery	Visceral trunk or lumbar trunk	Vesical ln. is absent in dog	22/16
Lumbar lc.					
24. Lumbar aortic lnn.	Dorsal, ventral and lateral to abdominal aorta and caudal vena cava	Deep parts of dorsal body wall, abdominal and pelvic cavities; urogenital organs; adrenals, efferent vessels from iliosacral lc.	Lumbar trunk or direct to cisterna chyli	Proper lumbar, renal, ovarian and testicular lnn. are absent in dog	24/6
Iliosacral lc.					
25. Medial iliac lnn.	At origins of deep circumflex iliac a. and v.	Pelvic wall, pelvic organs; possibly testis; afferent vessels from pelvic limb and lymph nodes thereof	Lumbar trunk or lumbar aortic lnn.	Lateral iliac, anorectal and uterine lnn. are absent in dog	24/8
26. Sacral lnn.	Between right and left internal iliac aa.	Rectum, genital organs and adjacency		Obturator ln. absent in dog	24/-
Superficial inguinal lc.					
27. Superficial inguinal lnn.	At the ramification of external pudendal a. and v. at level of inguinal teat or papilla	Skin of abdomen, scrotum and caudal pelvic aperture; tail; hind limb; caudal mammae	Medial iliac lnn.	Palpable; caudal epigastric, subiliac, accessory coxal lnn. and those of paralumbar fossa are absent in dog	14/11
Popliteal lc.					
28. Popliteal ln.	Between m. biceps femoris and m. semitendinosus on the hollow of the knee	Superficial and deep parts of pelvic limb distal to knee	Medial iliac and sacral lnn.	Palpable	33 and 34

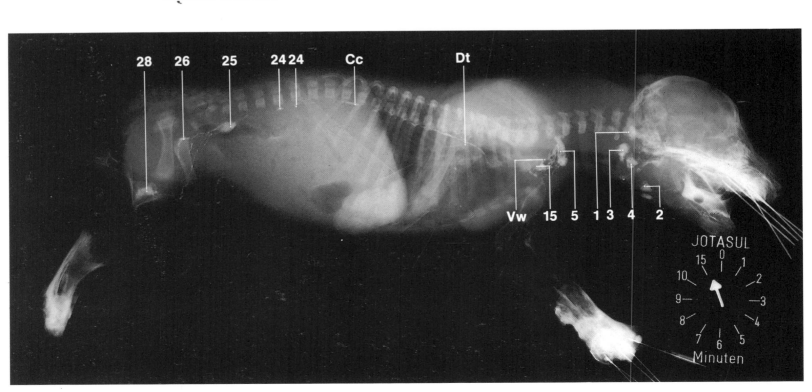

AN INTRODUCTION TO THE PHYSICAL-TECHNICAL BASICS OF X-RAY AND ULTRASOUND DIAGNOSTICS

Cordula Poulsen Nautrup

1 CONVENTIONAL AND DIRECT MAGNIFYING RADIOGRAPHY

1.1 X-rays

X-rays are electromagnetic waves with frequencies of approximately 10^{12} to 10^{14} MHz (megahertz).

1.2 Generating x-rays

When accelerated electrons hit an anode or target, part of the kinetic energy is transformed into bremsstrahlung (a continuous radiation spectrum of various wavelengths) and characteristic x-rays (a line spectrum). Most of the energy, however, is lost in the form of heat.

The emitted x-ray radiation, which is made up mainly of bremsstrahlung and, to an extent, of characteristic x-rays, is responsible for image formation. The voltage determines the most energetic radiation with the shortest wavelength. Potential differences useful for x-ray diagnostics and the resulting energies lie between 18 and 120 kV and 1 and 120 kV, resp. Soft x-rays of long wavelengths (low kV values) are opposed by harder and more energetic radiation of short wavelengths (high kV values).

1.3 The behavior of x-rays in material and the formation of radiographs

Materials absorb and scatter x-rays. Differences in absorption in an x-rayed patient, body part, or object produce a new radiation spectrum. All planes of a three-dimensional body are thereby presented on a single plane. A radiograph is thus a summative picture, in which areas lying one above or below another are superimposed (fig. 1).

1.3.1 Absorption in relation to the irradiated material

Absorption depends on the density and valence number of the irradiated material. Air or organs containing air with their low densities and resultingly low absorptions can be easily differentiated from other soft tissues with higher densities. Due to their higher densities and higher effective valence numbers, calcified or bony areas absorb x-rays better than soft tissues and are clearly contrasted from these.

1.3.2 Absorption in relation to the radiation quality

The quality of the x-rays further influences the level of absorption. Whereas with the use of soft x-rays absorption is predominant, absorption decreases and scattering increases with increasing radiation energy levels. In soft radiation exposures the large differences in contrast allow the differentiation between soft tissues and slightly calcified or ossified structures (fig. 3). In human medicine this technique is employed in mammography for radiological tumor diagnosis of the female breast. Medium hard to hard radiation is suited for the depiction of sufficiently calcified bone tissue (fig. 2).

1.3.3 Contrast media

Contrast media are necessary for the differentiation of vessels in soft tissues and bone (figs. 4, 5, 6) as well as for evaluating certain organ regions. Such media are classed as negative (less absorption than bone) or positive (higher absorption), according to their ability to absorb x-rays. The first group includes easily absorbed gases such as air, carbon dioxide, and nitrous oxide.

These are primarily used for presenting the alimentary canal and urinary bladder, often in the form of double contrast images, in which a positive contrast medium is combined with a negative. Positive contrast media have a higher density and a large valence number. In addition to barium sulfate, a water insoluble contrast medium for controlling passage in the esophagus and digestive tract, there are many water soluble media, usually containing iodine, for the depiction of vessels (angiography – figs. 4, 5, 6; arteriography, phlebography, lymphography) and nearly all of the body cavities (urography, bronchography, arthrography, myelography).

1.4 Image formation and recording

X-rays pass through the body film (photographic effect) or energize the luminescent chemicals of an intensifying screen or image amplifier (luminescence effect).

X-ray film images are negatives; highly-absorbent regions such as bone or calcifications appear white to light gray, whereas radiolucent areas, such as air-filled and parenchymatous organs, fat, muscle, cartilage, connective tissue, or fluids (blood, urine, liquid ingesta, etc.), appear dark (fig. 7). As a positive, the monitor image shows radiodense areas as black to dark gray, and radiolucent areas as white to light gray (fig. 8).

The analogue pictures of x-ray-closed circuit TV systems or radiographs can be digitalized and then reworked with a picture processor. Typically-used methods in medical diagnostics include digital measurements, multiple image integration for improving contrast, edge enhancement to improve the depth of field, and picture subtraction to bring out or isolate vessels (DSA: digital subtraction angiography).

Experimentally, color coding and pseudo-3D presentation after object shifting have proved useful for evaluating areas of low contrast (fig. 9). All methods of digital picture processing serve to increase the quality of x-ray film images with improved and more complete possibilities of evaluation, and at the same time to reduce the dangerous levels of radiation or the amounts of contrast media necessary.

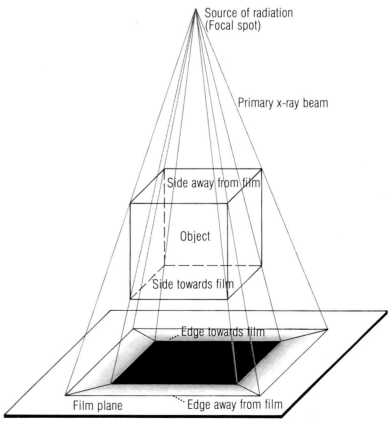

Fig. 1: Schematic drawing of the formation of a radiograph (summative picture).

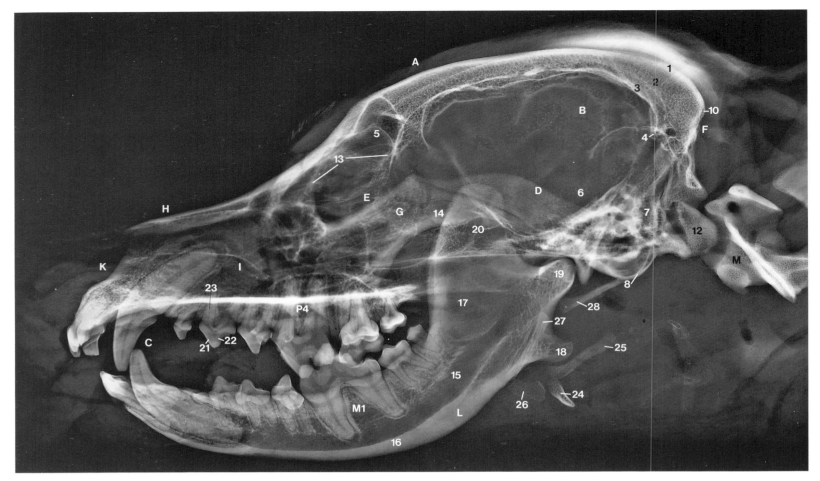

Fig. 2: Left cranial half of a adult dog. High-kV technique, 100 kV, lateromedial view — LM.

Anatomic structures (figs. 2,3)

Cranium	**I Maxilla**
1 External lamina	**K Premaxilla**
2 Spongy bone (diploe)	**L Mandible**
3 Internal lamina	15 Mandibular canal
4 Tentorial process	16 Mandibular body
	17 Mandibular ramus
Neurocranial bones	18 Mandibular angle
A Frontal bone	19 Condylar process
5 Frontal sinus	20 Coronoid process
B Parietal bone	
D Temporal bone	**Teeth**
6 Squamous part	C Canine tooth
7 Petrous part	P4 Fourth premolar
8 Tympanic part/tympanic bulla	M1 First molar
E Ethmoid bone	21 Enamel
F Occipital bone	22 Dentin
9 Squamous part	23 Dental cavity
10 External occipital crown	
11 Lateral part	**Hyoid bone**
12 Occipital condyle	24 Basihyoid
	25 Thyrohyoid
Face	26 Ceratohyoid
13 Orbits	27 Epihyoid
	28 Stylo and tympanohyoid
Facial bones	
G Zygomatic bone	**M Atlas**
14 Temporal process	
	N Axis
H Nasal bone	

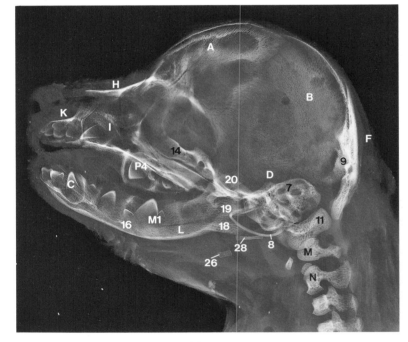

Fig. 3: Left cranial half of a six-day-old puppy. Low-kV technique, 40 kV, lateromedial view — LM.

1.5 The laws of projection

Picture formation with x-rays follows the laws of central projection; rays emanating from a more or less expansive, ideally point-formed focal spot diverge in straight lines.

1.5.1 Geometric magnification

In describing projections it differentiated must be made between the film-focal distance (FFD), focal-object distance (FOD), and the object-film distance (OFD). With a conventional image at a scale of 1 : 1 the FFD is nearly the same as the FOD, which means that with a large FFD the patient or area to be filmed is close to the film plane. Reductions in the FOD caused by raising the patient or object result in direct x-ray magnification (keep in mind the problem of geometric blurring, see

chapter 1.6.1, fig. 17). The factor of magnification, V, is the quotient of FFD to FOD, i. e., FFD : FOD (fig. 10).

1.5.2 Superpositioning

Two structures lying one above the other are projected into each other and can no longer be differentiated from each other. Cracks, fissures, or fractures, which run roughly perpendicular to the path of the x-rays remain hidden, due to superpositioning (figs. 11 A, 12). Only by rotating the object do the details run parallel to the path of the rays and then appear in radiographs (figs. 11 B, 13). Disturbing superpositioning can be reduced when examining anatomical preparations by preparing thin sections (0.1 — 1.0 mm). In vivo the superposition-free radiological depiction of structures of one plane is possible with the help of tomography.

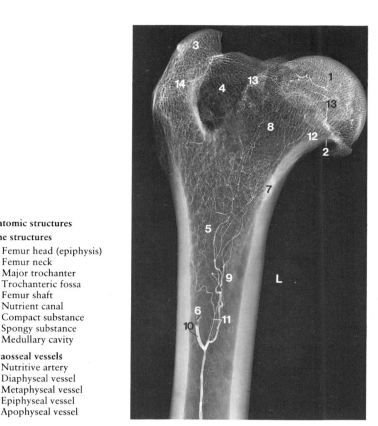

Fig 4: Radiograph of the proximal intraosseal vessels of the left femur filled with contrast medium. Mediolateral view — ML.

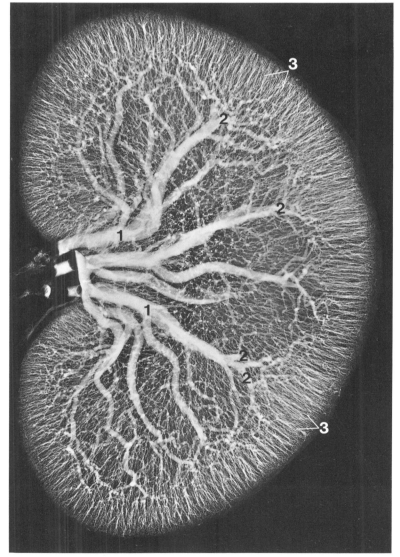

Fig. 5: Radiograph of the arterial kidney vessels filled with contrast medium.

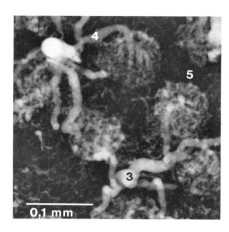

Fig. 6: Radiograph of the glomeruli in the renal cortex filled with contrast medium. (Tissue slice approximately 5 mm thick).

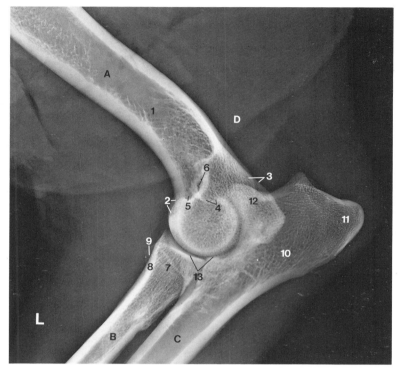

Fig. 7: Radiograph of the left elbow =negative. Lateromedial view — LM.

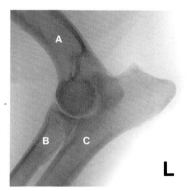

Fig. 8: Monitor image of the left elbow = positive. Lateromedial view — LM.

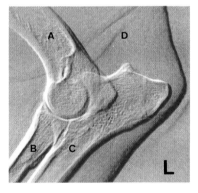

Fig. 9: Monitor image of the left elbow. Digital pseudo-3D-depiction after object shifting. Lateromedial view — LM.

1.5.3 Geometric distortion

The higher magnification in radiographs of areas further from the film, in contrast to areas near the film, is known as geometric distortion (figs. 1, 14 A, 15). An increase in the geometric distortion can occur by placing the object in the outer area of the x-ray beam (figs. 14 B, 16). Such an effect can also be obtained by positioning the object at an angle to the central x-ray beam. Geometric distortion can give an impression of three-dimensionality and simplify the categorizing of details to specific planes, especially with direct radiographic magnification. Exact measurements from radiographs, however, should only be made on images free of geometric distortion; the points to be measured should be located within the central beam and with larger FFDs should be close to the film plane.

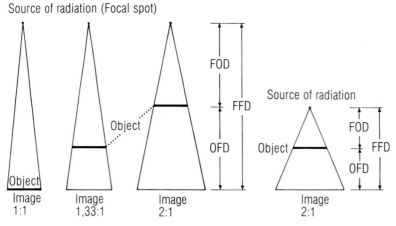

Fig. 10: Schematic drawing of a 1:1 scale image and various geometric magnifications.

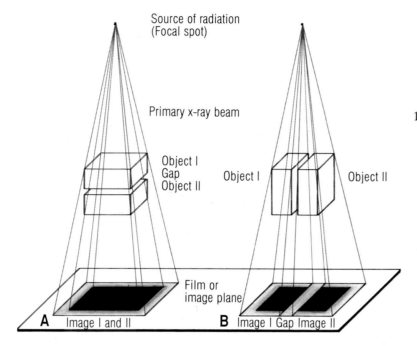

Fig. 11: Schematic drawing of superpositioning in the central projection; A: gap perpendicular, B: gap parallel to the beam.

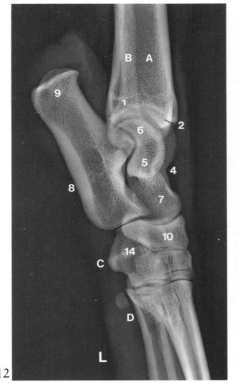
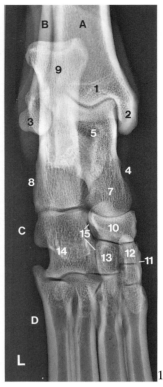

12

13

Fig. 12: Left tarsal joint. Mediolateral view — ML.

Fig. 13: Left tarsal joint. Dorsoplantar view — DPl. The typical hinge joint form of the tarsocrural joint and the intertarsal joints can be differentiated, in contrast to fig. 12.

Anatomic structures (figs. 12, 13)

A Tibia	7 Head
1 Tibial cochlea	8 Calcaneus
2 Medial malleolus	9 Calcaneal tuber
	10 Central tarsal bone
B Fibula	11 First tarsal bone
3 Lateral malleolus	12 Second tarsal bone
	13 Third tarsal bone
C Tarsal bones	14 Fourth tarsal bone
4 Talus	15 Intertarsal joints
5 Body of talus	
6 Trochlea of talus	**D Metatarsal bones**

1.6   The quality of radiographs

The ability to evaluate fine structures in radiographs or with an image amplifier and monitor presuppose sufficient picture quality. The following factors are responsible for the quality of radiographs:

sharpness or blurring,
contrast,
resolution of the image formation and recording system.

The functions sharpness and contrast are dependent on a number of various parameters, which are given by the geometry of the radiation source, the radiation, the object to be depicted, and the x-ray machine.

1.6.1   Sharpness or blurring

The recognition of details in radiographs can be limited or lost with blurring due to movement, geometric blurring, or film-screen blurring.

Blurring due to movement is the result of the patient's moving when insufficiently fixed, or the movement of organs (heart contractions, breathing movements, etc.). This can be avoided by stable positioning and using shorter exposure times (for example, cardiographs under 1 ms).

Geometric blurring (GB) is determined by the diameter of the focal spot and the relationship of the FFD to the FOD. Whereas a larger focal spot leads to significant blurring of the outer edges when the patient is raised above the film/image plane, an ideally point-formed radiation source gives sharp pictures without blurred edges in every area between the focal spot and film plane (figs. 18, 20). An object lying directly on the film plane is always sharply projected, regardless of the size of the focal spot. Whereas direct magnifications of up to 2.5x are possible with focal spots of 100 μm (mammography machines), direct magnifications of up to 200x with satisfactory sharpness are possible using microfocal spots of less than 10 μm (fig. 6).

Film-screen blurring depends on the type of screen used and the contact between the screen and film. X-ray or intensifying screens contain fluorescent crystals, which convert incoming x-rays into visible light. The photographic coating of the film in contact with the screen is blackened by the light emitted by the screen. Intensifying screens serve to reduce doses. The more sensitive the screen, i. e., the larger the amplification, and the shorter the necessary exposure time needed for blackening the film, the better the sharpness (fig. 23).

The three blurring parameters described here all influence each other at the same time. Due to the low dosage levels, reducing geometric blurring by using a smaller focal spot results in long exposure times and increases the danger of blurring due to movement. All three factors have to be tuned to each other and adapted to the situation.

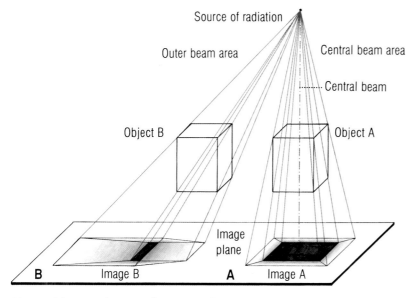

Fig. 14: Schematic drawing of geometric distortion. Geometric magnification of an object in the central beam path (A) and in the outer beam path (B).

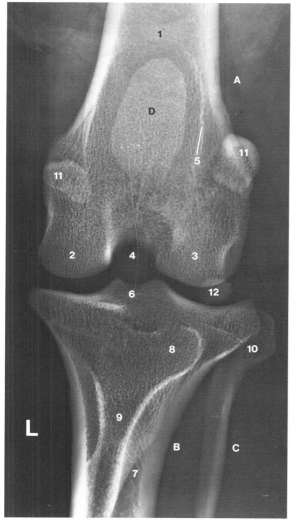

Fig. 15: Radiograph of the left knee joint within the central beam path. Craniocaudal view — CrCd = beam path.

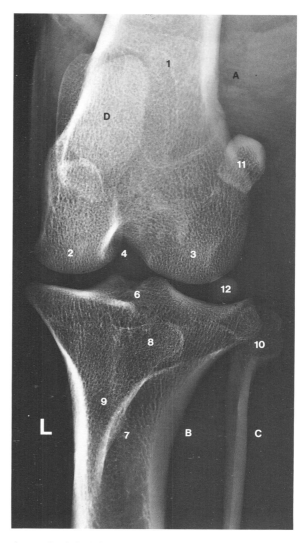

Fig. 16: Radiograph of the left knee joint in the outer beam path. Craniocaudal view, beam path 20° oblique to the craniocaudal beam path — Cr20L-CdMO.

Anatomic structures (figs. 15, 16)

A Femur	8 Tibial tuberosity
1 Femur shaft	9 Cranial margin
2 Medial condyle	
3 Lateral condyle	**C Fibula**
4 Intercondyloid fossa	10 Head of fibula
5 Femur trochlea	
	D Patella
B Tibia	11 Sesamoid bone of the gastrocnemius
6 Intercondylar eminence	muscle
7 Tibia shaft	12 Sesamoid bone of the popliteal
	muscle

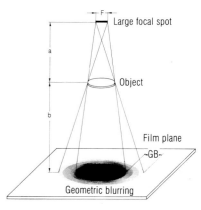

Fig. 17: Schematic drawing of geometric blurring (GB) caused by raising the object from the film plane when a large focal spot is used.

1.6.2 Contrast

Contrast means the difference between two intensities of blackening on the film or between two levels of brightness on the image amplifier. The contrast on the image recording system (x-ray film or monitor) corresponds to the differences in intensity after absorption of the primary beam by the patient or object. The contrast level depends mainly on the beam quality, the body region to be examined, and scatter radiation.

Soft x-rays give clearer differences in contrast than hard x-rays, for which the blackening gradations are smaller (see chapter 1.3.2). An area exposed with overly-hard radiation appears dull; it lacks bright whites and deep blacks.

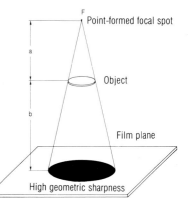

Fig. 18: Schematic drawing of sharp direct magnification when an ideally point-formed focal spot is used.

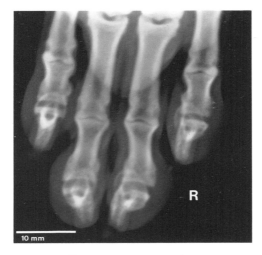

Fig. 19: Heavy geometric blurring in a 4x direct magnification made with a 400 μm focal spot (original photographically reduced for technical reasons of printing).

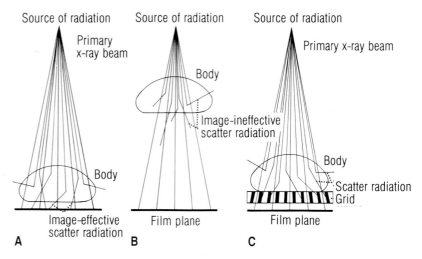

Fig. 21: Schematic drawings of the effect (A) and reduction (B, C) of scatter radiation (B- distance technique with large OFD, C-use of a grid).

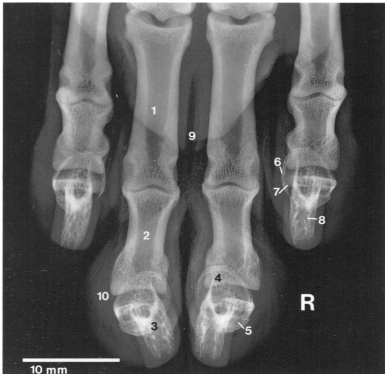

Fig. 20: High geometric sharpness in a 4x direct magnification made with a 5 μm microfocal spot. Middle and distal phalanges of the right rear paw. Dorsoplantar view – DPl. (original photographically reduced for technical reasons of printing).

Anatomic structures

Digital bones
1 Proximal phalanx
2 Middle phalanx
3 Distal phalanx
4 Extensor process
5 Flexor tuberosity

6 Unguicular crest
7 Unguicular groove
8 Unguicular process

Soft tissue
9 Metatarsal cushion
10 Toe pad

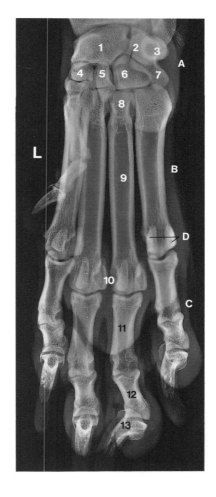

Fig. 22: Bones of the right front paw as seen with high resolution nonscreen x-ray film (Structurix D7, Agfa-Gevaert). Exposure values: 2.5 mm Al filter, FFD 80 cm, 60 kV, 50 mAs.

Clear differences in contrast are seen between bones and soft tissues and between vessels or organs filled with contrast media and the surroundings.

Scatter radiation, which originates in the irradiated material, reaches the recording surface multidirectionally and causes a diffuse blackening of the x-ray film or diffuse energizing of the illumination crystals of an x-ray screen or image amplifier (fig. 21 A). Fine differences in contrast, i. e., small differences in absorption in tissues, are covered up. Scatter radiation increases with increasing hardness of the primary beam. It is also dependent on the properties and size of the irradiated object. To reduce the scatter radiation the primary beam can be restricted with a diaphragm, a grid can be used (fig. 21 C), or the OFD can be increased (fig. 21 B).

Scatter radiation grids are made of lead strips, which are placed parallel to the primary beam and only allow these to pass, whereas multidirectional scatter radiation is absorbed. The disadvantage is the necessary increase in power, since in addition to the scatter radiation, a part of the primary beam is lost in the grid (fig. 21 C).

1.6.3 Resolution of the image formation and recording system

The resolving ability of the image formation or recording system determines the size of details, which can just be recognized. Periodical structures, such as the lead strips of a grid, which are smaller than the silver halide grains of the film or the illumination crystals of a screen or imaging device, cannot be distinguished from one another. The local resolution of nonscreen film is approximately 50 μm (fig. 22), for screen-film systems between 80−200 μm (fig. 23), and using image amplifiers and closed-circuit TV systems, between 200−420 μm. In order to recognize such small structures, the contrast between the structures and the surroundings has to be sufficient. This is generally not the case for biological material.

With conventional standard radiographs the screen-film system used generally limits the recognition of details to 100 μm.

All of the factors mentioned, which influence the quality of the picture, interact highly with each other. Small structures at the limit of resolution, which are affected by geometric blurring or show unclear edges as a result of superpositioning, can only be

Anatomic structures (figs. 22, 23)

A Carpal bones
1 Radial carpal bone
2 Ulnar carpal bone
3 Accessory carpal bone
4 First carpal bone
5 Second carpal bone
6 Third carpal bone
7 Fourth carpal bone

B Metacarpal bones
8 Base
9 Shaft
10 Head

C Digital bones
11 Proximal phalanx
12 Middle phalanx
13 Distal phalanx

D Sesamoid bones

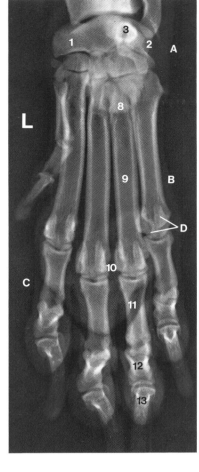

Fig. 23: Bones of the right front paw taken with a highly sensitive screen-film combination (Trimax T6 screen and XM film, 3M). Compared with fig. 22, lower resolution and increased screen blurring can be seen. Exposure values: 2.5 mm Al filter, FFD 80 cm, 60 kV, 1.5 mAs.

evaluated if the contrast is high enough. Reversely, with good sharpness and sufficient detail size the contrast is of less importance.

1.6.4 Labeling radiographs

The quality of radiographs can be further improved by comprehensive labeling, including exposure parameters, positioning, and patient identification.

Lead letters placed on the cassette at the time of exposure can be used to mark cranial, caudal, medial, lateral, right, and left. The terminology used for the views shown in radiographs generally follows guidelines set by the Nomenclature Committee of the American College of Veterinary Radiology.*) The positioning of the patient and the direction that the central ray of primary beam penetrates the body part, from the point of entry to the point of exit, are to be given using proper veterinary anatomic directional terms.

1.7 Radiation protection

In principle it can be said that x-rays exert dose-dependent, generally undesirable effects on biological tissues. Soft radiation is more likely to be damaging (especially to the skin), due to its greater absorption.

The requirements set on equipment and operation by state agencies are to be consciously followed. First and foremost is that the required image quality be obtained with the lowest amount of radiation possible.

2 PHYSICAL — TECHNICAL BASICS OF SONOGRAPHY OR ECHOGRAPHY

2.1 Ultrasound

Sound waves are longitudinal waves. The particles of a tissue are brought to vibrate by the sound. The direction of vibration corresponds to the direction of sonic propagation, which results in compression and rarefaction within the wave (fig. 24). Fre-

*) SMALLWOOD, J. E., M. J. SHIVELY, V. T. RENDANO, and R. E. HABEL: A standardized notation for radiographic projections used in veterinary medicine. Vet. Rad. 26, 2—9. (1985).

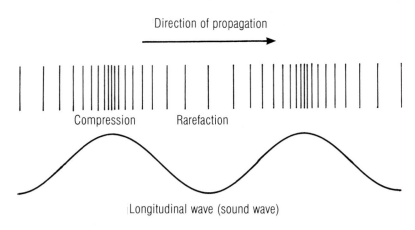

Fig. 24: Schematic drawing of the propagation of a longitudinal wave in tissue.

quencies above 20 kHz are known as ultrasound. The frequencies used in ultrasound diagnostics lie between 1—20 MHz (megahertz = 1000 kHz), and are considerably higher than the ultrasonic frequencies perceivable by animals, such as cats, dogs, or bats.

2.2 Behavior of sound waves in material

2.2.1 Wave propagation velocity and acoustic impedance

The velocity with which ultrasound travels in tissue (wave propagation velocity) depends on the ultrasound frequency and properties of the material. The speed averages 1540 m/s in soft tissues and liquids, is considerably lower in air, and distinctly higher in bone. The acoustic resistance or impedance of a tissue is the product of the speed of the sound beam in a given tissue and the density of the tissue.

2.2.2 Reflection, transmission, refraction, scattering, and absorption

When sound waves pass interfaces between two tissues with different wave propagation velocities or with different acoustic impedances (see chapter 2.2.1), a part of the sound beam is reflected (reflection — figs. 25, 26). The direction of reflection of a wave hitting an interface **perpendicularly** is the same as that of the incident wave. The nonreflected wave continues in the second medium without changing direction (transmission — fig. 25). If the interface is hit at an **angle,** the angle of reflection is the same as the incident angle. The transmitted wave is refracted (refraction — fig. 26).

Irregular acoustic interfaces cause reflection of the sound waves in all directions (scattering).

In addition, the sound beam is absorbed as it passes through tissues (absorption). The degree of absorption depends on the sound frequency and the quality of the insonated tissue. Higher frequencies are absorbed more strongly than low ones. When considering body tissues, bone and calcifications have high rates of absorption.

Fig. 25: Schematic drawing of the reflection and transmission of an incident sound wave hitting the interface perpendicularly.

Fig. 26: Schematic drawing of the reflection and refraction of an incident sound wave hitting the interface at an angle.

2.2.3 The effect of diagnostic ultrasound on biological tissues

The sound intensities used in the course of ultrasonic diagnostics of only a few mW/cm², as well as frequencies used, are considered safe, so that repeated pregnancy examinations are possible.

2.3 Generation and reception of ultrasound waves

The scanner head is both a transmitter and receiver. A piezoelectric crystal in the transducer or scanner is brought to vibrate by high-frequency electric pulses and converts them to ultrasound. At the same time reflected sound waves are transformed into corresponding electrical impulses, which can be made visible on a monitor. With an impulse duration of $1-2$ µs the relationship between the times of transmission and reception is approximately 1:500 to 1:1000.

2.4 Resolution

The axial resolution along the path of the ultrasound beam depends primarily on the frequency used. With frequencies of 1 and 10 MHz, structures of at best 1.5 mm and 100 µm, respectively, can be made out. High frequencies not only have a higher resolution, but also less penetration, since the rate of absorption is higher. 5.0 MHz has proven to be a universal frequency for cardiac and abdominal examinations of dogs.
The lateral resolution is less than the axial and is determined primarily by the geometry of the transducer.

2.5 Formation of two-dimensional ultrasonographs

2.5.1 Transducer-dependent form of the ultrasonographs

For the presentation of anatomical or pathological structures, the two-dimensional real-time technique is almost always used today. Depending on the ultrasound field scanned, one can differentiate between the sector scan and linear or parallel scan. A compromise between these two forms is the convex scan. Whereas the sector scanner produces a triangular image (figs. 27, 28), the image produced by a linear or parallel transducer is rectangular (figs. 29, 30). Only the sector scanner is suited for examination of the heart, due to the narrow intercostal acoustic window in the area of the cardiac notch of the lung. Abdominal examinations can also be made easily in dogs with sector transducers (fig. 28). Linear or convex scanners are also used for

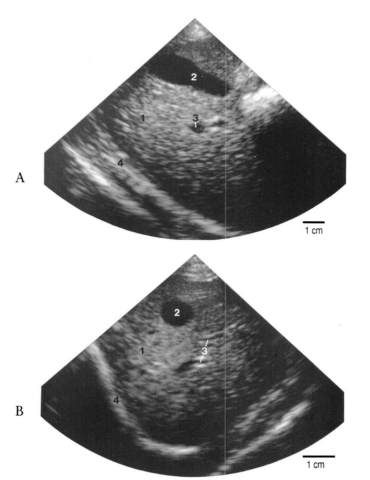

A

1 cm

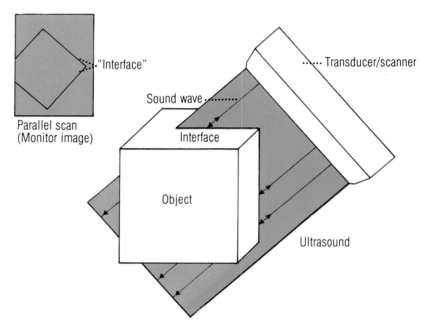

B

1 cm

Fig. 28: Sector sonographs of the liver with the gallbladder shown longitudinally (A) and transversely sectioned (B).

Anatomic structures

1 Liver, 2 Gallbladder, 3 Branch of the portal vene, 4 Diaphragm.

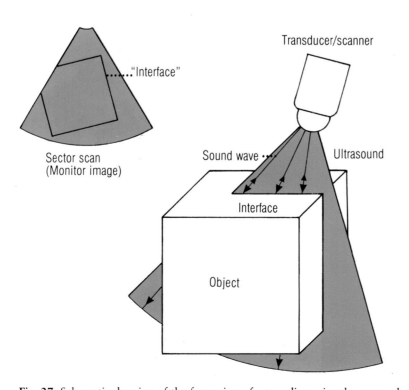

Fig. 27: Schematic drawing of the formation of a two-dimensional sonograph (section), using a sector scanner.

Fig. 29: Schematic drawing of the formation of a two-dimensional sonograph (section), using a linear transducer.

canine abdominal sonography, especially for diagnosing pregnancy, since they provide better resolution of structures close to the transducer (fig. 30).

2.5.2 Categorizing structures on the sonograph

In contrast to the radiograph (fig. 1), an ultrasonograph is a section, which shows the acoustic interfaces between various soft tissues (figs. 27, 29). Reflections of the ultrasound beam occur at each interface and appear on the monitor as echogenic regions, i.e., as bright areas (figs. 28, 30).
Fibrous wall structures of vessels and organs are seen as more or less hyperechogenic (white) areas, when hit perpendicularly with the ultrasound beam. Ultrasonographs of parenchymatous

organs (liver, spleen, renal cortex) and muscle (including the heart) are made up of numerous single echoes (white-spotted areas).

Nearly the entire beam intensity is reflected (strong white echo) at interfaces between soft tissues and structures containing air (lung), bone, or calcifications, as a result of the large difference in acoustic impedance. Since the remaining beam intensity is absorbed in penetrating bone or calcifications, tissues found behind them cannot be shown; a black anechoic area, acoustic shadowing, is seen on the monitor (fig. 30). The same is true for tissues covered by the lungs.

Homogenous tissue regions, such as renal parenchyma, or liquids, such as blood, bile, or urine, appear hypoechoic (constant dark gray) or anechoic (constant black).

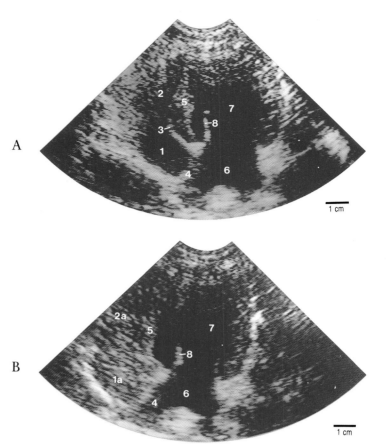

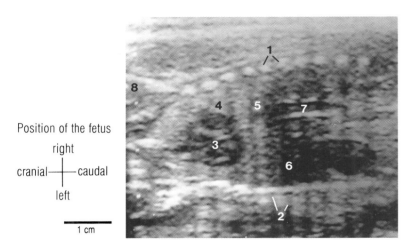

Position of the fetus

right

cranial —|— caudal

left

1 cm

Fig. 30: Linear scan of a Cairn Terrier, 59 d after conception. Horizontal section of a fetus, thorax, and intrathoracal abdomen with ribs and typical acoustic shadows.
(Grof, D., Wissdorf, H., Institute of Anatomy, Hannover Veterinary School).

Anatomic structures

1	Echos of the ribs	5	Diaphragm
2	Acoustic shadows of the ribs	6	Liver
3	Heart	7	Caudal vena cava
4	Lung (collapsed)	8	Forelimb

2.5.3 Contrast media

Ultrasound contrast media are used to improve presentations of vessels or heart chambers, or to evaluate heart defects, such as in the interventricular or interatrial septum (fig. 31). They contain extremely finely distributed gas bubbles as the echogenic substance. In addition to CO_2 or various well-mixed solutions, such as physiological saline, 5 % glucose solution, plasma expander, or patient blood, commercially prepared echocontrast media will be available in the future. The latter contain gas bubbles, which are bound to saccharide microparticles. Following intravenous injection the contrast medium can be traced to the pulmonary trunk. The presently available contrast media cannot pass the pulmonary capillaries.

2.5.4 Orientation on the ultrasonograph

By turning, tilting, or sliding the transducer, every sectional plane of the body can be reached in principle (fig. 28). The desired sonographic plane has to be correspondingly exactly selected or reproduced for measurements or repeated examinations. Regarding the localizations in ultrasonographs, areas near the transducer are seen in the upper part of the picture and areas distant to it in the lower part (figs. 27, 29). The right or left side of the picture corresponds to the left or right side, dorsal or ventral, or cranial or caudal side of the body.

2.6 Doppler echography

The waves reflected from stationary interfaces have a lower intensity, but the same frequency as the incident ultrasound waves (fig. 32 a). If, on the other hand, the waves strike a moving ob-

Fig. 31: Diastolic four-chamber view of the heart without (A) and with (B) contrast medium in the right atrium and ventricle.
(Tobias, R., Small Animal Clinic, Hannover Veterinary School).

Anatomic structures

1	Right atrium	4	Interatrial septum
1a	Contrast medium in the right atrium	5	Interventricular septum
2	Right ventricle	6	Left atrium
2a	Contrast medium in the right ventricle	7	Left ventricle
3	Opened tricuspid valve	8	Opened mitral valve

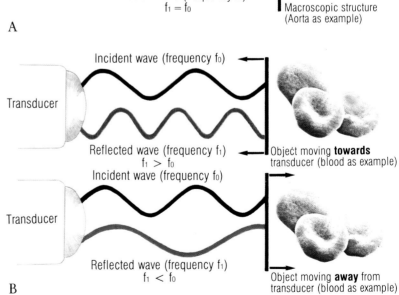

Fig. 32: Schematic representations of two-dimensional (A) and of doppler echography (B).

ject, such as an erythrocyte, the frequency of the reflected wave is altered in accordance with the direction and speed of the moving "interface" (doppler principle — fig. 32 b).

By definition, blood flowing towards the transducer is shown above the baseline (continuous or pulsed doppler) or in red

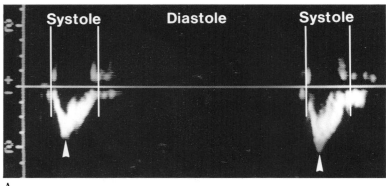

A

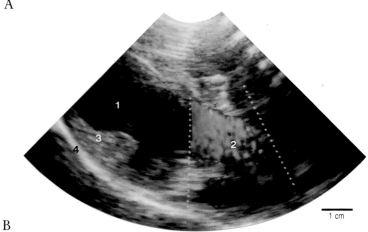

B

(color-coded doppler) — Reversely, i.e., blood shown below the base line or in blue is flowing away from the transducer. This presentation of the direction of blood flow is completely voluntary and has nothing to do with the type of vessel (artery or vein) or the oxygen content of the blood. Oxygen-rich blood in the ascending aorta, for example, is shown in blue (fig. 33 b) or below the baseline (fig. 33 a), since it flows away from the transducer.

The speed of the blood flow can be read directly from the y-axis with a conventional doppler and can be estimated from the color intensity with a color-coded doppler.

If the blood velocity exceeds the reception frequency of the transducer, the so-called aliasing phenomenon occurs, which corresponds to the stroboscopic effect in film. With color-coded dopplers this results in reversing of the colors and with conventional dopplers in reversed presentation, relative to the x-axis.

Fig. 33: Blood flow at the level of the aortic valve: pulsed doppler (A) and color-coded doppler (B).

Anatomic structures

Arrow shows maximum speed of the blood at the level of the aortic valve.
1 Left ventricle (outflow path)

2 Ascending aorta
3 Myocardium
4 Pericardium

2.7   Preparing the dog for ultrasonic examination

A prerequisite for quality ultrasonographs is sufficient contact of the transducer with the surface of the skin. For cardiae examination a wiping of the skin with alcohol and heavy application of contact gel is sufficient for most dogs. The same procedure can be used for abdominal sonography of dogs with little or medium amounts of hair. Animals with thick long fur and much underfur, however, have to be shaved in the area to be examined.

The positioning of the dog depends on the preference of the operator, the behavior of the animal, and the examination to be performed. Echocardiography in lateral recumbancy is proposed as well as cardial examinations made on sitting or standing dogs. Abdominal sonography is possible in lateral and dorsal recumbancy, as well as on standing animals.

SONOGRAPHIC ANATOMY

by CORDULA POULSEN NAUTRUP

AN INTRODUCTION TO SONOGRAPHIC ANATOMY

As with radiographic anatomy, sonographic anatomy has direct clinical references. Only a thorough knowledge of the normal ultrasonograph allows the successful clinical use of this method, which is relatively new to veterinary medicine. Several examples of ultrasonographs follow, showing various organs and organ systems. The most important structures in the ultrasonographs are depicted in sketches and labeled. The respective positions for the ultrasonographic examination are shown in accompanying schematic drawings.

HEART — COR

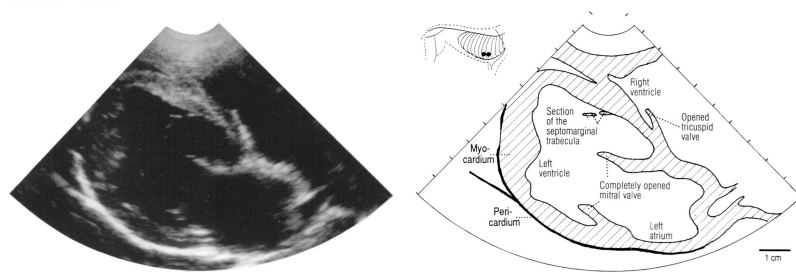

Fig. 1: Inflow tract of the left ventricle; late-diastolic longaxis section of the left ventricle.

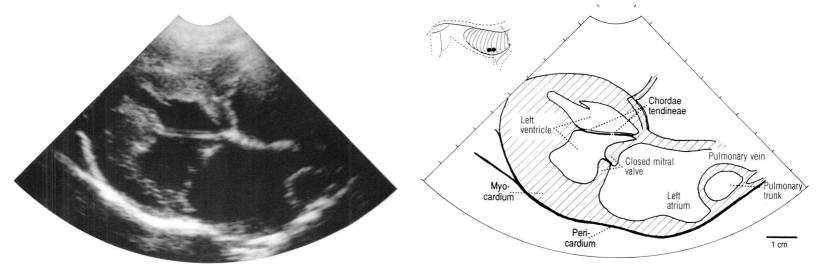

Fig. 2: Inflow tract of the left ventricle; late-systolic long-axis section of the left ventricle.

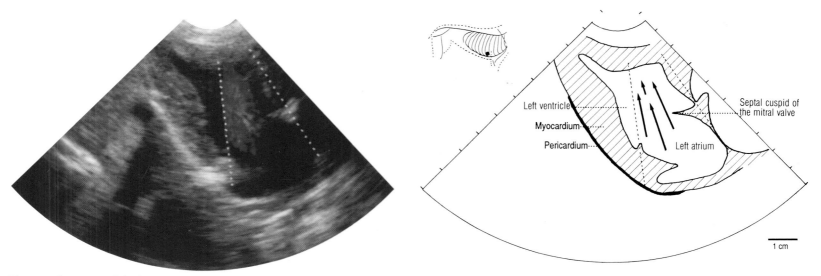

Fig. 3: Inflow tract of the left ventricle; long-axis section of the left ventricle with inflowing blood in the early diastolic phase.

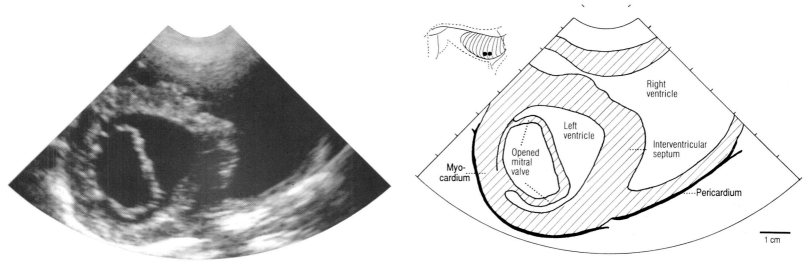

Fig. 4: Inflow tract of the left ventricle; diastolic short-axis section at the level of the mitral valve.

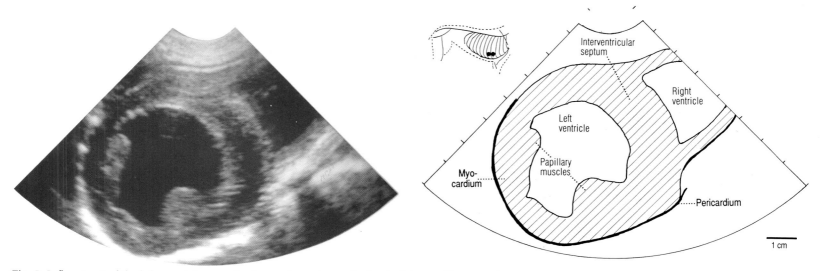

Fig. 5: Inflow tract of the left ventricle; systolic short-axis section at the level of the papillary muscles.

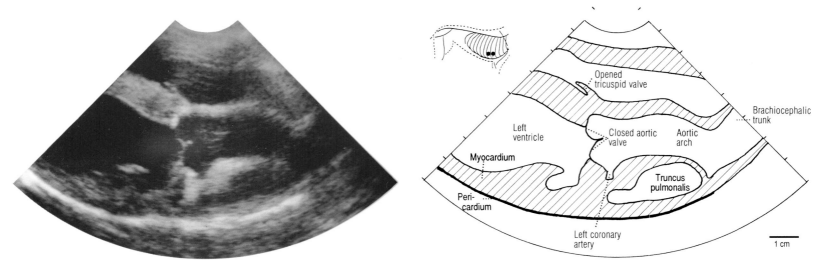

Fig. 6: Outflow tract of the left ventricle; diastolic long-axis section.

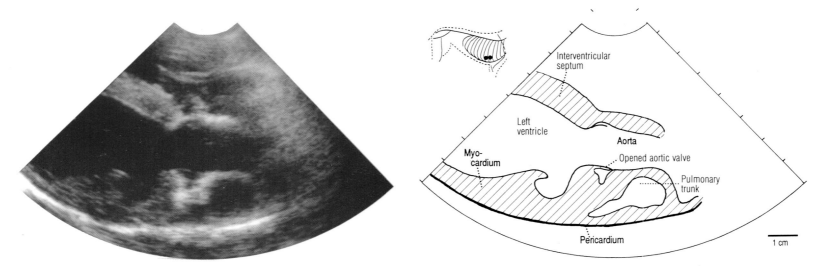

Fig. 7: Outflow tract of the left ventricle; systolic long-axis section.

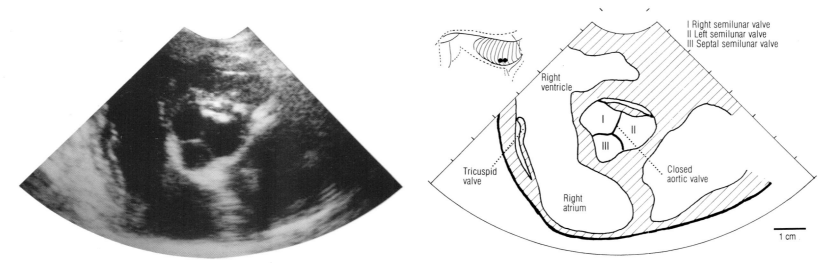

Fig. 8: Outflow tract of the left ventricle; diastolic short-axis section at the level of the aortic valve.

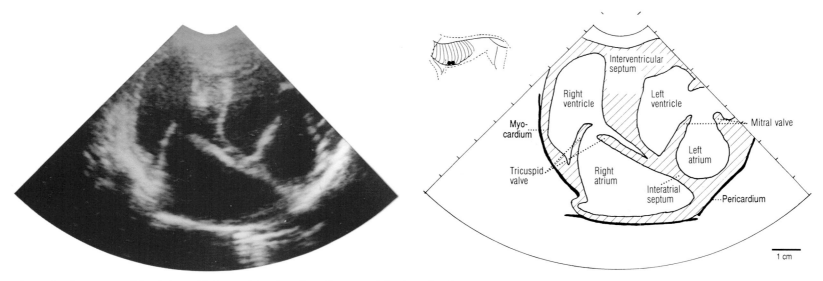

Fig. 9: Outflow tracts of the right and left ventricles; late-diastolic view of the four chambers.

79

LIVER – HEPAR

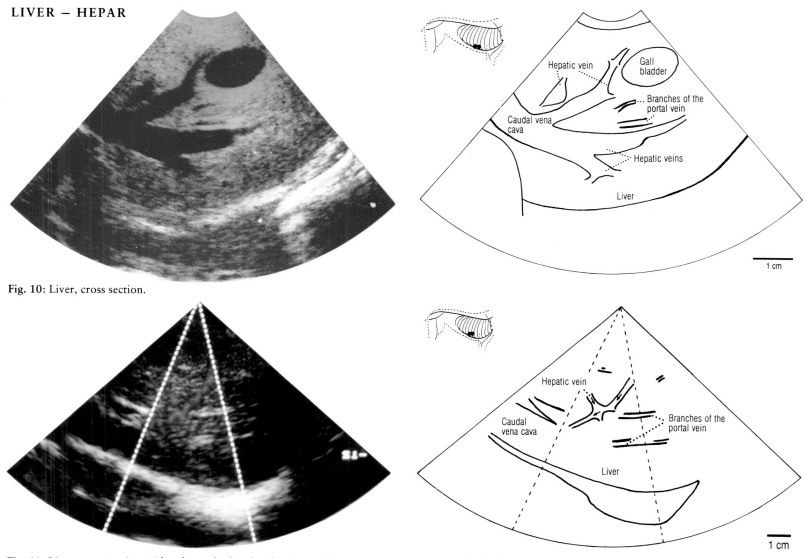

Fig. 10: Liver, cross section.

Fig. 11: Liver, cross section with color-code doppler depiction of the direction of blood flow in the hepatic veins.

SPLEEN – LIEN

OVARY – OVARIUM

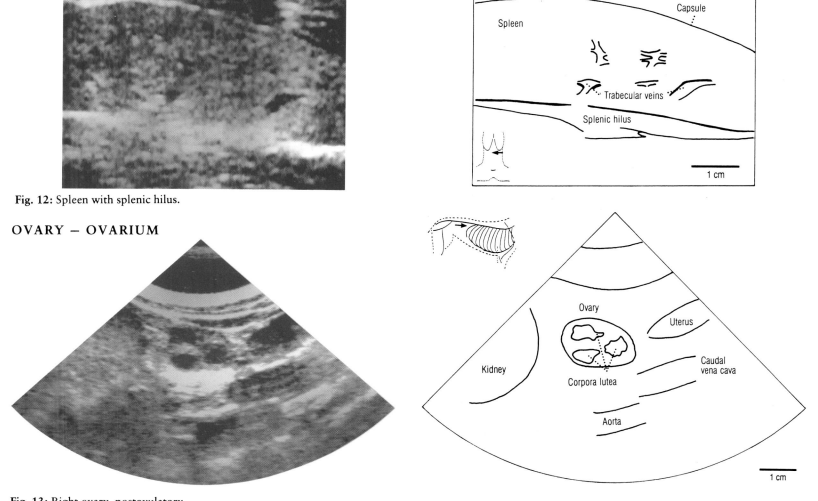

Fig. 12: Spleen with splenic hilus.

Fig. 13: Right ovary, postovulatory.
(Hayer, P., Günzel-Apel, A.-R., Lüerssen, D., Clinic of Andrology and AI and Small Animal Clinic, Hannover Veterinary School).

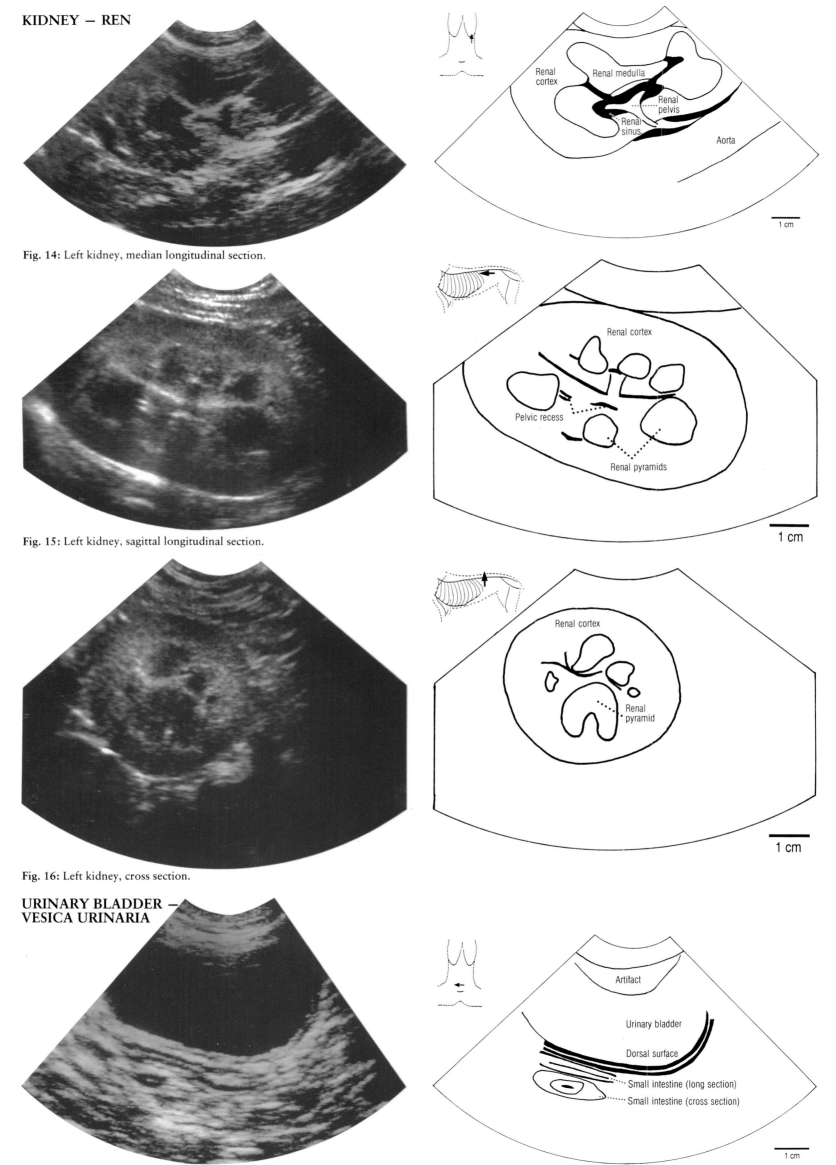

Fig. 14: Left kidney, median longitudinal section.

Renal cortex

Renal medulla

Renal pelvis

Renal sinus

Aorta

1 cm

Fig. 15: Left kidney, sagittal longitudinal section.

Renal cortex

Pelvic recess

Renal pyramids

1 cm

Fig. 16: Left kidney, cross section.

Renal cortex

Renal pyramid

1 cm

URINARY BLADDER — VESICA URINARIA

Artifact

Urinary bladder

Dorsal surface

Small intestine (long section)

Small intestine (cross section)

1 cm

Fig. 17: Filled urinary bladder, cross section.

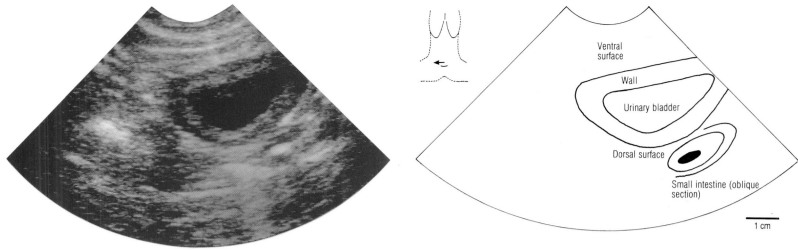

Fig. 18: Empty urinary bladder, cross section.

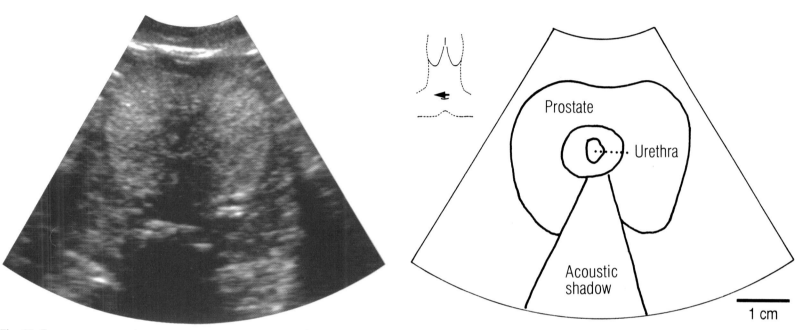

Fig. 19: Prostate, cross section.
(Lüerssen, D., Small Animal Clinic, Hannover Veterinary School).

TESTIS

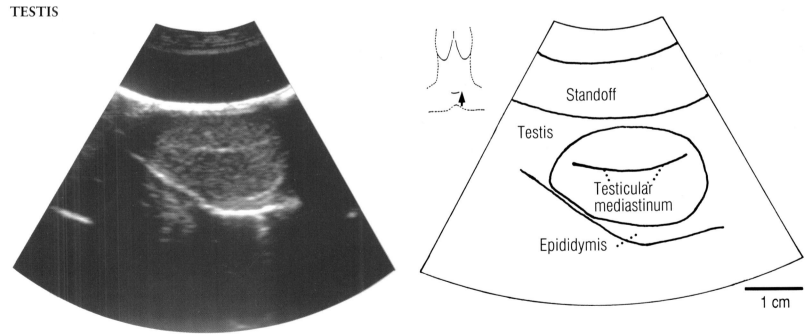

Fig. 20: Testis, longitudinal section.
(Lüerssen, D., Small Animal Clinic, Hannover Veterinary School).

Address of the author:
Dr. Cordula Poulsen Nautrup
Zentrum Anatomie
Medizinische Hochschule Hannover
Konstanty-Gutschow-Straße 8
D-30625 Hannover
Germany

Layout:
Susanne Fassbender, med. techn. assistant
Technical drawings:
Gertrud Baumgarten, technician

I would like to thank the FEINFOCUS Röntgen-Systeme Co. for allowing the use of their Microfocus x-ray machines. For the support with various ultrasound equipment, including color-coded doppler systems, I would like to thank Dr. W. Kästner, Hannover, and the following companies and people: aTL Advanced Technology Laboratories, especially Mr. R. Fischer, DYNAMIC IMAGING, especially Mrs. S. Wilson and Mr. T. Gerhards, as well as DIASONICS Sonotron, especially Dr. H. Schneider.

PELVIC LIMB – MEMBRUM PELVINUM

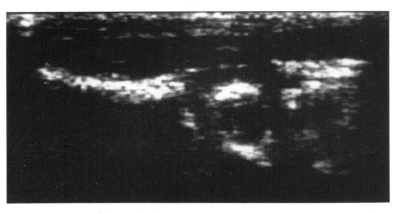

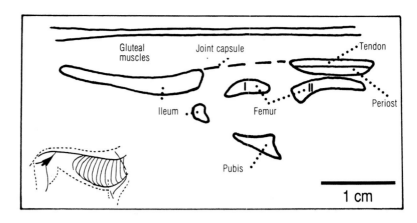

Fig. 21: Hip joint, longitudinal section.
(Kresken, J.-G., Köstlin, R.G., Clinic of Veterinary Surgery, Ludwig-Maximilian-University, Munich).

I (Head, ossified epiphysis)
II Neck

PREGNANCY

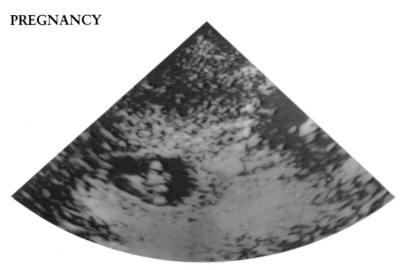

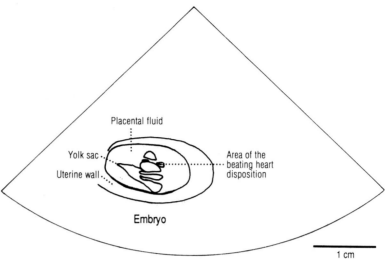

Fig. 22: Early pregnancy, embryo, 20d after conception.

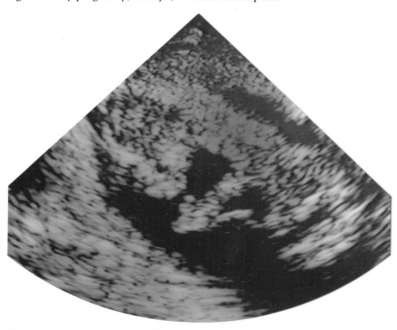

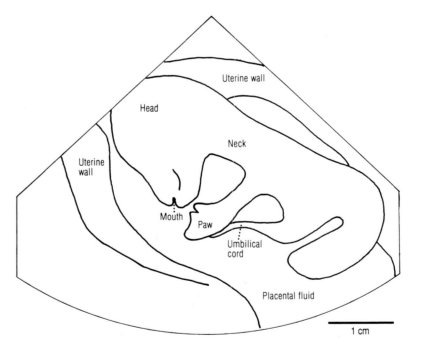

Fig. 23: Pregnancy, fetus, 38d after conception.

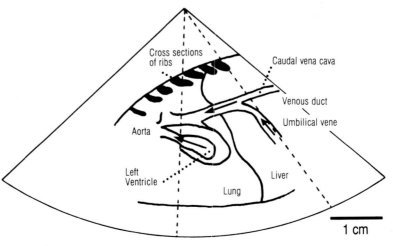

Fig. 24: Late pregnancy, fetus, 54d after conception; thoracic cavity and intrathoracal abdominal cavity, sagittal section, color-coded doppler depiction of the fetal circulatory system.

ADDITIONAL INFORMATION RELATING TO CLINICAL AND FUNCTIONAL ANATOMY

1. OSTEOLOGY: OSSIFICATION

1.1. **OSTEOGENESIS OF CARTILAGE** is deranged in the abnormal condition known as fetal chondrodysplasia or chondrodystrophy (chondros = cartilage; dys - defective, deficient; trophia - nutrition). A sharp decrease occurs in the longitudinal growth of long bones, following a disorganized proliferation of chondrocytes and an absence of their arrangement into columns. Chondrodystrophy occurs as a breed characteristic in the Dachshund and Pekingese among others, appearing as diaphyseal shortening and thickening of epiphyses of long bones (disproportionate nanism).

1.2. For **CHONDROCLASTIC ACTIVITY** to occur during bone growth, mineralization of cartilaginous ground substance is an essential prerequisite. Non-mineralized cartilage cannot be removed by chondroclastic activity and thus cannot be replaced by bone. With mineral deficiency and an imbalance in the optimum Ca/P ratio, mineralization in the **zona resorbens cartilaginea**, the erosion zone of large vesiculated cartilage, is limited. Infiltrating chondroclasts cannot destroy the cartilage and with its continued proliferation, one detects a thickening of the **epiphyseal cartilage** radiographically. Weight can no longer be supported.

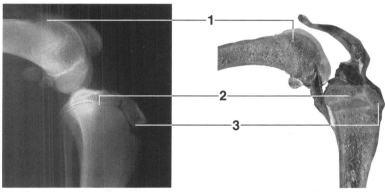

Distal epiphyseal cartilage of femur (1); proximal epiphyseal cartilage of tibia (2) and apophyseal cartilage of tibial tuberosity (3)

1.3. **CENTERS OF OSSIFICATION** are distinctly differentiated and well recognized radiographically, due to mineralization of cancellous substance as opposed to the (as yet) non-mineralization of the adjacent areas. The centers of ossification of different bones occur at precise times during an animal's growth. They are of value as an indication of maturity and can be used to estimate an animal's age.

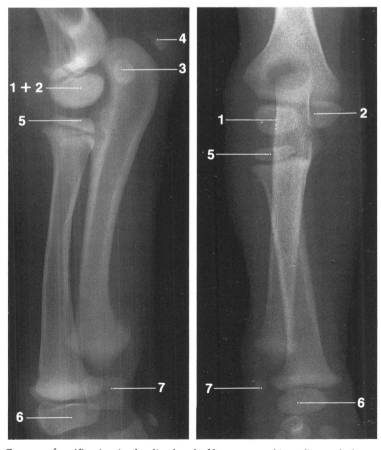

Centers of ossification in the distal end of humerus and in radius and ulna. 1 Capitulum of humerus; 2 Trochlea of humerus; 3 Medial epicondyle of humerus; 4 Tuber of olecranon; 5 Head of radius; 6 Distal epiphysis of radius; 7 Lateral styloid process.

1.4. **EPIPHYSIOLYSIS AND APOPHYSIOLYSIS** may be caused in young animals before the completion of growth of long bones by trauma, excessively heavy strain or abnormal movement. For example, this can occur to the tibial tuberosity in Osgood-Schlatter Disease (see 30.4.). After its surgical repositioning and fixation to the diaphysis, one can then undertake followup therapy. Normal epiphyseal cartilage and pathological epiphysiolysis can be differentiated radiographically.

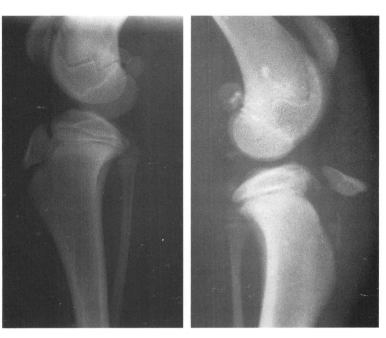

Left: Tibial tuberosity in normal position. Right: Detached tibial tuberosity.

1.5. **LONGITUDINAL GROWTH** of bone is promoted by growth hormone from the hypophysis and by thyroid hormones, and restricted by sex hormones. This bone growth terminates at the end of puberty. Premature removal of the gonads, as in castration or speying, leads to delayed ossification of the epiphyseal cartilage and in increase in final size of the animal (as in early castration of geldings and steers). In man and dog this is observed in tumors of the adenohypophysis causing acromegaly, a pronounced lengthening of the extremities.

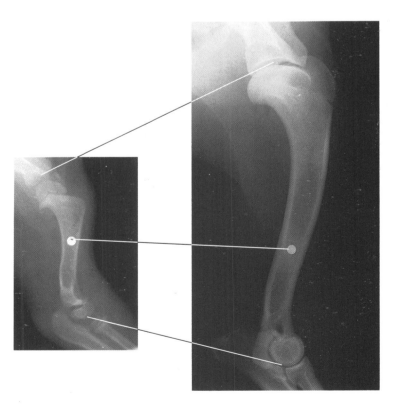

Objective assessment of longitudinal growth of humerus by radiography of the labelled diaphysis in a puppy (left), and after conclusion of growth (right). The proximal epiphysial cartilage has a definitely greater growth potential.

1.6. **BONE STRUCTURE** and **GROWTH** are influenced by hormones and vitamins. **Calcitonin** from the thyroid gland promotes bone growth through stimulation of osteoblasts and mineralization of osseous ground substance. An oversupply of **PARATHORMONE** from the parathyroid gland promotes osteolysis by activating osteoclasts. **Growth hormone** regulates metabolism in the epiphyseal disc and its deficiency in young animals produces dwarfism (nanism). **Sex hormones** accelerate the closure of growth plates. **Thyroxine** from the thyroid gland promotes the maturation of bone and its metabolism. Deficiencies of **vitamins A, C and D** act negatively on bone structure.

I.7. Bone **REGENERATION** after fracture is promoted by surgical union or reduction of the bone fragments concerned, followed by primary healing. An osteosynthesis of this nature is performed using different techniques such as pinning or the use of screws. In primary healing of a fracture, regeneration growth occurs when, after exact repositioning, the stable ends of the fracture are fixed to each other to avoid any movement. In the ideal instance, according to the principle of bone reconstruction, regeneration occurs where no free space is present on the line of fracture. Then in the region of fracture either no callus or only an insubstantial one, is present. Healing time is also shortened.

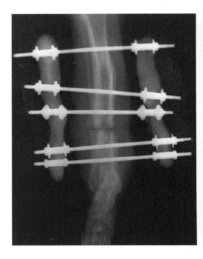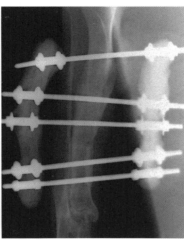

Transfixation (external screwing) of a fractured tibia. Left: Immediately following the operation (the fracture space is still open). Right: Closure of fracture space due to callus formation.

I.8. **REPAIR** by osseous tissue occurs with secondary healing of a fracture and resultant callus formation. In this case the fracture ends are not fixed and move with changes in tension and pressure. Secondary healing is achieved by the formation of a callus emanating from periosteum, perivascular connective tissue of associated blood vessels and the endosteum of the medullary cavity. Initially the union consists of connective tissue. Very quickly, however, connective tissue cells become osteogenic cells and osteoblasts. Variable amounts of cartilage are formed and then new mineralized osseous trabeculae are laid down. Thus, a primary callus is formed from cancellous bone and subsequently resculptured to give what is regarded by some as a secondary callus. At the conclusion of the healing of a fracture in this manner the callus of cancellous bone is replaced by definitive lamellar bone. The callus thickening is levelled or graded off so that one can hardly detect the site of fracture.

II. OSTEOLOGY: BONE STRUCTURE AND SHAPE

II.1. In **BONE RECONSTRUCTION** there can be local **atrophy** caused by persistent pressure due to degenerative processes. Local pathological atrophy may be caused by tumors or parasitic cysts, while general atrophy involving several bones stems from inactivity.

After fracture, autologous bone chip transplants are employed to compensate for overall loss of osseous and related substances, thus utilizing the physiological reconstruction of bone. At first the transplant serves in stabilization and disappears gradually. Its surviving blood vessels, however, connect with those of the surrounding transplantation bed and cause a proliferation of osteoblasts.

II.2. Muscle attachments having a large surface area, are fixed in the **PERIOSTEUM** by tendon fiber tracts which spread out at the site. Round tendons of fusiform muscles penetrate the periosteum and are anchored in the underlying compact bone. With intense muscle contraction, fragments of compact bone at the site of attachment can be torn away and this is referred to as a tear fracture. With continuous local mechanical insult periosteum reacts, forming cartilage and bone. Osseous protrusions or exostoses known as 'splints' are a result.

II.3. The **ENDOSTEUM** which lies on the boundary of the medullary cavity and lines the medullary cells in places, is thinner essentially than periosteum. During bone growth, bone removal by osteoclastic activi-

ty emanates mainly from the endosteum. Thus the thickening of compact bone induced by the periosteum, stops at the border of the medullary cavity, which is continually enlarging. The principle is external construction and internal destruction.

II.4. The **BLOOD VESSELS** of the long bones belong to four different systems which anastomose with one another to guarantee an elaborate blood supply. Anastomoses and collateral blood vessels ensure that the blood supply is maintained after numerous fractures, therapeutic osteosyntheses and ligation of individual blood vessels. The four systems of blood supply, in order of size, are as follows:

I. The largest blood vessels encompassing the first system are **nutrient vessels** which penetrate the medullary cavity by the **nutrient foramen** in the mid-length of the diaphysis, and subsequently through the **nutrient canal**. Finally they ramify in the sinusoids of the bone marrow and supply the compact bone from within outwards. The capillaries run longitudinally in the central (Haversian) canals of the osteons and supply the osseous tissue. Transverse connections between the osteons are known as the **perforating canals** of Volkmann, and through them venules, in the main, run to the periosteal veins.

II. In the vicinity of the epiphyseal line, the **metaphyseal artery and vein** enter the metaphyseal part of the diaphysis.

III. The **epiphyseal artery and vein** supply the extremities of the bone, that is covered by articular cartilage. (Epiphyseal vessels reach the epiphysis of the femur from the pelvis through the ligament of the head of femur.)
Before the completion of longitudinal bone growth, the epiphyseal vessels are separated from those of metaphysis and diaphysis by the epiphyseal cartilage. The vessels anastomose after the epiphysis and metaphysis have fused.

IV. Thin **periosteal vessels** and also those of the articular capsule penetrate the periosteum all over the osseous surface. They supply solely the superficial parts of the compact bone, while their branches penetrate the periphery of the compact bone. As described above, the large vessels establish a centrifugal system of blood circulation from the deeper parts of the bone.

II.5. The Nomina Anatomica Veterinaria of 1983 officially recognized a new **BONE SHAPE** namely the irregular bone. Vertebrae are now classed as irregular and not short bones (see 4.4.). On the basis of ossification type, short bones undergo endochondral ossification whereas vertebrae result from both endochondral and periosteal ossification.

II.6. Radiographically, **SESAMOID BONES** can be confused with pathological **free bodies (joint mice)**. These may become detached from pathological exostoses or from the synovial surface as in **osteochondrosis dissecans**. Then they move freely in the articular cavity.

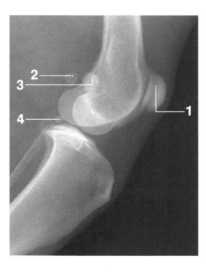

Sesamoid bones of the knee joint:1 Patella; 2 and 3 Sesamoid bones of the mm. gastrocnemii; 4 Sesamoid bone of the m. popliteus.

III. ARTHROLOGY

III.1. **PSEUDOARTHROSES** or 'False joints' are due to an absence of healing following instability or defective repositioning of the ends of a fracture. They attach to each other by connective tissue or cartilage. Abnormal movement, associated with instability, remains. A **nearthrosis** or new joint occurs after luxation, when tissues similar to articular cartilage and capsule form a new articular surface at the abnormal site.

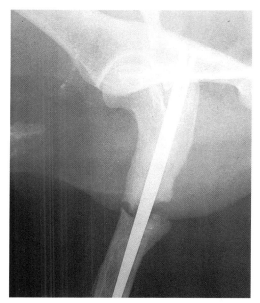

Pseudoarthrosis (left) after fracture and unstable osteosynthesis(intramedullary fixation or pinning) of femur.

Nearthrosis (below) following luxation and non-reduction of the hip joint. The nearthrosis on the ilium replaces the original articular surface.

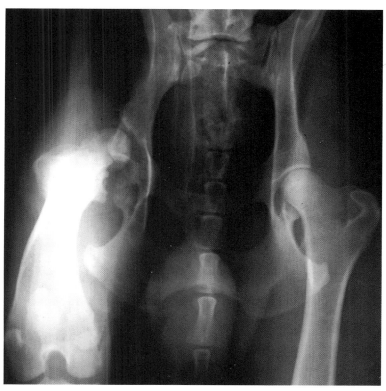

III.2. SYNOVIA may be interpreted as an ultrafiltrate of blood, modified by the secretion of specific synovial cells (histiocytic type B cells). Synovia nourishes the non-vascularized articular cartilage by perfusion, and its 'mucin' content (hyaluronic acid-protein complex) determines synovial viscosity. In hydrarthrosis, synovia is formed in excess following articular inflammation or severe stress.

III.3. The **ARTICULAR CAPSULE**, particularly on the palmar or plantar surface of the joint, is arranged in folds that disappear on extension. Articulations with a greater degree of movement have more extensive capsules. In contrast, stronger articulations with limited or no movement have tight capsules.

III.4. CAPSULAR LIGAMENTS become overstretched when strained and can tear in luxations. This may lead to degenerative arthroses.
A cruciate ligament rupture, involving the cranial ligament in 98% of affected cases, is the most frequent cause of stifle (knee) problems in dogs. The sequence of events is an instability of the joint with displacement of the tibia cranially with simultaneous fixation of the femur (the cranial drawer sign). Degenerative changes within the ligaments are mostly responsible. Purely traumatic tears are seldom a factor.

III.5. ARTICULAR CARTILAGE serves as an elastic cushion, permitting pressure changes and being capable of reversible deformation, and has a high water content which decreases with age. A perichondrium is absent on the surface of the hyaline cartilage, hence its regenerative capability is limited. In chondrosis dissecans, the articular cartilage is distinctly thickened in places and has necrosis at deeper levels. The pathological section is not capable of weight bearing. Overlying intact cartilage loses its foundation and can become detached. (Formation of

free bodies see II.6.). In young dogs this condition is found particularly in the shoulder joint but also quite often in the knee.

III.6. INNERVATION is derived from an intracapsular nerve plexus. From this, nerve fibers penetrate the synovial layer of the joint capsule but do not reach the stratum synoviale completely. Among other things torn ligaments cause articular instability, the resultant stretching of the capsule being very painful.

III.7. OUTPOCKETINGS OF THE JOINT CAPSULE are sites of choice for intracapsular injections.

III.8. SYNOVIAL BURSAE are distended during inflammation or bursitis due to an increase in synovia. Resultant pain is severe.

III.9. INTERVERTEBRAL ARTICULAR DISCS are subject to reconstruction associated with age. In chondroplastic (chondrodystrophic) breeds such as Dachshund, Pekingese and Scotch Terrier, this commences in their first year of life. The pulp nucleus and the inner connective tissue lamellae of the fibrous ring are replaced increasingly by cartilaginous tissue, in particular hyaline cartilage. Inadequate nutrition, derived from blood vessels at the periphery of the articular disc produces necrosis followed by dystrophic calcification. Due to these alterations the pulp nucleus loses its function as a 'water cushion' for maintaining distance, and as a tensile protection for the lamellae of the fibrous ring. Initially, necrosis and dystrophic calcification encroach on the inner part of the fibrous ring which finally tears dorsally. Necrotic and calcified portions of the pulp nucleus then prolapse into the vertebral canal. A ventral prolapse is rare and presents no symptoms. Prolapsed portions of pulp nucleus within the vertebral canal or in the vicinity of intervertebral foramina lead to pressure atrophy of the spinal cord or contusion of nerve roots. As a result, intense inflammation develops, with severe pain and eventual paralysis as sequelae (posterior paralysis in Dachshunds).

Due to calcification, the prolapsed portion of the pulp nucleus can be detected radiographically within the vertebral canal. Frequently, however, it is squeezed through the dorsal longitudinal ligament in the direction of the intervertebral foramen to the side. In the region of the thoracic vertebrae, a disc prolapse through the dorsal longitudinal ligament is impeded by the well developed intercapital ligament joining the heads of each rib pair immediately dorsal to the disc. Thus the condition is rarely seen at the thoracic level.

Surgery aims at removing the prolapsed disc material from the vertebral canal with resultant decompression of the spinal cord. This is facilitated by a laminectomy (see 4.5.). Besides this, after fenestration of the respective fibrous rings, remnants of the pulp nucleus of the disc remaining, and the pulp nuclei of adjacent discs, are removed as a prophylactic against further prolapse. In ventral fenestration an incision is made through the ventral longitudinal ligament, and fibrous ring and pulp nucleus are evulsed.

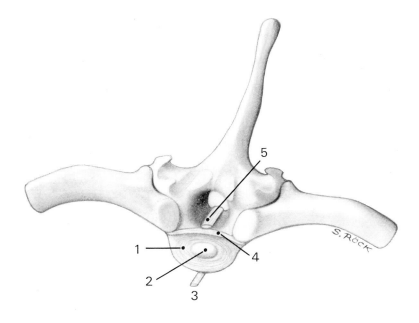

Tenth thoracic vertebra and adjacent intervertebral articular disc. 1 Fibrous ring; 2 Pulp nucleus; 3 Ventral longitudinal ligament; 4 Intercapital ligament; 5 Dorsal longitudinal ligament.

IV. MYOLOGY I

IV.1. The different **COLOURS**, not only of comparative muscle types but also of **individual skeletal muscles** may be studied on a dressed broiler. The 'white' largely anaerobic muscle consists mainly of white fiber types which are larger than the red and contract faster but also become fatigued faster. White muscle functions in the support and maintenance of stance of body as opposed to red muscle which is associated with movement. Red muscle is aerobic and contains high concentrations of myoglobin in its smaller red fiber types. As indicated above, red muscle has a slower contraction and rate of fatigue than white muscle and also has a better blood supply.

IV.2. For the most part **MYOGENESIS** proceeds from myotomes of the primitive vertebrae (consult an embryology text). From these, primary myoblasts differentiate in two different directions. A greater number develop further to secondary myoblasts, which, together with analogous cells form a long cell series, the myotube. This contains cell nuclei arranged centrally in series without a lumen, hence the term myotube is a misleading one. Inside the myotube, plasmalemmata between the secondary myoblasts disappear, resulting in a very long multinuclear cell, the syncytium. A small number of (primary) myoblasts become poorly differentiated satellite cells lying between the muscle cells and the external basal lamina (glycocalyx).

The significance of the satellite cells becomes clear in **muscle regeneration**. Providing the basal lamina remains intact in muscle cell damage, cell debris is removed by macrophages. The discontinuity between the defective ends of the muscle cells is closed over and continuity re-established by further differentiation of satellite cells and the synthesis of myofibrils. In larger defects such as tears of muscle fiber bundles, healing occurs with the production of scar tissue.

IV.3. **CONTRACTION OF SMOOTH MUSCLE** can also be caused by stretch stimuli and spontaneous discharge at the neuromuscular junction, as, for example, in lymph vessels and in the intestinal wall.

IV.4. The **QUALITY** and **CONSISTENCY OF TENDON COMPONENTS** and **MUSCLE CONNECTIVE TISSUE** determine the quality of meat. Another important factor is its maturation. Specific myofibrils relax and meat becomes tender due to biochemical changes particularly in glycogen.

V. MYOLOGY II

V.1. **MYASTHENIA GRAVIS**, as a serious muscle weakness, is characterized by rapid muscle fatigue (see also 16.6.). It can be either congenital or acquired. Antibodies are formed against the acetylcholine receptors, motor end-plates are defective and neuromuscular excitability is disturbed. Very often the striated muscle of the esophagus is affected leading to dilation (megaesophagus).

VI. NERVOUS SYSTEM

VI.1. The **NEUROGLIA** surrounds the nerve cell bodies and processes and separates these from blood vessels and their surrounding connective tissue. Peripheral glial cells, namely lemmocytes or Schwann cells, and satellite cells differ from central glial cells such as ependymal cells, astrocytes, oligodendrocytes and microgliocytes. Glial cells form a supporting structure, are significant for the nutrition of nerve cells, and maintain an optimal perineural milieu by eliminating neurotransmitter substances in the central nervous system. They also provide a protection for nerves by synthesizing the myelin sheaths (significance for nerve regeneration see VI.5.). Benign or malignant tumors (glioma) can have glial cells as their starting point. The neuroglia also forms the relatively large **neurophil** which is common to myelinated and non-myelinated nerve fibers and the continuously narrow intercellular spaces between them. In the grey substance of the central nervous system the neurophil separates the cell bodies from one another.

VI.2. The **FORMATION OF THE MYELIN SHEATH** commences prenatally and continues postnatally until maturation, an example of an important functional unification developing quicker than a less important one. The lamellar-shaped myelin sheath surrounds the axon, acting to a certain extent, as an insulator. In conjunction with the nodes of Ranvier, this permits a saltatory impulse transfer, that is the 'jumping across' of an impulse from node to node. Because of this, the velocity of conduction of the impulse is faster than that of the unmyelinated fiber. As a rule of thumb, the thicker the axon (dendrite), the thicker the myelin sheath, and the longer the internodal space, the faster the conduction of the impulse.

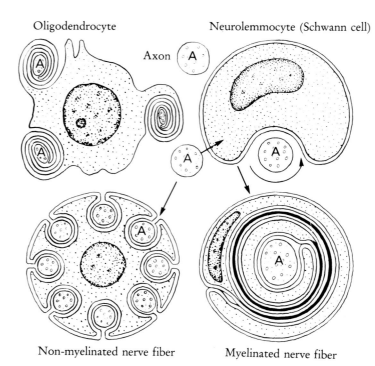

Oligodendrocyte   Neurolemmocyte (Schwann cell)

Axon

Non-myelinated nerve fiber   Myelinated nerve fiber

VI.3. The **NODES OF RANVIER** or nodes of the neurofibers lie at the breaks or intervals of the myelin sheath where the axon (dendrite) is slightly thickened. The internode or internodal space lies between two such nodes. Because myelin is absent the nodes of Ranvier are significant as the first sites at which local anaesthetic acts.

VI.4. **ENDO-, PERI- AND EPINEURIUM** protect the nerves from overstretching and anchor them to their surroundings by paraneurium.

VI.5. After transection or injury, **NERVE REGENERATION**, that is, the recovery of its capacity to conduct impulses, occurs initially on the proximal segment of the nerve concerned. The distal segment undergoes **degeneration**. Proximally, only a few internodal spaces degenerate, the nerve cell bodies remaining intact in most cases. The axons of the distal segment degenerate distally as far as their terminations. Proximally, regeneration begins with budding of the cells of Schwann. Likewise budding of the cells of Schwann of the distal segment occur and a defective bridging continuum is formed. The cells of Schwann, arranged in a bandlike series, one behind the other, act as a 'conduction tube' for the axonal buds of the proximal segment. These regenerate to the effector organ, very slowly at approximately 3 mm per day. The prerequisite for nerve regeneration is a proportionally short distance between proximal and distal segments of the severed nerve. Regeneration can be promoted by approximating the ends of such segments. Finally a new myelin sheath is formed. (In contrast, in a **neurectomy** to eliminate pain, an adequately long piece of nerve is ablated in order to prevent a natural union of the segments of the severed nerve. This procedure is debatable and seldom used now for treating sports injuries in horses.)

VII. VASCULAR SYSTEM

VII.1. **VEINS** are distinguished from arteries on the basis of numerous morphological characteristics. Generally they have a wider lumen and are more numerous. Several collateral veins, spoken of as concomitant veins, can accompany an artery in a common blood vessel-nerve pathway. The wide-lumened veins of a few organs such as liver, lung, spleen and skin, function as sites of blood storage. Thus heart stress, for example, is reduced due to an economic maintenance of venous function.

Venous return is promoted by the following different factors:

1. The heart has a suction action due to the change in level of its valves. Thus blood is sucked from the ostia of both venae cavae and the right atrium (see appendix 18.12.).
2. Parts of the animal body such as articulations and digital pads are compressed repeatedly due to movement, and thus exercise pressure on associated veins. Similarly with limited contraction of muscle, blood flow is propelled forward and this applies in particular to veins lying within muscle (deep) fascia. Furthermore, venous return is assisted by valves within the veins.
3. The arterial pulse wave with its rhythmic pressure variation can be transferred to concomitant veins.
4. With negative pressure changes within the pleural cavity during respiration, venous blood is sucked into the thorax. (When veins near the heart are injured because of negative pressure air can be aspirated with the danger of formation of emboli.)

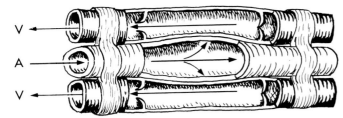

V
A
V

Effect of arterial pulsation on venous return

VIII. LYMPHATIC SYSTEM

VIII.1. **OPEN JUNCTIONS** are the morphological basis for **direct** or **indirect** lymphography (radiographic demonstration of the lymphatic system). With indirect lymphography, a dye is injected into the interstitial tissues. Due to the presence of open junctions, the dye particles (diameter approx. 0.1 um) reach the lymph capillaries labelling them, proximal lymph vessels and nodes. In so doing, the otherwise inapparent lymph vessels are located more readily in order to inject them with radiopaque material. With direct lymphography, a specific radiopaque substance is injected into a lymphatic vessel, the lymphatic pathways are identified radiographically, and any eventual pathological interruptions to lymph flow. For example, one may detect changes to lymph nodes and obliterating lymphangiopathy following degenerative-dystrophic endothelial changes.

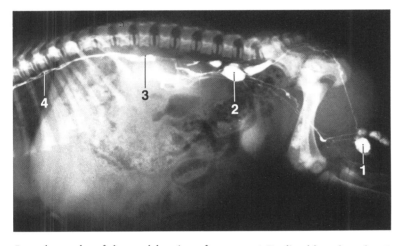

Lymphography of the caudal region of a puppy. 1 Popliteal lymph nodes; 2 Medial iliac lymph node; 3 Cisterna chyli; 4 Thoracic duct.

VIII.2. **LYMPH NODES** are of medical significance in that they are affected by **inflammation** within their drainage areas. Consequently, such areas should be known. Inflamed lymph nodes are enlarged, painful and hot. In **clinical diagnosis** it is important to realize that such nodes are more readily palpated, whereas normally only some superficial lymph nodes are palpated distinctly. Malignant tumor cells contained in the lymph flow are filtrated in the lymph nodes and are a focus for **metastases**. Because foreign material, including bacteria, is accumulated in lymph nodes, strategic nodes are incised routinely in **meat inspection** and consistency, colour and presence or otherwise of calcification checked.

VIII.3. Depending on their size **LYMPHOCYTES** are subdivided into small, medium-sized and large cell-types. B- and T-lymphocytes can be differentiated on functional grounds, but only by employing special methods can morphological differences be distinguished. **B-lymphocytes** stem from the bone marrow, are induced to differentiate in the avian cloacal bursa or equivalent organs, and discharge into the blood stream and lymph nodes (**B** stands for **bursa cloacae**). The **equivalent of the bursa** in mammals is not exactly known or defined. According to several authors the bone marrow is the equivalent and therefore, in an incorrect manner, **B** can refer to bone marrow. The B-lymphocytes are responsible for **humoral immunity.** This depends on the capability of **B**-lymphocytes, converted to plasma cells, to produce immunoglobulins (antibodies). The **T-lymphocytes** are produced in the bone marrow, differentiate further in the thymus (**T** stands for thymus), and are discharged into the thymus-dependent paracortex of lymph nodes and spleen. From the functional viewpoint they are differentiated into helper, killer, suppressor and memory cells. The short-lived **killer-cells** (living only a few days) are responsible for cellular immunity. They kill cells (bacteria, and transplant cells causing transplant rejection) and are themselves killed in the process, producing pus and suppuration. The **helper cells** regulate the transformation of B-lymphocytes to plasma cells and therefore control immunity. **Suppressor cells** check helper and killer cells and prevent excessive

immune reactions. The long-living **memory cells** (living more than a year) are responsible for the immune system reacting quickly with a secondary immune response, to a specific antigen confronting the immune system repeatedly.

2. SKIN

2.1. **CELL PROLIFERATION** and **KERATINIZATION** are governed by vitamin A and an 'epidermal growth factor'. Very fine (keratinogenic) filaments form the initial material within the cell allowing keratinization to take place. The process involves mainly three phases: 1. Intracellular keratinogenic filaments (tonofilaments) adhere to one another and, remaining inside the cell undergoing keratinization, are coated with a homogeneous protein. 2. The plasmalemma thickens. 3. Cementosomes (membrane coating vesicles or granula lamellosa), formed possibly from the Golgi apparatus, are discharged into the intercellular spaces by the process of exocytosis. Horn can be likened to a clinker wall, the horn cells being the bricks (or better, maybe units of softer ebonite), and the cementum or membrane coating material, the mortar or adhesive. Differences in horn quality are basically due to the quality of both the horn cells and the adhesive substance.

2.2. **AGING OF SKIN** is conditioned by atrophication of subcutaneous fat cushions, water loss, loss of elastic fibers and loss of tensile strength in those fibers remaining, as well as a decrease in skin thickness, gland and pigment cell numbers.

4. VERTEBRAL COLUMN

4.1. Excessive **LORDOSIS** and **KYPHOSIS** are pathological dorsoventral curvatures of the vertebral column. Lateral curvature is known as **SCOLIOSIS**. Pathological curvature of the spine can be caused by hemivertebrae and in rare cases, compression of the spinal cord can occur through a narrowing of the vertebral canal.

4.2. The **VERTEBRAL COLUMN** has many functions including protection, support and movement. The central nervous system requires **protection** since it has only a very limited healing capability after trauma. In the static position in quadrupeds, **support** and **weight-bearing** are comparable with similar functions in bridge construction. The **possibility of movement** between individual vertebrae is only small. The sum of the increments of movement, however, make a noticeable range of movements possible, particularly in dogs and cats. The importance of blood formation in the bodies of the vertebrae should be mentioned. The vertebral column, spinal cord and spinal nerves are arranged segmentally with the spinal nerves leaving the vertebral canal between analogous vertebrae. **Segmentation** can be restricted however through lack of development of two or more vertebrae to give a **block-formation.** (Sacralization of the last lumbar vertebra. A physiological fusion or synostosis occurs at the sacrum where the vertebral discs ossify.) **Spina bifida** is another congenital malformation of the vertebral column where the arches of the vertebrae do not close dorsally. This is often accompanied by an evulsion of a portion of the spinal cord to produce a **myelocele.** A **meningocele** involves only meninges.

4.3. **TERMINOLOGY** associated with **vertebrae** is derived from Latin in anatomical word usage and Greek in clinical usage. Thus the anatomical term vertebra, has, as its clinical equivalent, the word spondylos. For example, spondylosis is a degenerative disease of the intervertebral symphysis and adjacent vertebral bodies. Spondyloarthrosis is a degenerative disease of the intervertebral articulation.

4.4. The **BODY OF THE VERTEBRA** ossifies from three ossification centers, one diaphyseal and two epiphyseal, equivalent to the mode of ossification in long bones (see figure under VIII.1.). Traumatic epiphysiolysis may also occur in the vertebral body usually correlated with paraplegia. In this instance the prognosis is very unfavourable.

4.5. **VERTEBRAL MALFORMATIONS** include hemivertebra (mostly with the apex directed ventrally) and non-segmentation between vertebrae with ossification of the intervertebral discs.

4.6. The **VERTEBRAL ARCH** consists of a pedicle and a dorsal lamina. The lamina is removed surgically to provide access to the spinal cord and intervertebral disc and also to permit decompression of the cord (hemilaminectomy and laminectomy). The mamilloarticular processes are removed while intervertebral foramina, roots of spinal nerves and afferent and efferent blood vessels are preserved.

4.7. **ARTICULATIONS OF THE VERTEBRAL COLUMN** are classified as cartilaginous vertebral symphyses associated with intervertebral discs between the vertebral bodies, and synovial articulations between the vertebral processes. A vertebral disc is absent and a synovial articulation present between atlas and axis. Degenerative changes in the vertebral symphyses are known as spondyloses and

88

begin with osseous deposits on the ventral sides of the vertebral bodies and intervertebral discs. The resulting bridge-shaped exostoses can lead to ankylosis or stiffening of the vertebral column including the intervertebral articulations and discs.

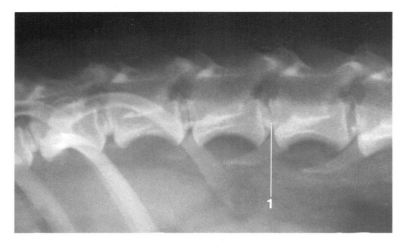

Calcified pulp nucleus (1, above) and spondylosis with formation of bridge shaped exostoses (2, below).

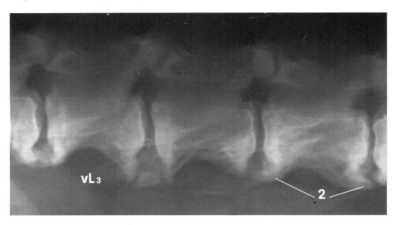

4.8. A subluxation of the atlantoaxial articulation can result from a congenital lack of the **DENS** of the **AXIS**, its postnatal detachment or fracture. Laminectomy is undertaken to stabilize the articulation and decompress the spinal cord, but usually damage to the dens is fatal. (This is the site of a broken neck in man.)

5. NECK AND THORAX

5.1. **PLATYSMA** is of clinical interest because wounds having a direction perpendicular to that of muscle fibers open wider than those where wound direction is parallel. In the first instance cutaneous muscle pulls the edges of the wound apart, in the second, muscle contraction brings the wound edges together.

6. VISCERA OF THE NECK

6.1. During surgery, entry can be made midventrally between the **MM. STERNOHYOIDEI** to reach cervical viscera or intervertebral discs of the cervical part of the vertebral column.

6.2. The **EXTERNAL JUGULAR VEIN** is a site of venipuncture during blood transfusion or when taking quantities of blood for storage. Only in exceptional cases is it used in infusion therapy. The vein is susceptible to trauma on account of its superficial position.

6.3. Because of their confluence at the venous angle, both the **JUGULAR TRUNK** and **THORACIC DUCT** are suitable for obtaining lymph at the site, using a fine needle. Such lymph is usually for experimental examination. Tears in the thoracic duct in the vicinity of its opening lead to chylothorax, the accumulation of lymph in the pleural cavity. Efforts can be made to ligate the duct, cream being fed before surgery to facilitate its indentification. The milky content of the thoracic duct is due to the resorption of long chain fatty acids into the intestinal lymphatic system.

6.4. Besides smooth muscle in its wall, the **ESOPHAGUS** also contains striated muscle innervated by the oesophageal ramus of the vagus nerve. It has been suggested that interruptions to the innervation are the origin of abnormal dilation of the esophagus. Stenosis (narrow-

ing) or obturation (occlusion) can be caused by tumors or foreign bodies or by developmental disorders due to a persistent right aorta (see 18.3). Foreign bodies generally lodge at sites where the esophagus is normally narrow namely at the cranial thoracic aperture or at the esophageal hiatus or on a level with the base of the heart. The relationship of esophagus to trachea is taken into account during surgery of the cervical esophagus. At the cranial aperture of the thorax, the esophagus lies on the left side dorsal to the trachea and this is the surgical site of preference (cupola of pleura - see 16.3.).

6.5. Reduction in tension of the **TRACHEA** produced by fibroelastic tissue and tracheal rings can produce collapse of the trachea. This phenomenon, also referred to as scabbard trachea appears most frequently in dwarfed dog breeds. Radiographically, tracheal rings so affected are flattened dorsoventrally and the trachea takes on a scabbard form, the origin of the syndrome of tracheal collapse. Tracheotomy may be undertaken in stenosis of the upper respiratory tract. After the mm. sternohyoidei and sternothyroidei have been removed to their respective sides, a tracheal entry is made between the tracheal rings on the ventromedian aspect of the trachea. A tracheal tube (with or without a T-piece) is inserted as an artificial respiratory route.

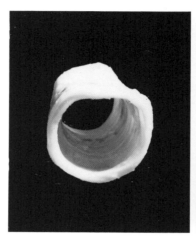

Transverse section of a normal trachea (left) and of a scabbard trachea of a Yorkshire Terrier (right).

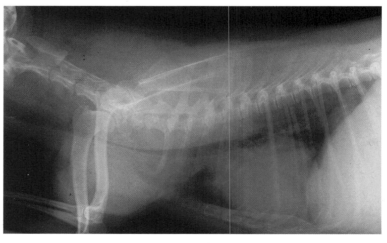

Radiograph of a scabbard trachea at the level of the cranial thoracic aperture.

6.6. The **THYROID GLAND** regulates metabolic processes due to the hormones thyroxine and tri-iodothyronine. As an endocrine gland discharging its hormone output into the blood stream, it has a relatively good blood supply. An abnormal enlargement of the gland, known as struma or goitre can accompany either hyper- or hypofunction. The condition is widespread in iodine deficient areas. The main parts of the thyroid gland develop dorsally from the endoderm of the root of the tongue, whereas parafollicular or C-cells stem from the ultimobranchial bodies. These arise primarily from the neural crest. Calcitonin, the hormone of the C-cells plays a part in calcium metabolism. The thyroglossal duct takes origin from the tongue and extends to the junction of larynx and trachea. The distal end of the duct develops into the right and left lobes of the thyroid.

The long rostral portion of the duct normally regresses. Occasionally, glandular remnants differentiate into accessory thyroid glands which may be present in all neighbouring tissues in contact with the embryonic primordium of the gland (tongue, larynx, trachea, and mediastinum at the base of heart). Changes in the thyroid gland and ongoing functional disturbances are also seen in the accessory gland tissue.

Remnants of the thyroglossal duct tend to produce cystic degeneration ventral to the tongue in the midline. In surgery of the thyroid gland, the recurrent laryngeal nerve is taken into account and preserved.

6.7. **PARATHYROID GLANDS** are essential to life. They should not be removed if at all possible, during surgery to the thyroid gland, otherwise hypocalcemic tetany can result. Where removal of the parathyroid glands is unavoidable because of non-detection within abnormal thyroid gland tissue, replacement therapy may be undertaken a few days postoperatively. Generally self-regulation results due to the presence of accessory parathyroid tissue. Normally parathyroid glands regulate calcium and phosphate metabolism through the output of parahormone. Parahormone excess causes **fibrous osteodystrophy** with increased decomposition of bone by osteoclasts and its replacement with connective tissue. In turn calcium from the osseous tissue is discharged into the blood. Parathyroid glands, in common with the thymus, develop from the third and fourth branchial arches. Generally, the external parathyroid gland (III) is situated craniolaterally on the lateral surface of the thyroid gland and the internal parathyroid gland (IV) on the caudomedial surface. In common with thymic tissue, parathyroids can be displaced to an abnormal position further caudally and within the thoracic cavity.

7. SKELETON OF THE THORACIC LIMB

7.1. During surgery, the **ACROMION** can be detached temporarily from its base by osteotomy, to create lateral access to the neck of the scapula. In so doing its relationship to the position of the **suprascapular nerve** is taken into account. At the end of surgery the acromion is reattached to the spine of the scapula, using orthopedic wire.

7.2. In the growing phase, the **HUMERUS** possesses an apophysis (greater and lesser tubercula) and an epiphysis (head of humerus) proximally. The associated centers of ossification coalesce in the middle of the growing phase and are separated from the diaphysis by a roof-like epiphyseal cartilage. Distally, the **trochlea** and the **capitulum of the humerus** form their own centers of ossification which fuse to form a common epiphysis after approximately two months. The distal epiphyseal cartilage has a distinctly limited growth-potential as compared to the proximal. In the main, the humerus grows in a proximal direction and this is clinically significant since disturbances to the proximal epiphyseal cartilage are more serious than those to the distal.

7.3. Of the bones of the **ANTEBRACHIUM**, or forearm, the **radius** has a proximal epiphysis, the epiphyseal cartilage of which has a very limited growth-potential. By contrast, the growth-potential of the distal epiphyseal cartilage is marked. The cartilage forms the boundary between the radial diaphysis and an undivided distal epiphysis. Proximally, the **ulna** has only one apophysis namely the olecranon, for insertion of the m. triceps brachii.

7.4. **FRACTURES OF THE OLECRANON** are due frequently to trauma and are often accompanied by a luxation of the elbow. Since the m. triceps brachii inserting on the olecranon tuber exerts very strong traction; 'open' reposition and very stable fixation of the fractured segments is necessary after fracture. An apophysiolysis seldom occurs.

7.5. In the German Shepherd and Saint Bernard, the **ANCONEAL PROCESS** is said to possess its own center of ossification and apophyseal cartilage only rarely. Its detachment, referred to as **isolated anconeal process,** can occur up to the age of nine months and results in lameness. Incongruency of the articular surfaces of the joint predispose to avulsion fractures. Some authors refer to changes occurring to the articulation as dysplasia of the elbow.

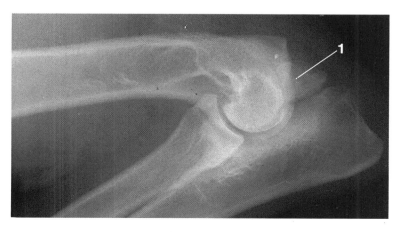

Isolated anconeal process (1) of a German Shepherd.

7.6. The **(DISTAL) ULNAR EPIPHYSEAL CARTILAGE** has a conical configuration and possesses a very marked growth-potential. Both **bones of the antebrachium** grow mainly distally. Regarding the thoracic limb one can also say that its largest long bones grow away from the elbow. Radius and ulna are tightly connected to each other distally by the radioulnar ligament and both must grow synchronously. Asynchronous disturbances lead to typical changes such as curvature of the radius in the Basset and Dachshund (distractio cubiti). One therapeutic possibility would be to transect the radioulnar ligament.

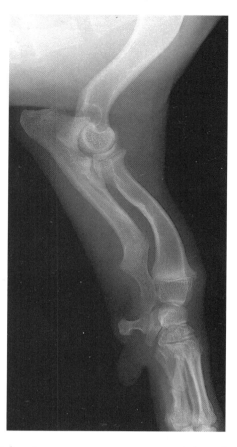

Distractio cubiti with step-formation of the elbow joint.

Fractures in the antebrachial region can cause premature epiphyseal closure and subsequent deformation of radius and ulna. Following fractures near the epiphyseal cartilage, deformations can also occur through **accelerated longitudinal bone growth.**

Inflammation limited to the region of fracture leads to hyperemia and better nourishment of the epiphyseal cartilage, which, as a result, proliferates with great intensity. Fractures of the medial or lateral styloid processes can cause joint instability in adult dogs.

8. NERVES OF THE THORACIC LIMB

8.1. The **NERVES OF THE BRACHIAL PLEXUS** (from nCv 5 or nCv 6 to nTv 2) can become paralysed through pressure, contusion or overstretching. In particular the 'short' nerves such as the axillary n. are in danger due to the short course between the exit of these nerves at the 'scalenus gap' and their entrance into the effector muscles. Rupture of the brachial plexus can occur with severe trauma (as in car accidents) where the roots of the nerves in the vicinity of the spinal cord are torn away. The nerves are irreparable because the resultant distal nerve segments retract and amputation neuromas are formed at the ends of the proximal segments. One may also detect Horner's syndrome (see 40.1.) in such cases. Due to tearing of the nerve roots and their connection with the sympathetic trunk the cervicothoracic ganglion may also be affected.

8.2. Following trauma, the **RADIAL NERVE** becomes damaged due to its course over the lateral supracondylar crest of the humerus. Radial paralysis accompanied by a functional deficit of the extensor muscles of carpus and digits is a sequel. The nerve is also endangered by fracture of the body of the humerus followed by excessive callus formation.

9. VEINS AND MUSCLES OF THE THORACIC LIMB

9.1. The **CEPHALIC VEIN** is accessible for venipuncture at the mid-length of the antebrachium. It is a convenient site for taking blood and giving intravenous injections.

9.2. The tendons of insertion of the **MM. SUPRASPINATUS, INFRASPINATUS, SUBSCAPULARIS** and **TERES MINOR** stabilize the shoulder joint and are collectively known as 'sleeve' muscles.

9.3. In cases of chronic inflammation (hygroma), the **SUBCUTANEOUS BURSA OF THE OLECRANON** is removed, following a skin incision.

9.4. The **M. ANCONEUS** covers the slight outpocketing of the caudo-proximal part of the elbow joint between the olecranon and the lateral supracondylar crest. With the joint fixed, the capsule of the joint is punctured there during intraarticular injections, the needle passing through the m. anconeus.

10. ANTEBRACHIAL MUSCLES

10.1. The **TENDONS OF THE EXTENSOR MUSCLES,** with the exception of the m. extensor carpi ulnaris, are surrounded by synovial tendon sheaths, at the level of the carpus. Sheath inflammation or tendovaginitis is often a sequel to bites, puncture wounds or tendon strain.

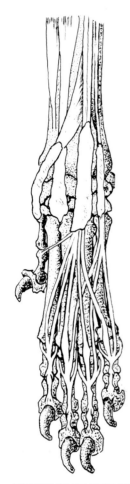

Tendon sheaths on the dorsal surface of the manus.

11. EFFERENT LYMPH FLOW FROM THE THORACIC LIMB

11.1. **SUPERFICIAL LYMPHATIC VESSELS** go to the superficial cervical lnn. and thus circumvent the axillary lnn. After removal of the former lymph nodes, a surgical procedure rarely undertaken, lymph flow from the thoracic limb is maintained but at a reduced quantity.

12. MUSCLES OF THE VERTEBRAL COLUMN

12.1. The **DORSAL VERTEBRAL MUSCLES** are pushed to the side to create surgical access to the vertebral column dorsally and midsagitally. Simultaneously the m. multifidus is detached from its insertions. This permits **dorsal hemilaminectomy** and facilitates the curettage of calcified material from the vertebral disc which had penetrated into the vertebral canal (see 4.5.).

13. RESPIRATORY MUSCLES

13.1. **RESPIRATORY MUSCLES** are transected during thoracotomy or opening of the thorax. Predetermined intercostal spaces are selected as sites of access to the different structures. For example, the eighth or ninth intercostal space is used to locate the diaphragm. The opening of the pleural cavity is made under general anasthesia, intubation and positive pressure respiration. When making surgical incisions through the intercostal spaces, intercostal vessels and nerves are also to be taken into account. Vein, artery and nerve lie craniocaudally in that order (see illustration p. 19A) and run along the caudal border of each rib. The ventral ends of the last ribs are also accompanied by an appropriate artery, vein, and nerve along their cranial borders. Therefore the surgical incision is made along the intercostal space at its midwidth and through the intercostal muscles. To permit greater access to the

thorax, rib resections may be made after the portions concerned are removed from their periosteal covering. To close the incision the thoracic wall is sutured layer by layer, avoiding puncture of intercostal vessels during the procedure.

13.2. **DIAPHRAGMATIC HERNIAE** can be congenital or acquired due to trauma. They are accompanied by displacement of abdominal viscera, mainly portions of small intestine, into the pleural cavity. During inspiration, the expansion of the lung is restricted to a greater or lesser extent by encroaching viscera from the abdominal cavity, which is sucked into the pleural cavity by negative pressure.

Congenital diaphragmatic hernia has its aperture generally at the esophageal hiatus where the esophagus is only loosely attached. The hernial sac which is regularly present is often the enlarged serosal cavity of the mediastinum (see p. 16). Where incomplete congenital closing of the ventral parts of the diaphragmatic insertion and pericardium occurs, abdominal viscera can also penetrate through these defects. A **diaphragmatic hernia acquired traumatically,** accompanies a related separation of a section of diaphragm and has no hernial sac. It causes a serious clinical picture. **Paralysis of phrenic nerves,** for example with high dosage of epidural anesthesia, leads to relaxation of the diaphragm, and in bilateral paralysis, to serious dyspnea. Hiccup is the result of a sudden closure of the rima glottidis due to a short spasm of the diaphragm following phrenic stimulation. Besides its large motor component, the **phrenic nerve** also possesses sensory fibers supplying diaphragmatic pleura and peritoneum.

14. THE BODY WALL

14.1. The **MAMMARY GLAND** is a **skin gland,** more specifically an extensively modified apocrine sweat gland or scent gland. Mammary tissue is embedded in a septum of superficial fascia of the trunk and to isolate it the surgeon may orientate himself without difficulty by locating the underlying deep fascia of the trunk. Only in the region of the inguinal and cranial abdominal mammae do blood vessels approach the mammae from their dorsal sides through the deep fascia. In a majority of bitches one has to take into account the **vaginal process** dorsal to the inguinal mamma. Its injury establishes an opening into the peritoneal cavity (see p. 15). Mammary **blood supply** from cranial and caudal epigastric vessels must be taken into account when removing tumors. Among other considerations hemorrhage from both sets of vessels is in opposite directions. **Efferent lymphatic vessels** and regional lymph nodes, namely axillary and inguinal superficial lymph nodes, determine the route taken by metastases in cases of mammary cancer. The first metastases appear predominantly in the regional lymph nodes. Efferent lymphatic vessels from the thoracic mammae also perforate the thoracic wall and reach the mediastinal and tracheo-

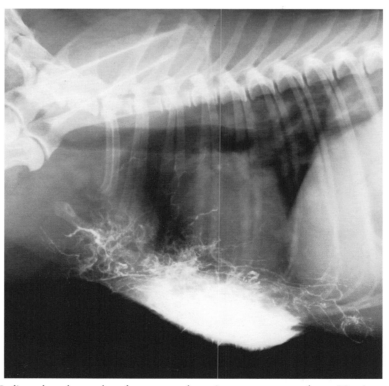

Indirect lymphography of tumorous thoracic mammary complexes 20 minutes after injection of radio-contrast material (Jotasul Schering). Lymph drainage cranially is in the direction of the axillary lymph nodes and caudally towards the superficial inguinal nodes. Perforating lymph vessels enter the thoracic cavity through the wall of thorax. (To confirm the anatomic details within the thorax, see radiograph upper page 94; radiograph by Professor Dr. D. Berens von Rautenfeld.)

bronchial lymph nodes. For this reason metastases appear frequently in the mediastinum and lung. In multiple malignant tumors one removes the diseased milk line of the affected side and the regional lymph nodes.

14.2. HERNIA OF THE ABDOMINAL WALL

HERNIA OF THE ABDOMINAL WALL leads to subcutaneous prolapse of viscera through the wall. This can occur at any site due to trauma or to predisposed weak sites such as the umbilicus in the young animal. In the fetus this affords passage to umbilical vessels and the urachus, the embryonic connection between urinary bladder and allantois. At an early stage, in fact, the umbilicus does serve as a temporary physiological umbilical hernia.

14.3. ABDOMINAL MUSCLES

ABDOMINAL MUSCLES are designed to diminish abdominal volume in an active manner. In so doing, **intraabdominal pressure** is raised in conjunction with contraction of the diaphragm. Indeed, in defecation, micturition, vomition and parturition, straight and transverse muscles take part in direct compression of the abdomen. Oblique abdominal muscles participate in oblique compression. In **difficult respiration** abdominal muscles and diaphragm function as antagonists. Due to contraction of abdominal musculature and relaxation of the diaphragm, the diaphragmatic cupola or dome is pushed cranially together with displaced abdominal viscera. As a result the pleural cavity, assisted by exspiration, is reduced.

15. THE INGUINAL REGION

15.1. INGUINAL HERNIA occurs frequently in males of the domestic species and also in the bitch. One can differentiate two types of hernia, the direct which is always acquired and the indirect which is either acquired of congenital. In **indirect inguinal hernia** the vaginal ring opening from the peritoneal cavity into the vaginal process acts as a hernial entrance and the process a hernial sac. And just as the vaginal process penetrates the abdominal wall and the inguinal space obliquely through the internal and external inguinal rings so also does the content of the hernia. Since the bitch very often has a vaginal process, that is, a potential natural hernial entrance, it too is predisposed to indirect hernia. Hernias of this type occur even more frequently in the bitch, because in the male dog, the vaginal process is occupied largely by the relatively hard spermatic cord. In the bitch, however, adipose tissue and the round ligament of the uterus are easily displaced by the herniated content advancing in the vaginal process. In **direct inguinal hernia** the hernial opening generally lies medial and adjacent to the vaginal ring and the herniated content goes directly, that is, by a short route, through the body wall. As a result, this is regularly lacerated. Where the herniated content reaches the scrotum, the condition is spoken of clinically as a scrotal hernia. After the reduction of the hernia and the return of content to the abdominal cavity, the external

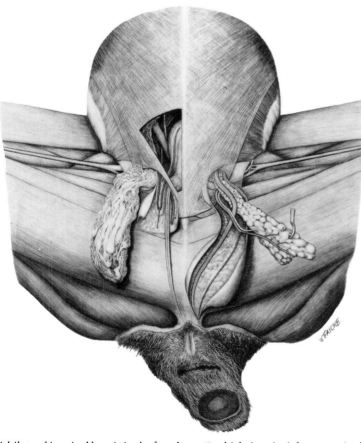

A bilateral inguinal hernia in the female cat (!) which, in principle, occurs in the same manner as in the bitch. Left side of illustration: direct hernia. Right side of illustration: indirect hernia into the opened vaginal process. (A window is placed in the body wall.)

inguinal ring is button sutured. Due to the castration required by the procedure, the vaginal process and its content are displaced slightly distal to the inguinal space.

15.2. In rare cases, the FEMORAL RING is the site of femoral hernia. Peritoneum and transverse fascia are everted as a hernial sac and accompanied by secondary laceration.

16. THORACIC CAVITY

16.1. The **PLEURA** lines a right and a left pleural cavity. In the live animal the **pleural cavity** is purely the width of a capillary space occupied by a serous liquid film. As a result adjacent parts of parietal and visceral pleura adhere to each other across the serous space without losing their capability to slide during respiratory movements of the lung. In formalized specimens the pleural cavity is enlarged artificially to a considerable extent. In vivo the cavity may contain varying quantities of different abnormal products such as air (**pneumothorax**), increased watery fluid (**hydrothorax**), fluid containing lymph or chyle (**chylothorax**) due to rupture of the thoracic duct, or fluid containing blood/purulent fluid (**hemothorax**). **Pneumothorax** can be either uni- or bilateral since the two pleural sacs are not completely separate in all cases. The elastic fibers of the lung draw it away from the thoracic wall towards the region of the hilus, and lung outline is lost radiographically. The mediastinum and the heart contained within it, are displaced towards the normal side. Furthermore, the heart, being attached to the aorta dorsally is also displaced towards the hilus of the lung.

With the removal of the apex of the heart from the sternum, the phrenicopericardiac ligament becomes tense and may be detected radio-

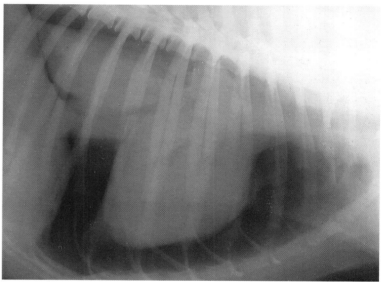

Pneumothorax

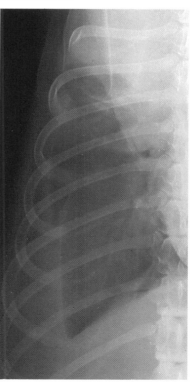

Hydrothorax (ventrodorsal radiograph). Due to presence of fluid, lung is pushed away from thoracic wall.

graphically. Lung volume on the normal side becomes restricted due to the displacement of the mediastinum and dyspnea becomes more pronounced particularly in ventilation pneumothorax (see below). Ongoing abnormal traction irritates the autonomic system and may cause disturbances such as circulatory collapse. One differentiates an **external pneumothorax** when air flows into the pleural cavity through an external wound (close it as quickly as possible!) and an **internal pneumothorax** when, for example, the flow is through a laceration of the lung. If the air circulates purely within the mediastinum, the condition is spoken of as a **pneumomediastinum.** This is also diagnosed radiographically since, air being a contrast medium, the intramediastinal organs and structures are each defined distinctly. A **ventilation pneumothorax** occurs when air is sucked through a wound during inspiration, and cannot escape in the exspiratory phase due to displacement or obstruction of the wound. An abnormal increase in fluid within the pleural cavity (**hydrothorax, hemothorax**) is detected radiographically, and the change in level of the fluid surface is also recognized between the standing and the laterally recumbent animal.

16.2. **SENSORY INNERVATION** of the parietal pleura is derived from the sensory part of the intercostal and phrenic nerves, making it susceptible to pain. Innervation of the mediastinum arises from the viscerosensory part of the vagus nerve. It is debated whether the pulmonary pleura has no innervation or whether only long-grade sensory nerves are involved.

16.3. During surgery at the **CRANIAL THORACIC APERTURE** one has to take into account the **pleural cupola.** Unintentional opening can cause pneumothorax.

16.4. The **PLEURAL RECESSES** (for example the costodiaphragmatic recess; not however, the mediastinal recess housing the accessory lobe of the right lung) provide a reserve spatial capacity not used in normal respiration. With deeper inspiration, particularly in yawning and in pathological lung enlargement such as emphysema, the caudal borders of the lung extend into the recesses at least partially. In rare cases of lung emphysema the caudal border of the lung field can reach almost to the line of insertion of the diaphragm. Normally in large dog breeds a four to six centimeter space occurs between the two.

16.5. The **PERICARDIAL CAVITY** is filled with only a small amount of serous fluid. A pathological increase and alteration of pericardial content is apparent radiographically in pneumopericardium (rarely in pneumothorax with injury to pericardium), hydropericardium (by disturbance of lymph drainage), and hemopericardium (when blood vessels rupture as in hemangiosarcoma with fatal pericardial tamponade).

16.6. The **THYMUS** is a primary lymphatic organ which, in the dog, consists of a thoracic portion only, having a right and a left lobe. Involution accompanied by fatty degeneration is never complete, with remnants persisting even into old age. Predominant opinion suggests that fetal stem cells migrate from bone marrow to the thymus where daughter cells differentiate into **thymocytes.** After differentiating into T-lymphocytes these colonize the **secondary lymphatic organs** such as lymph nodes and spleen and are responsible for cellular immunity. (They are efficient in transplant rejection.) **Atrophy of the thymus,** with disappearance of lymphocytes, can occur in nutritional disturbances, stress and wasting diseases. An abnormally **delayed regression** takes place in endocrine disorders and can be a sequel to early castration. A **thymic hyperplasia** with **overfunction** similar to that in man, also occurs rarely in young dogs with **myasthenia gravis,** a condition resulting in serious reduction of muscle strength. In man **thymectomy** (surgical removal) can effect a recovery.

A thymic hormone, **thymosine,** promotes the differentiation and maturity of T-lymphocytes. A blood-thymus barrier is present in the thymic cortex but its expression is only incomplete in the medulla. The barrier prevents antigens in the blood-stream from reaching the cortex of thymus and causing an antigen-antibody reaction.

17. LUNG

17.1. The **TERMINOLOGY OF THE LUNG** is derived from the Latin 'pulmo' meaning lung (for example pulmonary pleura), or from the Greek 'pneuma' air or breath (for example pneumonia meaning inflammation of the lungs).

17.2. The **LUNG FIELD** and its borders are determined by percussion.

17.3. The **LOBES OF THE LUNG** are further divided into **lung segments,** each being ventilated by a **segmental bronchus** at its center and supplied by an accompanying artery of the bronchoarterial type. The efferent venous supply is at the periphery of the lung segment, running separately to the artery. (In the ox a bronchovascular supply exists with artery, vein and segmental bronchus accompanying one another.) Lung segments can be considered as functional units since

segmental bronchi are not connected with one another. As a result they are diseased individually independent of the surrounding tissue. Radiographically an individual diseased lung segment appears as a local, sharply defined shadow.

17.4. **PULMONARY FUNCTION** is mainly concerned with gaseous exchange between blood and inspired air. Air is conveyed through a non-anastomotic bronchial system to alveoli where gaseous exchange occurs. This constitutes an enormous increase in the internal lung surface. Breathing is based on a change in volume of the thoracic cavity. During **inspiration** this is attained by a slight flattening of the dome of the diaphragm (diaphragmatic breathing) and by craniolateral displacement of the ribs (sternocostal breathing). Due to adhesive forces in the pleural space, the lung must follow the movements of the wall of the thoracic cavity and alveoli expand as a result. In the first instance **exspiration** is effected by perialveolar elastic nets constricting the alveoli passively. Narrowing of bronchi can cause exspiratory dyspnea as occurs in man when spastic muscle contraction leads to bronchial asthma. With inspiratory dyspnea, constriction generally occurs in the region of the upper respiratory tract as in swelling of the laryngeal mucous membrane (so-called edema of the glottis) or in 'scabbard' trachea with tracheal collapse (see 6.5.)

17.5. **PULMONARY BLOOD SUPPLY** is characterized by the separate course taken by arterial and venous branches at the segmental level. Arterial branches accompany the segmental bronchi while venous branches are situated at the periphery of the segment. Radiographically, the vascular pattern of the lung is outlined due to the strong absorption of X-rays by the air contained in the surrounding lung parenchyma. (Air is a contrast medium.) Absorption of rays does not occur in pneumothorax due to thickening of lung tissue more or less free of air.

17.6. The **ABSENCE** or only sparse viscerosensory innervation of lung and bronchial tree is why lung or bronchial tumors in man are recognized only at a late stage. Until then little or no pain occurs.

17.7. The **TRACHEOBRONCHIAL LYMPH NODES** are detected particularly in old urban dogs as the sites of deposition of coal dust due to their slate gray to black discoloration. This condition, known as anthracosis is due to the marked incorporation of coal dust in macrophages.

18. HEART

18.1. The **TERMINOLOGY OF THE HEART** is derived from the Latin 'cor cordis' meaning a heart, (for example basis cordis), or from the Greek 'cardia -ae', (for example epicardium). (The term cardia is also used for the entrance to the stomach.) Terms designating direction such as right atrium relate to the heart itself and not to the overall body.

18.2. A **PART OF THE SURFACE OF THE HEART** situated either directly adjacent to the thorax or indirectly so, due to the intervention of the lung between the two, defines the **field of the heart.** The position of the heart is ascertained by **percussion.** The right middle finger taps the left one placed on the body surface and where the heart is directly adjacent to the thorax a suppressed percussion sound (absolute cardiac dullness) is produced. Where the lung intervenes between them, a distinctly louder percussion sound (relative cardiac dullness) is detected. On laterolateral radiographs the heart lies with the apical end of the border of the right ventricle on the sternum. In a more oblique position of the longitudinal axis of the heart, the right heart lies more to the right and cranially, the left heart more to the left and caudally. On exspiration the latter is next to the diaphragm. The tips of both auricles are evident towards the left. In abnormal cases, the heart is displaced out of the thorax (ectopia cordis) either into the neck or abdomen. **Auscultation** of the different heart valves takes place at exactly defined sites.

18.3. The **LIGAMENTUM ARTERIOSUM** is the connective tissue remnant of the patent fetal **ductus arteriosus.** Through it, blood is conducted directly from the pulmonary trunk to the aorta by circumventing the lungs. Normally the ductus closes completely within a few weeks of birth to give the fibrous ligamentum. An abnormal persistent ductus arteriosus produces dramatic circulatory difficulties, since, depending on relative blood pressures, a portion from the aorta reaches the pulmonary trunk and circulates through the lung once again. Resultant 'machinery noises' are heard on auscultation. Other cardiac anomalies such as ventricular septal defects are often associated with persistent ductus arteriosus. Different surgical techniques are employed to close the patent ductus, during which, one should take into account the presence of the **left recurrent laryngeal nerve.** It bran-

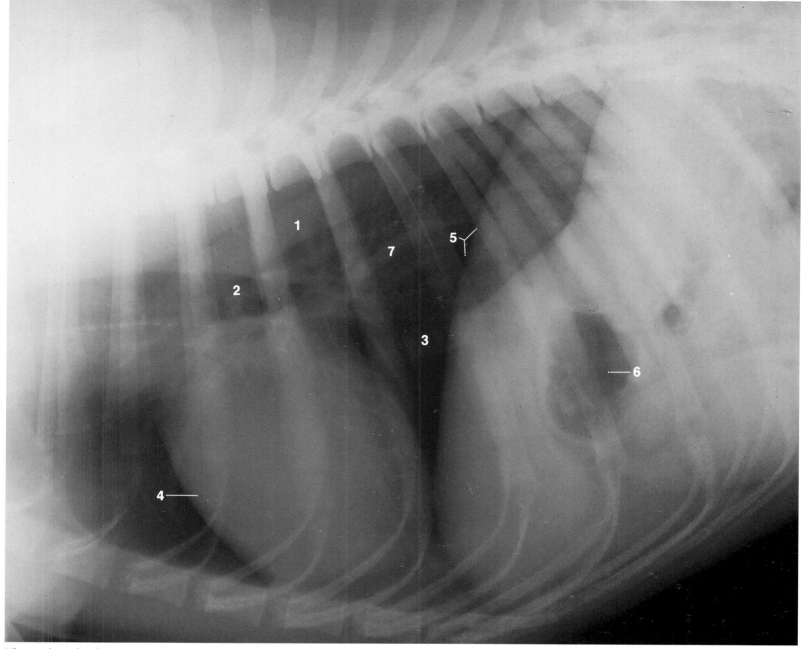

Thorax (lateral radiograph). 1 Aorta; 2 Trachea; 3 Caudal vena cava; 4 Border of right ventricle of heart; 5 Cupola or dome of diaphragm; 6 Gas in pyloric antrum; 7 'Lung pattern' due to pulmonary blood vessels.

ches from the vagus nerve at the ductus arteriosus (lig. arteriosum), and courses around the aortic arch.

In abnormal persistence of the right aorta, (refer to the embryology of the branchial arteries), the right aorta lies to the right of the esophagus instead of the normally present left aorta. In this abnormality, the distinctly elongated ductus arteriosus from the pulmonary trunk (which is situated regularly to the **left**) is directed dorsally and over the esophagus to the abnormal right aorta. As a result, the esophagus is pinched off by the ductus and is strongly dilated cranial to the constriction. This is the origin of one form of esophageal dilation (see 6.4.) and is easily detected radiographically after a barium meal. During surgery

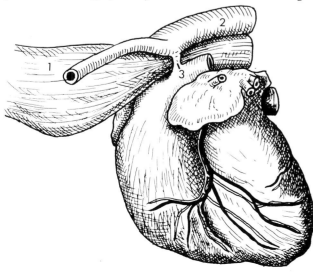

Persistence of right aorta. 1 Esophagus; 2 Descending aorta; 3 Ductus arteriosus. Details see p. 16A

the ductus arteriosus is ligated in two places and transected between them, because, generally, blood is still flowing through the duct.

18.4. In **TRICUSPID INSUFFICIENCY** the myocardium of right atrium and ventricle hypertrophies. During ventricular contraction (systole), a portion of the blood spurts back into the atrium, atrium and ventricle have to move greater quantities of blood, and muscle hypertrophies as a result.

18.5. The **VALVE OF THE PULMONARY TRUNK** often has a congenital (**pulmonary**) **stenosis or insufficiency** and the right ventricle hypertrophies. The pulmonary valve, like the aortic, is semilunar in type, its three valvulae attached to the inside of the vessel wall like swallow nests. Normally during right ventricular systole, the margins of the pulmonary valvulae are pushed to the inner wall of the pulmonary trunk. With ventricular diastole, the blood current tends to flow backwards and unfolds the valvulae, apposing them tightly. This ensures normal closure of the pulmonary valve.

The **heart valves** are attached to the **heart skeleton.** This is partly cartilaginous and in cattle even partly ossified. It consists of fibrous rings with fibrous triangles between them. The skeleton stabilizes the shape of the heart and serves as the insertion of atrial and ventricular myocardium. These approach the skeleton from their respective directions but never go beyond it. The closure of pulmonary and aortic semilunar valves produces the second (short) **heart sound.** The first (longer) sound is produced by tension vibrations of heart muscle in ventricular systole. Murmurs are abnormal noises resulting from stenosis (constriction), deficient closing (insufficiency) of heart valves, or persistent ductus arteriosus.

18.6. The **FOSSA OVALIS** is the thinnest site in the interatrial septum. It arises from the sudden closure of the fetal foramen ovale with the first respiratory movement at birth, following sudden changes in blood pressure relationships. In the fetus, blood flow through the foramen ovale bypasses the as yet functionless fetal lung. Oxygen rich blood from the placenta flows from the caudal vena cava and the right atrium through the foramen ovale to left atrium and left ventricle. It then proceeds by way of aorta, brachiocephalic trunk and common carotid artery to the head which receives oxygen-rich blood favourable for its development. The poorly oxygenated blood of the cranial vena cava reaches the right atrium where it crosses with the oxygen-rich blood from the placenta. The poorly oxygenated blood then proceeds through the right ventricle, pulmonary trunk, patent ductus arteriosus and descending aorta to the caudal part of the body. Because this receives poorly oxygenated blood, it lags in its development. A congenitally presistent foramen ovale causes severe circulatory disturbances.

18.7. The **EPICARDIUM** is the continuation of the pericardial serosal membrane on the heart surface. The smooth mesothelial covering permits a general slipping between the lubricated visceral and parietal laminae of the pericardium, thus providing a frictionless movement between heart and pericardium during heart action.

18.8. The **MYOCARDIUM** consists of a major part geared to work, and an addition part equipped for the conduction of stimuli (see 18.14.). The myocardium, the thickest constituent of the heart wall, consists of several muscle layers with different fiber directions which intermingle in a whorl-like manner at the heart apex. Just as skeletal muscle hypertrophies with additional work, so too does cardiac muscle. A proportionate harmonic cardiac hypertrophy involving all myocardium is evident with increased performance of the whole heart as, for example, in sportsmen. By contrast, in disproportionate hypertrophy only individual regions of the heart enlarge such as, for example, the right ventricle in pulmonary stenosis. Since a constant ratio persists between the numbers of heart muscle cells and the capillaries supplying them, hypertrophy of the heart is possible only to limited degree. When this limit is exceeded, hypertrophy leads to dilation of the heart. One can distinguish the different heart-wall thicknesses radiographically by using radiopaque substances. In addition, the dilated heart takes on a blunt-cone shape whereas the normal heart is more acutely conical. A dilated heart can no longer expel its blood volume completely in systole so that an observable amount of blood remains in diastole. This mixes with blood streaming into the chamber concerned.

18.9. ENDOCARDIUM merges with the intima of the large blood vessels and has a similar structure.

18.10. In its apical region the **INTERVENTRICULAR SEPTUM** consists essentially of heart muscle. A small membranous (connective tissue region) is present in the vicinity of the atrioventricular valves, and it is here that an interventricular foramen can be located. A foramen in the muscular septum is very rare. In septal defects blood flows from the left to the right ventricle in accordance with differences in blood pressure.

18.11. The **BICUSPID OR MITRAL VALVE,** just as with the other valves, may have a functional **insufficiency** and resultant hypertrophy of right atrium and ventricle. With a **stenosis** of the valve, the passage for the flow of blood is reduced markedly.

18.12. Insufficiency of the **AORTIC VALVE** leads to a reversal of blood flow into the left ventricle which hypertrophies on account of the increased work load. The **valvular mechanism** of all heart valves is the basic prerequisite for heart action with **ventricular muscle contraction** or **systole** and **relaxation** or **diastole**. Each of these consists of two phases and occurs synchronously in the left and right sides of the heart. In the first phase of **diastole,** the **relaxation phase,** all valves close and there is little or no pressure in the ventricle. In the second phase, the **phase of ventricular filling,** due to higher pressure in the atrium the atrioventricular cusps are pushed away by the blood flow, just as a door is opened by a gust of wind. In no case does valvular opening result from the movement of papillary muscles. In the first instance **ventricular filling** occurs due to a **lowering of the level** of the atrioventricular valves and blood is sucked from the atrium into the ventricle. Further factors contributing to ventricular filling are the **residual pressure in the venae cavae** and atrial contraction.

In the consecutive **systole** ventricular muscle contracts. In the first phase, the **contraction phase,** atrioventricular valves are closed passively because ventricular pressure is greater than that in the atrium. Because of this, the apex of the atrioventricular valve is pushed in the direction of the atrium and the chordae tendinae are now under tension to prevent the valve apex from being everted into the atrial region. Rather, the borders of the sail-like 'apex' of the atrioventricular valve are positioned so tightly together that the valve is impervious. In the second phase, the **expulsion phase,** a further rise in blood pressure drives blood into the large vessels namely aorta and pulmonary trunk.

Semilunar valves are pushed out of the way passively by the blood flow. At the end of the expulsion phase and the beginning of diastole, the blood pressure in the ventricles falls below that in the great vessels. The three pocket-like semilunar valvulae of each valve distend due to blood tending to flow backwards from the large vessels and this closes each valve.

With **external heart massage** heart action has to be maintained artificially for a limited time. In transient heart failure, for example in marked parasympathetic tonus conditioned by shock, strong artificial external pressure is exerted on the thoracic wall in unison with the heart beat. In turn the heart is compressed since it cannot avoid the external pressure transmitted through its surrounding pericardium. Due to this pressure aortic and pulmonary semilunar valves open passively and because of relaxation of cardiac muscle, blood is sucked from the venae cavae into right atrium. Thus, a direct flow of blood can be maintained artificially without ventricular contraction until heart beat is restored.

18.13. By strict anatomical criteria **CORONARY ARTERIES** are not end-arteries since they have narrow-lumened anastomoses. True end-arteries do not anastomose with one another, and supply only a completely defined tissue region. From the functional viewpoint, however, coronary arteries may be regarded as end-arteries. Their anastomoses are so narrow that, when a neighbouring artery is thrombosed, they cannot take over its blood supply fully. Necroses occur in the myocardium as a result, and this causes heart infarct or failure. With survival of the patient, cardiac muscle cells cannot regenerate and are substituted for by scar tissue of lower load-carrying capacity. (In comparison to man, heart-failure in dogs is of minor importance.) **Danger of thrombosis** in coronary arteries is increased by endocarditis in the left ventricle since detached portions of endocardium at the origin of the aorta can reach the coronary arteries directly. **Disturbances to blood flow** in coronary arteries cause angina pectoris in man, with severe cardiac pain radiating into the arm (see Zones of Head — 19.2.).
Blood supply to the heart itself occurs during diastole. In abnormally high frequency of the heartbeat, diastole is shortened and causes an undersupply of blood.

18.14. The **STIMULUS FORMATION AND CONDUCTION SYSTEM OF THE HEART** must function permanently to maintain heart action even during unconsciousness. The conduction system consists of specifically modified heart muscle cells which have lost their capability to contract. Stimulus formation occurs at the sinoatrial node (the pacemaker). If the pacemaker function of the sinuatrial node fails, the next part of the system, namely the atrioventricular node and the fasciculus of His, takes over as a substitute pacemaker but at an essentially reduced frequency. The overall system is a prerequisite to maintaining heart beat outside the animal body (for example the frog's heart).

In **heart transplants,** a portion of the conduction system, at least the atrioventricular fasciculus, is included. Thus heart function can be maintained in the recipient after inserting an artificial pacemaker. A knowledge of the conduction system makes it possible to understand an **electrocardiogram** or ECG. To formulate an ECG, resulting potentials of heart action are derived on the body surface, then amplified and recorded.

18.15. The **AUTONOMIC NERVOUS SYSTEM** modifies the frequency of stimulus formation (positively or negatively **chronotropic**), conduction velocity (positively or negatively **dromotropic**) and the strength of contraction of working muscle (positively or negatively **inotropic**). In this case sympathetic (positive) and parasympathetic (negative) divisions behave chiefly as antagonists. Baroreceptors (pressor receptors) are present in the right atrium and sinuses of the venae cavae.

19. AORTIC ARCH, AUTONOMIC NERVES

19.1. In its initial part the **AORTA** is curved. Its beginning widens into an onion shaped aortic bulb in which the coronary arteries arise from two of three bulges (the sinuses of the aorta containing the three aortic valvulae).
Adjacent to the vertebral column, the aortic arch continues as the descending aorta, known as the thoracic aorta cranial to the diaphragm and the abdominal aorta caudal to it (see illustration under 17.1). The aorta and the large arteries ramifying from it near the heart are recognized as **elastic arteries** due to their yellow colour. This is not so for coronary arteries.

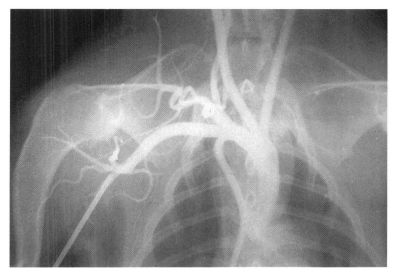

Arteriography of the branches of the aortic arch using radiopaque material injected into the **right** brachial artery, (above, a ventrodorsal radiograph, below, a lateral radiograph. For identification of blood vessels see p. 19A).

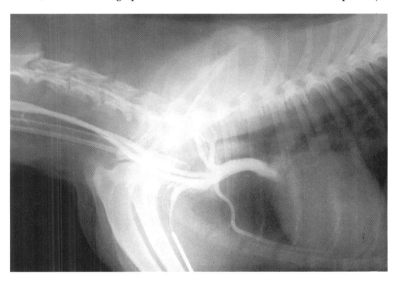

19.2. AUTONOMIC NERVES and cutaneous sensory nerves from the somatic nervous system arising from the same spinal segment form (conduction) arcs which affect each system reciprocally. Through cross-communications, cutaneous stimulus in a defined skin region (**zone of Head**) can influence appurtenant viscera. The 'saltation' of stimuli from the autonomic to the somatic nervous system and return, takes place in the spinal ganglia where small viscerosensory (autonomic) and large pseudounipolar (somatic) nerve cells are directly adjacent to one another. In illnesses or diseases of defined internal organs, pain is perceived in specific skin areas.

20. THE AUTONOMIC NERVOUS SYSTEM

20.1. The SACRAL PART OF THE PARASYMPATHETIC DIVISION reaches its effector organs by various routes still under debate. Furthermore, there are differing data on the radius of innervation and field of supply of the sacral part of the division and perhaps these are species specific. One veterinary text states the field of innervation is limited to the rectum and the nerves are supposed to extend along the insertion of mesorectum as far as the boundary with the descending colon. According to that, therefore the hypogastric nerve may be purely sympathetic. To add to the debate, it is stated that in man, the transverse colon is the boundary between supply from the sacral part of the parasympathetic division and the vagus nerve. New experimental results, however, speak of a still greater field of innervation of the sacral part, supposedly including the entire colon.

20.2. The OBTURATOR NERVE, particularly after a difficult parturition, can be crushed medial to the body of the ilium and craniodorsal to the entrance of the nerve into the obturator foramen. Obturator paralysis leading in turn to paralysis of adductor muscles occurs particularly in cows which have had large calves. The pelvic limbs of the cow concerned are no longer drawn towards the body and the animal cannot remain standing. In affected cattle in the frog-sitting position, tears occur in the adductor muscles.

21. STOMACH AND INTESTINES

21.1. The TERMINOLOGY used is derived from the Latin ventriculus - a small stomach. (The word is also used for the chambers of the heart and the side pockets of the laryngeal mucous membrane.) The Greek term, gaster - a stomach is also used, as in gastrosplenic ligament and gastritis or inflammation of the stomach.

21.2. GASTRIC FUNCTION concerns the transient storage of food and the regulation of the transport in small quantities into the small intestine. Hydrochloric acid acts as a disinfectant and in activating pepsin formed in the gastric glands. The enzyme then acts upon the food.

21.3. TORSION OF THE STOMACH is a dramatic dysfunction almost exclusive to large breeds of dogs and requires immediate veterinary intervention. Cranially the stomach is anchored only by the esophagus and its attachment to the esophageal hiatus. The dorsal mesentery of the stomach, by its nature, is displaced and is therefore unsuitable for stabilizing the stomach. In addition, the pyloric part of the stomach and the beginning of the duodenum are freely moveable so that the stomach is not fixed in its position. Thus acute gastric dilation can lead to a torsion or volvulus, generally caused by abnormal filling with swallowed air (aerophagy). A gastric torsion can be provoked experimentally by dilation of the stomach, but the reason the dog is incapable of expelling gas, under the circumstances, is as yet unclear. The mechanics with the beginning of gastric torsion is variable and in some cases also unclear.

In gastric torsion, gastric vessels, particularly the veins, are strangulated. With maximum over-distension, stomach rupture can occur, generally parallel to the greater curvature.

First of all the peritoneum ruptures, then the muscular tunic and finally the mucosa. During the process of gastric torsion, there is almost always a torsion of the greater omentum and associated splenic vessels. This leads to an interruption in splenic blood circulation and to splenic congestion, chiefly due to disturbance of efferent blood flow. A manifold enlargement of the spleen is a sequel, and in many cases a splenectomy or removal of the spleen is a necessary adjunct.

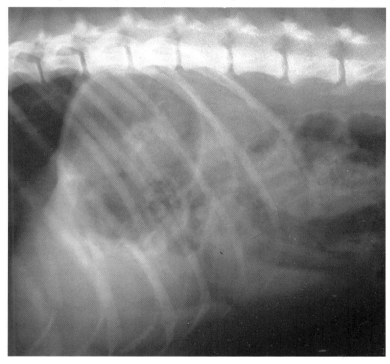

In viewing radiographs of the stomach, one must consider the position of the dog being radiographed. With the dog lying on its right side (upper radiograph) the gas pocket is situated on the left in the fundic region positioned dorsally. With the patient lying on its left side the gas pocket is present in the pyloric region which lies further to the right and ventrally (lower radiograph). Well defined folds of mucous membrane should be noted in the pyloric antrum (lower radiograph).

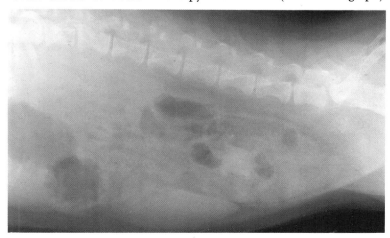

21.4. During GASTROTOMY or opening of the stomach, the surgical incision is made, if possible, on the parietal surface of the stomach midway between the two curvatures and along the long axis of the stomach. On account of muscle tone the incised mucous membrane is always strongly everted.

21.5. The **FUNCTION OF THE SMALL INTESTINE** consists essentially of the processes of secretion, digestion, resorption and transport. **Secretion** is maintained by the mucous producing **goblet cells** interspersed in the epithelium of the mucous membrane, from tubular **intestinal crypts** (kryptos-inconspicuous) or epithelial invaginations containing secretory cells, from **submucosal duodenal glands,** and particularly from the **pancreas** situated external to the intestine.

Digestion, that is the splitting of nutrients into resorbable units, takes place firstly in the intestinal lumen and then at the cell membranes of the brush border cells of the intestinal epithelium.

The brush border cells situated superficially, serve in resorption. To permit this, they have an enormous increase in surface area due to the dense erect microvilli each approximately one thousandth of a millimeter long. Transport of the resorbed nutrients takes place through the capillaries except for the long chain fatty acids which are transported in efferent lymphatic vessels.

Surface enlargement leading to a large resorption surface, is produced by recognizable macroscopic, microscopic and ultramicroscopic structures.

The macroscopic **circular folds** arise through folding of the mucosa and submucosa but not, however, of the muscular tunic. The microscopic **crypts** and particularly the **intestinal villi,** finger-like projection approximately 1 mm long, bear the **brush border cells.** These, in turn carry the microscopic microvilli covered by the ultramicroscopic glycocalyces.

Intestinal movements are classified as villous movements, pendulous movements and peristaltic waves. The parasympathetic division of the autonomic system propagates intestinal movements, while the sympathetic division inhibits them. Disturbances to intestinal transport can have different etiologies and lead to greater water resorption and constipation or coprostasis.

Villous movements, leading to a shortening of the intestinal villi, are due to the contraction of sparse smooth muscle cells. In this way the transport of lymph or chyle is promoted in the central lymph vessels. **Pendulous movements,** due to the contraction of the longitudinal muscle layer of the intestine, facilitate the mixing of intestinal content or chyme.

Peristaltic waves transport content further towards the colon. Initially the longitudinal muscle layer contracts and then, all importantly, the circular muscle.

21.6. **SOLITARY LYMPH NODULES** are scattered in the mucosal tunic. (They are also present in spleen and lymph nodes.) **Aggregated lymph nodules** (Peyers's patches) lie chiefly in the antimesenteric section of the wall of the ileum. The primary lymh nodules are spherical or elliptical collections of lymphocytes of diameter approximately 0.5 mm. The secondary nodules have a pale germinative center and are significant as the producers of B-lymphocytes.

21.7. **COLONIC FUNCTION** consists, essentially, in removing fluid to thicken the colonic content. It is mixed with mucous from the very numerous goblet cells and this promotes the further passage of content to the anus. The adluminal surface enlargement is due only to the presence of crypts. Intestinal villi are absent in the colon.

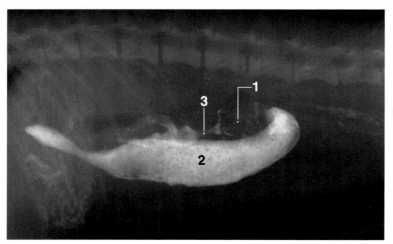

Radiograph of duodenum after administration of a radiopaque substance. Stenosis due to a foreign body ileus (produced by an acorn - 1). Prestenotic dilation of descending part of duodenum (2); post-stenotic constriction of ascending part of duodenum (3).

21.8. **DILATION OF THE COLON** with obstruction by colonic content (congenital megacolon or Hirschsprung disease) is due to an absence of regulated intestinal peristaltic movements. The syndrome is said to be due to a lack of intramural ganglion cells.

21.9. The **POSITION OF STOMACH AND INTESTINE** is characterized by a duodenum embracing a space opening cranially, and a colon bounding one opening caudally. This arises during the course of rotation of stomach and intestines (refer to an embryology text). The colonic length, which is hook-shaped, can be seen radiographically after radiopaque material is administered rectally. Clinically this procedure is adopted where colonic diseases are suspected (see radiograph).

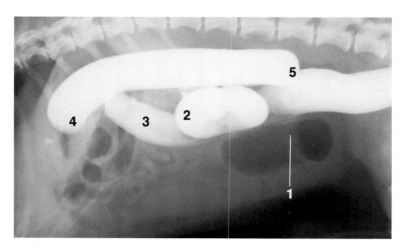

Radiograph of colon after rectal administration of a radiopaque substance. An ileus of the duodenum is detected due to a prestenotic gas pocket (1); Cecum (2); Ascending colon (3); Transverse colon (4); Descending colon (5) (kinked).

21.10. **ATRESIA OF ANUS, ATRESIA OF RECTUM,** and **ATRESIA OF BOTH ANUS AND RECTUM** are malformations occurring during ontogenesis. (Atresia is an absence of a normal opening resulting in this case from the failure of the closed embryonic anus or rectum to open.)

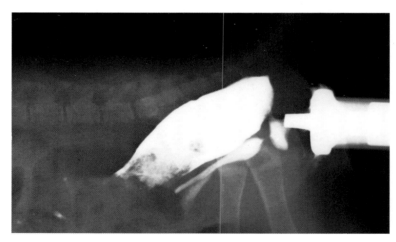

Atresia ani with a fistula to the vestibule of vagina. Application of contrast medium per vagina.

21.11. The **CIRCUMANAL GLANDS** are ductless hepatoid glands. The glands tend to undergo carcinogenic alteration or form fistulae.

21.12. The **PARANAL SINUSES** store secretion emanating from glands within their wall. This material, due to its deposition on feces during defecation, is a sexual attractant and defines the animal's territory. Normally the sinuses are emptied by pressure from the external sphincter of the anus, and feces of normal consistency acts as a support or abutment. A non-viscous, liquid fecal mass cannot serve as a site of abutment and the emptying of the sinus is prevented in such circumstances. This is due possibly to lack of pressure exerted on the liquid feces by the sphincter muscle during defecation. The retained secretion then decomposes, inflammation occurs, and secondary infection with subsequent abscess formation results. The abnormally full paranal sinuses are emptied by applying external digital pressure. With repeated abscess formation or persistent suppuration surgical removal of the sinuses may be necessary.

21.13. The DORSAL TAIL GLANDS are situated dorsal to the root of the tail. The glands are modified sebaceous and apocrine sweat glands, the secretion of which serves for species identification and intersexual communication. Fine 'wool hair' is absent in this region and a small baldy patch occurs in old male dogs due to loss of overlying hair.

22. PERITONEUM, LYMPHATIC SYSTEM OF STOMACH AND INTESTINES

22.1. The PERITONEUM consists of a serosa (similar to pleura and the internal lining of the pericardium), lined internally with serosal epithelium (mesothelium) as a consequence of development from mesoderm, and an underlying layer of lamina propria or connective tissue of the serosa. At many sites a subserosal lamina permits sliding as, for example, in the urinary bladder, which undergoes quick changes in circumference. Embedded in places in the subserosal lamina is subserosal fat or the internal panniculus adiposus. Depending on nutritional status and species, for example, this fat may be sparse in dogs and abundant in pigs. Serosal muscle occurs in places in the serosa covering the uterine tube and uterus.

22.2. The GREATER OMENTUM, as a duplication of peritoneum, possesses the same properties and functions. In addition, it is a fat depot and has an immune function due to its population of macrophages and lymphocytes. Where the greater omentum overlies areas of inflammation on adjacent organs, at such sites it becomes glued to the peritoneum lining the organ in question. This is seen frequently when dissecting fresh cadavers. After surgery to stomach and intestines, it is ingenious how the greater omentum positions itself over the site of suture and becomes anchored there.

22.3. The INTESTINAL LYMPH or chyle is more or less milky in colour on account of its fat content and its connection with digestion. Long chain fatty acids, in the form of chylomicrons are transported away initially in the central lymphatic vessel of the intestinal villus. This is assisted by the villus pump mechanism, and, as a result, lymph proceeds to the lymphatic capillaries. Short chain fatty acids and the products of protein and carbohydrate metabolism pass through the blood capillary net and the portal vein to the liver.

22.4. SPLENIC FUNCTION is involved with: blood storage, blood pressure regulation, blood processing due to phagocytosis of old erythrocytes, metabolism of hemoglobin breakdown products and the formation of immune-competent cells. In spite of its multifunctional role the spleen is not essential to the life of the animal since its functions can be undertaken by other organ systems. But in no circumstances is the spleen to be considered a superfluous organ and it should not be removed thoughtlessly.

Splenectomy, however, may become necessary due to a series of conditions such as gastric torsion, injuries and tumours (see 21.3.). With human splenectomy, it may be desirable to suture cubic-centimeter-sized implants of modified splenic tissue into the greater omentum. The spleen is adapted to a blood storage function (apparent from its size and shape) on account of the great numbers of smooth muscle cells in its capsule and the system of trabeculae proceeding from this. Due to the storage and removal of blood, the spleen acts in the regulation of blood pressure due to the discharge of the portal system into which the splenic veins are opening. A congested spleen is evident when the spleen is greatly enlarged following increased blood storage. This splenomegaly may necessitate a splenectomy which is made all the more difficult due to the quantity of blood present. When adrenalin is injected the smooth muscle contracts and blood is removed. In euthanasia a resultant congested spleen is said to be caused by a paralysis of the vasomotor centers of the brain. This is often seen in dog cadavers prepared for dissection. The white pulp with its splenic corpuscles and lymphatic periarterial sheaths forms the lymphocytes. The red pulp, named because of its high red blood cell content, is the site of processing of all erythrocytes. It is generally considered that blood flows in the red pulp partly through splenic sinuses (closed circulation) and partly in a free manner through the lymphoreticular tissue (open circulation). Because of animal species differences, this is not yet established in all such cases. Latero-lateral radiographs provide a good image of the spleen. Its ventral exremity bends to the right and across the midline, and may reach the right side of the body. Radiographically, this part of the spleen is featured as a triangular shadow, the triangle corresponding to the cross-sectional shape of the spleen.

23. LIVER, GALL BLADDER, PANCREAS

23.1. With the exception of its ventromedial part, the LIVER lies within the intrathoracic part of the abdominal cavity. Evenso, the ventromedial part cannot be palpated. In contrast, with pathological enlargement, the liver extends more or less caudal to the costal arch and this can be established by palpation, percussion (dull liver sounds) or radiography. Liver size, which, according to Evans/Christensen (1979), normally constitutes 3.8% of total body weight, is correlated with its numerous functions. These include its function as: 1. an endocrine gland in the synthesis of blood plasma proteins such as albumen, globulin and fibrinogen; 2. an exocrine gland in the production of bile and cholesterol; and in: 3. intermediate metabolism of fat, protein and carbohydrate; 4. storage of fat, glycogen and iron; 5. glycogenolysis and glyconeogenesis, and the maintenance of a physiological glucose concentration; and 6. detoxification of drugs and medicines. In the fetal stage the liver is relatively larger and occupies the predominant part of the abdominal cavity, its size being connected with its function in blood formation. A pathological diminution in liver size may be produced from pronounced scar tissue formation with accompanying shrinkage as for example in liver cirrhosis.

The classical liver lobules are still recognized as functional units macroscopically, having an approximate diameter of 1 mm and a height of 2 mm. Each lobule possesses: a central vein, seen as a dark point, with liver capillaries or sinusoids opening radially; trabeculae associated with liver cells radiating regularly; and a connective tissue capsule limiting the lobule peripherally. This fibrous perivascular capsule (of Glisson) is very distinct in the pig. According to another system of classification, a functional unit, the liver acinus, has at its center the smallest biliary ductule. It cannot be recognized macroscopically (consult an histology text).

Fatty degeneration of the liver which occurs following the effects of toxins or oxygen lack and leads to yellowish discoloration, is evident macroscopically in the progressive phase. Since blood in the liver capillaries flows from the periphery of the lobule to the central vein, toxins first of all affect the peripheral liver cells (hepatocytes) and these then undergo fatty degeneration. On the other hand with oxygen lack (hypoxia), fatty degeneration mostly occurs at the lobule center. Among other things liver atrophy may result from a congenitally persistent ductus venosus. Due to this, portal blood does not flow through the sinusoids but circumvents the parenchyma of the liver through the persistent ductus to flow directly into the caudal vena cava. The persistent duct as it were, acts as a portocaval shunt (see 45.3).

23.2. Bile is concentrated or thickened by water resorption in the GALL BLADDER. At the opening of the bile duct a sphincter muscle regulates bile flow into the duodenum. There, bile emulsifies fat, thus permitting an improvement in the effectiveness of digestive enzymes. Bile acids and pigments are resorbed in the small intestine (enterohepatic circulation) and subsequently are secreted in the bile once again. An excessive content of bile pigments in the blood appears as jaundice or icterus (pre-, intra-, and post-hepatic). In prehepatic icterus increased bile pigments result, for example, from a greater destruction of erythrocytes. Intrahepatic icterus is caused by damage to the liver cells while posthepatic icterus is caused by obstruction to the biliary duct system. Gall bladder and biliary ducts are demonstrated by contrast radiography (cholangiocystography), the radiopaque substance passing out of the liver and biliferous ductules. Soon after death autolysis of gall bladder and biliary duct system sets in, and the tissues in their vicinity are detected due to greenish imbibition of bile. This may be seen on autopsy.

23.3. The PANCREAS has an exocrine and an endocrine part. Its exocrine portion produces digestive enzymes for the breakdown of protein, carbohydrates and fat. The protein-splitting enzymes are present within the pancreas as inactive precursors and are activated to trypsin and chymotrypsin initially after their transport through pancreatic and accessory pancreatic ducts into the lumen of the small intestine. If activation occurs within the pancreas during pathological changes, self digestion takes place and results in pancreatic necrosis. This often leads to death after running a dramatic course. After death autodigestion takes place very quickly, the yellow-pinkish colour of the pancreas becomes discolored, turning a dark purplish, and its distinct lobulated structure is lost.

A partial pancreatectomy is necessary in acute necrosis or carcinogenic change. Normal tissue is separated from abnormal by the careful extraction of gland lobules from interlobar connective tissue, while preserving pancreatic ducts and blood vessels (see p. 23A). One should also take into account the anastomosis between cranial and caudal pancreaticoduodenal arteries along the right lobe of the pancreas.

The endocrine portion in the pancreatic endocrine cells or islets of Langerhans, regulates carbohydrate metabolism due to the hormones insulin and glucagon. Abnormal atrophy of the islets disrupts this process and produces, for example, a relative and absolute insulin deficiency leading to diabetes mellitus.

23.4. The architecture of the wall of the **PORTAL VEIN** is adapted to a comparatively high blood pressure. This is necessary for blood flow through the second (hepatic) capillary bed of the portal system. It is maintained by arteriovenous anastomoses in intestines and reinforced by the spleen, the venous blood of which drains into the portal vein. Characteristic of the wall of the portal vein are the fasciculi of longitudinal or adventitial muscles lying externally.

Due to arteriovenous anastomoses and the spleen (see below), adequate blood pressure is attained in the portal vein to permit blood flow through the capillary bed of the liver. This occurs even though the large part of the blood already flows through a capillary bed in the intestinal wall at the beginning of the venous limb of the circulation. The limited fall in blood pressure caused by this, is balanced by the blood-pressure-regulating function of the spleen (see 22.4.) and by blood circumventing the capillary bed through numerous arteriovenous anastomoses in the field of origin of the portal vein. A sufficient oxygen supply to the liver tissue is ensured by blood in the portal vein avoiding the capillary bed, since the narrow hepatic artery alone cannot supply such a gigantic organ. In summary, the portal vein collects the blood of unpaired veins from unpaired organs.

Venous flow from the caudal section of the rectum is extraportal, that is it is not included in the portal system and comes from paired middle and caudal rectal veins.. This blood does not reach the liver in the portal vein but rather flows through the general circulation of the body. This circumstance is used to best advantage by applying medicaments in suppository form. With resorption their full concentration is utilized.

24. ADRENAL GLANDS AND URINARY ORGANS

24.1. The **ADRENAL GLANDS** are retroperitoneal (as are kidneys, aorta, and caudal vena cava), each lying at the cranial extremity of its respective kidney. This close topographical relationship gives rise to the term suprarenal gland. As an endocrine gland, the adrenal has a function completely divergent from that of the kidney. Embryologically its cortex develops from mesoderm and produces steroid hormone. Its medulla develops from ectoderm (sympatheticoblasts) and produces adrenalin and noradrenalin. When removing the adrenal gland surgically, as for example with tumors, one should take into account the greater splanchnic nerve on its dorsal surface and the blood supply. The latter is derived from cranial, middle and caudal adrenal arteries.

24.2. With regard to the **KIDNEY** of the live animal, the **adipose capsule** has a soft consistency. As a result, the kidney appears to be 'swimming' in fat ventrolaterally.

Radiographically, the kidney cannot always be imaged satisfactorily without the use of a contrast medium. This is true, in particular for the left kidney, the caudal of the two. For better radiographic imaging, a radiopaque substance is administered intravenously and eliminated in the urine, the relevant parts of the urinary tract becoming visible as a result (elimination urography). In specific cases the retroperitoneal space and peritoneal cavity are filled with air before radiographing the kidney.

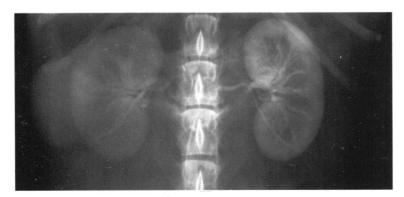

Radiograph of the renal arteries and their branches using retrograde aortography. Both kidneys are situated almost at the same level. This is not usual.

Congenital malformations, manifested as a failure in the connection between the nephron and the system of collecting tubules, result in cystic formation within the kidney. The organ is predisposed to this abnormality due to the different origins of nephron from the metanephrogenic blastema and collecting tubules from the ureteric buds.

The **blood vessels of the kidney** namely the renal artery and vein, penetrate the kidney substance via the respective interlobar vessels and then the arcuate arteries and veins. They serve in the primary formation of urine and in addition, supply the renal parenchyma and resorb non-obligatory urinary constituents from the renal tubules (loops of Henle). (The renal tubules of Henle loop reaching the **external part** of the renal medulla or pyramid are both thick and thin whereas only thin tubules are present in the internal part.)

The afferent glomerular arterioles which play a part in the formation of the primary urine, arise predominantly from interlobular arteries coursing radially within the kidney lobules to the kidney capsule. Rarely are the arterioles derived from arcuate arteries and veins. The afferent glomerular arterioles branch into capillary loops of the glomerulus in the **renal corpuscles.**

The efferent glomerular arterioles of the **subcapsular renal corpuscles** open into a cortical **capillary net** entangling the **renal tubule** of the cortex and connected to the **interlobular veins.** The renal medulla is supplied by **arteriolae rectae (spuriae)** coming from **juxtamedullary renal corpuscles,** and **arteriolae rectae (verae)** radiating from interlobular arteries, or rarely, arcuate arteries. Blood flow away from the kidney cortex is by way of **interlobular veins** and from the medulla via **venulae rectae.** All these, in turn, flow through **arcuate** and **interlobar veins** into the **renal vein.**

Renal arteries and veins

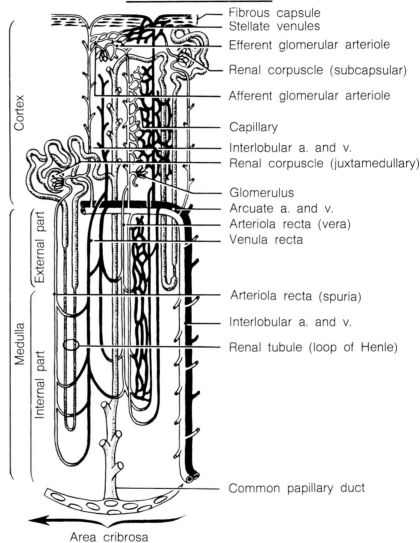

Schematic diagram of renal blood vessels.

24.3. The terminal part of the **URETER** opens into the urinary bladder by passing obliquely through the bladder wall. This prevents a reflux flow of urine towards the renal pelvis since increased pressure within the bladder closes the terminal part of the ureter at micturition. A unilateral duplication of the ureter occurs as a congenital abnormality of the ureteric bud and it divides prematurely before reaching the renal sinus. This can also accompany duplication of the renal pelvis. Ectopic openings of the ureters and accompanying incontinence may occur caudal to the sphincter muscles of the urethra (detected radiographically).

24.4. The **MUCOUS MEMBRANE OF THE URINARY BLADDER** is arranged in folds that disappear when the bladder is distended. The trigone which develops from mesoderm is constantly free of folds. The remaining part arises from endoderm. The smooth muscle of the urinary bladder has an external and an internal layer of longitudinal muscle with a circular muscle layer between them. These layers are not entirely separated from one another and merge in places. Muscle fasciculi shear away from the neck of the bladder to insert on the initial part of the urethra and with contraction of urinary bladder musculature these fasciculi widen the beginning of the urethra. The contraction of smooth muscle in the bladder wall is initiated by parasympathetic nerve fibers coming from pelvic nerves. Sympathetic fibers reach the urinary bladder in the hypogastric nerve.

The **blood supply** arises from the caudal vesical artery and vein which stem from the vaginal or prostatic artery and vein respectively, depending on sex. According to Evans/Christensen (1979), a functional cranial vesical artery arises from the umbilical artery in approximately 50% of cases. According to other workers the umbilical artery is generally obliterated completely, to give the round ligament of the bladder.

25. ARTERIAL AND VENOUS UTERINE RAMI

25.1. The **ARTERIAL AND THE VENOUS UTERINE RAMUS** from ovarian artery and vein respectively must be taken into account when performing surgery on ovary, uterine tube and uterine horn. The uterine rami can be regarded as anastomoses between ovarian vessels from aorta and caudal vena cava on the one hand, and vaginal vessels from the internal iliac artery and vein on the other. Thus the uterine horn derives its blood supply (uterine artery and vein) from two sources.

26. PERINEUM

26.1. The term **PERINEUM** is defined in several different ways. According to older anatomy texts the perineum lies between anus and vulva or scrotum, depending on sex, while Nickel/Schummer/Seiferle (1975) state that in the male, it extends between anus and bulb of penis. Nomina Anatomica Veterinaria (1983) takes the concept much further. It considers that the whole of the cutaneous-muscular closure of the caudal pelvic aperture in the vicinity of anal and urogenital canals is perineum. This also corresponds to clinical usage. An atrophy of the pelvic diaphragm, specifically the m. levator ani often leads to **perineal hernia** occurring in older male dogs. With surgical repair of the hernia, a castration is undertaken simultaneously to diminish hypertrophy of the prostate which is considered to be a predisposing factor.

The strength of the muscular closure of the caudal pelvic aperture may also depend on the alteration of sex hormone status with increasing age. The muscles concerned are often altered to such a degree that they can no longer be defined exactly. Portions of the m. levator ani which have undergone less modification are sutured to the m. sphincter ani externus. Surgical intervention in the perineal region may also be necessary with fistulae or carcinogenic change of the glands associated with the anus (see 21.10. - 21.12.). An **episiotomy**, an incision to widen the vulva, is performed to alleviate parturition or to remove vaginal tumors. The incision is made dorsally but does not extend to the m. sphincter ani externus.

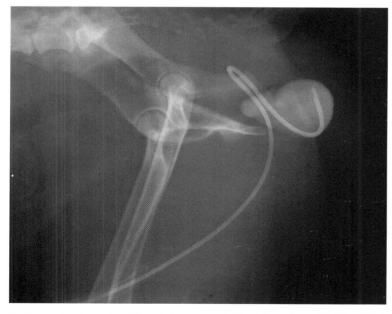

Radiograph of a perineal hernia in a male Collie. The urinary bladder filled with radiopaque material via a urinary catheter, is displaced into the hernial sac (retroflexion of the bladder).

27. PERITONEAL RELATIONSHIPS OF THE GENITAL ORGANS

27.1. The **PARAMETRIUM** contains an abundant, loose connective tissue between its two peritoneal laminae. Such connective tissue with its contained blood vessels favors the spread of inflammation. When a metritis encroaches upon the loose connective tissue of the broad ligaments, this is spoken of as parametritis.

27.2. By a process known as the **DESCENT OF THE TESTES,** the testes are displaced from their place of origin into the scrotum. The descent of the testes in the neonate is not fully completed until after parturition. When the descent is incomplete and the testis remains in the abdominal cavity or is lodged in the inguinal space it is known as a retained or concealed testis. The condition itself is spoken of as cryptorchidism from the Greek kryptos = hidden, and orchis = testis.

The **retained testis** has an abnormally long proper ligament of the testis and ligament of tail of epididymis. The testis tends to become cancerous, its temperature approaches body temperature, and spermiogenesis is deranged while the production of sex (steroid) hormones is almost undisturbed. For normal spermiogenesis to take place an essentially lower temperature is necessary, the internal temperature of the normal testis in the scrotum being in the vicinity of skin temperature. Considering its position in the scrotum from a physiological viewpoint, several other factors influence the **temperature of the testes** which is approximately 2°C less than body temperature. Two such factors may be mentioned. 1. The venous flow away from the testis is via the pampiniform plexus through the coiled loops of which runs the testicular artery. Before its intertesticular course, the testicular artery is cooled slightly due to the somewhat cooler blood within the surrounding pampiniform plexus. (In man and sheep, varicose changes in the pampiniform plexus can cause sterility.) 2. The m. cremaster (externus) and the smooth muscle of the tunica dartos of the scrotum contract to draw the testis closer to the warm body wall when ambient temperature is low, and relax in hotter temperatures, thus permitting the testis to slide into the cooler scrotum. The two muscles certainly function efficiently in species such as bull and ram which have a pendulous scrotum.

27.3. Depending on the different methods of castration the **TUNICS OF THE TESTIS** are dealt with in several ways. In castration with the covering vaginal process intact, it and its contained deferent duct are transected distal to the site of ligature. In a castration with an incised vaginal process, the 'naked' deferent duct and the spermatic funiculus or cord are ablated at the same site.

28. GENITAL ORGANS

28.1. The **TERMINOLOGY** used in the male relates to the Latin 'testis' or the Greek term 'orchis', as for example, in testicular, mesorchium, cryptorchid, cryptorchidism.

The testis is an endocrine gland and moreover produces the sperm. This, together with epididymal and prostatic secretions, comprises the ejaculate.

28.2. The efferent ductules of the **EPIDIDYMIS** (epi = upon, didymos = twins or metaphorically testes) develop from the transverse canaliculi of the mesonephros, and the duct of the epididymis from the Wolffian duct. From the division of epididymis into head, body and tail one may deduce the terms cranial and caudal extremities of testis.

28.3. The **DEFERENT DUCT** is the direct continuation of the duct of the epididymis. The surgical removal of a section of the deferent duct in sterilization is known as vasectomy because the duct was previously known as the vas deferens. In general, the surgical reunion of the two

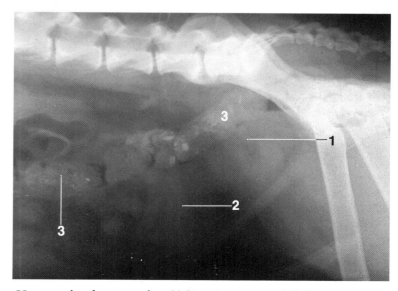

Hypertrophy of prostate of an old dog. 1 Prostate (tennis-ball siize); 2 Urinary bladder; 3 Coprostasis.

incised duct segments does not lead to recovery of male reproductive capability. Accompanying autonomic nerves are necessarily severed during the transection of the duct and testicular function impeded. Distal to the site of transection there is a deficit in peristalsis of the deferent duct indispensable to semen transport.

28.4. During ejaculation the **COLLICULUS SEMINALIS** closes the beginning of the urethra cranially due to its erectile-like tissue. Thus semen is discharged from the deferent duct distally through the urethra.

28.5. Particularly in older male dogs, the **PROSTATE** tends to undergo pathological changes including hypertrophy and cyst formation. In prostatic hypertrophy, pressure is exerted on the adjacent rectum and results in a pronounced and continuous urge to defecate. Less often there is difficulty in urinating. Frequently, continuous abdominal pressure transmitted to the pelvic cavity provokes a perineal hernia (see 26.1.). Especially is this so in old dogs, since both conditions, prostatic hypertrophy and perineal hernia, are promoted by hormonal changes associated with age. Successful treatment involves the removal of hormonal factors by castration or antiandrogen therapy.

28.6. **ERECTION** is regulated by the parasympathetic system and ejaculation by the sympathetic system. Erection of the penis is caused by intumescence of the erectile tissue of the corpus ridigum penis with arterial blood and the sponge-like corpus spongiosum penis with venous blood.

The following factors are involved in erection of the corpus rigidum: 1. In the dormant or quiescent phase, the contained intimal cushions close the helicine arteries and divert blood through arteriovenous anastomoses thus circumventing the cavernae of the corpus rigidum penis. At erection the cushions level out and blood flow in the helicine arteries is conducted to the cavernae. 2. Due to closure of the arteriovenous anastomoses increased blood supply is directed into the cavernae which, as a result, are filled tightly. 3. The cavernae are 'opened' due to relaxation of smooth muscle cells. 4. Veins are clamped down or pinched off due to their oblique passage through the tensed trabeculae. 5. Obstruction in the venous limb is intensified further by contraction of the m. ischiourethralis and the compression effect, on the male organ, of the corpus cavernosum (bulbus vestibuli) in the vestibule of the vagina of the bitch being served. In the male, the m. ischiourethralis which arises on the ischiatic arch, in common with the contralateral muscle, inserts onto a fibrous loop situated in the midline. Due to the contraction of the two muscles, blood flow is cut off from the dorsal vein of the penis which passes through the ring. Efferent blood flow is said to be delayed also by intimal cushions within the constricted veins. In recent investigations, however, the existence of arterial and venous intimal cushions for promoting erection is doubted.

The paired corpus rigidum penis becomes erect initially in the first phase of coitus. Only after penetration into the vagina does the unpaired corpus spongiosum penis complete its erection involving specifically the corpus glandis penis. Ejaculation commences within a minute of copulation. In the second phase of coitus the penis turns through 180° and at first cannot be detached from the bitch. An S-shaped torque of the penis delays further the efferent blood flow especially from the glans penis. This gives rise to the phenomenon known as the genital 'lock' or 'tie' (for blood supply to penis see p. 25). The **innervation of the penis** comes from parasympathetic, sympathetic, and sensory fibers. Parasympathetic fibers from the pelvic plexus bring about erection. In particular, sympathetic fibers innervate smooth muscle as well as the helicine arteries and are responsible for ejaculation. Sensory nerve endings are numerous on the glans penis and in the prepuce.

28.7. Having adopted a roof-like position, the **OS PENIS** limits the capability of the urethra to increase its diameter. As a result urinary calculi can block the penile or spongiose urethra directly caudal to the os penis. Dalmatians have a pronounced predisposition to form calculi consisting of urates. This is conditional to a congenital derangement of uric acid metabolism.

28.8. Like the testis, the **OVARY** has a twofold function, namely the preparation of oocytes for fertilization and the formation of ovarian steroid hormones.

Primary oocytes are situated in the **cortex of the ovary**, having been encased there postnatally. After a developmental stage they are distributed in different sized follicles. In the primary follicle the oocyte has a diameter of approximately 20 μm and is enveloped by follicular squamous epithelial cells, which become cuboidal subsequently. Initially at puberty a few primary follicles mature to become secondary follicles, the primary oocytes of which have a diameter of approximately 1/10

mm. Each is surrounded by a stratified follicular epithelium and a connective tissue covering, the external and internal theca respectively. The tertiary follicle has a follicular antrum or cavity containing follicular fluid between its soft yielding epithelial cells. At this stage the oocyte, with a diameter of approximately 2/10 mm, is supported by an elevation or cumulus of follicular epithelial cells and surrounded by a corona radiata. This is formed as a single layer from follicular epithelial cells. Then a few tertiary follicles grow to a mature size, are bladder-like and ready to rupture. They have a diameter of approximately 2 mm. The majority of follicles do not reach ovulation but perish through follicular atresia (atretos, without light or opening).

After the rupture of the follicle, a large endocrine gland about 2 mm in diameter, develops from the epithelial and connective tissue theca.

This is the **cycling corpus luteum** which regresses a few days after estrus and the termination of luteal synthesis. It is then known as a **corpus albicans** due to the associated whitish connective tissue. In the event of fertilization of the ovum, the cycling corpus luteum differentiates into a **corpus luteum of pregnancy** simultaneously with the development of the conceptus. The corpus luteum of pregnancy is active hormonally in the first half of gestation.

The **medulla of the ovary** is also known as the heterosexual part of the ovary since it is comparable embryonically to the primordium of the testis. The medullary cords correspond to embryonic testicular cords, but they disappear in mammals at an early stage. The medulla contains a thick reticulum of blood and lymph vessels and autonomic, mainly sympathetic, fibers.

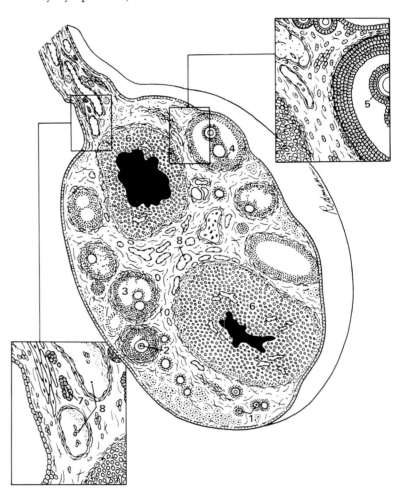

Longitudinal section of ovary. 1 Primary follicle; 2 Secondary follicle; 3 Tertiary follicle; 4 Cumulus of follicular epithelial cells; 5 Corona radiata; 6 Cycling corpus luteum; 7 Medullary cords; 8 Blood vessel.

28.9. The infundibulum of the **UTERINE TUBE** entraps the ovum after its ovulation. Inflammations of the tube are predominantly ascending, that is they emanate from the uterus and may lead to adhesion or occlusion of the labyrinthine tubal lumen. In such cases transport of the oocyte is either hindered or prevented. Sperm moving from the opposite direction against the flow of secretion within the tube may pass with little hindrance where closure is incomplete, and fertilize the ovum. Since the fertilized ovum cannot be transported further towards the uterus, it develops initially in the uterine tube which acts as a uterus however small it may be. The tube soon dehisces or bursts and in the human female, this leads to a **secondary abdominal pregnancy**. In the human female a **primary abdominal pregnancy** arises when the fertilized ovum is deposited directly into the abdominal cavity without having reached the uterine tube previously. A similar chain of events can lead to primary and secondary abdominal pregnancy in the bitch. In the vicinity of the extrauterine embryonic attachment or nidation, the peritoneum produces a proliferation rich in blood vessels. These, however, are insufficient for the development of the embryo and it dies at an early stage.

In comparison to the situation in human medicine abdominal pregnancy in the bitch is of minor veterinary importance.

28.10. The **UTERUS** allows nidation of the fertilized ovum to take place due to its glandular and richly vascular **endometrium.** The endometrial mucosa consists of non-ciliated secretory cells, ciliated cells and uterine glands. The forms of the last two structures change depending on the stage of the estrus cycle. Purulent endometritis with pus accumulation in the uterus is known as pyometra. This results from excessive and prolonged stimulation of the uterus by progesterone arising from ovarian changes such as cystic corpora lutea. Bacterial invasion does not initiate the disease process but is often associated with the pathological changes. During parturition the **myometrium** and the abdominal musculature act in the expulsion of the fetus. The **perimetrium** ensures that the uterus is capable of sliding within the abdomen and pelvis. During gestation smooth muscle cells of the myometrium lengthen approximately tenfold to about 1 mm and at parturition contract under the influence of oxytocin, a hormone of the hypothalamo-hypophyseal system.

28.11. An **UTERUS MASCULINUS** occurs in the male as a blind pocket in the vicinity of the colliculus seminalis. It is the remnant of the caudal end of the Mullerian duct.

28.12. The **CERVIX OF THE UTERUS** possesses a mucous plug in the cervical canal. This prevents ascending infection from the vagina and acts as a cervical seal during gestation. At estrus it liquifies and is discharged permitting passage of sperm through the cervical canal.

28.13. The **CLITORIS** may be enlarged considerably due to hormonal disorders of ovaries and adrenal glands (Cushing syndrome). At that stage the clitoris can be labelled a miniature replica of the penis with the reservation that in no case does the urethra terminate on it.

29. THE OSSEOUS PELVIC GIRDLE

29.1. The **OSSEOUS PELVIC GIRDLE** is often involved with **fractures,** particularly as a result of traffic accidents. In fractures where the fractured segments are displaced only slightly, spontaneous healing can take place in a quiet location without surgical intervention. The **muscles** adjacent to and attaching near the fracture site can promote healing due to their stabilizing influence but they may also increase displacement between the fractured ends due to contraction. In these cases spontaneous healing can have serious sequelae resulting for example in a narrowing of the birth canal.

29.2. Following trauma, each side of the pelvis may be separated from the other at the **PELVIC SYMPHYSIS.** Occasionally this is accompanied by dislocation of the sacroiliac articulation on the side of the injury. In the main, these interrelated injuries concern young dogs, the pelvic symphysis being still cartilaginous and ossifying only later in the adult state. Likewise, the tense connection at the sacroiliac joint is stronger and tighter in older dogs. In young individuals a very slight widening of the birth canal is possible due to cartilaginous symphyseal union. A ventromedian surgical approach to the pelvic symphysis is made through the symphyseal tendon (see p. 32). Surgical access to pelvic organs particularly in cases of prostatectomy necessitates splitting of the pelvic symphysis or removal of a portion of pubis.

29.3. **HIP DYSPLASIA** is a congenital condition present chiefly in the large breeds of dogs such as the German Shepherd, Newfoundland and Rottweiler. Depending on the degree of the condition, morphological changes are manifested as: a progressive flattening or bevelling of the acetabulum; incongruency between the head of femur and the articular or lunate surface of the acetabulum; and accompanying reactive tissue changes to the articular capsule and ligament of the head of the femur with the formation of collagen fibers. Furthermore, the head of the femur loses its spheroidal shape, the border of the acetabulum is reduced and the neck of the femur thickened. These anatomical changes promote luxations or subluxations of the hip joint (coxal articulation). Diagnosis is established unequivocally on the basis of radiographic findings and by the angular measurements of Norberg. In normal

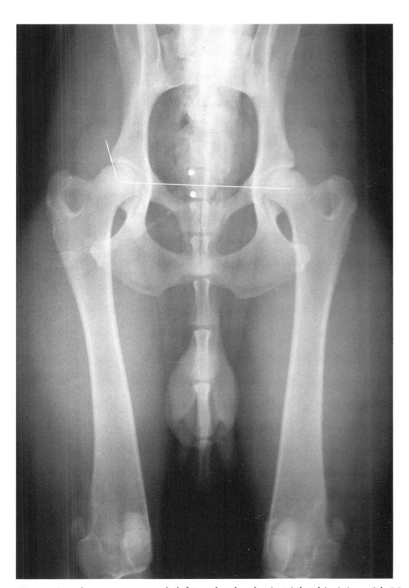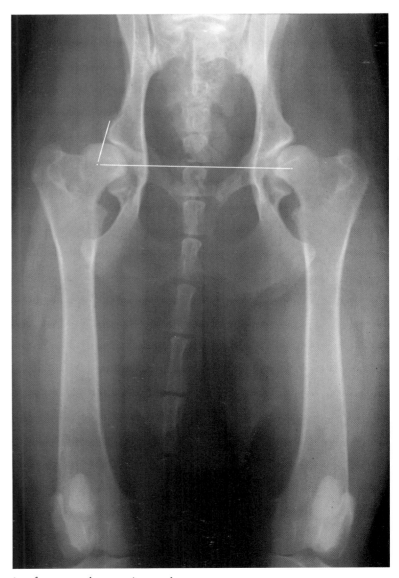

Comparison between a normal (left) and a dysplastic (right) hip joint with Norberg's reference angles superimposed.

dogs the line joining the midpoint of the head of one femur to that of the other forms an angle of approximately 105° with a line drawn to the cranial border of the acetabulum. In dogs affected with hip dysplasia the angle is less than 105° (see radiographs).

As a **surgical treatment** to relieve pain and delay further articular changes in the opening stages of the condition (low grade hip dysplasia) one performs a myectomy of the m. pectineus (et adductor longus) (see p. 32) or a tenotomy of its tendon of insertion.

Due to static changes in the hip joint brought about by the operation, the head of the femur attains a deeper position within the lunate surface of the acetabulum and the accompanying reduction in the tension of the articular capsule reduces pain. A deepening of the articular surface of the acetabulum is also achieved by osteotomy of the three pelvic bones and outer rotation of the acetabulum. In more advanced cases with concomitant luxation or subluxation, a resection of the head of the femur can alleviate the condition. Prosthetic hip replacements have been tested previously.

In creating a surgical access to the hip joint, because of its deep location, one cannot avoid the transection of muscles, tendons (tenotomy), parts of bone and the greater trochanter of the femur. Craniolateral access necessitates partial tenotomy of the m. gluteus profundus. A dorsolateral approach requires an osteotomy of the greater trochanter of the femur and the temporary displacement of the muscles inserting onto it. A ventral approach necessitates transection of the m. pectineus (et adductor longus) at its origin. (Alterations to head of femur, see below 30.1.)

30. PELVIC LIMB

30.1. HEAD OF FEMUR-Necroses are generally hereditary and aseptic in growing dogs, the smaller dog breeds such as Terriers and Miniature Poodles being those mostly affected. The etiology is presumed to be due to an insufficient supply of blood vessels. Progressive stages of the condition are detected radiographically, due to deformation of the head of the femur, indentation of articular cartilage and reparative processes such as compression of spongy substance and fiber formation in the endosteum. In the on-going condition, pathological changes such as endosteal tissue formation encroach on the neck of the femur. Good results are obtained by resectioning the head of the femur.

A **deformation of the head of the femur** also appears in the course of hip dysplasia (see under 29.3.).

In **fractures of the head of the femur,** also known as hip fractures of the neck of the femur, treatment must aim at avoiding an aseptic necrosis. The blood supply to the proximal end of the fracture should be interfered with as little as possible during surgery, while repositioning and fixation of the ends of the fracture should decrease disturbance to the blood supply from the distal part of femur. Blood supply comes chiefly from the periphery, more specifically from the lateral and medial circumflex arteries (see p. 34) the branches of which form a vascular plexus in the articular capsule. From there, fine branches go to the head of the femur. The epiphysis derives blood vessels from those of the pelvic bones which reach the femoral head through the ligament of the head of the femur. The diaphyseal vessels extend from the distal side through the neck of the femur and anastomose with epiphyseal vessels in the the adult. In young animals there is no anastomosis through the epiphyseal cartilage. Consequently in proximal fractures, epiphyseal blood supply may be deranged resulting in a necrosis of the head of the femur.

30.2. NECK OF THE FEMUR-Fractures can be treated in the same way as fractures of the femoral head (see above). In surgical treatment of such fractures, one should realize that the sciatic nerve runs over the m. gluteus profundus and both are situated dorsal to the neck of the femur. Fractures of the neck occur to a lesser extent in dogs than in man where the neck of the femur is essentially thinner. (In older humans the neck of the femur is vulnerable due to a particularly high incidence of osseous degenerative processes.) With a hip prosthesis the site of the condition is shortened and the prognosis improved considerably.

Normally the **head and neck of the femur** form an angle of about 130° with the body of the femur. Among other things a reduction of this angle (coxa vara) can be a consequence of fracture of the neck of femur. An increase in the angle (coxa valga) promotes subluxation.

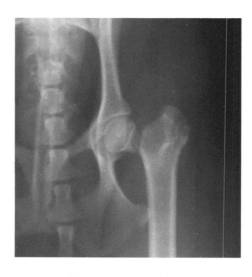

Fracture of the neck of femur in a Fox Terrier.

30.3. THE PATELLA-LUXATION (see 32.2.)

30.4. The TIBIAL TUBEROSITY (the insertion of the m. quadriceps femoris) and the cranial border of the tibia extending distally from it, form an apophysis with the tibia. During the growth period this is attached to the remainder of the tibia by an apophyseal cartilage. An abnormal detachment of the apophysis, favoured by the traction of the m. quadriceps, can lead to aseptic necrosis of the tuberosity following interruption to its blood supply (Osgood-Schlatter disease, see under 1.4.).

30.5. FIRST DIGIT. Where it is traumatized, an amputation is performed without generally posing any problems. Dewclaws are a breed characteristic and are duplicated in some breeds for example the Saint Bernhard.

32. MEDIAL ASPECT OF FEMUR and PATELLA

32.1. A myectomy or tenectomy may be performed on the **M. PECTINEUS (ET ADDUCTOR LONGUS)** in the treatment of hip dysplasia (see 29.3.).

32.2. PATELLA-LUXATIONS can be acquired or congenital. Generally the patella is displaced medially, no longer lying in the trochlea but on the medial condyle of femur. The axis of the limb turns laterally at the level of the knee. As an example of an acquired luxation of the patella

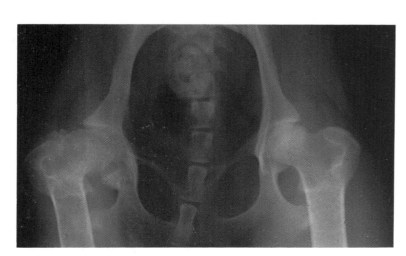

Head of femur-necrosis (left).

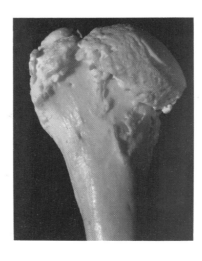

Deformation of the head of the femur (right, cranial view).

one should consider a laceration of the lateral femoropatellar ligament which pulls on the patella from the lateral sesamoid bone of the m. gastrocnemius. A much too flat trochlea of the femur and hypoplasia of the medial condyle of femur are quoted among other things as predisposing congenital factors contributing to congenital luxation. During surgical intervention the insertion of the patellar ligament can be displaced, the trochlea deepened or a metal prosthetic arch implanted to raise the level of the medial condyle.

When the cranial cruciate ligament is ruptured, a strip of fascia lata bounded by strips of patellar ligament or non-resorbable synthetic thread may be substituted. Depending on the surgical method used, the replacement tissue is guided through canals drilled to traverse the tibial tuberosity and the lateral condyle of the femur. The tissue may also run in an extraarticular manner between the sesamoid bones of the m. gastrocnemius and the tibial tuberosity.

33. THE SCIATIC, FIBULAR AND TIBIAL NERVES

33.1. In **PARALYSIS OF THE SCIATIC NERVE**, (for example in fractures of the pelvis), and the **FIBULAR** and **TIBIAL NERVES** following injections into the thigh muscles, different muscle combinations are involved. With sciatic involvement all muscles distal to the knee are paralysed. Involvement of the tibial nerve produces paralysis of the extensor muscles of tarsus, and injury to the fibular nerve paralysis of the flexors of tarsus.

34. BLOOD VESSELS OF THE PELVIC LIMB

34.1. The **MEDIAL** and **LATERAL CIRCUMFLEX FEMORAL ARTERIES** are the principal blood vessels supplying the head of the femur in the adult dog (see 30.1.).

35. HEAD

35.1. In its original form in cartilaginous fish, the **PRIMORDIAL CRANIUM** is a tubular cartilaginous brain capsule. In the course of phylogenesis the roof becomes more and more defective so that finally, only a basal cartilaginous cranium remains. The connective tissue defects are completed by osseous laminae arising directly from the connective tissue of the desmocranium and later fusing with the remaining cartilaginous primordial cranium at the base of the skull. Due to ossification emanating from numerous foci, the cartilaginous primordial cranium divides into several individual endochondral or replacement bones. These fuse partially with the membranous bones of the roof of the cranium to give **bones of a mixed nature** for example the temporal bone. In adults the only parts of the primordial cranium remaining are the nasal cartilages and the cartilaginous growth sutures at the base of the skull.

Fonticuli or fontanelle are connective tissue fissures which develop perinatally between the bones of the cranium. On account of the palpable pulse detected in them they are given the Latin name meaning a source. In the human neonate, it is possible to use the site for taking blood from the dorsal sagittal sinus. Due to peripheral growth of the bones limiting the cranium, the fonticuli are compressed early to form the cranial sutures. This process varies depending on the breed.

The **sutures** limit the individual bones of the skull and in the juvenile, function as growth gaps or seams. Their connective tissue proliferates strongly thus producing the initial material for osteogenesis. Skull sutures and fonticuli permit a small amount of reciprocal displacement to take place between the individual bones of the skull. Particularly in man, this is an advantage during parturition.

35.2. The **HYOID BONE** and sections of the larynx develop from the branchial skeleton. In the transition from an aquatic to a terrestrial life, the branchial apparatus of the organism underwent a functional alteration, in which its cartilaginous portions became integrated into the auditory and the voice producing apparatus. A part of the first branchial cartilage known as Meckel's cartilage, differentiates into incus and malleus, two of the ossicles of the middle ear. The remainder induces the development of the mandible and then disintegrates. The second branchial cartilage, Reichert's cartilage, gives rise to the stapes, the third of the ossicles of the middle ear, and the suspensory apparatus of the hyoid bone associated with the skull. The remaining parts of the hyoid apparatus and the larynx develop from the remaining branchial arches (refer to embryology text). (For radiographs, see under 41.7.).

36. SKULL

36.1. The **WALL OF THE SKULL** (the roof is the **calvaria**) consists of five layers.

I. The **pericranium** is fused with the connective tissue of the sutures and may be detached from extensive osseous surfaces lying between them. (Therefore subperiosteal bleeding does not pass to the outside at the sutures.)

II. The **external lamina** of the skull bone is a compact (cortical) osseous layer.

III. The **diploë** is a reduced layer of spongy bone with structural pecularities. Its small hollow spaces contain diploic veins concerned with thermoregulation. Due to intense blood flow through the diploë, temperature differences between external lamina and brain are equalized. (An over-demand on temperature regulation occurs in sunstroke.)

IV. The **internal lamina** is the internal cortical (compact) layer of cranial bone.

V. The **endocranium** is fused with the dura mater.

36.2. The **PARANASAL SINUSES** are cavities between the external and internal laminae into which respiratory mucous membrane extends from the nasal cavity. The sinuses are very small in neonates and increase considerably in size with age. Besides the frontal paranasal sinuses of the dog, many authors still recognize a **maxillary sinus.** This is not correct since no cavity develops within the maxilla. There is a laterally directed outpocketing of the maxilla and this should be called Recessus maxillaris. The roots of the maxillary cheek teeth border the recess so that in empyema of the root/s of these teeth the recess also becomes involved. Concerning the existence of an sphenoidal sinus there are also differing opinions. Nickel/Schummer/Seiferle (1984) state that it is not present in the dog whereas Evans/Christensen (1979) consider it present and packed with ethmoidal conchae. Functions ascribed to the paranasal sinuses are: Reduction in weight of the cranium; thermal isolation of the brain, eye and nasal cavity; a resonance space for voice formation; and an increased surface area of mucous membrane and resultant secretory surface. Inflammation of the paranasal sinuses (sinusitis) arises in general from the nasal cavity. Recovery from infection of the paranasal sinuses is impeded due to their narrow routes of access. Following swelling of the mucous membrane, the entrances are obstructed and their aeration ceases. Since mucous membrane resorbs air, (keeping in mind self-recovery after pneumothorax), pressure is also reduced within the sinus. In turn, this produces venous congestion, fluid accumulation and bacterial invasion in the frontal sinuses. In this event they are trephined, flushed out and the connections to the respective nasal cavities widened. Generally a suppurative sinusitis is recognized by a uni- or bilateral nasal discharge and the chronic condition can lead to a 'meltdown' of the internal lamina of the adjacent bone and development of a meningoencephalitis.

In long-standing processes of this nature **radiography** may demonstrate the degree of inflammation and fluid accumulation, new bone formation and osseous destruction.

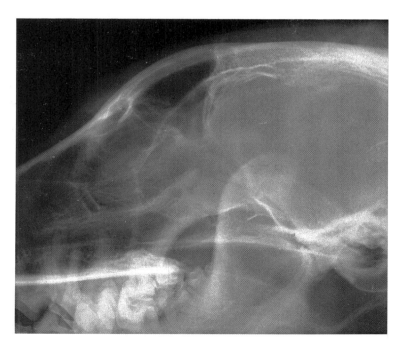

Radiography of the frontal sinus of an adult dog.

04

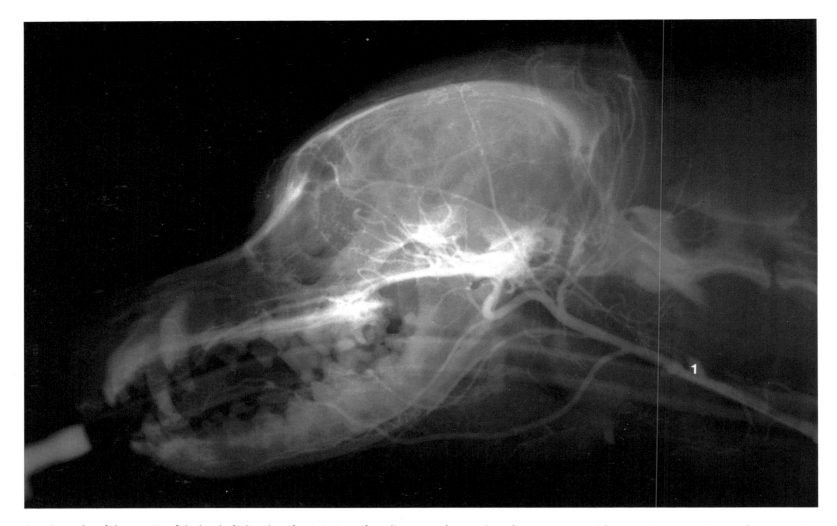

Arteriography of the arteries of the head of a live dog after injection of a radiopaque substance into the common carotid artery. Due to traction on the artery, the internal carotid artery and the carotic sinus are displaced caudally (1). For identification of arteries see p. 37A.

36.3. The **SELLA TURCICA** is occupied completely by the hypophysis and its surrounding subarachnoid space. Therefore by indirect radiography one can deduce the size of the radiolucent hypophysis or pituitary gland. The sella turcica enlarges with a pathological increase in size of the pituitary gland because osseous tissue gives way under the increased pressure. Osseous tissue on the inside of the sella is destroyed through osteoclastic activity and is built up on the outside through the action of osteoblasts. Particularly is this true in man where the sella turcica is deep and almost closed off dorsally by the diaphragma sellae.

36.4. Externally the **FACIAL BONES** have osseous features, some of which are easily palpated as for example the osseous orbit, zygomatic bone and infraorbital foramen. Using these, one can orientate the skull during clinical examination and surgery. Deeply situated, non-palpable parts of the skull are identified radiographically.

36.5. In primitive species the **TEMPOROMANDIBULAR ARTICULATION** is formed by incus and malleus and derivatives of the first branchial cartilage (of Meckel). This can be regarded as the primary articulation. In the course of evolution a change of function occurs with the transformation of incus and malleus to ear ossicles, and a secondary articulation, the one remaining, is constructed. Officially this is condylar in type and permits a wide 'angle of opening' of the mouth. It has a pronounced instability, however, with the possibility of a 'physiological luxation' occurring, as for example, with yawning and maximum opening of the mouth.

The **articular disc** is a biconcave, smooth fibrocartilage dividing the joint cavity into two levels or 'storeys', one situated on top of the other.

37. THE FACE

37.1. The **SUBLINGUAL VEIN** is easily accessible on the ventral side of the tongue. In an emergency situation it is used for intravenous injections of stimulant, for example, in the anesthetized dog.

37.2. The **ARTERIES OF THE HEAD** are identified by radiography and compared with the lower illustration on p. 37A.

Arteriography (see illustration) is a special method of demonstrating specific arteries and their branches after the intraarterial injection of radiopaque material. An injection into the vertebral artery and/or internal carotid artery is performed to demonstrate arteries at the base of the brain. The external arteries of the head are demonstrated by an injection of contrast medium into the external carotid or the common carotid artery. On the basis of the course of the artery one can detect the three dimensional formation of new blood vessels and estimate any interruptions to blood flow.

37.3. **FACIAL PARALYSIS** always leads to different degrees of pathological atrophy distal to the site of damage, and may appear either bilaterally or unilaterally. With **central facial paralysis** the lesion is situated in the medulla oblongata in the vicinity of the facial nucleus. With a deficit to the mimetic muscles, the muscles of eyelids and forehead remain intact because the nuclei of these muscles only, receive impulses from both cerebral hemispheres. With **peripheral facial paralysis** the lesion lies either at the facial nucleus or along the course of the facial nerve.

In **complete peripheral paralysis** efferent nerve branches may be injured within the petrous temporal bone with corresponding degrees of injury. Damage to the chorda tympani, the nerve to the stapedial muscle and the greater petrosal nerve produces respectively disturbances in taste sensation and salivary secretion, hearing, and lacrimal secretion. Derangement of salivary secretion is significant in man. Paralysis of mimetic muscles produces muscle flaccidity, resulting for example, in a drooping of the eyelid. More particularly in this case a nervous deficit to the muscles of the eyelid causes it to be neither completely closed (by the m. orbicularis oculi) nor completely opened (by the m. levator anguli oculi medialis). The intact m. levator palpebrae superioris, supplied by the oculomotor nerve and the smooth m. tarsalis also prevent the open eyelids from closing. When the eyelids do not close completely, lacrimal fluids are not distributed over the cornea and this leads to corneal desiccation, cloudiness and opacity, and eye infection. As stated above, damage to the greater petrosal nerve may derange lacrimal secretion and produce the same result. The corneal reflex (see 40.1.) is weakened with facial paralysis. With a nervous defect to the m. buccinator, food collects in the oral vestibule of the cheek and cannot be pushed between the dental arches.

38. FACIAL AND MANDIBULAR MUSCLES

38.1. During plastic surgery the **M. SCUTULOAURICULARIS** is displaced dorsomedially to erect the auricular cartilage when it is drooping or pendulous. The muscle is sutured to the m. interscutularis and the scutiform cartilage.

38.2. To create surgical access to the dorsolateral surface of the cranium, the M. TEMPORALIS is detached from the temporal fossa. This is a necessary initial procedure in the treatment of fractures and in the removal of subdural and subarachnoid hematomata in the region of the frontal and parietal bones. (The latter surgical procedure is seldom resorted to.)

39. MANDIBULAR NERVE

39.1. PARALYSIS OF THE MANDIBULAR NERVE following trigeminal neuritis of unknown etiology leads to a deficit of the muscles of mastication and a drooping downwards of the mandible.

39.2. The MENTAL NERVES are blocked with local anesthetic at the palpable sites of exit, namely the mental foramina. They supply sensory innervation to the lower lip. (Sites of anesthesia, see illustration below 43.17.)

40. EYELIDS AND LACRIMAL APPARATUS

40.1. The UPPER AND LOWER EYELIDS limit the palpebral aperture and in rare congenital cases remain closed. They may be regarded as fibromuscular 'laminae', the inner surfaces of which bear a non-keratinized stratified squamous epithelium that turns into a prismatic epithelium within the conjunctival sac and finally extends to sclera and then cornea. This epithelium is the conjunctiva. Both external and internal lid surfaces merge with each other at the free border of the eyelid. On sagittal section this border is approximately rectangular and bears eyelashes on its external and internal margins. These serve as a protection from foreign material. The internal 'skeleton' of the eyelid is the tarsus, a curved plate of collagen fibers adapted to the curvature of the eye.

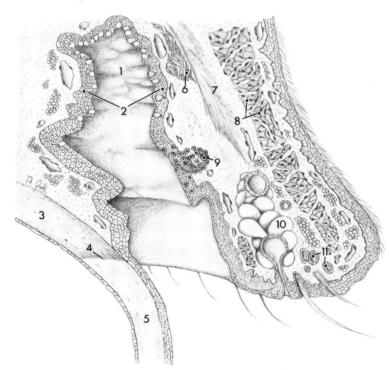

1 Superior fornix of conjunctiva; 2 Conjunctiva (epithelium with goblet cells); 3 Sclera; 4 Limbus of cornea; 5 Cornea; 6 M. tarsalis superioris; 7 Superior tarsus; 8 M. orbicularis oculi; 9 Conjunctival lymph nodules; 10 Tarsal gland; 11 Ciliary gland.

The **muscular basis of the eyelid** is formed by the m. orbicularis oculi (innervated by the facial nerve), the m. levator palpebrae superioris (oculomotor n.) in the upper eyelid, and the smooth m. tarsalis (sympathetic innervation). A drooping of the upper eyelid may have different origins such as damage to the oculomotor nerve, facial paralysis (see 37.3.) and sympathetic paralysis. Horner's syndrome, due to damage of sympathetic fibers of nT1, is recognized by several clinical signs. These include: Ipsilateral myosis or constriction of the pupil; narrowing of the palpebral fissure due to ptosis or drooping of the upper eyelid; extrusion of the nictitating membrane; and enopthalmus, an ocular bulb situated within the orbital cavity in an abnormally deep manner.

The **glands of the eyelids** include: the tarsal glands or sebaceous glands of Meibom opening on the border of the eyelid; and the ciliary glands or apocrine sudoriferous glands of Moll. Obstruction of the tarsal glands produces a painless nodule formation on the border of the eyelid and inflammation of the ciliary glands produces a very painful 'sty'.

Lacrimal fluid is distributed over the surface of the cornea due to **blinking of the eyelid.** Closure of the eyelid is brought about reflexly by tangential contact between lid and cornea (blink or lid-corneal reflex). The afferent nerve fibers of the reflex arc run in the long ciliary nerve (VI, see p. 40), and the efferent course in the facial nerve to the m. orbicularis oculi. Defects in the positioning of the borders of the eyelids occur in a two-fold manner. An infolding of the lid margins or entropion irritates the cornea mechanically and an outfolding or ectropion may occur in facial paralysis where the conjunctival sac is exposed. Such defects of the free borders of the eyelids lead to corneal and conjunctival inflammations. In case of the generally treatment is unsuccessful and eventual surgical correction is required.

40.2. The THIRD EYELID or nictitating membrane lies at the medial angle of the eye and is supported by the contained cartilage. Pressure applied to the eye ball pushes the membrane across the corneal surface. The same clinical sign arises in tetanus where there is contraction of the m. retractor bulbi, and in inflammation of the m. temporalis in the German Shepherd (eosinophilic myositis). Numerous lymph nodules (so-called lymph follicles) are present on the external and particularly the internal surface of the eyelid. They enlarge in follicular conjunctivitis and are removed surgically.

The **conjunctiva** extends from the posterior surface of the eyelid to the **fornix of the conjunctiva**, around the sclera and, as the bulbar conjunctiva, onto the corneal surface. The prismatic epithelium of the conjunctival sac is capable of resorbing fluids as in the application of eyedrops and certain drugs. Goblet cells of the conjunctiva produce a mucous secretion to reduce friction during blinking of the eyelids.

40.3. The LACRIMAL GLANDS and GLANDS OF THE THIRD EYELID secrete serous lacrimal fluid. This has a nutritive and moistening function on the cornea, regulating its physiological tumefaction and transparency. The fluid also acts as a cleansing agent of conjunctiva and conjunctival sac as a protection against conjunctivitis. This is achieved by antibacterial substances such as lysozyme.

An **abnormal exhaustion** or drying up of lacrimal flow leads to keratoconjunctivitis sicca, that is a desiccated cornea and inflammation of the conjunctiva. After failure of conservative treatment such as artificial lacrimal fluid preparations, a surgical operation was designed to reposition the parotid duct into the lateral commissure of the affected eye. This method has been practiced more frequently in the past than it is today.

40.4. Occlusion of the LACRIMAL PUNCTA or the NASOLACRIMAL DUCT may be cleared by catheterization or flushing. In the former instance entry is made through the puncta. The nasolacrimal duct occasionally has two nasal openings, the caudal or additional one opening deep within the nasal cavity at the rostral end of the osseous lacrimal canal. With a blockage of the nasolacrimal duct, fluid escapes from either punctum when the other is used as the site of injection or flushing. In such instances the nasolacrimal duct is flushed through by inserting the catheter through the nostril and into the rostral ostium of the duct.

40.5. ANESTHESIA OF THE EYE is attained by injection of local anesthetic within the periorbita or into the orbital fissure. The (sensory) ophthalmic nerve (V1) and motor nerves to eye muscles are desensitized, the latter to avoid a retraction of the ocular bulb during the operation (see illustration below 43.17.).

41. NOSE AND LARYNX

41.1. In brachycephalic dog breeds such as Pekingese the NARES or nostrils have a tendency to be stenotic since there is insufficient cartilaginous support for the nostrils. In a similar manner to a one-way valve, closure of the nostrils occur during inspiration with resultant difficult breathing or dyspnea. In such dog breeds facial bones are also relatively short and nasal cartilages are not covered at the level of the tip of the nose. During clinical examination the nasal cartilages may also be deformed while keeping the muzzle closed, and the air flow is then interrupted unintentionally.

41.2. The NASAL CAVITY is included in the 'respiratory passageways' in which inspired air varies in temperature and humidity, and is cleaned of foreign material during transit. The equalization of the temperature of inspired air (30°C) is attained through turbulence of airflow and the very good blood supply to the nasal cavity. Extensive humidification of inspired air is made possible by a warm, moist, nasal mucous membrane, while foreign bodies such as fine particles and insects are delayed in the hair bordering the nares. The finest particles and bacteria and spores remain adherent to the moist mucous membrane and are removed in the first instance by the action of cilia, or expelled by coughing and sneezing. The nasal cavity fulfils the further important function of regulating body temperature.

Inflammation of the nasal mucous membrane may be caused by viruses, bacteria or mycoplasms, indicated by frequent sneezing, nasal discharge and tissue thickening. Tissue thickening and tumors causing deformation and destruction of nasal conchae or septum may be demonstrated radiographically.

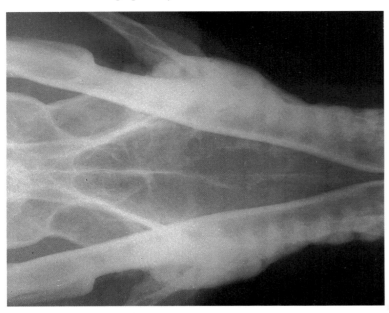

A ventrodorsal radiograph of the nasal cavity with partial destruction of nasal septum and hard palate by a tumor.

41.3. The **LARYNGEAL MUCOUS MEMBRANE** may swell markedly in the region of vestibule, for example in allergic reactions, and in extreme cases block the passage of air. The usual term 'edema of the glottis' used for this condition is incorrect since the glottis and apparatus of phonation is not swollen. More particularly, the epiglottis is edematous.

Besides the vestibular and vocal folds, another **fold of mucous membrane of the larynx**, the **aryepiglottic fold**, is present. This reaches from epiglottis to arytenoid cartilage and forms the lateral boundary of the aditus of the larynx. The **piriform recess** of the pharynx lies lateral to the aryepiglottic fold and the aditus. Dorsally, the cuneiform and corniculate processes of the arytenoid cartilage form the cartilaginous supports of the aryepiglottic fold of its respective side. A collapse of the cuneiform process occurs in small brachycephalic dog breeds and is often associated with a stenosis of the nares. Serious dyspnea can result during inspiration since the affected cuneiform process is displaced into the lumen of larynx and compresses or restricts it.

41.4. The **VOCAL LIGAMENT** extends from the vocal process of the arytenoid cartilage to the inner surface of the thyroid cartilage ventrally. It consists, predominantly, of elastic fibers, and in common with the m. vocalis is a part of the vocal cord.

41.5. The **RIMA GLOTTIDIS** is bounded ventrolaterally by the vocal folds where it is termed the intermembranous part of the rima, and dorsally by arytenoid cartilages at the intercartilaginous part of the rima. The glottis functions in voice formation, produced by vibrations of the margins of the vocal folds which depend in turn on sound frequency. Laryngeal musculature can vary the tension of the vibrating margins of the vocal folds in a voluntary manner and, in common with the strength of the air current, influence vibration, frequency and amplitude. In this way the volume and pitch of sound are regulated. Age, sex, and steroid hormone content, for example testosterone, influence length and thickness of the margins of the vocal folds and thus the pitch of the voice. With castration for example, the longitudinal growth of the vocal ligaments ceases and this results in a higher register of the voice. Sound produced at the glottis is modified in spaces such as nasal or oral cavities and paranasal sinuses which act as resonance chambers. Functionally the glottis is associated with coughing, where the rima glottidis is closed initially. Then, due to an exspiratory impulse, the rima is pushed open under greatly increased pressure and air is expelled very fast. Coughing cleans the air passage, foreign material and mucus being discharged with the impulse of air.

41.6. In brachycephalic dog breeds, the **LATERAL VENTRICLES OF THE LARYNX** may prolapse into the lumen of the larynx due to

negative pressure, and are then ablated surgically. They are also removed when dealing with paralysis of the recurrent nerve (roaring, see 41.8.).

41.7. The **LARYNGEAL CARTILAGES** form the skeleton of the larynx. The **elastic epiglottic cartilage** is situated cranially to protect the aditus of the larynx, its curvature facilitating this. The remaining laryngeal cartilages are hyaline and in older dogs areas of calcification may be seen radiographically. The cartilages are connected by articulations one under the other and in this way possess a certain freedom of movement. The **cricothyroid articulation** lies between the caudal cornu of the thyroid cartilage and the lateral surface of the cricoid cartilage. The **cricoarytenoid articulation** lies between the arytenoid cartilage and the cranial border of the cricoid cartilage.

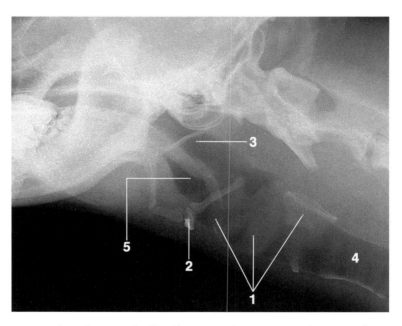

Laryngeal cartilages (1); basihyoid (2); nasopharynx (3); trachea (4); and pars digestoria of pharynx (5). (For details of hyoid apparatus see p. 41A.)

41.8. The **LARYNGEAL MUSCLES** connect the thyroid, cricoid and paired arytenoid cartilages with one another. According to their function one differentiates '**muscles of position**' inserting onto the arytenoid cartilage and through their movement altering the extent of the rima of the glottis, and '**tension muscles**' which vary the tension of the vibrating wall of the larynx. The 'muscles of position' include the **m. cricoarytenoideus dorsalis** that draws the muscular process of the arytenoid cartilage caudodorsally and medially, and the vocal process and attached vocal fold of its respective side laterally. This widens or dilates the rima of the glottis. The **m. cricoarytenoideus lateralis** is likewise a muscle of position and functions as an antagonist to the m. cricoarytenoideus dorsalis. It draws the muscular process of the arytenoid cartilage ventrally and laterally and swivels the vocal process medially. In so doing it positions the margin of the vocal fold against that of the other side. This therefore narrows or constricts the rima of the glottis. The 'tension muscles' include the **m. cricothyroideus**, the only laryngeal muscle to derive motor innervation from the cranial laryngeal nerve.

In **paralysis of the recurrent laryngeal nerve** causing stenotic inspiratory noises termed 'roaring', those muscles supplied by the nerve atrophy, the first being the m. cricoarytenoideus dorsalis. In contrast to 'roaring' in horses, where the left recurrent laryngeal nerve is mostly affected, the condition in dogs is rare and chiefly bilateral. Treatment involves resection of the vocal folds or the mucous membrane lining the lateral ventricles of the larynx.

41.9. In the midline, the **VESTIBULE OF THE MOUTH** is divided partly into two by the superior and inferior labial frenula. The **philtrum** extends as a groove from the upper lip onto the **planum nasale**, or nasal plate, thus bisecting it. **Cheiloschisis** or **harelip** is a congenital defect of the upper lip lying between the embryonic nasomedial process and the maxillary process. A maxillary cleft divides the maxilla between the canine and incisor teeth.

41.10. A cleft **PALATE** occurs as a congenital defect or after trauma. The latter condition is seen in cats having fallen from a height. A **median cleft palate** or palatoschisis is due to a failure or incomplete union of the embryonic palatine processes. Oral and nasal cavities are no longer separate and material to be swallowed enters the nasal cavity. A combination of several defects can involve **a cleft lip, palate and maxilla.** The cause of these defects has been ascribed to genetic, toxic and hormonal factors, as well as enviromental influences such as hypo- and hypervitaminoses. Treatment involves surgery. Further **malformations** involving alterations to or shortening of maxilla or mandible, are known as: **agnathia** (gnathos - a jaw), an absence of maxilla or mandible; **brachygnathia,** a shortening of maxilla or mandible; and **prognathia,** a lengthening of either bone.

In contrast to the hard palate, the **soft palate** has no osseous foundation and is supported in particular by striated muscles and the palatine aponeurosis. The mucous membrane on the oral surface is cutaneous in type, whereas that on the nasopharyngeal surface is a respiratory epithelium. This is capable of sliding and swells markedly when inflamed. In small and middle sized dog breeds the soft palate is long and relaxed and as a result, problems may occur with breathing, and increase with age. The soft palate can be shortened surgically. To perform an hypophysectomy as for example in the treatment of Cushing's syndrome, surgical access is gained through soft palate and sphenoid bone.

41.11. The **PHARYNGEAL OSTIUM OF THE AUDITORY TUBE** is surrounded by an elevation formed partly from supporting cartilage and partly from adjacent muscle tissue. The auditory tube serves as a pressure equalizer between middle ear and tympanic cavity on the one hand, and the external ear on the other. Only in this way can the tympanic membrane separating the two, vibrate optimally. By being subjected to great differences in altitude, pressure equilibrium is disturbed, producing 'pressure in the ears' and temporary difficulty in hearing. Pressure equilibrium is restored by swallowing, the wall of the auditory tube being opened by the contracting m. tensor veli palatini which inserts onto it. In pharyngitis accompanied by tubal catarrh, the opening mechanism at the ostium fails to work on account of the marked swelling of the mucous membrane. Thus, due to resorption of air out of the tympanic cavity and a lack of air current in the auditory tube, air pressures are no longer balanced. Ear ache and disturbance in hearing result.

41.12. **TONSILS** belong to the lymphatic organs. The paired palatine tonsil is the most important of those of the lymphatic pharyngeal ring in the dog. In comparative anatomy it is classified as nonfollicular and housed in a fossa. The palatine tonsil has a characteristic stratified squamous epithelial covering. In the lamina propria mucosae deep to the epithelium there is lymphoreticular tissue containing typical primary and secondary lymph nodules. Numerous lymphocytes from the lymphoreticular tissue penetrate the epithelial covering, thereby 'reticulating' it. At a deeper level the lymphoreticular tissue is bounded by a distinct tonsillar capsule of connective tissue.

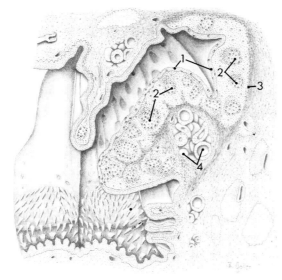

Palatine tonsil in tonsillar sinus. 1 Epithelial covering; 2 Lymph nodules; 3 Tonsillar capsule; 4 Mucous glands.

The tonsil **functions** presumably in defence against infections penetrating through nose and mouth. Infection is recognized and the formation of specific antibodies takes place. Like other lymphatic tissue, tonsils are best developed in young animals. This is necessary since the young animal has to sustain infections against which the adult has already gained its immunity.

With increasing age tonsillar tissue regresses. **Tonsillectomy** is indicated in chronically recurring tonsillitis or pharyngitis or in carcinoma of the squamous epithelial covering of the tonsil. The entire tonsillar tissue is enucleated from its underlying tonsillar capsule, keeping in mind that complications may occur due to hemorrhage from nume-

rous branches of the lingual artery. The palatine tonsil should not be removed without due consideration since it fulfils an important immune function. Nevertheless an excessive production of antigen-antibody complexes in the course of recurring tonsillitis may lead to a considerable detrimental effect on kidneys and heart.

42. **MUSCLES OF PHARYNX**

42.1. Of the **PHARYNGEAL MUSCLES,** the pharyngeal constrictors arise on cranium, hyoid apparatus and larynx, and form a 'sling mechanism' opening ventrally. They insert dorsally on the pharynx along the median pharyngeal raphe which is a longitudinal tendinous band attached to the base of the cranium.

42.2. The **MUSCLES OF THE SOFT PALATE** radiate in the palatine aponeurosis, thus forming the connective tissue skeleton of the soft palate.

Their **function** is to seal off the nasopharynx during swallowing and equalize pressure between the middle and external ear (see 41.11.). During **swallowing** food is passed from the oral cavity through the pharynx and into the esophagus, the nasopharynx and aditus of larynx being closed off simultaneously. In the initial **voluntary phase of swallowing** the muscles of the floor of the mouth become tense and the tongue is pushed against the hard palate like a piston. As a result, food material is pressed caudally and passes over the convexity at the base of the tongue to reach the narrow oropharynx in the vicinity of the intrapharyngeal ostium. There the final **involuntary phase** of swallowing begins with the contraction of the constrictor muscles of the pharynx. Food material passes to left and right, past the epiglottis and through the left and right piriform recesses into the esophagus.

The bolus then reaches the stomach by peristaltic movement along the esophagus. In the narrow oropharynx, food must be secured from going in three 'wrong' directions. 1 **A reversal back into the oral cavity** is prevented through pressure of the tongue against the soft palate. 2 **Entrance into the nasopharynx** is obstructed due to elevation and tension of the soft palate. 3 To prevent the **uncovering of the aditus of the larynx,** the elastic epiglottic cartilage is deflected over the aditus in a passive manner. This occurs with forward movement of the larynx and upward projection of the base of the tongue.

42.3. The **SYMPATHETIC DIVISION** of the autonomic system supplying the head is derived from the cranial cervical ganglion, with postganglionic fibers accompanying the large blood vessels, particularly the internal carotid artery (internal carotid plexus), the external carotid artery (external carotid plexus), and the internal jugular vein (the jugular nerve). These postganglionic fibers proceed to the effector organs. **Stimulation of the sympathetic division** causes dilation of the pupil (mydriasis) due to strong contraction of the m. dilator pupillae, and a widening of the palpebral aperture due to strong contraction of the smooth m. tarsalis. In **Horner's syndrome** the sympathetic supply of the head may be deranged or interrupted along its considerable length, for example by goitre, enlarged lymph nodes, tumors of the mediastinum or prolapse of intervertebral discs. This is evident with ptosis (a narrow palpebral aperture), myosis (a constricted pupil), prolapse of the third eyelid and enophthalmos (a retracted ocular bulb).

42.4. The **PARASYMPATHETIC DIVISION** of the autonomic system supplying the head is included in cranial nerves III, VII, IX and X. Preganglionic parasympathetic fibers of the **oculomotor nerve (III)** proceed in the orbit to the ciliary ganglion from where postganglionic fibers reach the mm. ciliares (accommodation) and the m. sphincter pupillae (miosis). Parasympathetic fibers associated with the **facial nerve (VII)** run through the pterygoid canal to the pterygopalatine ganglion. After synapsing there, postganglionic fibers proceed to meninges, nasal and lacrimal glands, and the nasal mucous membrane. Another parasympathetic part of the facial nerve reaches the salivary glands in the chorda tympani. The **parasympathetic fibers** included in the **vagus nerve (X)** run predominantly within the large body cavities, and generally have their first synapses in the intramural ganglia of the effector organs. The proximal vagal ganglion, situated in the jugular foramen contains the nerve cell bodies of somatic afferent sensory fibers. The nerve cell bodies of the viscerosensory fibers are in the distal vagal (nodose) ganglion. In general, it is true to say that the synapsing of somatic sensory fibers occurs neither in the proximal ganglion nor in the distal. The nerve cell bodies of the motor autonomic fibers are situated in the rhombencephalon.

42.5. The **AURICULAR CARTILAGE** is involved in the surgical treatment of chronically recurring otitis externa. A section of the cartilage in the shape of a blunt V is removed from the external auditory canal (see p. 42A), keeping in mind the abundant blood supply to the area. Blunt dissection is used to retract the parotid gland from the lateral aspect of the external auditory canal. Then skin and remaining auricular cartilage are sutured at the transition between horizontal and perpendicular parts of the canal. This provides a drainage route for secretion from the wound, greater access for air, and a better opportunity for healing to occur. In rare cases, deep seated conditions, such as neoplasms of the horizontal part of the external auditory canal, require the removal of the whole canal, both perpendicular and horizontal. The free part of the auricular cartilage is preserved in such circumstances.

42.6. The **PINNA** may also exhibit an extensive **hematoma** on its inner concave surface and less frequently on its external surface. In some cases the origin is obscure, while others are due to shaking of the head coincidental with otitis externa, or from trauma resulting from a blow. The hematoma is due to hemorrhage from smaller branches of auricular blood vessels of which the finer rami run through small apertures in the auricular cartilage to the concave surface. Treatment involves drainage or surgical incision.

Ear wax which has undergone bacterial decomposition is a predisposing factor to otitis externa. The yellow-brown material is a mixture of secretion from modified apocrine sweat glands, sebaceous glands and desquamated keratinized cells.

43. TONGUE AND DENTITION

43.1. The **ROOT OF THE TONGUE** is noticeably large and torose in brachycephalic dog breeds and can hinder the restricted airflow and breathing.

43.2. The **APEX OF THE TONGUE** contains the **lyssa** rostrally and in the midline. Centuries ago it was wrongly considered to be a worm causing rabies, lyssa being the Greek term for the disease. In fact, the lyssa is a morphological pecularity in the apex of the tongue of carnivors, easily demonstrated on cross section. It is a connective tissue tube approximately 4 cm long containing fat, striated muscle and cartilage cells.

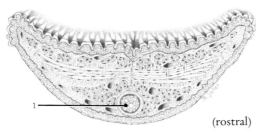

(rostral)

Cross section of apex of tongue with lyssa (1) situated midsagittally.

43.3. **TASTE RECEPTORS** are a constituent part of the taste buds chiefly present in the ring-shaped grooves of vallate papillae, on the foliate and to a lesser extent on the fungiform papillae. In addition, in man and probably the domestic mammals, taste buds are also present at the root of the tongue and the base of the epiglottis. These are innervated by sensory fibers of the vagus nerve (X). Taste buds are barrel-shaped and consist of approximately thirty cells (taste cells, supporting cells and basal cells). The 'barrel opening', that is the gustatory pore, is situated on the surface of the mucous membrane. The taste or gustatory cells are secondary sense cells which synapse with sensory nerve fibers. The moat-like groove surrounding the central part of the vallate papilla contains some 150 taste buds. **Serous gustatory glands** (of Ebner), opening on the floor of the groove, probably dissolve material to be tasted and wash it away subsequently. Serous and seromucous glands, regarded as 'minor' or 'intrinsic' salivary glands are present on the caudal third of the tongue.

43.4. Due to the three dimensional arrangement of longitudinal, transverse and perpendicular muscle fibers, the **intrinsic muscles of the tongue** have a characteristic structure that is organ specific. Here the typical envelopment of bundles of muscle fibers with endo-, peri-, and epimysium as well as deep fascia, is modified drastically. The striated muscle fibers also have occasional branchings which are not typical.

Functionally, the three dimensional arrangement of the muscle fibers of tongue permits an enormous range of movement which is of particular advantage in the prehension and assimilation of food. In mammals the tongue can form a tube and by contraction of perpendicular and transverse muscle fibers become thinner and actually longer. This is a unique function of muscle.

43.5. The **SALIVARY GLANDS** secrete saliva into the vestibule or the oral cavity proper, the several glands being grouped together as the **major salivary glands.** Of these, in the dog, the parotid gland secretes a serous, and the mandibular and sublingual glands a seromucous saliva. The **regulation** of salivary flow is under the control of the **parasympathetic division** and to a lesser extent the sympathetic, and parasympathomimetic drugs stimulate its secretion. The flow of saliva occurs before and during the intake of food. **Salivary calculi** and **dental tartar** are salivary sediments containing high levels of calcium and in the case of tartar, also decomposed and foul-smelling cell detritus. Salivary calculi lie in salivary ducts and can obstruct them. In turn this may lead to an accumulation of secretion, retention cysts and pressure atrophy of the salivary gland concerned. Laceration of the salivary ducts gives rise to cysts (see 43.7.).

43.6. The **PAROTID GLAND** is traversed by blood vessels and the facial nerve. Therefore pathological changes such as inflammation or tumors, can cause facial paralysis. The parotid duct, running superficially across the m. masseter may be transplanted into the lateral angle of the eye to provide a 'substitute' for lacrimal fluid in cases of keratoconjunctivitis sicca (see 40.3.).

43.7. The **SUBLINGUAL GLAND** is clinically significant, injury to its duct causing the formation of a sublingual cyst. This lies on the floor of the oral cavity and is known as a ranula because of its similarity to a frog's abdomen. When situated cervically it is also known as a honey cyst. A diagnosis as to whether the right or left duct is affected, can be confirmed by contrast radiography. Radiopaque material is injected through the sublingual caruncle. Since the monostomatic sublingual and the mandibular gland are tightly adjacent, and the surgical excision of the sublingual gland without injury to the neighbouring mandibular gland is not possible, both are removed in common.

43.8. **SUPERNUMERARY TEETH (POLYDONTIA)** arise from additional teeth primordia. **Reduction in the number of teeth (oligodontia)** is observed particularly in brachycephalic dog breeds where the last (caudal) teeth of the dental arch are absent. Where embryonic teeth primordia are absent, **diastases** or separations occur between teeth. With tooth loss due to trauma or breaking off of a tooth, the osseous alveolus closes over after a few weeks. This, however, is not true for the alveolus of the canine tooth in the old dog, and after lodgment of food particles an inflammation of the nasal cavity can erupt. Within the dental arch there is self-regulation with a tendency to restore the continuity of the dentition. In the **normal position of the teeth** upper incisors are somewhat rostral to the lower ones to give a physiological cutting or shearing dentition. A faulty cheek dentition occurs where maxilla and mandible are situated exactly one above the other. **Anomalies of position** appear due to rotation of a tooth around its long axis (torsion) and around a transverse axis in brachycephalic breeds.

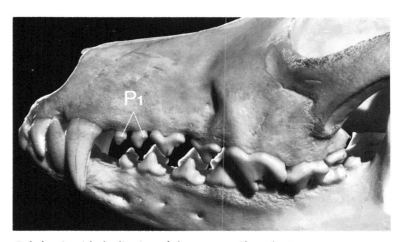

Polydontia with duplication of the upper wolf tooth (P1).

43.9. **INCISOR TEETH** are triangular-shaped, their wear and tear depending on diet and mode of behaviour, as for example with retrievers.

43.10. In the dog, the **CANINE TOOTH** is used as a weapon and the very long root makes an extraction difficult. Transverse fractures occur, resulting in exposure of the pulp cavity.

43.11. **PREMOLAR** and **MOLAR TEETH** both have a cutting and a crushing function.

43.12. **ROOT NUMBERS** of the teeth exhibit considerable variation with several having undergone fusion. In hindsight they must be considered after dental extractions where portions of root remain in the alveoli. Anomalous positioning of roots, particularly those separating divergently, can make tooth extraction difficult.

43.13. The upper **SECTORIAL** or carnassial tooth has three roots, the lower, two. Inflammation and suppuration of the upper sectorial radiate to surroundings. Fistulae from the caudal of the roots have their openings ventral to the orbit under the eye.

43.14. **DENTAL ENAMEL** is the hardest substance of the animal body, essentially harder than the ingredients of the diet. Enamel producing cells also form a cuticle covering the enamel, and after the eruption of the tooth this is worn away within a short time. Consequently enamel cannot regenerate if, for example, it becomes demineralized and degenerates due to dental caries caused by acid-producing bacteria. Enamel hypoplasia with the formation of groove-like enamel defects occurs on the incisors in conjunction with distemper. This predisposes to caries. Caries unassociated with infections, attacks the last premolar and first molar tooth of the upper dental arch in particular. Due to the disintegration of enamel the exposed dentin is discolored a very dark brown and subsequently caries can produce an opening into the pulp cavity. In the dog dental caries generally begins on the occlusial surface of the tooth, rarely at the boundary between cement and enamel which is the major site in man. The **pulp cavity** contains blood vessels and nerves surrounded by loose connective tissue thus suggesting its mesenchymal origin.

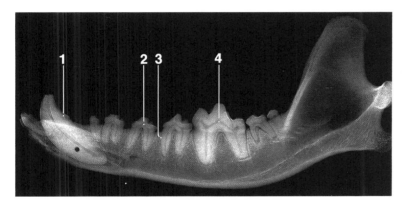

Dental series of mandible with enamel (1), dentin (2), cement (3) and pulp cavity (4). For identification of teeth see p. 43A.

43.15. The **GINGIVA** or gum is that portion of the mucous membrane of the oral cavity which covers the alveolar processes of the jaw bones. It projects like a collar with the outer marginal epithelium directed upwards over the base of the crown of the tooth. Its inner marginal epithelium is directed down deeper and is attached to the neck of the tooth, covering the periodontal membrane (periodontium) superficially. When the inner marginal epithelium detaches from the tooth, **gingival pockets** or crevices arise which cause inflammation of the peridontium (peridontitis) due to bacterial attack. A tooth being extracted must be separated from the gum taking into account gingival innervation from the buccal and sublingual nerves.

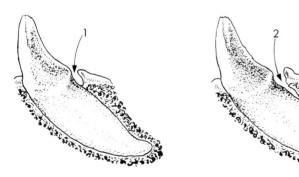

Gingival pocket (1) due to detachment of the inner marginal epithelium and bone (osseous pocket 2) from the tooth.

43.16. To attain local **ANESTHESIA OF THE UPPER CHEEK TEETH** one administers a nerve block to the infraorbital nerve. A fine gauge needle is used to enter the infraorbital canal through the infraorbital foramen and then local anesthetic is deposited around the nerve. The foramen is palpable at the level of the third or fourth premolar tooth. Another procedure requires local anesthetic to be injected around the infraorbital nerve in the pterygopalatine fossa caudal to its entrance into the infraorbital canal.

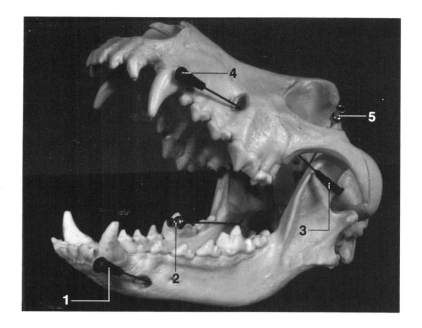

Sites of injection of local anesthetic in the head region: 1 Mental foramina for anesthesia of the mental nerve; 2 Mandibular foramen for the inferior alveolar nerve; 3 Pterygopalatine fossa for the maxillary nerve; 4 Infraorbital foramen for the infraorbital nerve; 5 Infraorbital fissure for nerves situated within the orbit.

43.17. To attain local **ANESTHESIA OF THE LOWER CHEEK TEETH** the inferior alveolar nerve is blocked by an injection at the mandibular foramen, the entrance to the mandibular canal. The mandibular foramen is palpated within the oral cavity approximately 2 cm caudal to the last molar tooth. Similarly, to anesthesize the mental nerves, one injects into the mental foramina. This also permits surgery to the lower lip and mandible.

44. SPINAL CORD AND MENINGES

44.1. **EPIDURAL ANESTHESIA** necessitates the injection of a local anesthetic into the epidural space. Injection sites are selected to avoid injury to the spinal cord and injection into the subarachnoidal space. Anesthetic diffuses through the tissues enveloping the roots of the spinal nerves, and blocks nerve transmission. Depending on quantity and type of anesthetic used and the injection site chosen, one speaks of a cranial or high block when the segmental nerves cranial to the sciatic nerve are blocked. Blocking of segmental nerves caudal to the sciatic nerve is referred to as a caudal, low or deep block. In high epidural anesthesia the motor nerves of the pelvic limb namely femoral, obturator and sciatic nerves are anesthetized and the animal is incapable of standing. In low epidural anesthesia sensory nerves, namely pudendal and caudal cutaneous femoral nerves and motor nerves to tail and pelvic diaphragm are blocked. Furthermore, with a low epidural block the animal remains standing, an important feature in large animal surgery. The lumbosacral interarcuate space is generally selected as the injection site in high epidural anesthesia (lumbosacral injection). The site lies caudal to the last lumbar spinous process and between the sacral tubera, the needle being directed perpendicular to the skin surface. The location of the needle in the anticipated epidural space is confirmed by aspiration. If it lies in the subarachnoid space, cerebro-

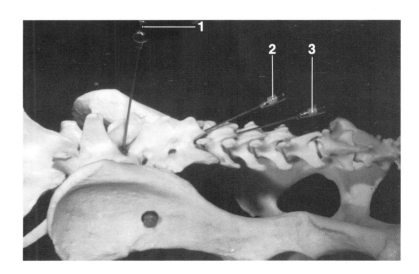

Injection sites for epidural anesthesia in: 1 Lumbosacral interarcuate space; 2 Sacrococcygeal interarcuate space; 3 Intercoccygeal interarcuate space.

spinal fluid is aspirated, while blood is aspirated if the internal vertebral (venous) plexus is punctured. If either situation results, the needle is withdrawn slightly and is aspirated again to check its position. The sacrococcygeal interarcuate space or intercoccygeal interarcuate spaces I to II are the injection sites of choice for low epidural anesthesia.

These 'lower' injection sites are also used for high epidural anesthesia if a sufficient dose of local anesthetic is given. (The needle is inserted cranioventrally at an angle to the skin surface. See illustration.)

44.2. **LESIONS OF THE SPINAL CORD** produce, apart from anything else, derangements of movement, eventually involving flaccid paralysis and deficit or weakening of spinal reflexes.. Compression of the spinal cord occurs with slight space occupying lesions because it has little space within its stiff meningeal envelop to permit deformation. **Compression** of the spinal cord is caused by many conditions including

prolapse of an intervertebral disc, fractures and subluxations. Laminectomy or hemilaminectomy is a recommended procedure for decompression of the cord.

44.3. **MYELOGRAPHY** is a radiographic method for the examination of the spinal cord. Alterations to the vertebral column affecting the spinal cord are demonstrated on an ordinary radiograph (flat plate). Recognition of primary changes to the spinal cord such as occur with tumors, is facilitated however, by using a radiopaque substance as a contrast medium. This is injected into the cisterna cerebromedullaris (magna in clinical terms) through the atlantooccipital space, with the articulation flexed. Inexperience can lead to injury of the spinal cord when using this procedure. A stoppage of flow of contrast medium indicates a constriction of the subarachnoid space. In the case illustrated below, this is caused by a prolapse of the intervertebral disc.

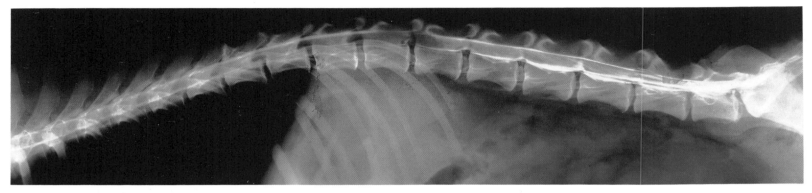

Myelography after injection of contrast medium into the subarachnoid space cranially, through the atlantooccipital space into the cisterna cerebromedullaris (magna) and caudally, through the lumbosacral space. An abnormal peripheral enlargement is present between the thoracic and the lumbar vertebral column at the interruption of subarachnoid filling with contrast medium. This has led to compression of the subarachnoid space.

45. BRAIN

45.1. **MENINGITIS** (inflammation of the cerebral and spinal meninges) may spread from pathological conditions in adjacent areas for example with ulceration of the nasal cavity, or with transmission by the vascular route. Severe pain results since the meninges have an abundant innervation. Stiffness in the cervical region is apparent, the animal avoiding movement and hence traction on the meninges.

45.2. **ENCEPHALITIS** or inflammation of the brain, often connected with hydrocephalus, may be of unknown etiology or appear in conjunction with viral infections such as distemper and rabies, or in parasitic diseases such as toxoplasmosis.

In **canine distemper,** inflammatory changes are produced initially in the white matter of brain and spinal cord. Symptoms caused by lesions in the brain stem are motor derangement and dullness, whereas lesions in cerebellum and spinal cord cause ataxia and paresis respectively. The **rabies** virus generally spreads from a bite in the tissue of the dog to spinal nerves along the central nervous system. There, the virus causes inflammation of the nerve cell bodies, with death finally supervening. There is no treatment for rabies.

Toxoplasmosis often causes inflammation of spinal cord and nerves and lesions in cerebrum and cerebellum.

45.3. **METABOLIC ENCEPHALOPATHY** is observed in conjunction with portocaval shunts (see 23.4.). Due to an abnormal anastomosis between portal vein and caudal vena cava within the liver, a portion of portal blood, depending on pressure ratios, flows directly into the caudal vena cava. As a result, insufficient filtration of portal blood occurs in the liver and cerebral tissues are damaged noticeably due to toxic metabolic products.

45.4. **BRAIN TUMORS** may emanate from glial cells of the ectodermal portion of the brain, or from cerebral meninges and blood vessels, that is the mesodermal portion of the central nervous system. Tumors may also stem from other organ systems by metastases, reaching the brain by way of the vascular system. Tumors damage brain tissue through compression or through its conversion to cancerous tissue. Subsequent to an abnormal accumulation of fluid, increased pressure in the brain leads to disturbances of permeability and cerebral edema.

45.5. With lesions in the brain stem, the **RETICULAR FORMATION** brings about alterations in muscle tone and disturbances to movement. Changes in consciousness may range from apathy to coma.

45.6. The **CEREBELLUM** regulates and coordinates motor function by forming a link between sensory and motor limbs of the peripheral nervous system. It receives information from the spinal cord via ascending pathways, and impulses from the centers of balance, sight and hearing and from motor areas of the cerebral cortex. Above all, cerebellar lesions produce disturbances of movement known as cerebellar ataxia.

45.7. The ventral portion of the **MIDBRAIN** or **MESENCEPHALON** transmits long nerve pathways travelling between cerebrum and spinal cord. Such autonomic centers or systems as those controlling respiration and circulation are situated in the tegmentum, and complicated motor output from the midbrain is mediated, for example via such nerves as those to the ocular muscles. The rostral colliculus contains the Edinger-Westphal nucleus, controlling the pupillary reflex, while the subcortical integration centers for hearing are situated in the caudal colliculus.

45.8. The **THALAMUS** is a subcortical center for the integration of sensory sensation such as temperature, pain and deep sensory sensation. It also contains a coordination center for the sense of sight. When affected by pathological conditions the severe pain resulting cannot be relieved by analgesics.

45.9. The **EPIPHYSIS** or pineal gland represses the development of the gonads, an epiphyseal deficit producing a precocious sexual maturity. The epiphysis is said to control circadian rhythm.

45.10. In the **HYPOTHALAMUS** one locates the supraoptic and the paraventricular nucleus which synthesize the hormones of the neurohypophysis, namely oxytocin and adiuretin. Other hypothalamic nuclei synthesize releasing hormones which regulate in turn the secretion of hormones from the adenohypophysis. The hypothalamus also contains important autonomic centers for regulating body temperature, water balance, hunger and thirst.

45.11. **HYPOPHYSEAL (PITUITARY) TUMORS** always cause changes in hormone status (see 45.10.) and also neurological symptoms. On comparative grounds the diaphragma sellae is underdeveloped and often the growth of a tumor in the direction of the hypothalamus is unhindered.

46. THE CEREBRUM OR TELENCEPHALON

46.1. The **CEREBRUM** is arranged in lobes such as frontal, parietal, temporal and occipital as well as the insular region. From the functional viewpoint the cerebral cortex is arranged somatotopically with definite areas of the body and their sensory and motor innervations represented on defined areas of cortex. A map of this reciprocal action between cerebrum and the dependent parts of the body is known as an homunculus, or little man, illustrated lying on the cerebral surface. The form of the homunculus is disproportionate and distorted, since biologically important functions and their respective parts of the body, such as hand and head, are represented as being disproportionally large. In turn these are represented also by correspondingly large areas of brain.

46.2. The **CORPUS STRIATUM** is a central relay for motor innervation particularly involved with slow movement. Within limits it is comparable to the cerebellum.

46.3. The **INTERNAL CAPSULE** contains the initial portion of the pyramidal pathway which is strongly compressed. Approximately 4/5 of its fibers cross from one side to the other at the pyramidal decussation situated between medulla oblongata and spinal cord. Lesions of the internal capsule emanate from adjacent blood vessels and in man cause the complex of symptoms termed stroke or apoplexy. This occurs rarely in dogs.

46.4. The **ARTERIES OF THE BRAIN** are end arteries supplying well-defined fields, and having no anastomoses with adjacent arteries coursing parallel to them. A deficit in blood supply, due to emboli or extensive hemorrhage from a ruptured blood vessel, generally causes sharply defined infarcts and scarring (see illustration under 37.2.).

The **blood-brain barrier** lies in the region of the capillary bed (for morphological detail refer to histology text). In particular the capillaries associated with the barrier are non-permeable, with many substances contained in the blood unable to reach the central nervous system. On the one hand brain tissue is protected from damaging blood constituents; on the other, drugs required for treatment of central nervous conditions reach the barrier and are found in markedly reduced concentrations in brain tissue on its deeper side.

The **barrier between blood and cerebrospinal fluid** may be interpreted as a strongly modified variant of the blood-brain barrier.

47. SINUSES OF THE DURA MATER AND CEREBROSPINAL FLUID

47.1. The **SINUSES OF THE DURA MATER** are involved above all in the thermoregulation of the head. Blood within the diploic veins is cooled at the roof of the cranium, and also in the nasal cavity, by the passage of inspired air. From both regions cooled blood flows within the sinuses to produce a cooling system for the brain. That blood from the nasal cavity is transmitted dorsally to the ventral venous system of the brain through the ophthalmic (venous) plexus is of clinical significance. For example, in tumors of the nasal cavity carcinogenic cells may be displaced and sediment out in the venous system because of the slow or almost absent blood flow. This then gives rise to new tumor growth.

Sinus venography is the radiographic demonstration of the venous sinuses of the dura mater. Contrast medium is injected into the angular vein of the eye and reaches the venous sinuses of the brain through the ophthalmic plexus. Interruptions to the flow of injected material may be produced by space occupying lesions in the cerebral region.

47.2. **ENCEPHALOGRAPHY** or radiographic demonstration of the brain, is obtained by injecting a suitable contrast medium into the cerebellomedullary cistern to delineate the surface configuration of the brain. One specific method is **ventriculography.** A lateral ventricle (see text illustration p. 47) is punctured through the cerebrum, contained cerebrospinal fluid withdrawn, and the ventricle filled with air which serves as a negative contrast medium (pneumoencephalography). Ventriculography permits size and symmetry of the ventricles to be estimated. In encephalography, contrast medium can also be injected into the internal carotid or vertebral arteries, while injection into the common carotid artery permits all arteries of the head to be demonstrated. **Computer tomography** is the most efficient but also the most cost-intensive method of examination. It is restricted to use in human medicine. No radiographic film is used and X-ray absorption values are calculated by computer and reproduced on a computer tomogram.

47.3. **CEREBROSPINAL FLUID** or CSF is formed in considerable quantities (approximately 350 ml per day in the dog). Furthermore a balance must exist between its production and resorption since the total quantity of fluid in ventricles of brain, central canal and subarachnoid space remains constant. Sites of resorption are considered to be: the subarachnoid granulations (see p. 45A —10), the continuations of the subarachnoid space over the beginnings of spinal nerves, and the arachnoid tissue associated with cerebral blood vessels. Hydrocephalus internus occurs where there is an imbalance between production and resorption, due either to an increased production of fluid or a stoppage of its outflow. Thus the brain ventricles are enlarged, brain tissue compressed progressively and the head is disproportionately large. This results in the bones of the skull being abnormally thin. Pathological increase in circumference of the narrow regions of the 'internal' fluid spaces is of importance. It involves the interventricular foramen, the mesencephalic aqueduct and the lateral apertures of the fourth ventricle connecting 'internal' and 'external' fluid spaces.

Besides its nutritive and thermoregulatory functions the CSF has a mechanical role chiefly as a protective envelope of fluid. This inhibits brain movement during violent blows to the head.

Lesions may occur at the site of the blow and after some delay, on the opposite side also. With violent rotation, shearing forces occur at the sites of passage of blood vessels from the skull, leading in turn, to hemorrhage at the confluence of cerebral veins with the sinuses of the dura mater. The composition of cerebrospinal fluid (CSF) is altered during infections resulting in encephalites and meningites. In the first instance protein content and lymphocyte and macrophage populations increase, and in several infections the causal agent is isolated directly from the fluid. Therefore it is important to be able to obtain CSF for diagnostic purposes. The procedure involves the insertion of a needle through the atlantooccipital space into the cerebellomedullary cistern and the aspiration of cerebrospinal fluid.

49. HORMONE SYNTHESIS IN THE HYPOTHALAMO-HYPOPHYSEAL SYSTEM

49.1.

HYPOTHA-LAMUS	NH-EFFEC-TOR HOR-MONE	TARGET ORGAN CELLS	PRINCIPAL ACTION
Oxytocin Vasopressin = Adiuretin	Neurosecretory transport to NH	Myoepithelium, smooth muscle, kidneys, blood vessels	Milk flow, parturition, renal water resorption, increase in blood pressure
	AH-EFFEC-TOR-HOR-MONE		
RH and RIH	Somatotrophic-hormone (STH)	Increased union of cells	Body growth
RH and RIH	Prolactin	Mammary gland	Milk secretion
	AH-REGULATOR HORMONE		
RH	Thyroid stimulating hormone (TSH)	Thyroid gland	Metabolism
	Gonadotrophin: Follicle stimulating hormone (FSH)	Sertoli cells and follicular epithelial cells	Testicular and follicular maturation
RH	Luteinizing hormone (LH - female) (ICSH - male)	Ovarian and testicular interstitial cells, corpus luteum	Estrogen androgen and synthesis of progesterone
RH	Adrenocorticotrophic hormone (ACTH)-	Cortex of adrenal gland	Synthesis of steroid hormone
RH and RIH	Melatropin of pars intermedia Melanocyte stimulating hormone	Pigment cells	Pigmentation

RIH = Release Inhibiting Hormone; RH = Releasing Hormone; NH = Neurohypophysis; AH = Adenohypophysis; ICSH = Interstitial Cell Stimulating Hormone.

12

Peripheral endocrine glands, such as thyroid, adrenal cortex and gonads, influence the hypothalamo-hypophyseal system through a feedback mechanism. As an example of a negative feedback mechanism, consider the following: A stress situation produces the release of ACTH-RH in the hypothalamus and a rise in ACTH production in the adenohypophysis. This is followed by an increased output of glucocorticoids from the adrenal cortex. Through negative feedback on the hypothalamus release of ACTH-RH, ACTH, and glucocorticoids are subsequently reduced. As an example of a positive feedback, consider a high rate of hormone synthesis in the ovary (estrogen in interstitial cells and follicular epithelial cells) enhancing the release of hypothalamo-hypophyseal hormones such as LH-RH and LH to permit termination of ovulation.

49.2. **RENIN** is released due to a fall of arterial blood pressure in the kidneys. It converts angiotensinogen of blood plasma into angiotensin I. In different organs such as the lung, this is converted into angiotensin II which stimulates the cells of the zona glomerulosa of the adrenal gland to synthesize aldosterone. This mineralocorticoid produces an increase in intravascular fluid volume and an elevation in blood pressure due to the resorption of sodium and water. The rise in blood pressure diminishes the synthesis of renin once again.

50. SENSE ORGANS (EYE)

50.1. The crystal clear **CORNEA** retains its transparency due to a defined state of swelling or tumefaction. This is maintained externally by lacrimal fluid and internally by aqueous humor. Light is refracted only on the outer surface of the cornea which is therefore compared to the front (fixed) lens of a camera.

The healthy cornea does not possess blood vessels and since it is nourished by perfusion, metabolic diseases such as diabetes mellitus may cause corneal opacity. In inflammation of the cornea or keratitis, blood vessels 'sprout' within the cornea and hinder sight. In superficial inflammatory processes these 'buds' are associated with the vessels of the conjunctiva, whereas with inflammation involving deeper layers, vessels penetrate from the greater arterial circle of the iris. Corneal opacity occurs soon after death because the normal state of hydration is lost through alteration of the outer (Bowman's) and inner (Descemet's) limiting membrane.

50.2. On the **INNER SURFACE OF THE IRIS** one finds the shimmering pigment epithelium responsible for the blue or gray eye colour present for example in individual huskies. A brown eye colour is the result of greater quantities of pigment in the stroma of the iris. Pigment-free consolidations of connective tissue in the iris appear white. A partial or totally white iris, however, is rare in the dog.

It is generally agreed that the **smooth muscles** of the iris develop from ectoderm. The m. sphincter pupillae is innervated by the parasympathetic autonomic division. A narrow pupil (myosis) is produced by administering a myotic such as pilocarpine, by direct stimulation of the parasympathetic division, or by a sympathetic paralysis. Contraction of the m. dilator pupillae is induced by administering a sympatheticomimetic drug and the pupil dilates (mydriasis) as a result.

50.3. An **OPACITY OF THE LENS** is known as a cataract. Surgery involves extraction of the lens. The lens capsule is generally removed (intracapsular extraction) because the condition can emanate from there also. With a luxation of the lens into the anterior chamber of the eye an extraction of the lens becomes essential.

50.4. The **RETINA** is examined using an ophthalmoscope. The pupil is dilated with a mydriatic and a stream of light projected onto the fundus of the eye using a mirror. The clinician or surgeon inspects the fundus of the eye through a central aperture in the mirror, and is able to diagnose such conditions as inflammation, atrophy and detachment of the retina. As a further example, increased intracranial pressure resulting from hydrocephalus, meningitis or tumors, may lead to an abnormal rostral distension of the optic disc over the surface of the retina. Of the photoreceptors in the dog, the rods definitely predominate over the cones in the proportion of 95:5, whereas in man the numbers are equal. Since cones are responsible for colour perception this seems to be less evident in the dog. Furthermore, as the cones of the macula are responsible for sharp vision, it would also seem that the latter is less well developed in the dog.

50.5. In **GLAUCOMA** the retina is compressed due to an increase in intraocular pressure. This leads to retinal atrophy and blindness. Cyclodialysis as a method of treatment, aims at draining aqueous humor from the iridocorneal angle.

REFERENCES

Adams, D. R., 1986: Canine Anatomy. Iowa State University Press, Ames

Amman, E., E. Seiferle und G. Pelloni, 1978: Atlas zur chirurgisch-topographischen Anatomie des Hundes. Paul Parey, Berlin, Hambg.

Bargmann, W., 1977: Histologie und Mikroskopische Anatomie des Menschen, 7. Aufl. Georg Thieme, Stuttgart

Barone, R., 1976: Anatomie Comparée des Mammiferes Domestiques; T. 1 – Osteologie; T. 2 – Arthrologie et Myologie; T. 3 – Splanchnologie, Foetus et ses Annexes. Viget Freres, Paris

Baum, H., 1917: Die im injizierten Zustand makroskopisch erkennbaren Lymphgefäße der Skelettknochen des Hundes. Anat. Anz. 50:521-539

Baum, H., 1917: Die Lymphgefäße der Haut des Hundes. Anat. Anz. 5O:1-15

Baum, H. und O. Zietzschmann, 1936: Handbuch der Anatomie des Hundes, 2. Aufl., Paul Parey, Berlin

Benninghoff, A. und K. Goerttler, 1980, Hrsg. Ferner H. und J. Staubesand: Lehrbuch der Anatomie des Menschen, Bd. I, 13. Aufl., Bd. II 1979, 12. Aufl., Urban und Schwarzenberg, München, Wien, Baltimore

Berg, R., 1982: Angewandte und topographische Anatomie der Haustiere. 2. Aufl., Gustav Fischer, Jena

Bloom, W. and D. W. Fawcett, 1975: A Textbook of Histology, 10. Ed. W. B. Saunders Comp., Philadelphia, London, Toronto

Bojrab, M. J., 1981: Praxis der Kleintierchirurgie. Enke, Stuttgart

Bolz, W., 1985, Hrsg.: O. Dietz: Lehrbuch der Allgemeinen Chirurgie für Tierärzte. 5. Aufl., Enke, Stuttgart

Bradley, O. Ch., 1948: Topographical Anatomy of the Dog. 5. Ed., Oliver and Boyd, Edinburgh, London

Bucher, O., 1980: Cytologie, Histologie und mikroskopische Anatomie des Menschen, 10. Aufl., Hans Huber, Bern, Stuttgart, Wien

Budras, K.-D., 1972: Zur Homologisierung der Mm. adductores und des M. pectineus der Haussäugetiere. Zbl. Vet. Med., C, 1:73–91

Budras, K.-D., F. Preuß, W. Traeder und E. Henschel, 1972: Der Leistenspalt und die Leistenringe unserer Haussäugetiere in neuer Sicht. Berl. Münchn. tierärztl. Wschr. 85:427 431

Budras, K.-D. und D. Seifert, 1972: Die Muskelinsertionsareale des Beckens von Hund und Katze, zugleich ein Beitrag zur Homologisierung der Lineae glutaeae unserer Haussäugetiere. Anat. Anz. 132:423–434

Budras, K.-D. und A. Wünsche, 1972: Arcus inguinalis und Fibrae reflexae des Hundes. Gegenbaurs morph. Jb. 117:408–419

Chandler, E. H., J. B. Sutton and D. J. Thompson, 1984: Canine Medicine and Therapeutics. 2. Ed., Blackwell, Oxford, London

Christensen, G. C., 1954: Angioarchitecture of the canine penis and the process of erection. Amer. J. Anat. 95:227–261

Christoph, H.-J., 1973: Klinik der Hundekrankheiten. Fischer, Stuttg.

Dämmrich, K., 1981: Zur Pathologie der degenerativen Erkrankungen der Wirbelsäule bei Hunden. Kleintierpraxis 26:467-476

Dahme, E. und E. Weiss, 1983: Grundriß der speziellen pathologischen Anatomie der Haustiere. 3. Aufl., Enke, Stuttgart

Dellmann, H.-D. and E. M. Brown, 1976: Textbook of Veterinary Histology. Lea and Febiger, Philadelphia

De Lahunta, A., 1983: Veterinary Neuroanatomy and Clinical Neurology. 2. Ed., W. B. Saunders Comp., Philadelphia

De Lahunta, A. and R. E. Habel, 1986: Applied Veterinary Anatomy. W. B. Saunders Comp., Philadelphia

Dobler, Chr., 1967: Papillarkörper und Kapillaren der Hundekralle, Schweine- und Ziegenklaue. Morph. Jb. 113:382-428

Donat, K., 1971: Die Fixierung der Clavicula bei Katze und Hund. Anat. Anz. 128:365–374

Donat, K., 1972: Der M. cucullaris und seine Abkömmlinge (M. trapezius und M. sternocleidomastoideus) bei den Haussäugetieren. Anat. Anz. 131:286–297

Ellenberger, W. und H. Baum, 1943: Handbuch der vergleichenden Anatomie der Haustiere. 18. Aufl., Springer, Berlin

Evans, H. E. and A. de Lahunta, 1980: Miller's Guide to the Dissection of the Dog. 2. Ed., W. B. Saunders Comp., Philadelphia, London, Toronto

Evans, H. E. and G. C. Christensen, 1979: Miller's Anatomy of the Dog. 2. Ed., W. B. Saunders Comp., Philadelphia, London, Toronto

Faller, A., 1980: Anatomie in Stichworten. Ferdinand Enke, Stuttgart

Franke, H.-R., 1970: Zur Anatomie des Organum vomeronasale des Hundes. Diss. med. vet., Freie Universität Berlin

Gelatt, K. N., 1981: Veterinary ophthalmology. Lea and Febiger, Philadelphia

Gerneke, W. H., 1986: Veterinary Histology. Dept. Anat. Faculty of Vet. Sci., Univ. of Pretoria

Getty, R., 1975: Sisson and Grossman's Anatomy of the Domestic Animals. Vol. 2 – Porcine, Carnivore, Aves. 5. Ed., W. B. Saunders Comp., Philadelphia, London, Toronto

Getty, R., H. L. Foust, E. T. Presley and M. C. Miller, 1956: Macroscopic anatomy of the ear of the dog. Amer. J. Vet. Res. 17: 364–375

Gräning, W., 1937: Beitrag zur vergleichenden Anatomie der Muskulatur von Harnblase und Harnröhre. Z. Anat. und Entwicklgesch. 106:226–250

Grandage, J., 1972: The erect dog penis. Vet. Rec. 91:141–147

Habel, R. E., 1985: Applied Veterinary Anatomy. Pub. by author, Ithaca, N. Y.

Henning, Ch., 1965: Zur Kenntnis des M. retractor ani et penis s. clitoridis et constrictor recti (M. retractor cloacae) beim Hund. Anat. Anz. 117:201–215

Henning, P., 1965: Der M. piriformis und die Nn. clunium medii des Hundes. Zbl. Vet. Med., A, 12:263–275

Henschel, E. und W. Gastinger, 1963: Beitrag zur Arteriographie der Aa. carotis und vertebralis beim Hund. Berl. Münchn. tierärztl. Wschr. 76:241-243

Henschel, E., 1971: Zur Anatomie und Klinik der wachsenden Unterarmknochen mit Vergleichen zwischen der Distractio cubiti des Hundes und der Madelungschen Deformität des Menschen. Arch. Experim. Vet. med. 26:741–787

Henschel, E., 1980: Zur Röntgenanatomie des Schädels. Kleintierpraxis 25:287–292

Henschel, E., 1983: Das Hüftgelenk von Hund und Katze – eine Enarthrosis? tierärztl. prax. 11:345–348

Hentschel, E. und G. Wagner, 1976: Tiernamen und zoologische Fachwörter unter Berücksichtigung allgemeinbiologischer, anatomischer und physiologischer Termini. Fischer, Stuttgart, New York

Hoerlein, B. F., 1978: Canine Neurology. Diagnosis and Treatment. 3. Ed., W. B. Saunders Comp., Philadelphia, London, Toronto

Hyrtl, J., 1880: Onomatologia Anatomica. Braumüller, Wien

International Committee on Gross Anatomical Nomenclature, 1983: Nomina Anatomica, 3. Ed., Nomina Histologica, 2. Ed., Ithaca, N. Y.

Jacoby, S., 1968: Über einige Abkömmlinge des M. sphincter marsupii beim Hunde. Anat. Anz. 122:234–338

Kadletz, M., 1932: Anatomischer Atlas der Extremitätengelenke von Pferd und Hund. Urban und Schwarzenberg, Berlin, Wien

Kahle, W., H. Leonhardt und W. Platzer, 1979: Taschenatlas der Anatomie für Studium und Praxis, Bd. 1–3, 3. Aufl., Thieme, Stuttg.

Kealy, J. K., 1981: Röntgendiagnostik bei Hund und Katze. Enke, Stuttgart

King, A. S., 1978: A Guide to the Physiological and Clinical Anatomy of the Thorax. 4 Ed., Dept. Vet. Anat., University of Liverpool, Liverpool L69 3BX

King, A. S. and V. A. Riley, 1980: A Guide to the Physiological and Clinical Anatomy of the Head. 4. Ed., Dept. Vet. Anat., University of Liverpool, Liverpool L69 3BX

Knoche, H., 1979: Lehrbuch der Histologie. Springer, Berlin, Heidelberg, New York

Koch, T. und R. Berg, 1981 - 1985: Lehrbuch der Veterinär-Anatomie. Bd. 1-3, Gustav Fischer, Jena

Krölling, O. und H. Grau, 1960: Lehrbuch der Histologie und vergleichenden mikroskopischen Anatomie der Haustiere. Paul Parey, Berlin, Hamburg

Krstić, R. V., 1978: Die Gewebe des Menschen und der Säugetiere. Springer, Berlin, Heidelberg, New York

Krstić, R. V., 1984: Illustrated Encyclopedia of Human Histology. Springer, Berlin, Heidelberg, New York, Tokyo

Krüger, G., 1961: Veterinärmedizinische Terminologie. 2. Aufl., Hirzel, Leipzig

Laue, E., 1987: Makroskopische, licht- und elektronenmikroskopische Untersuchungen über das Lymphgefäßsystem des Pferdes vom Huf bis zum Karpalgelenk. Diss. med. vet., Freie Universität Berlin

Leonhardt, H., 1985: Histologie, Zytologie und Mikroanatomie des Menschen. 7. Aufl., Thieme, Stuttgart

Lippert, H., 1982: Lehrbuch der Anatomie nach dem Gegenstandskatalog. Urban und Schwarzenberg, München, Wien, Baltimore

Lohse, C. L., D. M. Hyde and D. R. Benson, 1985: Comparative Development of Thoracic Intervertebral Discs and Intra-Articular Ligaments in Human, Monkey, Mouse, and Cat. Acta anat. 122: 220–228

Menge, H., 1954: Menge-Güthling: Enzyklopädisches Wörterbuch der lateinischen und deutschen Sprache. Langenscheidt KG, Berlin

Nickel, R., A. Schummer und E. Seiferle, 1982–1984: Lehrbuch der Anatomie der Haustiere. Bd. 1–4, Paul Parey, Berlin, Hamburg

Niemand, H. G., 1980: Praktikum der Hundeklinik. 4. Aufl., Paul Parey, Berlin, Hamburg

Nitschke, Th., 1966: Zur Frage der Vena profunda glandis des Rüden. Zbl. Vet. Med., A, 13:474–476

14

Nitschke, Th., 1966: Der M. compressor venae dorsalis penis s. clitoridis des Hundes. Anat. Anz. 118:193−208

Nitschke, Th., 1970: Diaphragma pelvis, Clitoris und Vestibulum vaginae der Hündin. Anat. Anz. 127:76−125

Nitschke, Th., 1976: Die Rami orbitales des Ganglion pterygopalatinum des Hundes − zugleich ein Beitrag über die Innervation der Tränendrüse. Anat. Anz. 139:58−70

Preuß, F., 1967: Anleitung zur topographischen Ganztierpräparation des Hundes. Selbstverlag, Berlin

Preuß, F., 1976 und 1979: Medizinische Histologie. Teil 1 − Zellen und einfache Zellverbände; Teil 2 − Zusammengesetzte Zellverbände. Paul Parey, Berlin, Hamburg

Preuß, F., K.-D. Budras und E. Henschel, 1967: Processus vaginalis, Hernia inguinalis paravaginalis und Prolapsus omenti bei einer weiblichen Katze. Berl. Münchn. tierärztl. Wschr. 80:371-373

Preuß, F. und E. Henschel, 1968: Praktikum der angewandten Veterinäranatomie. Teil I: Hund, Schwein, Geflügel. Institutsaufl., Freie Universität Berlin

Preuß, F., W. Müller und K. Donat, 1966: Die Rectus-Scheide des Hundes. Berl. Münchn. tierärztl. Wschr. 79:331−333

Rauch, R., 1963: Beitrag zur arteriellen Versorgung der Bauch- und Beckenhöhle bei Katze und Hund. Zbl. Vet. Med., A, 10:397−429

Rohen, W. J., 1977: Topographische Anatomie. 6. Aufl., Schattauer, Stuttgart, New York

Rohen, W. J., 1977: Funktionelle Anatomie des Menschen. 3. Aufl., Schattauer, Stuttgart, New York

Rohen, W. J., 1978: Funktionelle Anatomie des Nervensystems. 3. Aufl., Schattauer, Stuttgart, New York

v. Sandersleben, J., K. Dämmrich und E. Dahme, 1985: Pathologische Histologie der Haustiere. 2. Aufl., Fischer, Stuttgart

Schebitz, H. und W. Brass, 1975: Allgemeine Chirurgie für Tierärzte und Studierende. Paul Parey, Berlin, Hamburg

Schebitz, H. und W. Brass, 1985: Operationen an Hund und Katze. Paul Parey, Berlin, Hamburg

Scholtysik, G., 1962: Die normale Zwischenwirbelscheibe des Hundes normaler und chondrodystrophischer Rassen. Diss. med. vet., Freie Universität Berlin

Schwarz, R., J. M. W. Le Roux, R. Schaller et al., 1979: Micromorphology of the Skin (Epidermis, Dermis, Subcutis) of the Dog. Onderstepoort J. vet. Res. 46:105-109

Seifert, D. und K.-D. Budras, 1979: Wachspräparate und Kunstharzeinbettungen als Anschauungsmaterial für den topographisch-anatomischen Präparierunterricht. Verh. anat. Ges. 73:1207−1212

Siegel, E. T., 1982: Endokrine Krankheiten des Hundes. Paul Parey, Berlin, Hamburg

Smollich, A. und G. Michel, 1985: Mikroskopische Anatomie der Haustiere. Gustav Fischer, Jena

Van de Velde, M. und R. Fankhauser, 1986: Einführung in die veterinärmedizinische Neurologie. Paul Parey, Berlin, Hamburg

Weiss, L. and R. O. Greep, 1973: Histology. 4. Ed., McGraw-Hill Book Comp., New York

Wenzel-Hora, B. J., D. Berens v. Rautenfels und H. M. Siefert, 1982: Direkte und indirekte Lymphographie am Hund. tierärztl. prax. 10:521-529

Wünsche, A. und K.-D. Budras, 1972: Der M. cremaster externus resp. compressor mammae des Hundes. Zbl. Vet. Med. C, 1:138−148

Zapp, A., 1967: Die Corpora capsularia und die Mm. interossei der Hundefüße. Diss. med. vet., Freie Universität Berlin

Zietzschmann, O., 1928: Über den Processus vaginalis der Hündin. Deutsche tierärztl. Wschr. 20−22

20